Sustainable Construction Technologies & Advancements in Civil Engineering (ScTACE-2020)

Held online due to COVID-19

IOP Conference Series: Materials Science and Engineering
Volume 1006

Bhimavaram, India
9 - 10 October 2020

Editor:

Pala Gireesh Kumar

ISBN: 978-1-7138-3318-5
ISSN: 1757-8981

Printed from e-media with permission by:

Curran Associates, Inc.
57 Morehouse Lane
Red Hook, NY 12571

Some format issues inherent in the e-media version may also appear in this print version.

Institute of Physics
Dirac House, Temple Back
Bristol BS1 6BE UK

Phone: 44 1 17 929 7481
Fax: 44 1 17 920 0979

techtracking@iop.org

Additional copies of this publication are available from:

Curran Associates, Inc.
57 Morehouse Lane
Red Hook, NY 12571 USA
Phone: 845-758-0400
Fax: 845-758-2633
Email: curran@proceedings.com
Web: www.proceedings.com

TABLE OF CONTENTS

IOP Conf. Series: Materials Science and Engineering 1006 (2020) 011001 doi:10.1088/1757-899X/1006/1/011001

PREFACE

We were delighted with a high level of satisfaction and aspiration, indeed it's a great privilege to recapitulate the two day National Virtual Conference. We are conscious on how the Corona pandemic has made a global impact and continued to hit majority of the sectors, with education being one of the most affected. At such a time the extended lockdown period has forced the education sector in India to rapidly evolve. Visibly, over the last few months education has changed dramatically with a considerable rise in e-learning, whereby teaching is executed remotely and on digital platforms. To succor people and to set up for success during this unique time, the Department of Civil Engineering of Shri Vishnu Engineering College for Women (Autonomous) organized a 2 day National Virtual Conference on Sustainable Construction Technologies & Advancements in Civil Engineering (ScTACE-2020) in collaboration with Indian Concrete Institute – Student Chapter. The programme engendered a technically rewarding experience under the superintendence of Department of Civil Engineering on 9th and 10th October 2020 supported on Zoom platform, for productivity a way to create more meaning, stability and empowerment through all the chaos of change each of us face.

The utmost goal and facet of ScTACE-2020 is to bring the academicians, engineers, industry researchers with one accord to discourse and share their grounding, indignation about most aspects of sustainable solutions and deliberate over the challenges encountered in the explication of the solutions. The two day programme was excogitated with an even handed to the presenters as well as listeners with a bodacious array of 11 keynote and 2 invited guest speakers with prowess from IISc, IIT's, NIT, renowned Universities and Industry from India and overseas. The symposium beheld a noteworthy representation with 350 entrants from variegated institutions, organizations across the country and overseas made this 1st conference of its kind a truly conference in spirit. The varied genres of trending technologies or materials in constituent to the sustainability and the series of interactive sessions promoted the indefatigable spirit of 250 learners and encouraged their determination in adding an extra skill to their forte. 65 papers (49 papers were selected and shortlisted for publication upon reviewer's decision) were presented with divaricated 12 parallel panels, formed to be the heart of the conference and provided ample opportunity for discussion that brought new perspectives as a stepping stone in the enhancement of knowledge and skill of those working in the field of civil engineering. The proceedings provided a written record of the synergy with 130 long and short papers with 55% of acceptance rate publishing in the IOP Conference Series: Materials Science and Engineering owing to the Peer Review Policy that represented a solid framework which emerge a new interaction in the fullness of time. The event ended with the unveiling of the conference sovereign leaving everyone's experience with ScTACE-2020, a fruitful and long lasting one.

All in all the plenary lectures, progress and presentations bridged the gap making it possible for the people to stay productive and learn or to get au courant with newfangled technologies or skills. We divulge our unfeigned delight to all the external chairpersons for their admirable serenity and accorded stupendous time for colloquy, elucidation of doubts mooted by the partakers. We take the entitlement in acclaiming all the keynote and guest expertise for their credence and immense deliberation that aided the budding graduates and professionals to hone their skills in the art of civil engineering and pursue their long-procrastinated passions during the lockdown. We are also honor-bound to profess our acknowledgement to the management of Shri Vishnu Engineering College for Women for its verve in lending a hand and believing thatthere should be no reason for people to stop learning new skills in the competitive times we live in. We would like to express our gratitude and appreciation for all the reviewers who helped us maintain the capacity constraints and high quality of manuscripts included in the proceedings. Equally or exceedingly we bestow our commendation to the support team, committee members, and student coordinators for their stance and idiosyncrasy in handling the contenders and making this conference viable.

We promise to bring forth the same bright picture and charming landscape with the aforementioned support, participation and will continue to extend the success for a long time.

Convener/Chief Editor ScTACE-2020

Dr. Pala Gireesh Kumar
Associate Professor & HOD
Shri Vishnu Engineering College for Women (Autonomous)

ScTACE 2020

Peer review declaration

All papers published in this volume of *IOP Conference Series: Materials Science and Engineering* have been peer reviewed through processes administered by the Editors. Reviews were conducted by expert referees to the professional and scientific standards expected of a proceedings journal published by IOP Publishing.

- **Type of peer review:** Double-blind
- **Describe criteria used by Reviewers when accepting/declining papers. Was there the opportunity to resubmit articles after revisions?**
  A basic plagiarism check is done as a primary filtration and the papers with less than 15% plagiarism are sent to Reviewers and further, they are suggested and briefed out to keep a check on novelty and continuity of flow along with the clarity of thought and correctness concerning the data. Upon satisfying the above-mentioned criteria, the papers were submitted for final publication with IOP. The papers, which failed to meet the proposed criteria, were sent to the respective authors along with comments from reviewers and a second chance was given to resubmit their work addressing the comments.

- **Conference submission management system:** EasyChair

- **Number of submissions received:** 130

- **Number of submissions sent for review:** 130

- **Number of submissions accepted:**49
  (NOTE: Out of 130 submissions, 65 papers were shortlisted for ScTACE 2020, out of which only 49 papers were considered for publication with IOP, and the remaining 16 papers were considered under the category of the presentation without publication)

- **Acceptance Rate (Number of Submissions Accepted / Number of Submissions Received X 100):** 38%

- **Average number of reviews per paper:** 2

- **Total number of reviewers involved:** 25

- **Any additional info on review process (i.e., plagiarism check system):** Every submitted paper has gone through a plagiarism check using the software **Turnitin**. And every key factor like novelty, continuity of flow, clarity, and correctness of thought was taken into consideration and based upon these; a strict acceptance process was adopted. Papers meeting the high proposed standards were accepted for publication with IOP (49) and the papers with basic proposed standards were considered for presentation without publication only (16).

IOP Conf. Series: Materials Science and Engineering 1006 (2020) 011003 doi:10.1088/1757-899X/1006/1/011003

- **Contact person for queries:** Dr Pala Gireesh Kumar, Assoc Prof. & HOD-CE, Convener, ScTACE 2020, SVECW(A). (Email ID: gireeshcivil@svecw.edu.in)

Design and Analysis of Filler Slab

Mahananda R K[1], Vikas Mendi[2] and Raveesh R M[3]

[1]PG student, Dept. of Civil Engineering, RV College of Engineering, India
[2]Assistant Professor, Dept. of Civil Engineering, RV College of Engineering, India
[3]Ph.D. Scholar, Dept. of Water Resources and Ocean Engineering, National Institute of Technology Karnataka, India

Email: vikasm@rvce.edu.in

Abstract. Filler slab technology is an innovative and cost effective technology where the dead load of slab is reduced by replacing the concrete with filler material. The concept behind the use of filler-slab technology is to reduce a substantial portion of concrete in the tension zone, since all the concrete in the tension zone does not contribute to the tensile properties. This concrete is replaced with lightweight, inert and inexpensive filler without compromising with the quality and structural stability of the structure. Two-way slab is designed; the filler blocks are placed between the reinforcement spacing by providing a cover of 20mm. The filler materials are granite dust and foundry sand. This filler slab is analyzed using STAAD.Pro and ANSYS software. Filler slab is compared with the conventional slab of same size. This study describes the Structural behavior and cost effectiveness of filler slab when compared to the standard slab.

1. Introduction

As per the annual reports of Census 2011, around two-thirds of the total Indian population cannot afford their shelter [1]. Affordability is one major reason for people to go shelter less. If houses can be constructed economically by using innovative techniques to reduce the consumption of materials, more people can afford to construct their own houses.

Concrete being the highest consumed material on earth after water (Cement Industry Federation, Australia), is produced and used at a large scale in the construction industry. Ideas to reduce the usage of concrete without compromising the quality of construction will not only lead to cost-effective buildings but also reduce carbon emissions when looked at a large scale.

Many innovative techniques and economical methods are being proposed and filler slab technology is one such innovative and cost-effective technology where a dead load of the slab is reduced by replacing the concrete. Concrete is good in withstanding compressive forces and steel is good in withstanding tensile forces. The main aim behind the use of filler-slab technology is to condense a substantial portion of concrete below the neutral axis since all concrete in the tension zone does not add to the tensile properties. This concrete is replaced with lightweight, inert and inexpensive filler without neglecting the quality and structural strength of the structure [2].

By reducing the quantity and weight of concrete, the slab becomes less expensive, but the strength has to remain the same as the conventional slab. In many of the areas, to reduce the heat impact in the building during hot weather at an economic cost, the filler-slab technology acts as an excellent thermal insulator and brings down the temperature inside the building.

IOP Conf. Series: Materials Science and Engineering 1006 (2020) 012001 doi:10.1088/1757-899X/1006/1/012001

2. Need for alternatives

Construction segment in India is accountable for major input of energy resulting in the major emission of CO_2 (22%) into the atmosphere [3]. To reduce the harmful effect, sustainable technology is required. This technology is used to remove a substantial portion of concrete in the tension zone and replace it with lightweight, inert and inexpensive filler. This particular technology can not only reduce the cost of the materials but will also reduce the amount of carbon emission by lowering the use of energy-consuming materials. Overall, it is concluded that 30% of the carbon emission is reduced [4].

3. Filler material selection

The filler materials are selected upon satisfying the following conditions:

- The material must be non-reactive (inert) in nature.
- The limitation must be applied in terms of water absorption as hydration of concrete is affected.
- It must be lightweight, which will result in the reduction of the overall weight of the structure.
- It must be cost-effective when compared to the cost of concrete replaced.
- The size and thickness of the material must satisfy the reinforcement spacing and the size of the slab.
- Aesthetic views must be considered to avoid ugly faces of the ceiling.

4. Materials used

Various materials such as Cement (53 Grade), M-sand, Coarse aggregate and reinforcement steel are used in this study. Cement is used as a binder which binds, sets and hardens the other materials in contact. M-sand is simply nothing but crushed granite rocks into finer particles. Its size is finer than coarse aggregate and coarser than silt. M-sand is generally used as a substitute for river sand due to its cost expense. Coarse aggregate are gravel stones of bigger size than M-sand. These are usually obtained by crushing granite rocks into smaller dimension. The size to be used in concrete should generally lie between 10 - 20mm. Reinforcement steel are hot round bars, provided with deformation patters to give a better bonding between the steel and concrete. These steel are used in concrete to provide tensile strength.

4.1 Filler materials procured

To maintain a sustainable environment, the materials to be chosen were to be of a waste or discarded material, which can help in reusing the waste. After researching many materials like the mine waste and many other waste, three filler materials were short listed which satisfied the criteria and were chosen as filler materials for design and analysis of a filler slab. These materials were chosen as they were easily available/accessible, locally available in abundant quantities and the process of manufacturing the filler blocks with these materials can be done at site with minimum cost. Granite dust, Foundry sand and discarded plastic are the materials chosen for this experimentation.

Fig. 1 Granite dust, Foundry sand, solid plastic

4.2 Cost estimation of plastic blocks

Cost estimation is one important parameter which is to be considered as it should cost less than that of the concrete being replaced. Upon visiting various Industries, looking out for the best price we found an Industry located near Magadi road where the price of each plastic portion was Rs19 of size 80mm x 80mm x 20mm. The combination of 36 of such plastic portions will make one single block as shown in Fig. 2.4. All these plastic portions can be attached by using epoxy resin to make it a block. The filler slab requires 25 of such blocks as per the design of the slab, which results in 900 plastic portions. Therefore, the total cost of the filler material is Rs 17,100/- which is not economic when compared to the amount of concrete being replaced.

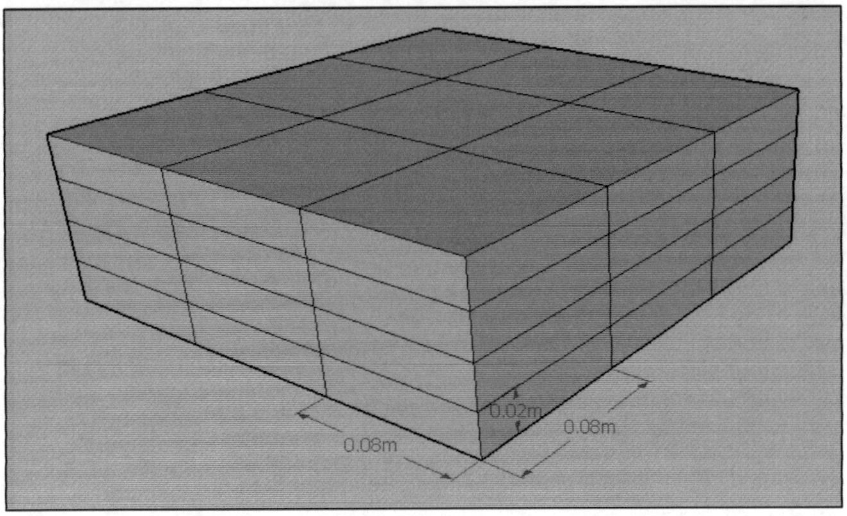

Fig. 2 Plastic block arrangement filler slab

5. STAAD Pro

STAAD Pro is inclusive structural designing software that reports all features of structural engineering including model development, verification, analysis, design and review of results which was being originally developed by Research Engineers International in 1997. It supports over 90 international steel, concrete, timber & aluminium design codes. This software is basically designed for civil structures in designing and analyzing purpose. This software is constructed in such a way designing of 2D and 3D modelling can be done and the structure can be defined for analysis.

6. ANSYS software

It is a global software designed for engineering simulation to check the behaviour upon testing. It also acts as a mechanical software that can used to simulate the models of structures, fluid dynamics or to design any other machine components. This software is helpful in many fields in terms of simulation. The version used is ANSYS (2019) R1.

7. Methodology

The following steps are being followed in the project and a flow chart is being shown in Fig. 3:

1. Experimental procedure in finding the properties of the basic materials and the filler materials.
2. Casting and testing of granite dust cylinders by using various measures of stabilizer to find the optimum stabilizer content.
3. Designing two way slabs and introducing filler materials in the tension zone between the reinforcement spacing by proving optimum cover.
4. Using the basic properties obtained through experimentation and other literatures, modelling is done using STAAD.Pro and ANSYS software.
5. Results are extracted upon analyses. These results are compared with the conventional slab.
6. Cost comparison between the filler slab and conventional slab is done.

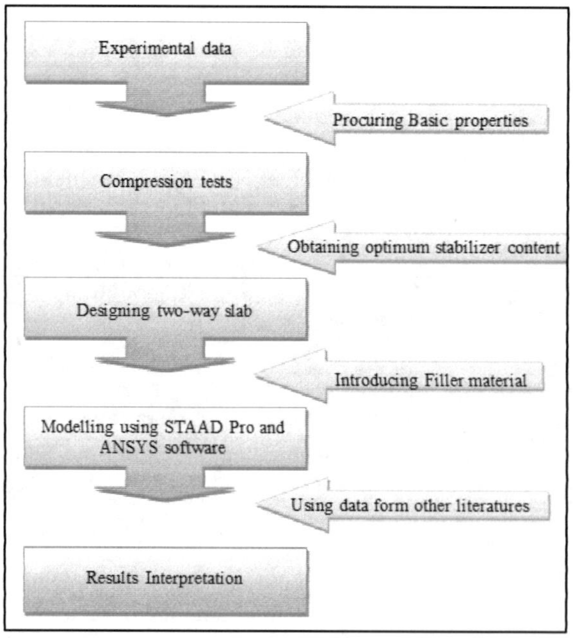

Fig. 3: Steps being followed during the project

7.1 Construction of Filler slab

The shuttering of the slab is placed and held stiff. This is followed by the reinforcing bars are placed over the shuttering as per design of the slab. These bars are tied using binding wire at every cross section of the bars and cover is provided at the bottom of the bars by using cover clocks.. Filler blocks are then placed over the shuttering proving a cover of 20mm between the reinforcement as shown in Fig. 2.6. Concrete of M20 grade is prepared and poured. Compaction is done using needle vibrator to get rid of air voids. On completing, deshuttering should be done after 7 days and curing must be done for 28 days to attain its strength.

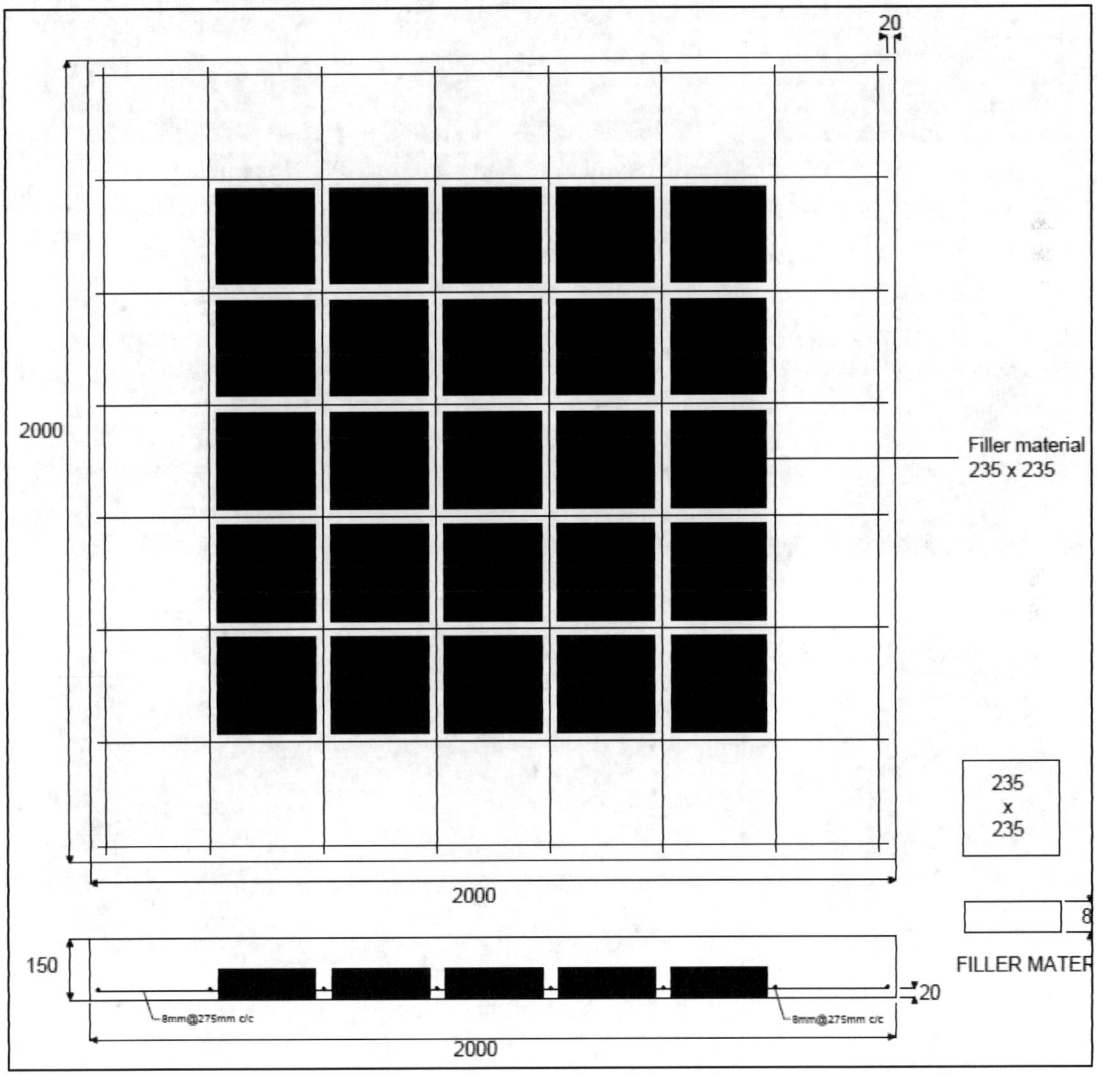

Fig. 4 Typical Filler slab design

IOP Conf. Series: Materials Science and Engineering 1006 (2020) 012001 doi:10.1088/1757-899X/1006/1/012001

7.2 Experimental tests carried out to find basic properties

Specific gravity test	**Moisture content test**	**Cone penetration test**
	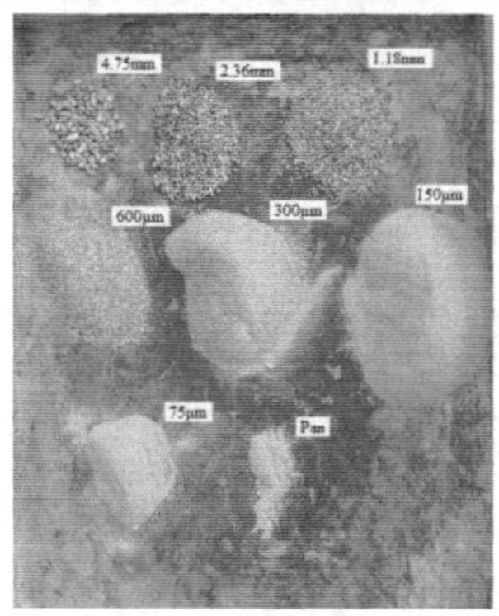	
Bulking test	**Sieve analysis**	**Proctor test**
	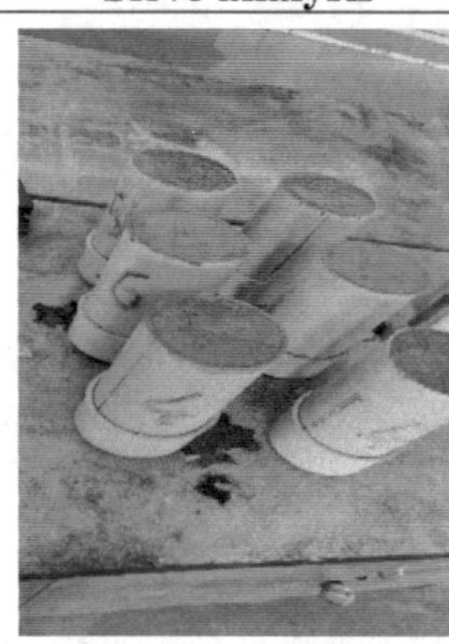	
Mixing of various composition	**Casting of cylinders done using PVC moulds**	**Compression test of cylinder**

Fig. 5 Basic tests conducted

7.3 STAAD.Pro modelling

The modelling of slab was done using STAAD.Pro V8i (SELECT series 6). The following steps were followed in the analysis of the slab. Similar steps were followed for conventional slab and filler slabs using various filler materials:

1. Geometry of the slab
2. Meshing
3. Support conditions
4. Material properties
5. Design parameters

7.4 ANSYS modelling

The modelling of slab was done using a ANSYS (2019) R1. The following steps were followed in the analysis of the slab. Similar steps were followed for conventional slab and filler slabs using various filler materials:

1. Boundary conditions
2. Engineering properties (Concrete and Steel)

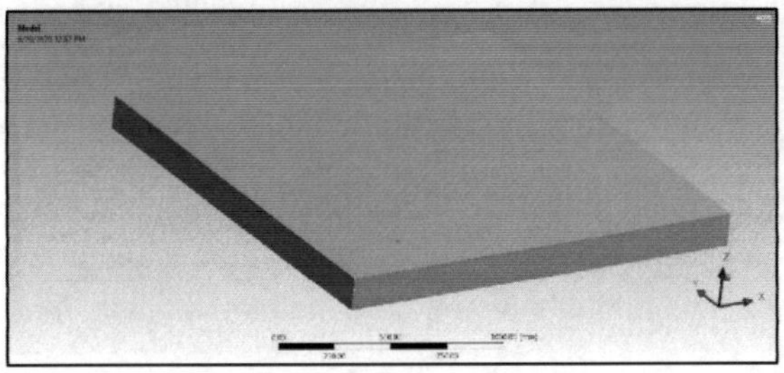

Fig. 6(a) Geometry conditions of conventional slab

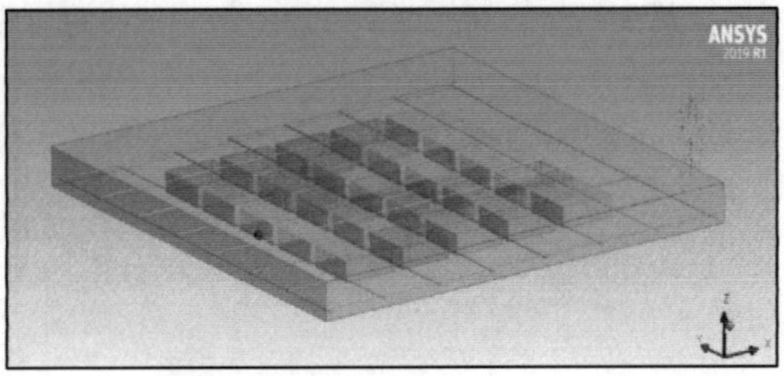

Fig. 6(b) Geometry conditions of the filler slab

8. Results and discussion

The results of properties of filler materials and the compressive strength of filler material has been tabulated. The cost comparison between the conventional slab and filler slab is estimated. Also, the results from STAAD.Pro and ANSYS software are interpreted.

8.1 Properties of filler material

The basic properties of the filler material were found experimentally and the results are shown in Table 1.

Table 1 Filler material properties

Property	Granite dust	Foundry sand	Range as per Code	Acceptable or Not acceptable
Specific Gravity [5]	2.32	2.05	2 – 2.5	Acceptable
Moisture content [5]	1.45%	Nil	-	-
Cone penetration test (liquid limit) [6]	16%	5.04%	-	-
Bulking [5]	66.67%	18.75%	30 - 40	Not acceptable
Fineness modulus (Sieve analysis) [5]	1.62 (very fine sand)	2.3 (Medium sand)	2 – 3.5	Acceptable
Maximum Dry density (Proctor test) [7]	1.19 g/cc	1.75 g/cc	Around 1.182	Acceptable

8.2 Properties of cement

Properties of Ordinary Portland cement (OPC) 53 grade was found experimentally satisfying the standards as per BIS specifications IS 12269 (2013) [7]. The properties of cement are presented in Table 2.

Table 2 Properties of 53 grade Cement

Property	Cement	Range as per Code	Acceptable or Not acceptable
Specific gravity [9]	3.14	Around 3.15	Acceptable
Normal consistency [9]	31% by weight of cement	26 - 33	Acceptable

8.3 Compression tests of granite dust cylinders

The compression test of the filler material (Granite dust) with various dose of stabilizer was experimented in the laboratory to check its optimum stabilizer and the results are as shown in Table 3. A graph was being plotted for its compressive strength with various dosages of stabilizer as shown in Fig. 7

Table 3 Compression test results

Cement (%)	Water (%)	W/C ratio	Weight (kg)	Density (kg/m³)	Curing days	Load (kN)	Compressive strength (MPa)
4	16	4	4.1	2204.30	7	5	0.58
4	16	4	4.06	2136.84	14	8	0.92
4	16	4	4.18	2200.00	28	9	1.04
6	16	2.67	4.19	2327.78	7	9	1.04
6	16	2.67	4.21	2263.44	14	11	1.27
6	16	2.67	4.17	2241.94	28	14	1.62
8	16	2	4.17	2194.74	7	14	1.62
8	16	2	4.19	2252.69	14	17	1.96
8	16	2	4.18	2200.00	28	20	2.31
10	16	1.6	4.1	2204.30	7	23	2.66
10	16	1.6	4.12	2168.42	14	26	3.00
10	16	1.6	4.15	2231.18	28	29	3.35
12	16	1.3	4.16	2236.56	7	28	3.23
12	16	1.3	4.15	2184.21	14	37	4.27
12	16	1.3	4.18	2309.39	28	41	4.74

\

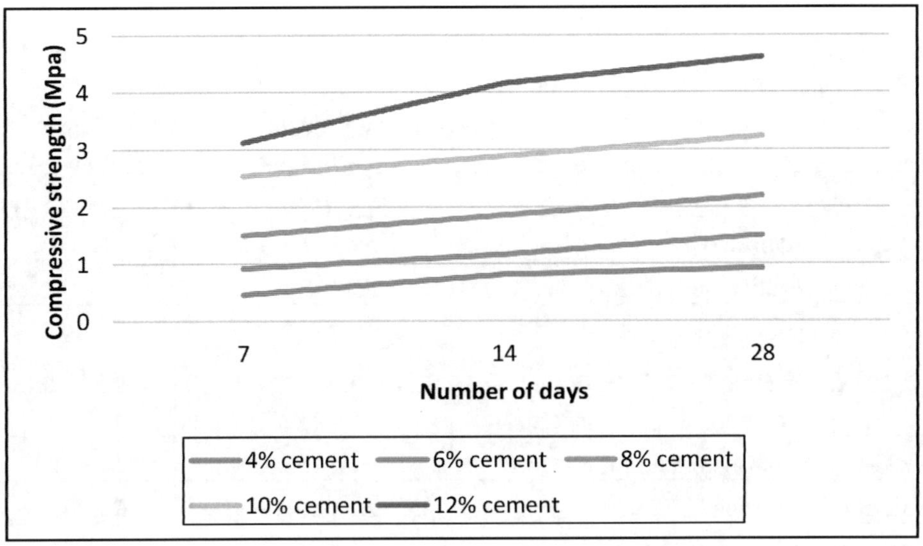

Fig. 7 Graph of compressive strength with various dose of stabilizer

The compressive strength of granite dust with various dosages of cement was found experimentally. As the percentage of cement increases, the strength also increases and the stabilizer content must also satisfy the economic parameters. Therefore, the optimum dosage percentage was found to be 10% using the increase in strength for every 2% increase in dosage as shown in table 4.

Table 4 Increase in strength of Granite dust

% cement	7 days	14 days	28 days	% increase in	% increase in	% increase in
	Strength (Mpa)	strength (Mpa)	Strength (Mpa)	7 days strength	14 days strength	28 days strength
4	0.46	0.81	0.92			
6	0.92	1.16	1.5	50.00	30.17	38.67
8	1.5	1.85	2.19	38.67	37.30	31.51
10	2.54	2.89	3.23	40.94	35.99	32.20
12	3.12	4.16	4.62	18.59	30.53	30.09

8.4 Cost analysis of slab

On calculating the amount of concrete, the amount of steel, the cost for shuttering and the cement required for the making of filler material, cost analysis was done in comparison with the conventional slab as shown in table 5 and it was observed that the total cost required to construct conventional slab was Rs.6150 and the cost required to construct filler slab was Rs 5923.

Table 5 Cost comparison of slab

	Conventional slab (2m x 2m)	**Filler slab (2m x 2m)**
Volume of concrete (m^3)	0.6	0.4848
Cement required (kg)	229.9	185.7
Fine aggregate required (kg)	405.8	327.9
Coarse aggregate required (kg)	667.2	539
Cost of cement for concrete	Rs 2299	Rs 1857
Cost of Fine aggregate for concrete	Rs 259.7	Rs 209.9
Cost of coarse aggregate for concrete	Rs 400.3	Rs 323.4
Cost of shuttering	Rs 2000	Rs 2000
Weight of steel (kg)	12.64	12.64
Cost of steel	Rs 632	Rs 632
Cement required of filler material (kg)	-	36.3
Cost of cement for filler material	-	Rs 363
Total cost	Rs 5591	Rs 5385
10% over head	Rs 559.1	Rs 538.5
Final cost	Rs 6150.1	Rs 5923.5

8.5 STAAD analysis

8.5.1 Conventional slab

Analysis of Conventional slab was done using STAAD.Pro software to study the behaviour of slab. Results obtained are mentioned in below.

The stress that is being experienced upon applying the maximum load is the Absolute stress. Stress contour is being obtained as shown in Fig.8(a) resulting the maximum stress of 0.2696 N/mm^2 acting along the diagonal and minimum stress of 0.093 N/mm^2 as shown in Fig.8(a) acting at the mid-corners of the slab.

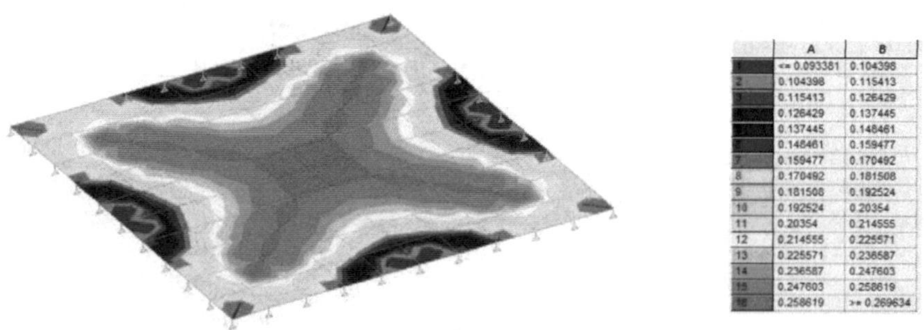

Fig. 8(a) Absolute stress contour

These are the forces the resists flexural activity along both direction. The shear force contour along X-direction and Y-direction are obtained as shown in Fig.8(b) resulting the maximum value to be 0.027 N/mm^2 and the minimum value to be -0.027 N/mm^2.

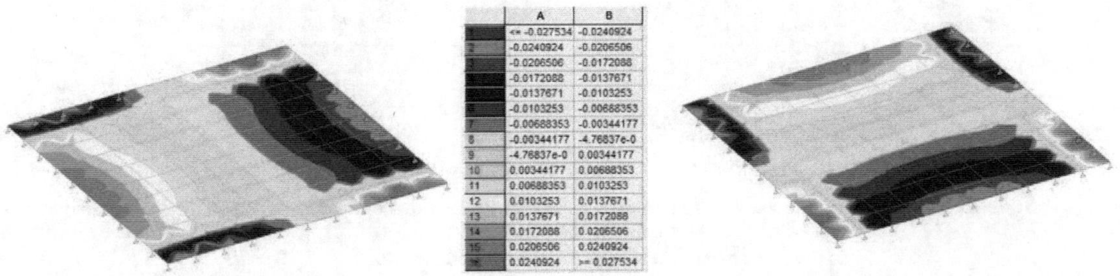

Fig. 8(b) Shear force contour along X-direction (left) and Y-direction (right)

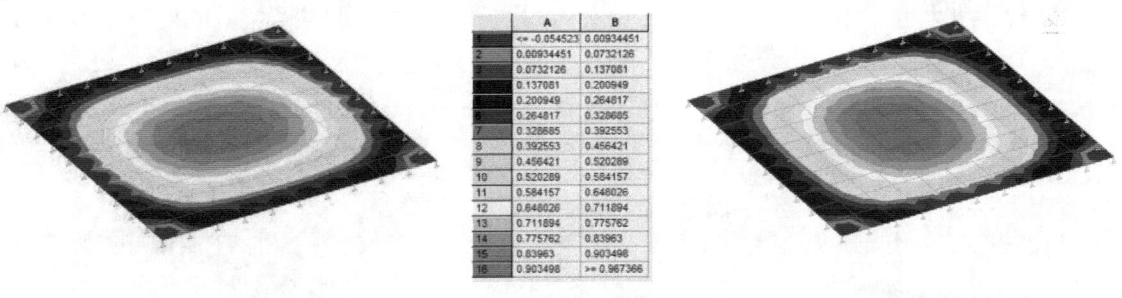

Fig. 8(c) Bending moment contour along X-direction and Y-direction

The slab is allowed to deflect upon applying the required load and was observed to have maximum deflection at the centre of the slab as shown in Fig.8(d). The slab experienced maximum deflection of 0.11mm

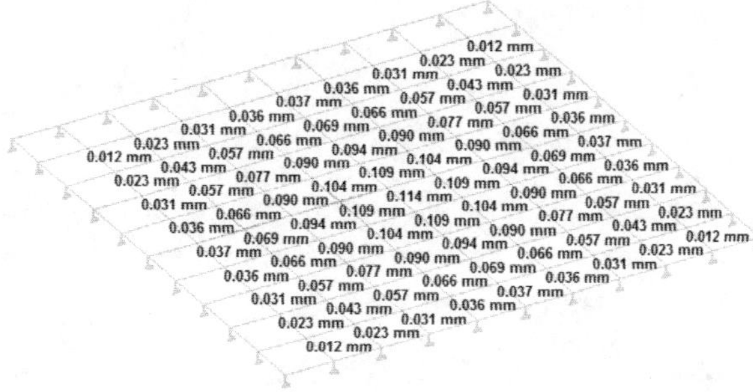

Fig. 8(d) Deflection of slab

8.5.2 Analysis of Filler slab (Granite dust)

The stress that is being experienced upon applying the maximum load is the Absolute stress. Stress contour is being obtained as shown in Fig.9(a) resulting the maximum stress of 0.262 N/mm^2 acting along the diagonal and minimum stress of 0.090 N/mm^2 as shown in Fig.9(a) acting at the mid-corners of the slab.

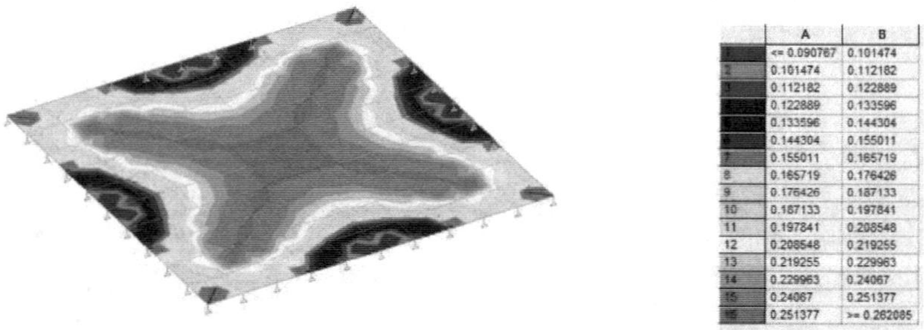

Fig. 9(a) Absolute stress contour

These are the forces the resists flexural activity along both direction. The shear force contour along X-direction and Y-direction are obtained as shown in Fig.9(b) resulting the maximum value to be 0.026 N/mm^2 and the minimum value to be -0.026 N/mm^2 as shown in Fig.9(b).

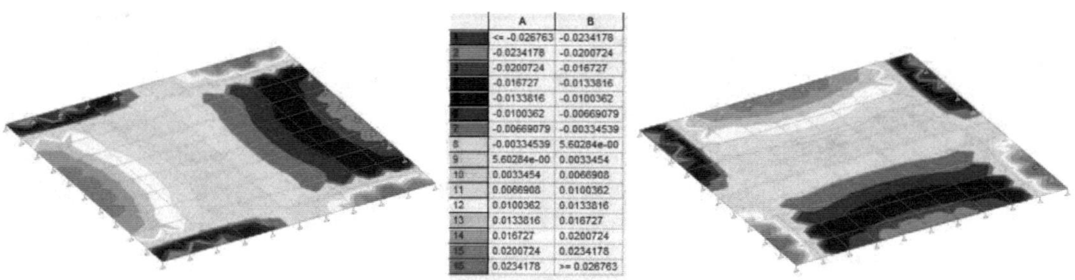

Fig. 9(b) Shear force contour along X-direction (left) and Y-direction (right)

Since, the designed two-way slab was simply supported, maximum bending moment occurs at the mid span. The bending moment contour along X-direction and Y-direction are was obtained as shown in Fig. 9(c) resulting the maximum value to be 0.940 kNm/m and the minimum value to be -0.052 kNm/m

.

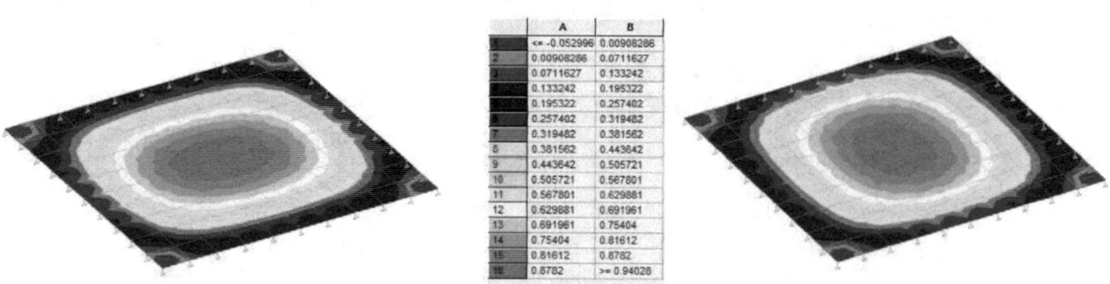

Fig. 9(c) Bending moment contour along X-direction and Y-direction

The slab is allowed to deflect upon applying the required load and was observed to have maximum deflection at the centre of the slab as shown in Fig.9(d). The slab experienced maximum deflection of 0.059mm.

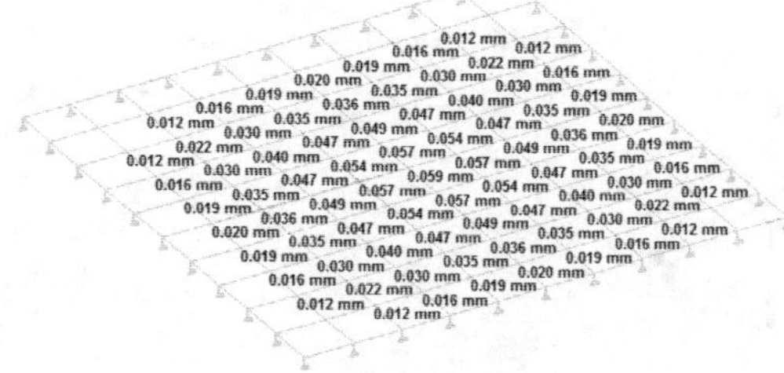

Fig. 9(d) Deflection of slab

8.5.3 Analysis of Filler slab (Foundry sand)

The stress that is being experienced upon applying the maximum load is the Absolute stress. Stress contour is being obtained as shown in Fig.10(a) resulting the maximum stress of 0.249 N/mm^2 acting along the diagonal and minimum stress of 0.086 N/mm^2 as shown in Fig.10(a) acting at the mid-corners of the slab.

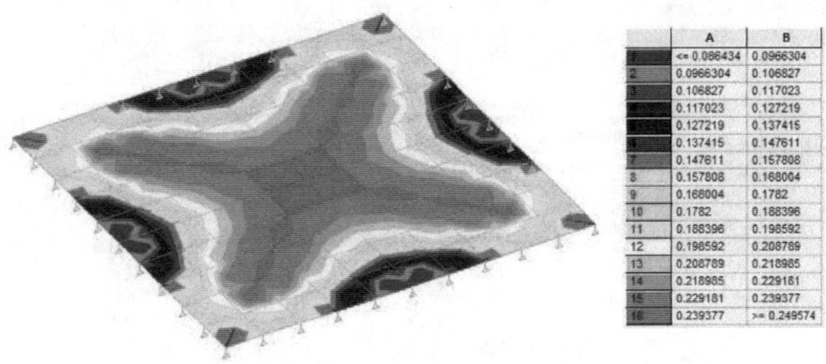

	A	B
1	<= 0.086434	0.0966304
2	0.0966304	0.106827
3	0.106827	0.117023
4	0.117023	0.127219
5	0.127219	0.137415
6	0.137415	0.147611
7	0.147611	0.157808
8	0.157808	0.168004
9	0.168004	0.1782
10	0.1782	0.188396
11	0.188396	0.198592
12	0.198592	0.208789
13	0.208789	0.218985
14	0.218985	0.229181
15	0.229181	0.239377
16	0.239377	>= 0.249574

Fig. 10(a) Absolute stress contour

These are the forces resists flexural activity along both direction. The shear force contour along X-direction and Y-direction are obtained as shown in Fig.10(b) resulting the maximum value to be 0.025 N/mm^2 and the minimum value to be -0.025 N/mm^2.

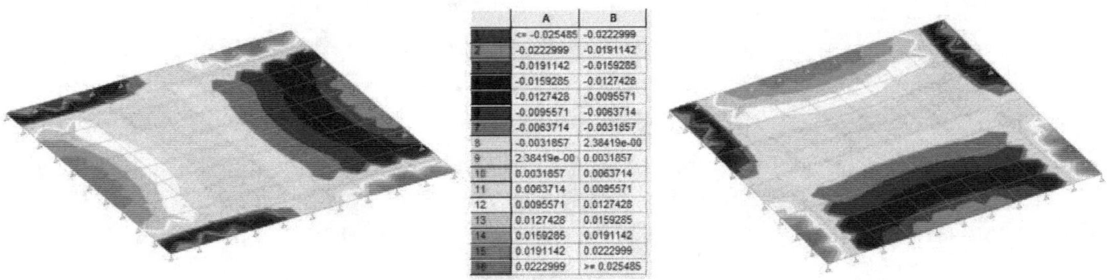

	A	B
1	<= -0.025485	-0.0222999
2	-0.0222999	-0.0191142
3	-0.0191142	-0.0159285
4	-0.0159285	-0.0127428
5	-0.0127428	-0.0095571
6	-0.0095571	-0.0063714
7	-0.0063714	-0.0031857
8	-0.0031857	2.38419e-00
9	2.38419e-00	0.0031857
10	0.0031857	0.0063714
11	0.0063714	0.0095571
12	0.0095571	0.0127428
13	0.0127428	0.0159285
14	0.0159285	0.0191142
15	0.0191142	0.0222999
16	0.0222999	>= 0.025485

Fig. 10(b) Shear force contour along X-direction (left) and Y-direction (right)

Since, the designed two-way slab was simply supported, maximum bending moment occurs at the mid span. The bending moment contour along Y-direction was obtained as shown in Fig.10(c) resulting the maximum value to be 0.895 kNm/m and the minimum value to be -0.050 kNm/m.

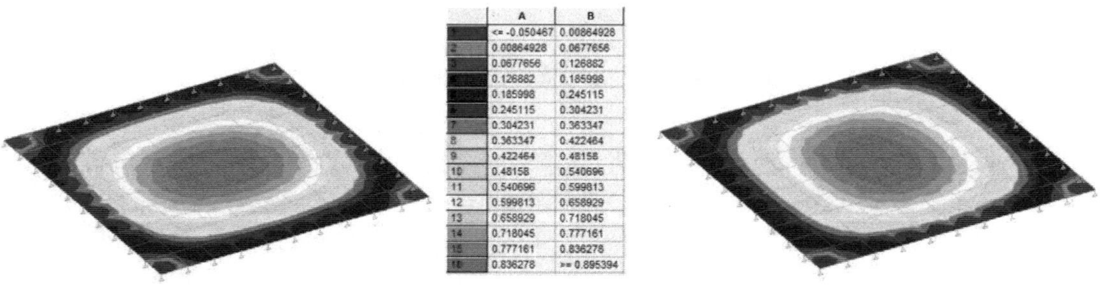

Fig. 10(c) Bending moment contour along X-direction and Y-direction

The slab is allowed to deflect upon applying the required load and was observed to have maximum deflection at the centre of the slab as shown in Fig.10(d). The slab experienced maximum deflection of 0.056mm.

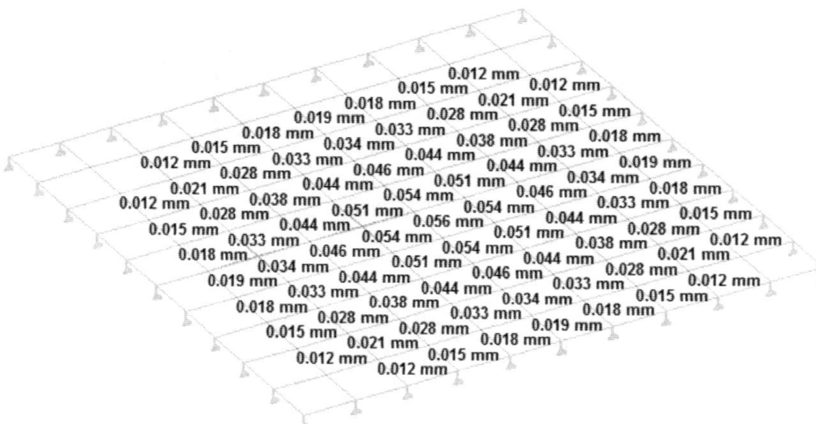

Fig. 10(d) Deflection of slab

8.6 ANSYS software

8.6.1 Analysis of Conventional slab

The stress that is being experienced upon applying the maximum load is the Shear stress. Stress contour is being obtained as shown in Fig.11(a) resulting the maximum stress of 0.25 N/mm^2 acting along the diagonal and minimum stress of -0.35 N/mm^2 acting at the mid-corners of the slab. On calculating the readings obtained, it is observed that there was a maximum shear force of 32.5kN which is under the limit as per IS 456 [8] and minimum shear force acting -45.5kN.

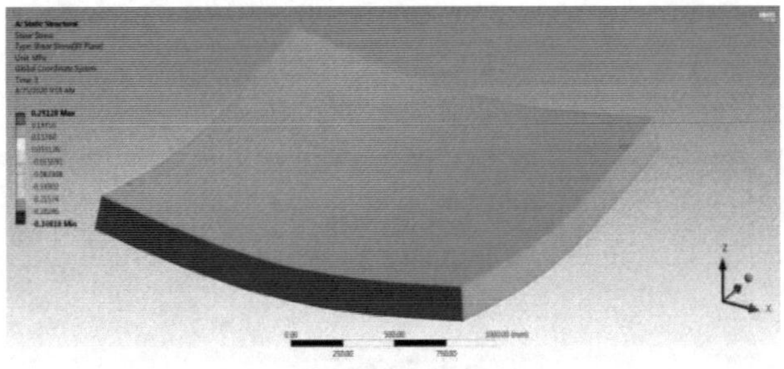

Fig. 11(a) Shear stress of slab

It is a one dimensional stress which is usually a stress for its plasticity. The equivalent stress contour is obtained as shown in Fig.11(b) resulting the maximum value to be 2.44 N/mm^2 and the minimum value to be 0.008 N/mm^2 and it is observed that minimum stress is experienced on all four corners of the slab.

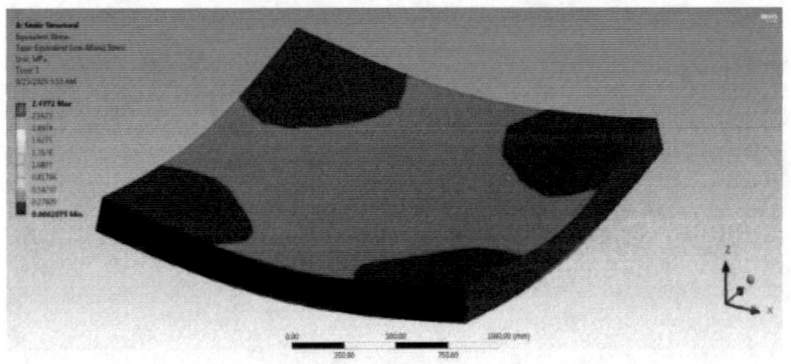

Fig. 11(b) Equivalent stress contour

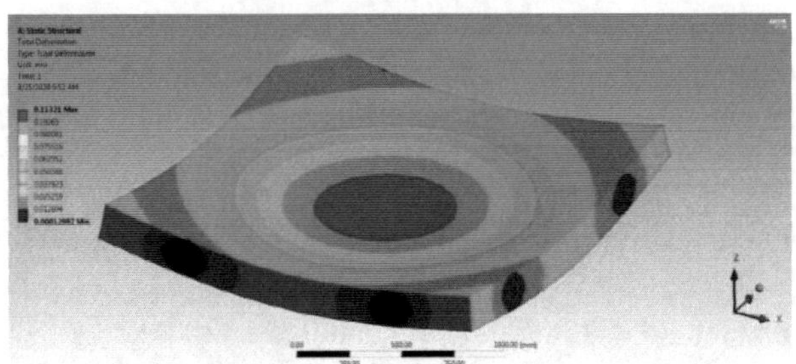

Fig. 11(C) Deformation contour

8.6.2 Analysis of filler slab (granite dust)

The stress that is being experienced upon applying the maximum load is the Shear stress. Stress contour is being obtained as shown in Fig.12(a) resulting the maximum stress of 0.264N/mm^2 acting at the opposite corners of slab and minimum stress of -0.33 N/mm^2 acting along the diagonal.

On observing the readings obtained, it is observed that there was a maximum shear force of 34.32kN and minimum shear force acting -42.9kN.

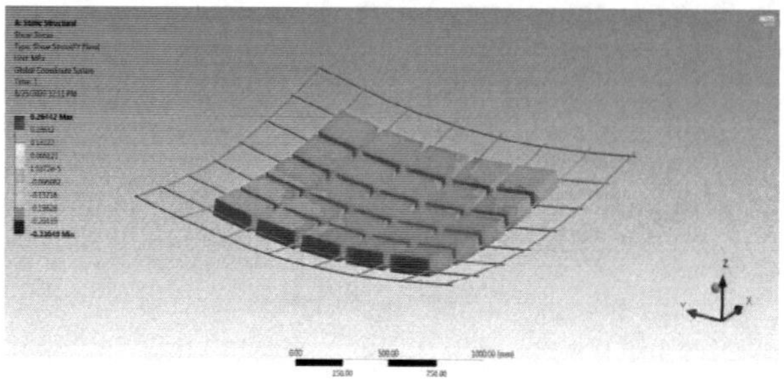

Fig. 12(a) Shear stress of slab

It is a one dimensional stress which is usually a stress for its plasticity. The equivalent stress contour is obtained as shown in Fig.12(b) resulting the maximum value to be 2.93 N/mm^2 and the minimum value to be 0.001 N/mm^2 and it is observed that minimum stress is experienced on all four corners of the slab as well as on the filler material.

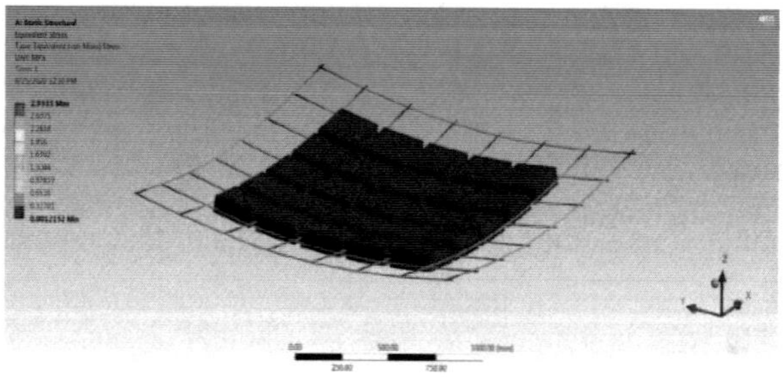

Fig. 12(b) Equivalent stress contour

The slab is allowed to deflect upon applying the service load and was observed to have maximum deflection at the centre of the slab as shown in Fig.12(c). The slab experienced maximum deflection of 0.13mm.

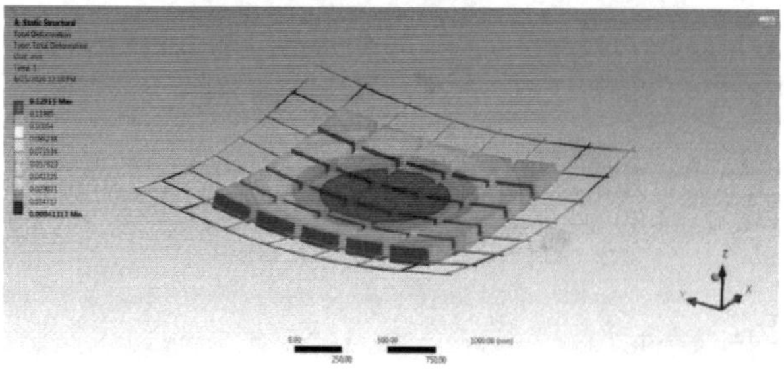

Fig. 12(c) Deformation contour

8.6.3 Analysis of filler slab (foundry sand)

The stress that is being experienced upon applying the maximum load is the Shear stress. Stress contour is being obtained as shown in Fig.13(a) resulting the maximum stress of 0.26 N/mm^2 acting at the opposite corners of slab and minimum stress of -0.325 N/mm^2 acting along the diagonal.

On observing the readings obtained, it is observed that there was a maximum shear force of 33.8kN and minimum shear force acting -42.25kN.

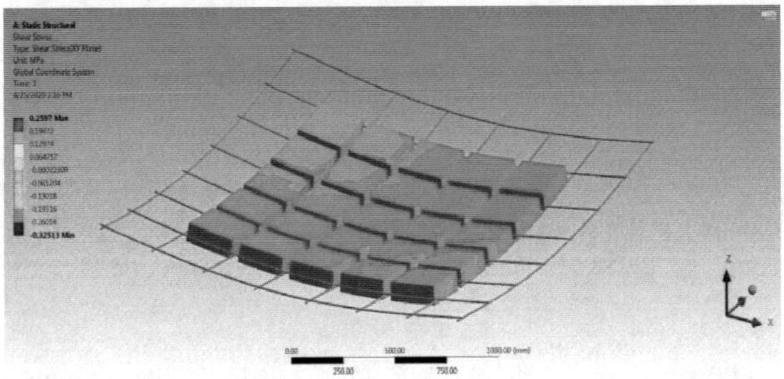

Fig. 13(a) Shear stress of slab

It is a one dimensional stress which is usually a stress for its plasticity. The equivalent stress contour is obtained as shown in Fig.13(b) resulting the maximum value to be 2.91 N/mm^2 and the minimum value to be 0.001 N/mm^2 and it is observed that minimum stress is experienced on all four corners of the slab as well as on the filler material.

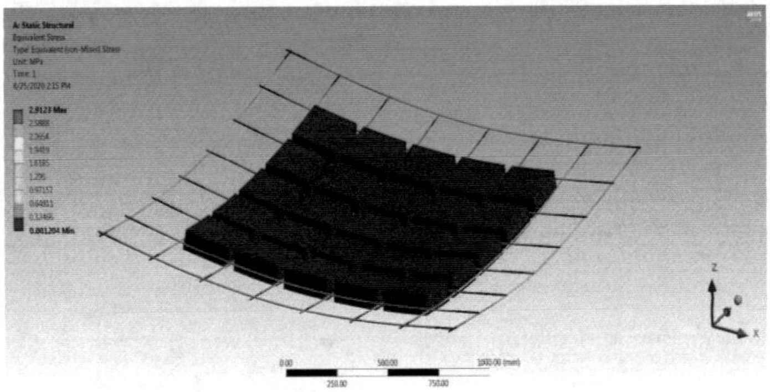

Fig. 13(b) Equivalent stress contour

The slab is allowed to deflect upon applying the service load and was observed to have maximum deflection at the centre of the slab as shown in Fig.13(c). The slab experienced maximum deflection of 0.125mm.

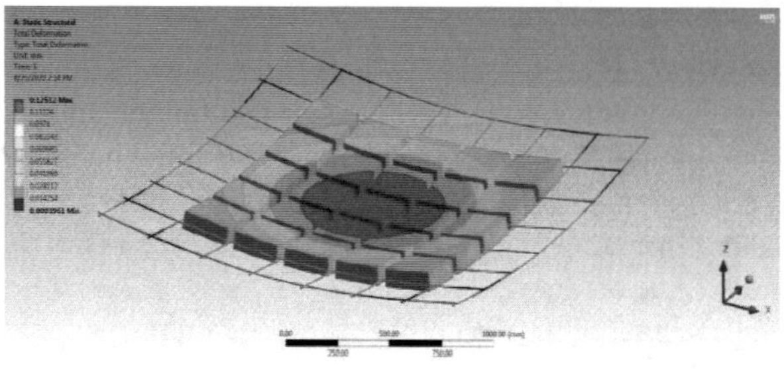

Fig. 13(c) Deformation contour

9. Conclusions

Based on the results obtained from the analysis done using ANSYS software, a comparison is done between the filler slab and conventional slab. It has been explained by following terms:

- All the basic tests conducted experimentally satisfies the respective Indian standards.
- Construction of filler slab was found to be economical by saving 4% when compared to the RCC slab of the same size.
- On the application of load, the shear stress on the Filler slab using Granite dust as the filler material was increased by 5.3% and the shear stress on the Filler slab using Foundry sand as the filler material was increased by 3.9% when compared to the conventional slab.
- Upon applying load, the shear force on the Filler slab using Granite dust as the filler material was increased by 5.3% and the shear force on the Filler slab using Foundry sand as the filler material was increased by 3.9% when compared to the conventional slab.
- Upon analysis, the Equivalent stress on the Filler slab using Granite dust as the filler material was increased by 16.72% and the Equivalent stress on the Filler slab using Foundry sand as the filler material was increased by 16.15% when compared to the conventional slab.
- On testing for its deflection, the deflection of the Filler slab using Granite dust as the filler material was increased by 15.4% and the deflection of the Filler slab using Foundry sand as the filler material was increased by 12% when compared to the conventional slab while the deflection of the conventional slab was cross verified using STAAD.Pro software.
- In this study, it is observed that there is no much difference in terms of Deflection, Shear stress, Shear force and Equivalent stress. On the other hand, adopting this technique can reduce the cost by 4%.

References

[1] Office of the Registrar General & Census Commissioner, India, Ministry of Home Affairs, Government of India, www.censusindia.gov.in, CensusInfo, 2011 (Accessed on 12-04-2020).
[2] Harry V and Pinto P 2014 *Journal of Information, Knowledge, and Research* **14(2)** 975–6744.
[3] El-Dieb A S, El-Ghareeb M A, Abdel-Rahman M A H and Nasr E S A 2018 *Journal of Buildings and Engineering* **15** 61–69.
[4] Morel J C, Mesbah A, Oggero M and Walker P 2001 *Buildings and Environment* **36(10)** 1119–1126.
[5] IS 2386-3 1963 Methods of test for aggregates for concrete.
[6] IS 2386-1 1963 Methods of test for aggregates for concrete.
[7] IS 2720-5 1985 Methods of test for soils.
[8] IS 875-1 1987 Code of practice for Design loads.
[9] IS 12269 2013 Ordinary Portland cement, 53 grade – Specification.

IOP Conf. Series: Materials Science and Engineering 1006 (2020) 012002 doi:10.1088/1757-899X/1006/1/012002

Analysis of Traffic Congestion Impacts of Urban Road Network under Indian Condition

S R Samal[1], P Gireesh Kumar[2], J Cyril Santhosh[3], and M Santhakumar[4]

[1]Assistant Professor, KIIT Deemed to be University, Bhubaneswar, India
[2]Associate Professor and Head, Department of Civil Engineering, Shri Vishnu Engineering College for Women, Bhimavaram, West Godavari District- 534202, India
[3]Assistant Professor, Coimbatore Institute of Technology, Coimbatore, India
[4]Professor, National Institute of Technology, Tiruchirappalli, India

E-mail: satya.samalfce@kiit.ac.in

Abstract. Traffic congestion has been an area of major bother across the globe. The existing infrastructure is not able to cope with the new traffic demand. Furthermore the restriction of the space and outside activities influencing the traffic congestion. The emerging country like India, where the traffic conditions comprise heterogeneous traffic with no lane discipline, further creates more complicated scenarios for the researcher. A substantial portion of working hours is getting wasted on the roads because of traffic congestion, which imposes the negative effect on the overall economy. There has been numerous of literature and studies for analysis of the traffic congestion and its impacts. However, the result has not been much satisfying. In the present study congestion forecast is aimed under mixed traffic with no lane discipline towards identifying the inherent viability of the diversified traffic situation and presents better recommendations in controlling and evading these prolonged traffic jams. The urban highway systems were considered as a study area. Required particulars were collected by a License Plate Matching method using video graphic survey for the day rush timing considered from 8:00 AM to 10:00 AM and off peak hours to estimate the travel time of a distinct class of motorized vehicles for selected sections of the urban roadway. Congestion indices for both the up and down traffic of a particular road were evaluated from the data collected from the video recording. Traffic congestion impacts were analyzed and possible mitigation measures have been suggested. The study mainly focused on traffic jam indices with regarded to travel time reliability measures to observe the functional effectiveness of the urban road network.

1. Introduction

Congested traffic, leads a significant threat that hinders the growth of a country. Due to the growth in population and motorized vehicles, there is huge congestion on the roadway. Increasing population and vehicles, the transhumance of rural inhabitants to metropolitan cities hunting for superior opportunities are the fundamental reason of the traffic jam. Traffic Congestion may be categorized into two types, i.e. recurring type, which is primarily taken place due to capacity and behavioral issues or non-recurring type, which is because of accidents, construction, or emergencies. The recurring and non recurring congestion, having different causes, but having indistinguishable effects. Traffic jam accelerates additional hindrance, causing the unpredictability to traveler engendering to personage tension and risky transportation circumstances. Defective traffic signaling structures, scarcity of the workforce, limited roadway width and overtaking proneness of motorists extended the road congestion. The valuable working hours of the road users getting wasted on the road, which secondarily adversely influenced the overall economy. Traffic congestion not only affecting the human being, but also elevates the pollution.

Traffic congestion also influences accessibility and mobility. Traffic congestion increases travel time and fuel costs, which adversely effecting organizations and employees distributing goods and services. Traffic congestion is normally described as the traffic demand surpassing the roadway capacity. Traffic density (the number of vehicles in a specified length of a roadway) is one of the most frequently used congestion indicators. The travel time reliability measures such as level of service, roadway congestion index and lane-mile duration index can also be considered to identify the traffic congestion parameters. For evaluating the traffic congestion in the urban roadway, the speed performance index is one of the indicators. There is extensive mental as well as physical stress on drivers because of traffic congestion and increases their aggression which might induce unfortunate incidents like road crashes. Every day by millions of people across the globe are directly affected by the traffic congestion. Traffic congestion originates significant air pollution and noise pollution consequently aggravate the whole atmosphere surrounding. Congestion pricing techniques as a measure of congestion mitigation have been suggested by the majority of the researcher. The impact of traffic congestion can be related to Fuel expenditure cost, Transit cost, Health associated matters and Environment concerned matter.

The current traffic framework is inadequate to encounter the traffic demand due to an exponential increase in population as well vehicle. This research aims to analyze the traffic congestion impacts of the urban road network of a smart city in India. Congestion indices have been evaluated to examine the operational efficiency of the road network. In order to plan the trips productively and to choose the routes suitable travel time reliability measure has a significant part to the travelers. Finally, the study also examines the traffic congestion impacts and its mitigation measures.

2. Literature review

Qingyuet al. (2007) analyzed qualitative and quantitative of the external costs that the urban traffic congestion pricing. Described the external costs of traffic congestion in terms of extra travel time costs, environmental pollution costs, traffic accident costs, and fuel consumption costs.

Lyman, and Bertini (2008) described congestion measures and travel time reliability measures. Discussed different travel time reliability measures, but considered the buffer index as important measures. They suggested that adding travel time reliability measure can be a superior comprehension for the transportation network.

Padiath et al. (2009) illustrated the traffic congestion indicator in terms of traffic density. In order to establish the practicality, few approaches for density forecast under similar traffic circumstances are tired under the mixed traffic scenario.

Uddin (2009) addressed the issues of increasing population and vehicles, especially passenger cars with some statistics.. He also found that vehicles in India are distributed unevenly He identified that the transhumance of rural inhabitants to metropolitan cities hunting for superior opportunities is the one of the primary reasons of the urban traffic congestion.

Zhiliet al. (2009) forecasted the significance of congestion charge as per the locality, land use attributes, and present travel environment. Evaluated the travel situations under congestion pricing depends on travel demand model and given recommendations for executing the congestion charge.

McGroarty (2010) focused on causes of recurring congestion, which is mainly occurring due to capacity and behavioural issues and non-recurring congestion, which is primarily occurring due to events which creates unanticipated congestion. Though the occurrences are different, but the impacts are similar

Mahmud et al. (2012) found that restricted resources invested in the growth of transportation infrastructure, a quick jump up in transport demand, the intervenes of enormous non-motorized

vehicles on roads, lack of implementation of appropriate and realistic traffic governance strategy are the critical transport issues in almost all the urban areas.
Rao, and Rao (2012) focused mostly on urban arterial congestion. The study executed based on measurement metrics such as speed, travel time/delay and volume and level of service. They have adopted the travel time reliability measures to identify the traffic congestion parameters.
Dubey, & Borkar (2015) focused on detection techniques of the traffic congestion. They have described the major problem incurred during the detection of the traffic congestion on the urban road network. They have suggested to adopt GPS based tecnique for the betterment of the result in accordance with the traffic safety.
He et al. (2016) have adopted performance index to determine the existing road network conditions of congestion in urban road networks. For evaluation of traffic congestion, they have adopted the speed performance index as the evaluation indicator of urban road traffic system.
Samal, and Das (2020) have taken the buffer index to predict the congestion level in the specified road network. They have established regression modelling to see the differences in observed value and predicted value. Some of the recommendations have been suggested.

3. Data collection and analysis

3.1 Travel time estimation
Utilizing the common analytical methods or by application of computer software mean travel times are evaluated from independent travel times. The time indispensable to travel a path connecting any two points of interest called as travel time. For a comparatively short span travel time can be determined by accepting that the mean speed at a specific point (spot speed) is constant. The acceptance of steady speed over a brief roadway portion is relevant to continuous flow provisions with stable flow traffic pattern. For measurement of congestion indices travel time is in important parameter.

3.2 Travel time reliability
The measure of travel time reliability is the replacement of the volume-to-capacity proportion. It assists to govern the acceptability of the assistance furnished by the urban network The V/O ratio concentrates especially to construct additional roadways to accommodate the traffic demand, whereas travel time reliability measures are concentrated regarding superior administration of the current structure.

3.3 License plate matching techniques
It is one type of travel time data collection technique. Registration number matching approaches comprise of gathering vehicle registration number and entry time at different points, complementing the registration number between successive points, and determining travel times through the differences in entry times.
Four fundamental techniques are employed for gathering the information's about registration number matching approaches.

- **Physical Method:** With the help of conventional pen and paper or tape recorder gathering the information's about registration number and physically extracting the required data.
- **Portable Computer:** Gathering the information's about registration number in the working areas by utilizing movable computers that voluntarily extract the required data.
- **Video graphic technique with Physical Method:** By making use of video cameras gathering the information's about registration number in the working areas and physically extracting the required data.
- **Video graphic technique accompanied by Character Identification:** Gathering the information's about registration number in the working areas by video graphic technique

and voluntarily extracts the required data by utilizing computerized registration number character Identification.

The present study employed the Video graphic technique with Physical Method, i.e. by utilizing video cameras gathering the information's about registration number in the working areas and physically extracting the required data.

3.4 Congestion indices

To evaluate the serviceability and overall performance of the present urban roadway, congestion indices are the significant variables.. Congestion indices are the estimate of road network serviceability and overall performance. Different congestion indices are

- Travel Time Index (TTI)
- Buffer Index (BTI)
- Planning Time Index (PTI)

$$Travel\ Time\ Index = \frac{Mean\ Travel\ time}{Off\ peak\ travel\ time}$$

$$Buffer\ Time\ Index = \frac{95th\ percentile\ travel\ time - Mean\ Travel\ time}{Mean\ travel\ time}$$

$$Planning\ Time\ Index = \frac{95th\ percentile\ travel\ time}{Off\ peak\ travel\ time}$$

The present study has adopted Buffer Time Index to evaluate the congestion index for the selected urban road networks. BTI represents the traveler need to put additional time in their mean travel time to make sure on time arrival most of the time.

3.5 Selection of the study area

One of the significant and the first step of the research work is the selection of the study area. The location of study area decides the requirement of research activities to be done in consideration of the existing transport framework, traffic behaviour and traveller information. The urban road network has been taken as the study area for the evaluation of congestion indices.

Table 1: Lane attributes of the chosen roadway

Sequence No.	Category of road	Lane attribute
1.	Arterial	6 lane divided
2.	Arterial	6 lane divided

The present study focused on the two locations, where different data's were collected. The traffic volume count was executed during morning rush hours (from 8.00 A.M.to 10.00 A.M) and morning off-peak hours (from 2.00 P.M. to 3.00 P.M.) employing video graphic survey in the chosen road locations. According to Indian Road Congress106 -1990, “Guidelines for Capacity of Urban roads in Plain areas", the volume count is analyzed.

Table 2: Recommended Passenger Car Unit (PCU) factors for different category of vehicles in urban road networks

Sequence number	Category of vehicles	Equivalent PCU Factors Percentage Composition of different Vehicles	
		5%	10% and above
1	Two wheeler, Motorcycle or Scooter	0.5	0.75
2	Passenger car, pick up van	1.0	1.0
3	Auto Rickshaws	1.2	2.0
4	Light Commercial Vehicle	1.4	2.0
5	Truck or Bus	2.2	3.7
6	Agricultural Tractor Trailer	4.0	5.0
7	Cycle	0.4	0.5
8	Cycle Rickshaw	1.5	2.0
9	Horse Drawn Vehicle	1.5	2.0
10	Hand cart	2.0	3.0

Table 3: Buffer Time Index (%) of the chosen urban roadway

Road Number	Direction	Peak average travel time (Sec)	95th percentile travel time (Sec)	BUFFER TIME INDEX (%)
Selected Road 1	1-2	12.70	16.50	30
	2-1	13.10	16.60	27
Selected Road 2	1-2	15.10	20.30	35
	2-1	14.20	18.40	30

Among the different congestion indices, the present study mainly focused on the Buffer Time Index, which gives acceptable outcomes in the mixed traffic environment. According to the constitution of the vehicle and lane assignment, the travel time and congestion indices are varied. Smaller the Buffer Index, better the serviceability.

4. Impacts of traffic congestion

The impact of congestion can be studied in three ways, such as impact on economy, impact on the environment and impact on health. For the current research work the relevant information has been assembled from various activities, e.g. group conversation, questionnaire review, physical monitoring, individual Interaction, journals and articles, website articles, and preceding works on the relevant matter to study the impact of traffic congestion on various parameters.

4.1 Impact on economy

People are forced to wasting money in different methods due to traffic congestion, such as wasting man hour, additional transportation cost, additional fuel consumption, vehicle operating and maintenance cost and diverse cost. From the collected data the study found that travellers are wasting 50% of their traffic hourly during the journey, e.g. if the traveller requires 1 hour during off peak hours for the journey and the same length of the road will take 1.30 hours to the traveller during the peak hours. The extra time, which they can transform towards precious working hour that could contribute to the overall economy.

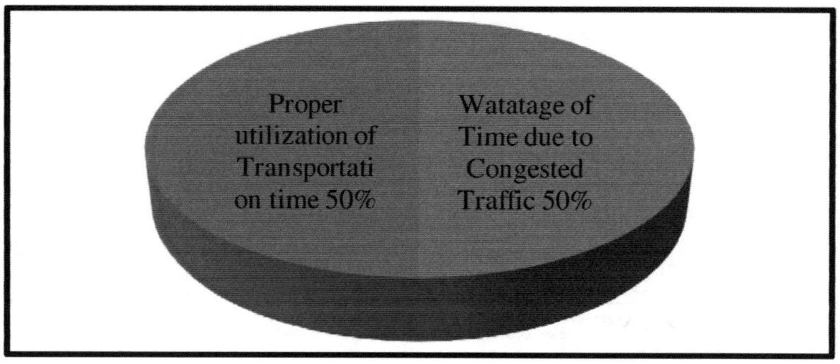

Figure 1: Wastage of time due to traffic congestion

4.2 Impact on health

Because of congestion people are forced to spend extra time on the road, which directly affecting the health of the individuals. The impact of health due to congestion cannot be neglected. The data have been collected as mentioned sources to study the impact of congestion on health related issues. From the collected response we found that, 81% individuals are getting affected because of traffic congestion, Whereas 16% individuals gave a response that their health was not getting affected due to traffic congestion and 3% response did not have any conclusive answer.

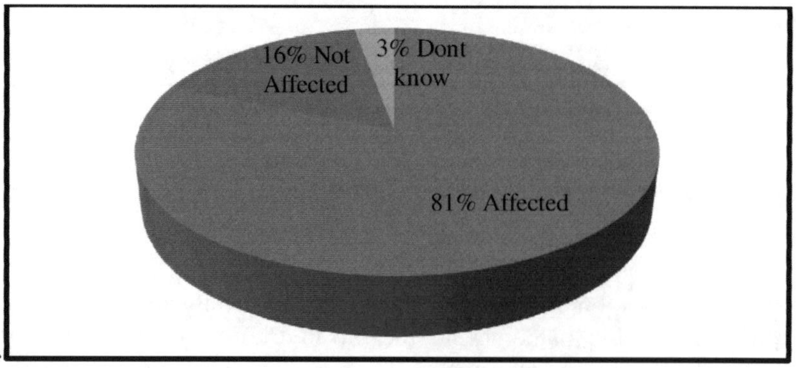

Figure 2: Percentage of people suffered due to traffic congestion

The study has categorized the symptoms of suffering in different ways, such as respiratory complication, neuralgia related difficulties, psychological distress, trouble in hearing, unpredicted sweating, fatigue, suffocation, dust allergy, difficulties in visibility, digestion problem, dehydration etc.

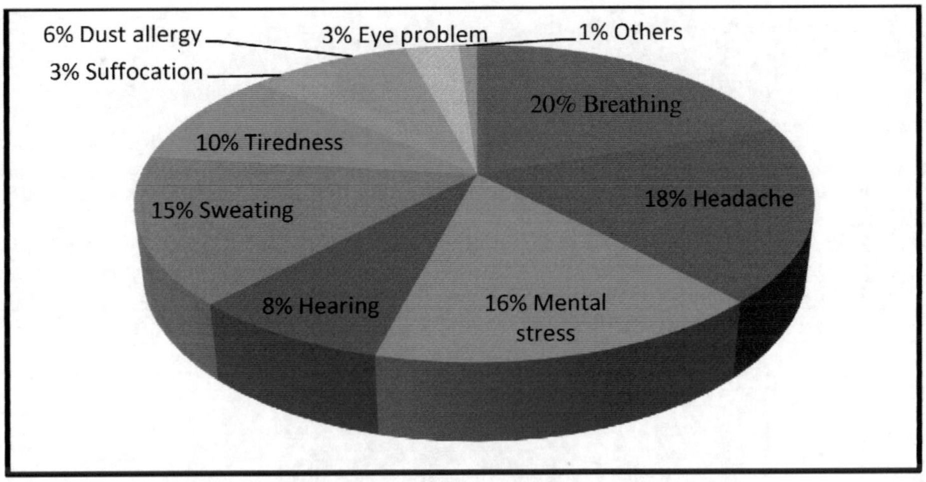

Figure 3: Percentage of categorized symptoms suffered due to traffic congestion

It can be seen that the majority of the suffering is related to breathing (20%), headache (18%), mental stress (16%) and sweating (15%), which can be considered as high level of distresses. Whereas hearing (8%), tiredness (10%) and dust allergy (6%) considered as a moderate level of distresses. Low level of distresses is suffocation (3%), Eye problem (3%) and others e.g. digestion problem, dehydration etc.

4.3 Impact on environment

The environment plays an important role in traffic congestion. The environment is getting polluted on the whole by sound pollution and air pollution. During the traffic congestion the roadway is of its full capacity. Everyone wants to be free from congestion. The sound created from vehicle as well as horns by the operator really creates the worst environment. Not only the people who suffer from traffic congestion, but also the neighbours people are also disrupted. When vehicles put a stop for a prolonged period of time on the road due to congestion, the gases like SO2, CO, CO2, NO2 emitted, which is lightweight compared to air yet too harmful. The impact of environment can be can be categorized in different aspects, e.g. disturbance due to the horns, disturbance due to noise pollution, irritation, ill temper, having problem with concentration and sleeping, trouble in hearing and honking of drivers in traffic congestions.

5. Suggested mitigation measures

The present study mainly focused on the improvement of the existing infrastructure which does not require any huge investment. The data collected from various sources, e.g. general people, experts, researchers, academicians, engineers, research article, etc.

Table 4: Traffic congestion mitigation measures

Activities	Perception Percentages (%)
Awareness building,	3
Construction of flyovers	11
Public transportation uses	7
Strict traffic law Implementation	16
Rescheduling of Office/School timing.	4
Subway	3
Removing Non Motorized Vehicle (NMV)	2
Decentralization	16
Controlling road side activities	16
Car free days and discourage private cars	1
High Parking charge	3
Increasing pedestrian facilities	2
Adequate parking facilities	16

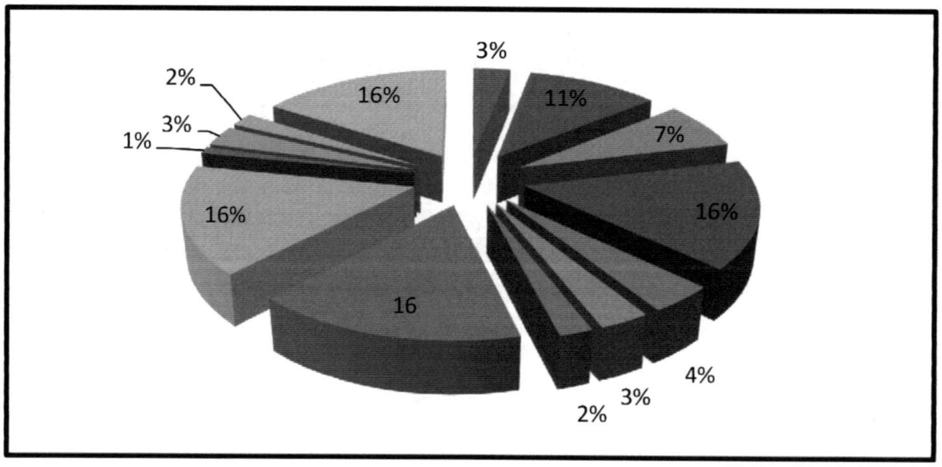

Figure 4: Percentage of factors Suggested for the mitigation measures

The present study tried to collect the information regarding the implementation of congestion mitigation techniques. New constructions are not given priority as the present study mainly focused on improvement of existing infrastructure. Very interesting facts are collected from the data as most of people are aware about the congestion, but they are not following the traffic rules. Majority of people gave a strong information about decentralization of goverment organization, Important sectors e.g hospitals, colleges, indursdries etc. From the most of the litereture we observed that more imporatnce given to Removing non motorized Vehicle (NMV), Car free days, High Parking charge, Increasing pedestrian facilities to counteract the congestion, where as in this study we have collected less response in this regards. The reason was the due to the increase of two wheeler the amount non motorized Vehicle (NMV) are getting absolute day by day. Even though there will be implementation of High Parking charge, Increasing pedestrian facilities, still without strict traffic law all of these factors cannot be implemented.

6. Conclusions and recommendations

Following conclusions can be drawn from the present study and analysis of traffic congestion data were collected from urban road networks in India and recommendations are also provided for the mitigation regarding the traffic congestion in the studied area. According to the vehicle composition and lane distribution, there is the variation of travel time and congestion indices. Buffer Time Index displays adequate outcome in mixed traffic conditions among the other indices. Several factors were considered to assess the congestion indices as well as analyze the influence of traffic congestion on the overall economy, environment including health. Congestion cannot be eliminated totally due to the increase in population as well as vehicle, specially two wheelers and passenger cars. Among the suggested congestion mitigation measures, the majority of the people have suggested that the congestion can be minimized by Implementation of strict traffic law, decentralization of important organization, controlling road side activities and adequate parking facilities.

There has to be proper coordination between the expert / researcher those are involved in the planning parts and the engineers those are basically involved in the execution part. Heterogeneous traffic conditions prevailing in developing countries like India are more complicated. The impact of traffic congestion adversely affects the economy, health, environment and traffic safety, which is the major concern for the researcher to come up with the comprehensive and new dimensions to tackle these challenges. Lane discipline, though a major factor no work is described to measure the congestion indices with account of Lane discipline in mixed traffic environment. Although the number of non-motorized vehicle are not so high, but it plays a significant part in city network and its impact on traffic jam eventually may not be ignored.

References

[1] Aftabuzzaman M 2007 September Measuring traffic congestion-a critical review *In 30th Australasian Transport Research Forum.*
[2] Bigazzi A Y 2011 Traffic congestion mitigation as an emissions reduction strategy.
[3] de Palma A & Lindsey R 2011 Traffic congestion pricing methodologies and technologies *Transportation Research Part C: Emerging Technologies*, **19** **(6)** 1377-1399.
[4] Downs A 2000 Stuck in traffic: Coping with peak-hour traffic congestion *Brookings Institution Press.*
[5] Dubey P P & Borkar P 2015 February Review on techniques for traffic jam detection and congestion avoidance *2nd International Conference on Electronics and Communication Systems (ICECS)* (pp 434-440) IEEE.
[6] He F, Yan X, Liu Y and Ma L 2016 A traffic congestion assessment method for urban road networks based on speed performance index *Procedia engineering* **137** (pp 425-433).
[7] Lyman K and Bertini R L 2008 Using travel time reliability measures to improve regional transportation planning and operation *Journal of the Transportation Research Board* 2046 (1-10).
[8] Mahmud K, Gope K & Chowdhury S M R 2012 Possible causes & solutions of traffic jam and their impact on the economy of Dhaka City *J Mgmt & Sustainability* **2** (112).
[9] McGroarty J 2010 Recurring and non-recurring congestion: causes. Impacts and solutions.
[10] Padiath A, Vanajakshi L, Subramanian S C & Manda H 2009 October Prediction of traffic density for congestion analysis under Indian traffic conditions *12th International IEEE Conference on Intelligent Transportation Systems* (pp. 1-6) IEEE.
[11] Qingyu L, Zhicai J, Baofeng S and Hongfei J 2007 Method research on measuring the external costs of urban traffic congestion *Journal of Transportation Systems Engineering and Information Technology* **7** (9 – 12).
[12] Rao A M, and Rao K R 2012 Measuring urban traffic congestion-a review *International Journal for Traffic and Transport Engineering 2* (**4**).
[13] Samal S R and Das A K 2020 Evaluation of traffic congestion parameters under heterogeneous traffic condition: a case study on bhubaneswar city *Transportation Research* (pp. 675-684) *Springer* (Singapore).
[14] Uddin A 2009 Traffic congestion in Indian cities: Challenges of a rising power Kyoto of the cities, Naples.
[15] Van Woensel T and Cruz F R 2009 A stochastic approach to traffic congestion costs *Computers and Operations Research 36* (**6**) 1731-1739.
[16] Zhili L, Chunyan L and Cheng L 2009 Traffic impact analysis of congestion charge in mega cities *Journal of Transportation Systems Engineering and Information Technology* **9** (57 – 61).

IOP Conf. Series: Materials Science and Engineering 1006 (2020) 012003 doi:10.1088/1757-899X/1006/1/012003

Preparation of Inundation Maps by Developing Unsteady Flow Model of Pamba River- Decision Support System for Disaster Management

V. Johnpaul[1], C R. Jayakumar[2], R. Sindhu[3], N. Balasundaram[4] and S. Solai Mathi[5]

[1,3,5] Assistant Professor, Department of Civil Engineering, Karpagam Academy of Higher Education, Coimbatore-21, Tamil Nadu, India.
[2]PG Student, Department of Civil Engineering, Karpagam Academy of Higher Education, Coimbatore-21, Tamil Nadu, India.
[4] Professor, Department of Civil Engineering, Karpagam Academy of Higher Education, Coimbatore-21, Tamil Nadu, India.

E-Mail : johnpaulv2490@gmail.com, jkcr76@gmail.com

Abstract. Nowadays extreme rainfall events are more frequent across our country. Controlled releases from reservoirs would be necessitated to ensure the safety of dams and people downstream during extreme inflow. Mitigation measures are to be planned and preparedness actions to be taken effectively to reduce the impact of such catastrophe at downstream flood plains of reservoirs. Advance information on the extent of the area likely to be in undated consequent to releases from reservoirs is vital in developing an effective action plan for the timely evacuation of people. In the present study, it has been attempted to develop an unsteady flow model of the Pamba River in Kerala& the flood plain downstream for determining the area likely to be inundated corresponding to a largely controlled release in the tune of 900m3/s. The unsteady flow model is developed in HEC-RAS to study the flow behavior of flood wave caused by the flow hydrograph with peak 900m3/s as it propagates to the downstream plains. From the studies, it is found that the transformation of flow profile of flood wave can be better estimated using the HEC-RAS 2D flow model and the results can be used as important information in the decision-making process for Disaster Management Authorities.

1. Introduction

Flood is one of the most frequent, pervasive and devastating natural hazards in the world nowadays. One of the major challenges during a flood is to get an overall view of the incident with the accurate extent of the affected area and to predict the possible developments.

In August-2018 flood occurred in Kerala, the major rivers Pamba, Periyar, and Chalakkudy were flooded heavily consequent to the unprecedented extreme rainfall. The extreme rainfall in the basins caused an abnormal rise of water level in the rivers & banks. Most of the rivers were over flown and lead to large scale submergence in the flood plains. District Administration as well there lie authorities had to put a lot of effort to evacuate the people to safer places due to lack of estimation on the area likely to be inundated and flood hazard parameters consequent to the extremeevent.

As a case study, River Pamba, the third-longest river of Kerala, was selected to estimate the extent of the area likely to be submerged & flood parameters consequent to a large control release from the upstream Pamba reservoir with better precision by developing a 2D unsteady flow model of the river reach and its flood plains in HEC-RAS.

1.1 Location of River Basin
Pamba river basin is located in the southern part of Kerala. The river Pamba traverses through the districts of Pathanamthitta and Alappuzha before it empties into the Arabian Sea. Major settlements like Ranni, Kozhenchery, Aranmula, Chengannur, etc. are located on the banks of Pamba. The location map is shown in Figure 1 below.

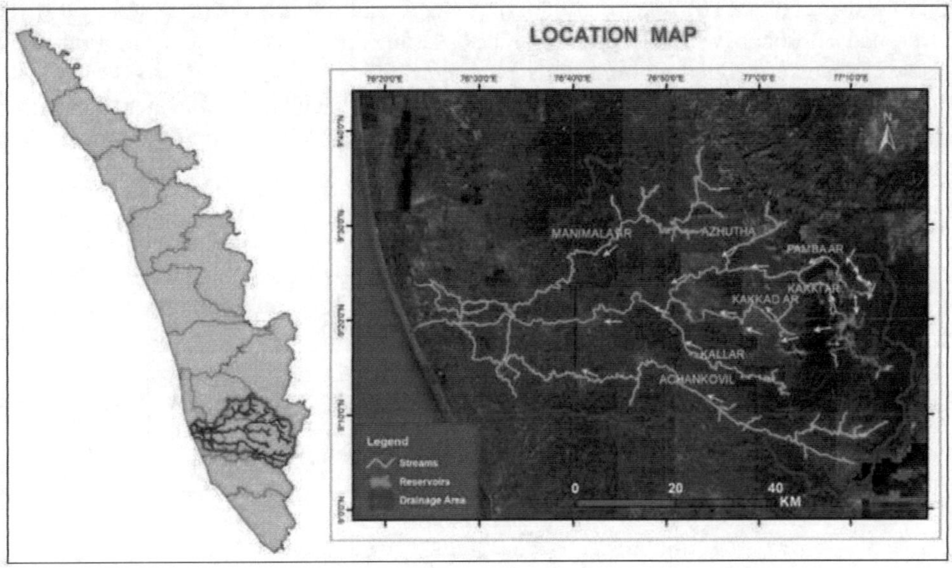

Figure 1. Location Map of Pamba Basin

1.2 Water utilization
Water resources of Pamba are being utilized for Irrigation, Water supply, and Power Generation. Dams and Diversion structures are constructed in the mainstream as well as its tributaries at various locations as part of Irrigation, Water Supply, and Power Generation projects.

Pamba Dam is one among them and is constructed in the upper reaches of the basin creating a reservoir as part of the Sabarigiri Hydro Electric Project (300 MW). Pamba reservoir receives water from a catchment of about 73 Sq.Km. Water from the reservoir is diverted to Power. In the caseof excessive inflow to the reservoir, the spilled water flows to the river downstream. Excessive releases from the reservoir encroach the flood plain downstream and pose a threat to the people downstream.

2. Materials and methods

2.1 Materials
The hydraulic modeling exercise was done using HEC-RAS. HEC-RAS is the Hydraulic Engineering Center's River Analysis System software developed by the US Army Corps of Engineers. The 2D Flow modeling approach is adopted here. Arc GIS was used for processing the output maps. The data used for the modeling are, Digital elevation model (DEM), land use, and

rainfall data. The JAXADEM of 30 m resolution was used to generate the terrain. The height accuracy of the 30-meter DEM is 5 meters (root means quareerror). The 'PMP Atlas of West Flowing Rivers of Western Ghats' prepared jointly by Central Water Commission and India Meteorological Department, Government of India and published in February 2015, Flood Estimation report published by CWC, Guidelines for Mapping Flood Risks Associated with dams published by CWC, etc. were used for the study. Glob Cover 2009 v 2.3 data set developed by the European Space Agency (ESA) was used to estimate the Roughness coefficients of the model. The geo-referenced village/ taluk/ district maps were used to identify the areas likely to be inundated.

2.2 Methods

The method adopted in the unsteady flow modeling is the hydraulic routing of the hydrograph corresponding to large control release by solving St.Venant's equation of unsteady flow using a finite difference approach with HEC-RAS. Input Parameters: Basic inputs, Terrain and LULC (Manning's N) data were added through RAS Mapper.

Geometric Data: Storage Area and 2D flow area with cell size 100m are created in the Geometric data file. Unsteady Flow Data: In this analysis, the upstream boundary condition provided is flow hydrograph shown in Figure 2 below.

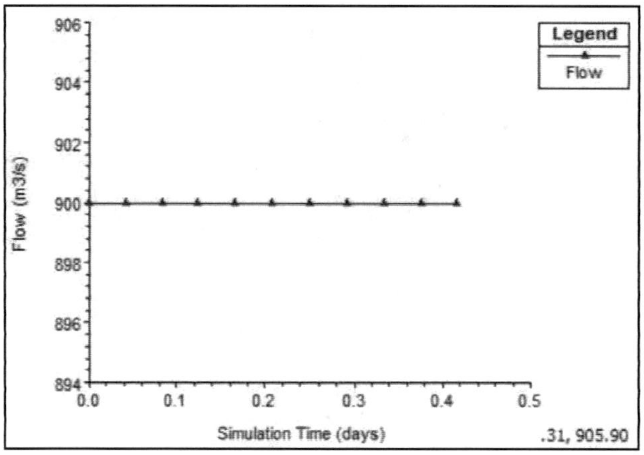

Figure 2. Input Hydrograph at the upstream boundary

The downstream boundary condition provided is normal depth.
Unsteady Flow Analysis: The unsteady flow is simulated for 10 hours with a computational interval of 1 second. However, the output intervals were provided as 5 seconds.

3. Results and discussion

3.1 Results

The outputs of the unsteady flow model are output hydrographs, inundation maps as Raster files depicting various flood hazard parameters viz. The velocity of flow, Depth of flow, Flood arrival time, etc. and extent of inundation as Shape file. The Inundation areas depicting flood hazard parameters Velocity of flow, Depth of flow, Flood arrival time etc. were processed in Arc GIS. Taluk boundaries in geo-referenced maps were overlaid on the inundation maps to analyze the hazard parameters in each Taluk and output maps are prepared. The outputmaps depicting Flood arrival time, Depth of flow & Velocityof flow are shown as Figure 3, Figure 4, and Figure 5 below.

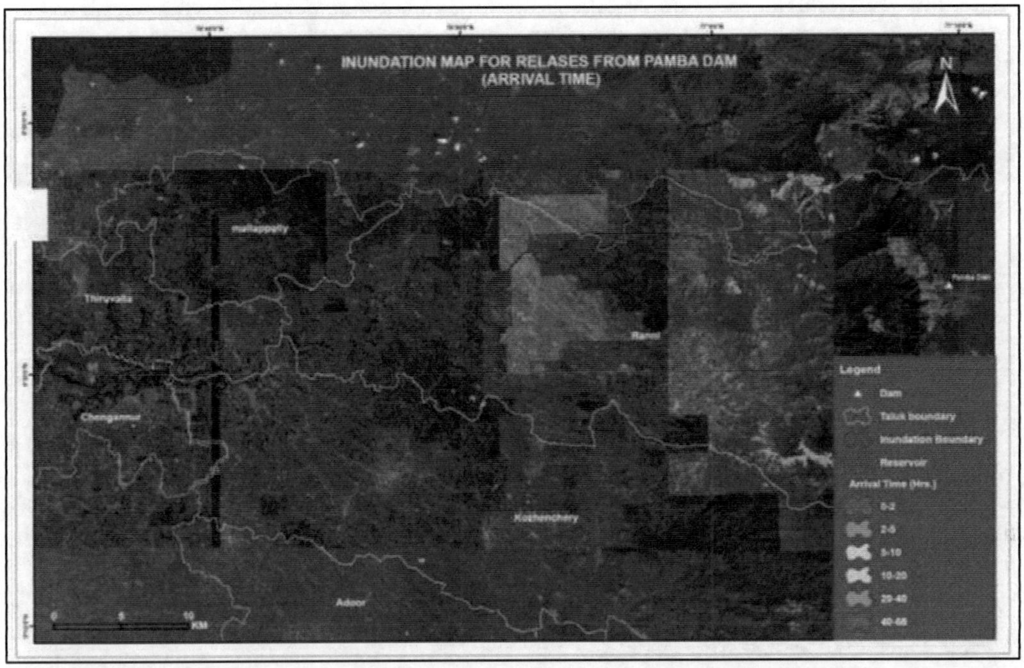

Figure 3. Inundation map depicting Flood arrival time

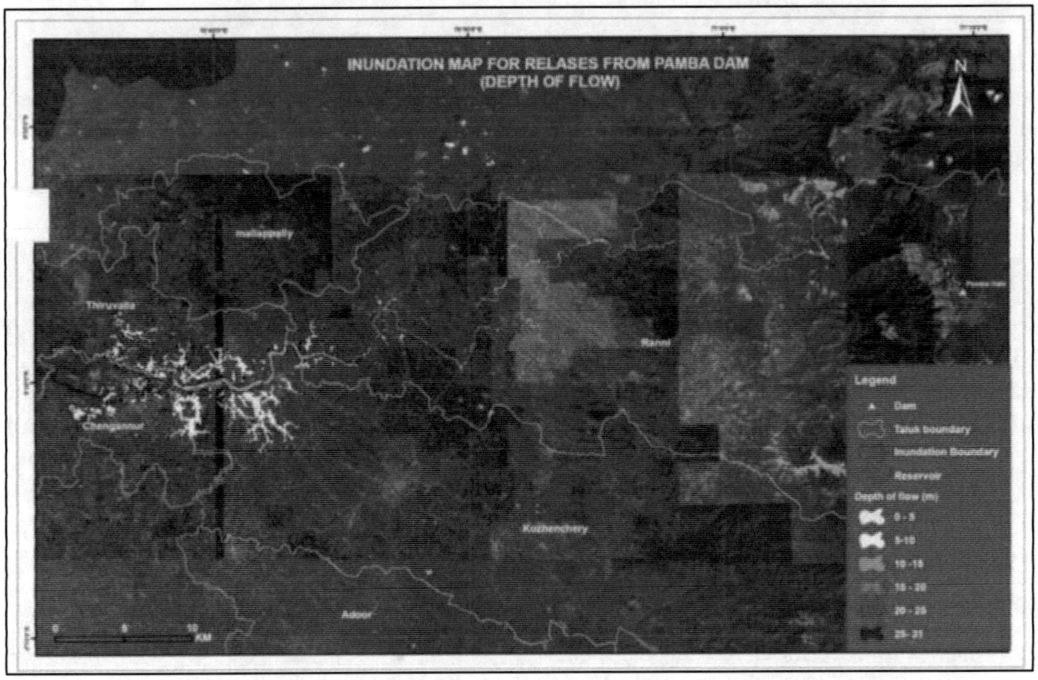

Figure 4. Inundation map depicting Depth of flow

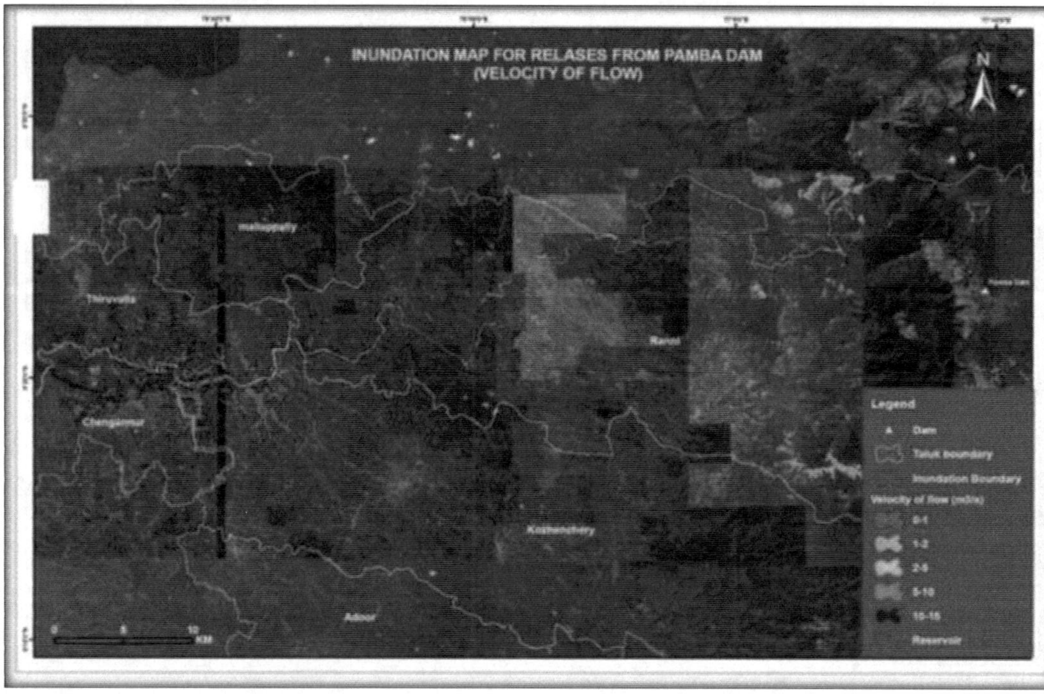

Figure 5. Inundation map depicting Velocity of flow

3.2 Discussion

On analyzing the output maps it can be seen that four Taluks are affected consequent to a release of 900m3/s from the reservoir. Extent of inundation, Velocity of flow, Depth of flow, Flood arrival time at any point in question can be easily determined from the geo-referenced output maps.

4. Conclusion

This study asserts that by developing an unsteady model of a river system and flood plain downstream, the effect of flooding due to varying releases from the reservoirs can be better analyzed using HEC- RAS and the outputs can be used as a guide for Disaster Management Authorities in decision-making process including identification of the extent of inundation, flood hazard parameters in various locations, etc. at any point in question. This helps in the early evacuation of people from the areas likely to be flooded consequent to large releases from the reservoirs and can thereby be safeguarding the life of the people and infrastructure downstream.

References

[1] Ajin R S et. al. 2013 Flood hazard assessment of Vamanapuram River Basin, Kerala, India: An approach using Remote Sensing & GIS techniques *J.Advances in Applied Science Research* ***4(3)*** 263-274.

[2] Amitangshu Acharyaa et. al. 2019 When the river talks to its people: Local knowledge-based flood forecasting in Gandak River basin, India *J.Environmental Development* ***3*** 55–67.

[3] C F Fratiniet et. al. 2012 Three Points Approach (3PA) for urban flood risk management: A tool to support climate change adaptation through trans disciplinarity and multi functionality *Urban Water Journal* ***9*** **(5)** 317-331.

[4] D Dhana Lakshmi et. al. 2019 Assessment of heavy precipitation events associated with floods due to strong moisture transport during summer monsoon over India *Journal of Atmospheric and Solar- Terrestrial Physics* ***189*** 123-140

[5] H A P Hapuarachchi et. al. 2011 A review of advances in flash flood forecasting: Hydrological Processes **25** 2771-2784.

[6] J Bartholmes et. al. 2005 Coupling meteorological and hydrological models for flood forecasting Hydrology and Earth System Sciences Discussions *European Geosciences Union* **9(4)** 333-346.

[7] K H V Durga Rao et. al. 2014 Kedarnath flash floods: a hydrological and hydraulic simulation study *Current Science* **106** 598-600.
[8] K Jenkins et. al. 2017 Assessing surface water flood risk and management strategies under future climate change: Insights from an Agent-Based Model *Science Total Environment* **595** 159-168.
[9] Mehebub Sahana et. al. 2019 Vulnerability to storm surge flood using remote sensing and GIS techniques: A study on Sundarban Biosphere Reserve, India *Remote Sensing Applications: Society and Environment* **13** 106-120.
[10] Nektarios N Kourgialas et. al. 2011 Flood management and a GIS modeling method to assess flood-hazard areas – A case study *Hydrol Sci. J.* **56(2)** 212–225.
[11] Rajesh Kumar et. al. 2018 Flood hazard assessment of August 20, 2016 floods in Satna District, Madhya Pradesh, India *Remote Sensing Applications: Society and Environment* **11** 104-118.
[12] R Resmiet et. al. 2016 Stable isotope characteristics of precipitation of Pamba River basin, Kerala, India *J Earth System science* **125** 1481–1493.
[13] Md Rejaur Rahman et. al. 2018 Thakur Detecting, mapping and analyzing of flood water propagation using synthetic aperture radar (SAR) satellite data and GIS: A case study from the Kendrapara District of Orissa State of India *Egyptian J Remote Sensing and Space Science* **21** S37-S41.
[14] M Woodward et. al. 2011 Real Options in flood risk management decision making *Journal of Flood Risk Management* **4** 339-349.
[15] Rupal K Waghwala et. al. 2019 Flood risk assessment and resilience strategies for flood risk management: A case study of Surat City *Inter. J. Disaster Risk Reduction* **40**.
[16] Satish Nagalapalli et. al. 2018 An appraisal of flood events using IMD, CRU, and CCSM4-derived meteorological data sets over the Vaigai river basin, Tamil Nadu *J. Sustainable Water Resources Management* 1–14.
[17] Sumit Das 2019 Geospatial mapping of flood susceptibility and hydro geomorphic response to the floods in Ulhas basin, India, *J. Remote Sensing Applications: Society and Environment* **14** 60-74.
[18] Tauseef Ahmad et. al. 2018 Flood hazard vulnerability assessment in Kashmir Valley, India using geospatial approach: *J. Physics and Chemistry of the Earth* **105** 59-71.

Site Suitability Assessment for Neelambur Panchayat using GIS and AHP Techniques

V Navin Ganesh[1], E Sachin[2], R Ravikumar[3], S Sanjay Kumar[4] and S Vijayakumar[5]

[1]Assistant Professor (Senior Grade), Department of Civil Engineering, PSG Institute of Technology and Applied Research, Coimbatore, Tamil Nadu, India.
[2,3,4,5]UG Student, Department of Civil Engineering, PSG Institute of Technology and Applied Research, Coimbatore, Tamil Nadu, India.

E-mail: navinganesh@psgitech.ac.in

Abstract. The rapid urbanization in the fast growing cities is posing a major problem for the planning authorities. Most of this urbanization is experienced in the peri-urban areas which refer to the settlements beyond, about or around cities. They are characterized by the changing economic and employment structure along with rapid population growth, changing spatial development patterns, unregulated growth, haphazard developments, low quality housing and inadequate infrastructure facilities. Proper site suitability assessment involving detailed investigation of the land parcels with respect to the available resources required for the urban settlements which will lead to sustainable growth of the peri-urban area is required. In the current research study, site suitability assessment has been carried out for Neelambur, one of the peri-urban areas of Coimbatore city. The factors considered for identifying the land parcels suitable for urban settlement included the land-use of the study area, proximity of sites to the nearest road, proximity of sites to the nearest facilities of importance and slope of the study area. The Analytical Hierarchy Process based Weighted Overlay Method was used in Geographic Information System to identify the land parcels suitable for urban settlements in order to ensure a sustainable development in the study area.The results indicated that 5.29 sq.km of the study area was classified as highly suitable for urban settlements. The results of the current study can be used by the local planning authorities in establishing development control norms for ensuring the sustainable growth of Neelambur.

1. Introduction

In today's world, the rapid population growth is creating numerous challenges for the goal of achieving sustainable development [1]. The rapid population growth has resulted in the major cities around the world facing the problem of rapid urbanization. The rapid urbanization is a result of the lack of basic social, employment, economic and environmental sustainability and opportunities in the rural areas. This has created a lot of stress on the already congested cities. The reports indicate that 55 % of the world's population is living in the urban areas. In the context of India, 34 % of the population is living in urban areas which are projected to increase up to 40 % in 2030 and 53 % in 2050 [2]. Most of these urbanization settlements are expected to come up in the peri-urban areas of major cities. The peri-urban areas as referred to the settlements beyond the city limits are in a transition from rural to urban characteristics due to the spillover of people from the nearby core city areas and semi-urban areas. If the development of these peri-urban areas are not monitored and controlled, it may lead to the degradation of these areas thereby affecting the sustainability of not only those areas but also the connected city [3-5]. This highlights the need of the hour which is the identification and evaluation of suitable sites for the development of urban settlements which is also a part of the general land-use planning [6]. Many researchers have attempted the GIS and AHP based land suitability assessment for the identification of proper sites for urban settlements and services planning in order to ensure a sustainable growth of those areas [7-14] The present research work aims to identify the land parcels that are suitable for urban settlements using the AHP based weighted overlay method in GIS for Neelambur, one of the peri-urban area of Coimbatore city. The need for taking up the urbanization issue of Coimbatore city arises from the fact that Coimbatore being the 15th largest urban agglomeration in India has in the recent years has experienced rapid, haphazard and ungoverned urbanization in the peri-urban areas of the city [15].

2. Materials and methods

2.1 Study area

Neelambur, the study area chosen for the current research work is a village panchayat and one of the peri-urban areas of Coimbatore city. It is situated in the administrative block of Sulur in the Coimbatore district of Tamil Nadu state in India. Neelambur is situated within the geo-coordinates 11.06N and 77.10 E. As per the census 2011, 8,382 was the total population of Neelambur. Major industries specializing in the fields of textile, manufacturing and metal casting foundries are situated in Neelambur which attracts lot of migration of people from nearby districts and villages due to the enormous employment potential and opportunities available. The study area is also covered by agricultural lands which makes agriculture a prominent economic activity of people in the study area. The presence of many educational institutions also makes the study area a potential area for migration and settlement of people from nearby villages and districts [16]. The study area map is depicted in figure 1 having a total area 19.12 sq.km.

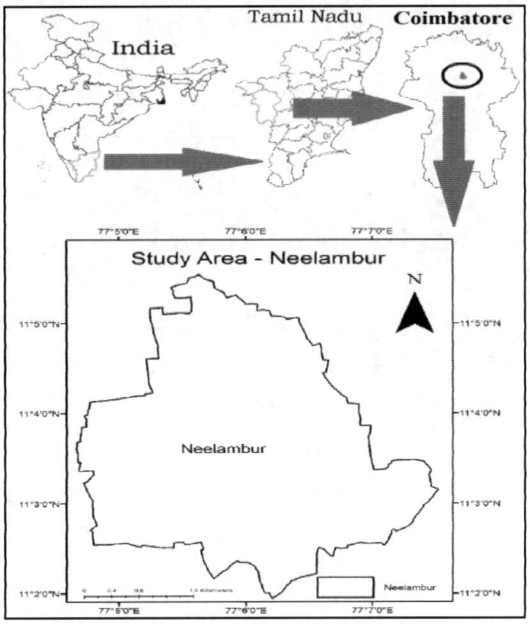

Figure 1. Study area map

2.2 Research methodology adopted

The administrative boundary map of Neelambur was used as the base map in ArcGIS for the current research. The base map was geo-referenced using the geo-coordinates obtained in the field using hand-held GPS in the geo-reference tool. The geo-referenced base map of the study area was used for the preparation of four different thematic layers of the study area namely LULC map, slope map, road map and facility. Once the thematic layers were prepared, the maps were rasterized based on the sub-criteria of the factors considered. Then all the rasterized criteria maps were processed using the weighted overlay tool in ArcGIS by assigning criteria weights derived from the AHP analysis and sub-criteria ranks to generate the site suitability maps. The research methodology adopted in the current study is indicated in figure 2.

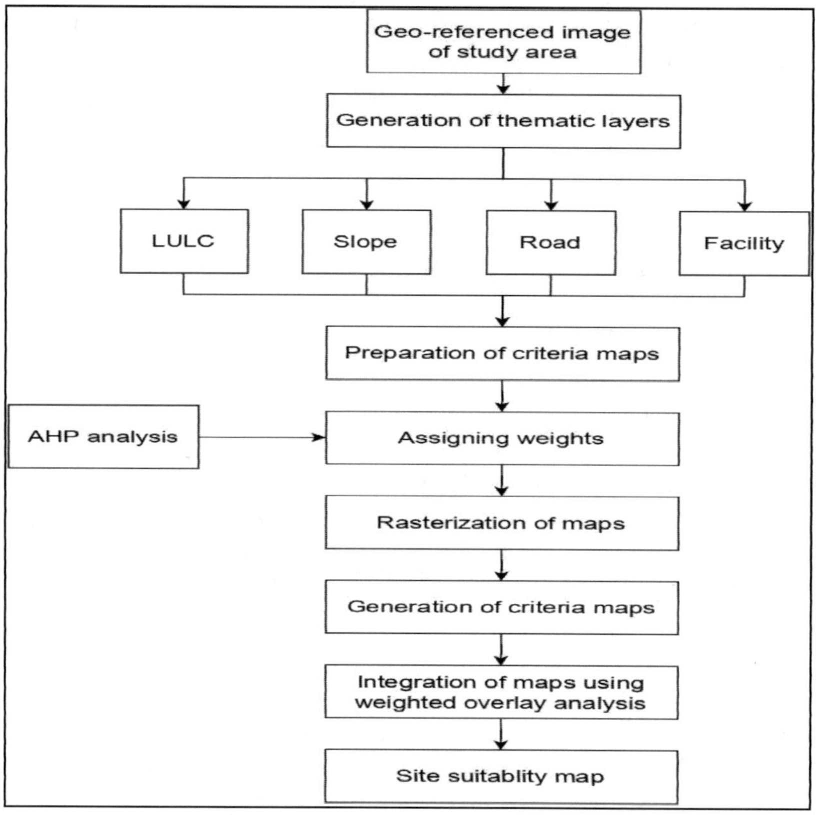

Figure 2. Research methodology adopted

2.3 AHP Analysis

One of the powerful tools under the Multi Criteria Decision Making (MCDM) technique is the AHP (Analytical Hierarchy Process). It uses the pair-wise comparison matrix to determine the relative weights for the criteria which are considered in the study. An AHP pair-wise comparison matrix has been prepared in the current study based on the four factors considered namely land-use land classification, slope, proximity to road and proximity to facility and the criteria weights have been determined based on the pair-wise comparison matrix as shown in table 1, developed based on the evaluation scale of Saaty[17]. In the table, the relative importance of one factor over another factor is indicated using the numbers 1 to 9. The equal importance of two parameters is represented by the value 1, whereas the value 9 indicates the extreme importance of one factor over the other. The Consistency Ratio (CR) arrived for the AHP pair-wise comparison matrix developed was 0.05, which is less than the normal limiting values of 0.1. It indicates that the AHP pair-wise comparison matrix developed for the current research has high correctness and consistency.

IOP Conf. Series: Materials Science and Engineering 1006 (2020) 012004 doi:10.1088/1757-899X/1006/1/012004

Table 1. AHP pair-wise comparison matrix

Criteria	LULC	Slope	Proximity to road	Proximity to facility	AHP Weights
LULC	1	1/5	1/3	1/3	0.08
Slope	5	1	5	5	0.60
Proximity to road	3	1/5	1	1	0.16
Proximity to facility	3	1/5	1	1	0.16
Total Weight					1.00

2.4 GIS thematic Layers

The four thematic layers prepared for the current research included the land-use land classification, slope map, road map and facility map. The LULC map of the study area was prepared using the maximum likelihood based supervised image classification techniques in ArcGIS. The Lands at 8 OLI/TIRS C1 Level-1 satellite image with the image acquisition date of 11.01.2020 with path number 144 and row number 052, downloaded from the USGS earth explorer website was used for the preparation of LULC map. The prepared LULC map indicated that the study area Neelambur was covered by 3.30 sq.km of built-up land. 4.47 sq.km of the study area was covered by crop land, 10.70sq.kmwas covered by fallow land and 0.65sq.km was covered by water bodies like lakes and ponds. The LULC map of the study area is represented in figure 3.

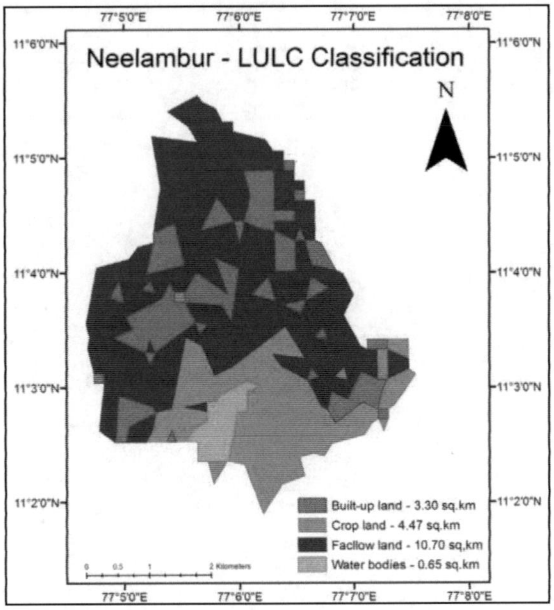

Figure 3.LULC map

The slope map was prepared using the slope tool in ArcGIS for which the SRTM DEM image downloaded from the USGS earth explorer website was used. The slope map was classified into three categories based on the slope value in degrees. Class 1 consisted of areas with slope in the range 2 degree to 9 degree. Class 2 consisted of areas with slope in the range of 1 degree to 2 degree and the Class 3 consisted of the areas where slope varied between 0 to 1 degree. The slope map of the study area is indicated in figure 4.

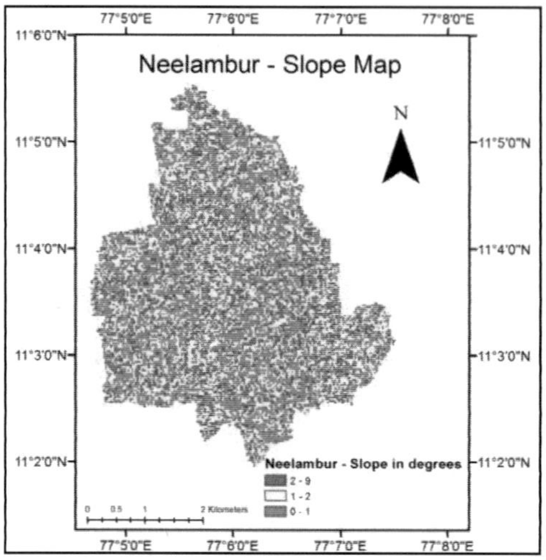

Figure 4. Slope map

The roads within the boundary of the study area were digitized using the administrative boundary map and also using the Google maps data. The different categories of roads such as National Highways, State Highways, District roads and Village roads were considered as a single category for the purpose of digitization. Once the road map was prepared, the Euclidean distance tool in ArcGIS was used to identify the land parcels that were within the range of 200 m, 400 m and 600 m from the edge of the roads. The road map along with the Euclidean distance is represented in figure 5.

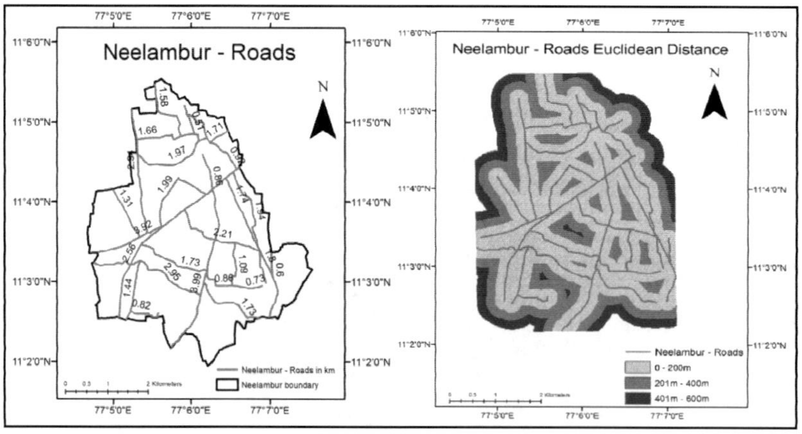

Figure 5. Road map with Euclidean distance

Since any urban settlement will be developing around a major industry, educational institution, tourism spot, etc., the geo-location of important facilities in the study area including the major industries, educational institutions, hospitals, government office buildings were recorded using the hand-held GPS and those points were digitized in the GIS system. Based on the location of points, the Euclidean distance tool was used to identify the land parcels within the ranges of 200 m, 400 m, 600 m, 800 m and 1000 m from the facility points. The facility map along with the Euclidean distance is represented in figure 6.

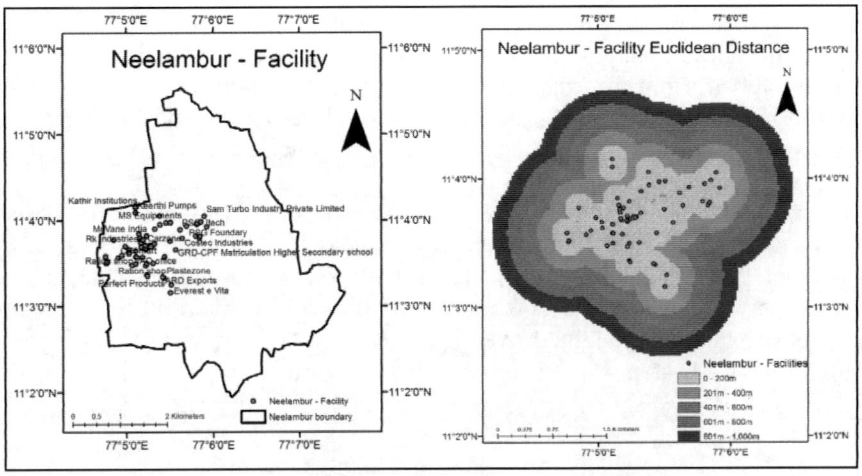

Figure 6. Facility map Euclidean distance

3. Results and discussions

The individual thematic layers prepared in ArcGIS were rasterized based on the sub-criteria determined for each factor. After rasterization, the individual thematic layers were given as input in the weighted overlay tool of ArcGIS where the AHP criteria weights for factors and the ranks for the sub-criteria were assigned. The sub-criteria and their corresponding ranks are given in table 2.

Table 2. Criteria weights and sub-criteria ranks assigned for weighted overlay analysis

Criteria	AHP Weights	Sub-criteria	Rank
LULC	8	Fallow land	5
		Crop land	3
		Built-up land	1
		Water body	1
Slope (degrees)	60	2-9	5
		1-2	4
		0-1	3
Proximity to road (m)	16	0-200	5
		200-400	4
		400-600	3
Proximity to facility (m)	16	0-200	5
		200-400	4
		400-600	3
		600-800	2
		800-1000	1

The LULC criteria were divided into four sub-categories namely fallow land, crop land, built-up land and water body. Since, the urban settlements can be allowed to develop on the fallow land; the fallow land was assigned the highest rank of 5, whereas the built-up land and water body cannot be used for the urban settlements resulting in assigning the rank of 1 for them. The most suitable site for urban settlements should be located at higher grounds than the surrounding areas resulting in assigning highest rank of 5 for slope value between 2 degree and 9 degree. Flat grounds are not preferred owing to the drainage problems resulting in assigning 3 for slope value between 0 degree and 1 degree. The urban settlements are preferred to be closer to the existing roads. The land parcels which are progressively away from the roads are not preferred for developing new settlements as access to roads and the site being located in the close proximity to roads is an important factor in the development of urban settlements. This results in the assigning of the highest rank of 5 for land parcels falling within 200 m from the edge of the road. The sites falling within 200 m and 400 m from the edge of roads is assigned a rank of 4 and those land parcels falling within 400 m and 600 m are assigned a rank of 3. The location of major facilities close to the land parcels selected for the development of urban settlement is an important factor in making that land parcels a highly suitable site. The highest rank of 5 is assigned to those land parcels which are falling within 200 m of any important facility. A rank of 4 is assigned for the sites falling between 200 m and 400 m from the facility, rank of 3 is assigned for sites falling between 400 m and 600 m, rank of 2 is assigned for sites falling between 600 m and 800 m and rank of 1 is assigned for the sites falling between 800 m and 1000 m. Once, the criterion weights and sub-criteria ranks are assigned, the weighted overlay analysis in GIS system produces the output site suitability map for the study area which is represented in figure 7.

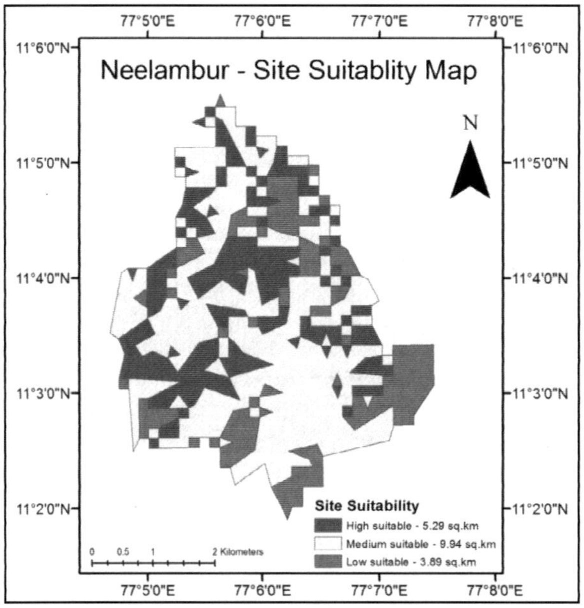

Figure 7. Weighted overlay analysis result

The output site suitability map obtained in the current study indicated that 5.29 sq.km of the considered study area Neelambur is classified to be having land parcels which are highly suitable for urban settlements. The results also showed that 9.94 sq.km of the study area was classified as having medium suitable sites and 3.89 sq.km was classified as having low suitable sites for urban settlements. More high suitable sites are concentrated on the central and western parts of the study areas owing to the high accessibility and close proximity to the roads. The presence of more facilities in the form of industries, hospitals, educational institutions, government offices, etc., is also a major reason for high concentration of high suitable sites in these areas. The low suitable sites are highly concentrated on the south west and southern parts of the study area owing to the lack of accessibility to roads and also due to the presence of more water bodies and agricultural lands. The medium suitable sited are distributed over the entire study area owing to the wide distribution of fallow lands. These medium suitable sites if properly reclaimed and developed based on stringent development control regulations can offer a better solution to the problem of lack of suitable urban settlements in the study area which can cater to the high migration of working people from nearby villages and districts.

4. Conclusions

The research work presented in this paper aimed at identifying the land parcels suitable for urban settlements which will ensure the sustainable growth of the study area Neelambur. The research methods used in the current study applied the Weighted Overlay procedure in the Geographic Information System technique based on the criteria weights derived using the Analytical Hierarchy Process to identify the land parcels that were suitable for urban settlements over the long run thereby ensuring the sustainability of the study area. The final results of this study indicated that only 27.7 % of the study area was suitable for development into urban settlements, whereas 20.3 % of the study area was not suitable for further developments. The results also showed that 52 % of the study area was medium suitable for the development of urban settlements indicating the need for development control exercise in these areas. The study presented a novel and minimalistic approach for identifying suitable areas for developing urban settlements based on GIS and AHP techniques. The results of the study when taken up and applied in the field by the respective administrative and planning authorities with the necessary development control regulations can ensure the sustainable development of the peri-urban area Neelambur.

References

[1] Department of Economic and Social Affairs, Population Division 2019 *World Population Prospects 2019 Highlights* (New York: United Nations)

[2] Department of Economic and Social Affairs, Population Division 2019 *World Urbanization Prospects 2018 Highlights* (New York: United Nations)

[3] Kolbl R and Haller R 2008 *Periurban* – A comparison between India and Western countries *Association for European Transport and Contributors*

[4] WP 7 deliverable Dissemination 2006 A comparison between the periurban developments in the EU & US and India (PERIURBAN)

[5] Dutta V 2012 Land Use Dynamics and Peri-urban Growth Characteristics *Environment and Urbanization* (ASIA) **3** 277-301

[6] Qiu D and Zhang J 2011 Urban residential land suitability index system and its comprehensive evaluation – A case study of Wenzhou *Procedia Engineering* **21** 439-445

[7] Parry J A, Ganaie S A and Sultan Bhat M 2018 GIS based land suitability analysis using AHP model for urban services planning in Srinagar and Jammu urban centers of J&K, India *Journal of Urban Management* **7** 46-56

[8] Duc T T 2006 Using GIS and AHP Technique for Land-use Suitability Analysis *Proceedings of International Symposium on Geoinformatics for Spatial Infrastructure Development in Earth and Allied Sciences*

[9] Mukhopadhaya S 2016 GIS-based Site Suitability Analysis: Case Study for Professional College in Dehradun *Journal of Civil Engineering and Environmental Technology* **3** 60-64

[10] Sarath M, Saran S and Ramana K V 2018 Site Suitability Analysis for Industries using GIS and Multi Criteria Decision Making *ISPRS Annals of the Photogrammetry, Remote Sensing and Spatial Information Sciences* **4** 447-454

[11] Liu R, Zhang Ke, Zhang Z and Borthwick A G L 2014 Land-use suitability analysis for urban development in Beijing *Journal of Environmental Management* **145** 170-179

[12] Santosh C, Krishnaiah C and Deshbhandari P G 2018 Site suitability analysis for urban development using GIS based multicriteria evaluation technique: a case study in ChikodiTaluk, Belagavi District, Karnataka, India *IOP Conference Series: Earth and Environmental Science* **169**

[13] Dong J, Zhuang D, Xu X and Ying L 2008 Integrated Evaluation of Urban Development Suitability Based on Remote Sensing and GIS Techniques – A Case Study in Jingjinji Area, China *Sensors* **8** 5975-5986

[14] Anugya, Kumar V and Jain K 2017 Site Suitability Evaluation for Urban Development Using Remote Sensing, GIS and Analytic Hierarchy Process (AHP) *Proceedings of International Conference on Computer Vision and Image Processing, Advances in Intelligent Systems and Computing* 377-388

[15] Yagya A 2017 Sustaining urban growth through ecological restoration: The Case of Coimbatore (New Delhi: Observer Research Foundation)

[16] Navin Ganesh V, Pricilla D, Rajkumar R and Vishnuvardhan K 2020 Spatial Assessment of Ground Water Quality for Neelambur using GIS and AHP Techniques *Journal of Critical Reviews* **7** 1387-1395

[17] Saaty T L 1980 The Analytic Hierarchy Process (New York: McGraw Hill)

IOP Conf. Series: Materials Science and Engineering 1006 (2020) 012005 doi:10.1088/1757-899X/1006/1/012005

Use of Eco-Coolers in Indoor Cooling

Vikas Mendi[1], Anvit Kumar[2], Bhavani Singh Yadav[3], Divyanshu Verma[4], Ankit R J[5]

[1]Assistant Professor, Department of Civil Engineering, R.V College of Engineering, Bengaluru-560059
[2,3,4,5]U.G Student, Department of Civil Engineering, R.V College of Engineering, Bengaluru-560059

E-mail: vikasm@rvce.edu.in

Abstract. With the growing global warming and rise in pollution levels across the world in addition to several other factors such as deforestation, urbanization and ozone layer depletion, there has been a continuous increase in global rise in annual average temperature. Eco-cooler is a cheap and eco-friendly device made from non-biodegradable waste which can be used to reduce the indoor temperature of the building thus giving a comfortable living experience. Although a conventional design with symmetrical hole design is adopted in some under-developed countries as a commercial way, no study is being conducted for the improvement in performance of the Eco-cooler. Hence the present study has the main objective of evaluating different factors of dependency of an Eco-cooler, establishing them as a cheap and effective way of reducing indoor temperature and increasing its efficiency.

1. Introduction

Eco-cooler is a cheap and eco-friendly device made from non-biodegradable waste which can be used to reduce the indoor temperature of the building thus giving a comfortable living experience. Although a conventional design with symmetrical hole design is adopted in some underdeveloped countries as a commercial way, no study is being conducted for the improvement in performance of the Eco-cooler. Hence the present study had the main objective of evaluating different factors of dependency of an Eco-cooler, establishing them as a cheap and effective way of reducing indoor temperature and increasing its efficiency. The basic science behind the working is the hot air that enters the bottle through the cut end is compressed gradually at the neck section of the bottle, which in turn turns the air cooler before it enters the room.

1.1. Temperature trends

The present global scenario has seen a drastic change in the world temperature with the average temperature increasing day-by-day. The effects of global warming have taken a significant toll on the daily lives of the people. In the modern world where the climate change is becoming inevitable, the world overall temperature is rising at the rate of 0.8°C/decade.13 of the 15 warmest years overall in the world where in the last 15 years. The last decade was also the warmest. Stats show that in India alone the temperature change trend is very disturbing. Annual mean temperature in India is increasing rapidly. It will breach the 1.5°C marks if it continues to follow the current trend in the next decade. As recorded 2016 was the warmest year of the history. Even during monsoon season, there is an increase of 1°C temp per century. According to the research conducted, it was established that the outdoor temperature should be around 22° -24°C for a comfortable living experience. But due to the drastic increase in temperature over the past years, it has become absolute necessary to provide a cheap and effective way of reduction of indoor temperature with sustainability principle in mind.

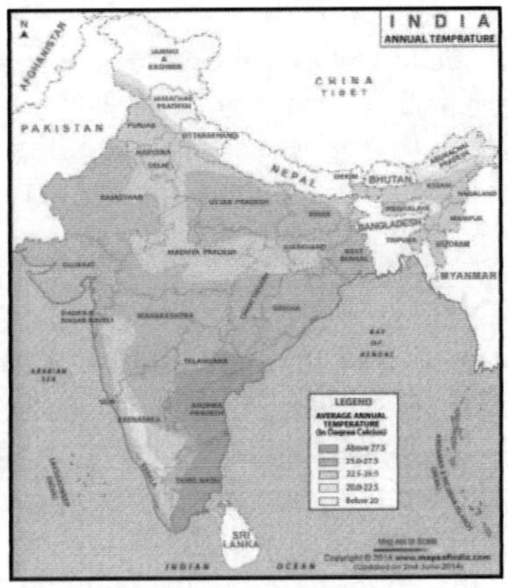

Figure-1: Temperature trends in India

1.2 Working principle

A survey of the room in which the Eco-cooler needs to be placed is done and the window facing the maximum anticipated air flow is chosen. The ends having larger diameter of the bottles are placed facing out of the window thus permitting maximum air in while not holding further radiation to return within the area. Now, once the air passes through the outer side of the Eco-cooler, the air gets compressed going through the larger diameter section area to the smaller diameter section near the throat.

The compression, due to bottleneck, can increase speed of the air, consecutively decreasing its pressure. The compressed gas undergoes speedy enlargement before long once it leaves the throat of the bottleneck. This speedy enlargement decreases the temp. of the air current and some form of air mass creation takes place that successively attracts the encircling air into the stream.

This Venturi result- the reduction in fluid pressure that results once a fluid flows through a constricted section of a pipe governs the Eco-cooler's operating. Since the pressure decreases because the speed will increase at the outlet of nozzle in an Eco-cooler, the air commencing of the eco cooler is way cooler than the air getting into (Gay-Lussac's Law).

1.3 Objective

The major objectives of the present study are as follow:

a. To analysis the cooling capacity and efficiency of the conventional Eco-cooler.
b. To analysis the wind effect on the cooling capacity of the Eco-cooler.
c. To study the effect of using water as an aid in increasing the efficiency of the Eco-cooler.
d. To analysis the effect of increasing the hole area per surface area on the plywood and adding more plastic bottles on the efficiency of the Eco-cooler.

2. Literature review

A deep literature survey was conducted to get a detailed knowledge about the scope, past studies conducted in the field and need for the Eco-cooler. Following are the observations based on cumulative topics.

2.1 Indoor air quality and window positioning

There were specific studies focused on the effectiveness of passive or indirect cooling techniques and its implementation to reduce energy consumption by a building while having the benefit of an improved thermal performance simultaneously. Residential buildings in Dubai were taken for this purpose and a case building was selected. An energy simulation software – IES was used to associate the data obtained with the efficiency of the applied cooling strategies. Solar shading was also used and observed using Sun Cast analysis which was included in the IES software. About 23% reduction in energy consumption was observed with the implementation of the cooling methods. Also, the present scenario for the demand for energy and growing health issues have raised an alarm and hence, natural ventilation is being considered in many areas to be an effective alternative. This mostly depends on the size and orientation of the window openings considering the building is naturally ventilated. Hence, the focus in the research was on single-sided and cross ventilated-buildings and its effects on indoor thermal comfort. Computational Fluid Dynamics was used to study the indoor airflow for a 3D model. Further, it was checked for grid independence and experimental validation was successful on a reduced scale at a wind tunnel and with the network model. Following on the idea that natural ventilation is imperative and open windows are the necessity for having good ventilation. Even then, the knowledge of the efficiency of an open window and its performance remains pretty much unclear. The experiments for two types of windows and their performance in rooms but in an isolated manner were showcased. It showed that airflow inside rooms can be described by the traditional theory as is done for jets. Also, layered flow and semi-empirical flow element models were developed for the estimation of thermal comfort parameters in the occupied region i.e., where human activity is to be present.

2.2 Ventilation effectiveness as indoor room temperature

A study focused on the measurable values that ventilation in domestic buildings can achieve because it is tedious to estimate it using conventional methods and may be costly. The study works on a simple equilibrium-state, heat-balance calculation approach that is backed with monitoring data to model the effectiveness of reducing ventilation rates to the minimum standards for individual homes. In this way, it was easy to differentiate energy consumption and assess how much reduction could occur with this as well as develop better models for further analysis. To understand the temperature variations, it was important to develop better and realistic models that can represent the temperature models and predict heat regulation in an effective manner so that the people can have comfort inside as well energy conservation can be achieved. The continuity, momentum and concentration equations were input in CFD simulations and various models were designed such as the linear ventilation model, low dimensional model, and artificial neural networks to establish the role of ventilation. After the input of the required data, the simulation showed that linear ventilation can be used to predict the indoor environment and heating, ventilation & air conditioning control and bring down air-conditioning costs by about 32%.

2.3 Climate change global warming and greenhouse effect

A research discussing the Greenhouse Gases (GHG) and its effect, the Kyoto Protocol and about the renewable energy sources that are available in today's world was done. Climate change is directly a consequence of Greenhouse Gas emission. This also told us about the climate change actions taken throughout the USA, Europe, and Asia. It was concluded that GHG Emissions are a problem to the current world's climate and stated the 4 general measures that can be taken to reduce GHGs so that the market is also not being affected with the measures that will be taken in this process. The burning of fossil fuels emitting CO_2 is the reason for the increase in mean surface temperature and it will increase the temperature by 2-3°C by the middle of next century. Also, the harmful gases like chlorofluoromethanes, nitrous oxide, and aerosols enhance the rate of global warming. The authors insisted to shift towards the use of nuclear energy instead of fossil fuels. As the CO_2 spreads globally, every country must adopt a different way of energy consumption to control the CO_2 emission. Global warming is mainly because of deforestation and degraded forests. Restoring degraded forests has a huge potential for averting global climate change.

Due to global warming and increased energy cost, reducing power consumption by computers has become important. Software controls the flow of processing hence indirectly affects the energy consumption and its life cycle energy affects the CO_2 emission. Software like Linux 3.2 Kernel and other software of different sizes were taken to analyse the total contribution and the energy consumption associated with their production towards green-house gas emissions. The results have been discussed with the consideration of the sizes of the software ranging from 10 kilobytes to 1 megabyte. They are generic in nature. The result shows that as the software grows, the energy consumed at the production stage will be more than the total lifecycle energy. Linux 3.2 Kernel, the total lifecycle energy of a single copy founded exceeding 51GWh (31 kilotons of CO_2 emission)/year.

Research also showed that primary energy demands account for 25% in addition to 11% usage of solid fuels and oils for the purpose of primary home heating in residential sectors of Ireland. Replacing oil and solid fuel usage with air source heat pump (ASHP) technology could offer household cost savings, reductions in emissions, and reduced health impacts. Using this technique, 600 Euros annually can be saved for each household. Not only that, but there can also be a reduction in global CO_2 emission by 4.3 million tons per annum. The study also verified the importance of thermal insulation as an effective way of keeping home hot in winter and cool in summer. Ultimately the analysis clearly suggests that ASHPs can play an important role in managing residential emissions.

2.4 Factors responsible for working and construction of eco-cooler

To establish theoretical foundations regarding fluid flow under the venturi effect, a special model called the venturi tube was taken under observation where the effects can be emphasized. A three-dimensional model of the venturi tube was constructed with Solid Edge V20 and analysed using ANSYS CFX for highlighting the fluid flow inside. It was found that this model can provide a solution to determine the fluid flow rate in a hydraulic or pneumatic installation using manometers.

Venturi shaped roofs can be made for better ventilation and air circulation and research were conducted to validate the same. A specific venturi shaped roof in the University of Putra Malaysia was considered and a case study and CFD simulations were performed. The results showed that the venturi effect could be seen in the specific roof and hence a correct choice of ventilation technique and design has been applied for this case. However, the result can be improved by improving the ratio of contraction and introducing a slight change in the design inspired by Bronsema's roof design. The Eco-cooler was used for this and this solution can prove effective as the temperature reductions were about 3 degrees Centigrade. Also, the results suggested that the use of a bottle with a 4-inch diameter at the base and 1-inch diameter at the brim will yield better efficiency than other bottles of varying dimensions. Since wind velocity plays an import role in the performance of the Eco-cooler, with an objective of making a detailed wind map of India, long-term data on an hourly basis on wind speed from 70 meteorological centres of India Meteorological Department have been collected and analysed. Using the Gumbel probability paper approach, the extreme value quantities have been derived. A design basis wind speed for each site for a return period of 50 years was evaluated. Following analysis of the collected data after appropriate processing, the site-specific changes in the design wind speeds at the contemporary wind zone map for the design of buildings/structures were highlighted and revisions to the previous map were suggested.

2.5 Gaps identified in literature survey

After going through the literature survey, following gaps were identified:

a. No analysis is done on the cooling capacity and efficiency of the Eco-cooler.

b. Wind effect is not taken into consideration for the cooling capacity of the Eco-cooler.

c. No attempt is made to study the effect of using water as an aid in increasing the efficiency of the Eco-cooler.
d. No analysis is done on the effect of increasing the hole area per surface area on the plywood and adding more plastic bottles on the efficiency of the Eco-cooler.
e. No record of varying material during construction of Eco-cooler to check the effect on the cooling capacity.
f. Possibility of addition of phase transition material as an aid to increase efficiency is neglected.

3. Construction and experimental setup

3.1 Construction of eco-cooler

The guidelines followed for the construction of Eco-cooler for experimentation purposes are as follow:

a. Measure the dimensions of the window on which Eco-cooler will be mounted.
b. Take plywood of the same dimension and 6mm thick (transparent or opaque).
c. Make required inlet diameter cuts on the wood at 10cm c/c distance.
d. Gather reusable bottles and cut them so that ratio of inner to outer diameter is 1:4.
e. Silicone Gel is used to adhere the half-cut bottles in positions.
f. Ensure no further movement occurs by applying aid to the Silicon gel.

Install the Eco-cooler on different windows of the room (based on wind movement direction and requirement)

3.2 Estimation of unit rate of eco-cooler

Considering the window of 3X3 feet for which the Eco-cooler is supposed to be designed.

- Size of Eco-cooler: 4X4 feet (1/2 feet each side for fixing)
- Plywood typical cost being Rs. 80/- square feet (good quality BWP)

BWP Plywood is an acronym for Boiling Waterproof Plywood which is confirming IS 710 specifications and is Marine Grade Exterior Wood.

- Cost of plywood: 80 X 16 = Rs. 1280/-
- Cost of Silicone for one unit (Transparent) = Rs. 200/-

Silicone is the best-known sealant available in the market which bonds almost all types of materials be it wood, metal, or plastic. Silicone is the most used sealant in the civil engineering industry.

- Assuming 50 recycled bottles to be used of suitable diameter of nozzle depending on design.
- Cost of each recycled bottle = Rs. 2/bottle.
- Total cost of recycled bottle = 2X50 = Rs. 100
- Miscellaneous cost for making holes in the plywood = Rs. 200

This involves the cost incurred to drill the holes of required specified and designed diameter by using the drill gun generally which is available with the carpenters and other such workers.

- Total Cost of Eco-cooler for window of size 3X3 = Rs. 1780/-

- Cost of Eco-cooler per square feet = Rs. 197.78/- = Rs. 200/-

To cool the same room more easily by using a modern method of reducing the temperature in a room it costs up to Rs. 20000/- which comes up with added on maintenance, electricity bills every month and the other problems that are being caused by the gases and chemicals being emitted by this method.

3.3 Experimental Setup
Following experimental procedure is followed for obtaining observations for different experimental conditions:

a. Mount the Eco-cooler on the window in the direction of the maximum wind flow or opposite to it depending on the experiment.
b. Install 6 temperature sensors (Ambient Weather WS-10-X4 Wireless Indoor/Outdoor 8-Channel Thermo-Hygrometer) (one on each wall, one on roof and one on the floor).
c. In the interval of 15 minutes, note down the temperature change.
d. If the temperature remains constant for 4 consecutive readings, the device has delivered the maximum efficiency.
e. Add a water wall on the diametrically opposite position to the mounted Eco-cooler (if unavailable, use large containers containing water placed near the mounted Eco-cooler).
f. Note the change in temperature in the intervals of 15 minutes.
g. Introduce change in design of the Eco-cooler and note down the effect of the induced change.

4. Observations and results
The following conditions were used to collect observational data and compute them in order to obtain results:

4.1 Assessing the effect of wind on eco-cooler's performance
Before mounting the Eco-cooler on the actual window, table fan was used to create the wind flow, thus, accumulating the effect of wind velocity on the performance of the Eco-cooler. The observation obtained is recorded as follows:

Table-1: Eco-cooler fitted on table fan

Parameters	9th March	10th March	11th March	12th March
Inlet air velocity	0.2 m/s	0.3 m/s	0.4 m/s	0.5 m/s
Room temp.	24.3° C	24.0° C	24.8° C	24.4° C
Outlet air velocity	1.1 m/s	1.2 m/s	1.2 m/s	1.3 m/s
Outlet air temperature	24.3° C	24.0° C	24.7° C	24.2° C

The data collected shows a directly proportional relationship stating that more the wind velocity, better will be the efficiency and performance of the Eco-cooler. Therefore, it is evident that wind velocity plays an important role in temperature reduction and cooling capacity of an Eco-cooler.

4.2 Eco-cooler (design 1) under normal performance criteria.

The constructed Eco-cooler was mounted on a window with maximum wind flow. Wind velocity at the start of the experiment was noted and change in room temperature every 15 minutes was noted.

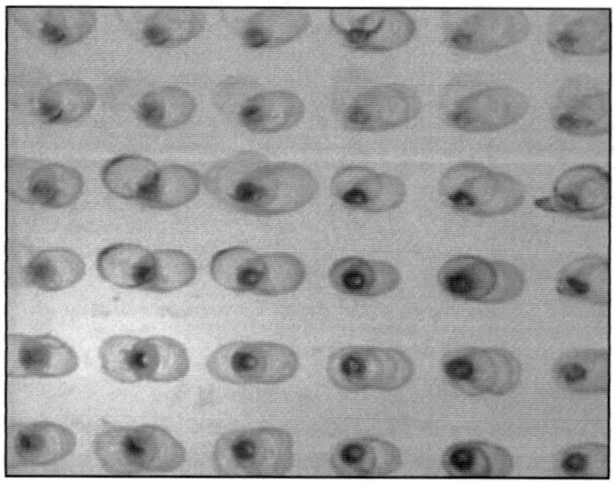

Figure-2: Design 1 - Square pattern arrangement

Table-2: Observation on cooling capacity (wind velocity at start 3.2 m/s)

Time interval (min)	Initial temp. (° C)	Final temp. (° C)	Reduction in temp.(° C)	Cooling capacity (° C/hr.)
0-15	26.8	26.7	0.1	0.4
15-30	26.7	26.5	0.2	0.8
30-45	26.5	26.2	0.3	1.2
45-60	26.2	25.9	0.1	0.4
60-75	25.9	25.9	0.2	0.8
75-90	25.9	25.9	0.0	0.0

Calculations:

- Overall Reduction in temperature (R) = 26.8 °C - 25.9 °C = 0.9°C
- Cooling Capacity = 0.9 °C ÷ 1.25 hr = 0.72 °C/hr
- Efficiency = (0.9 °C ÷ 26.8 °C) x 100 = 3.36%

4.3 Eco-cooler (design 1) with water aid

The same mounted Eco-cooler is now aided by keeping a large water container which will theoretically absorb the heat and thermal energy from the entering air thus giving a better cooling experience. (Wind velocity at the start 2.9 m/s).

IOP Conf. Series: Materials Science and Engineering 1006 (2020) 012005 doi:10.1088/1757-899X/1006/1/012005

Table-3: Observation of cooling capacity

Time interval (min)	Initial temp. (° C)	Final temp. (° C)	Reduction in temp. (° C)	Cooling capacity (° C/hr.)
0-15	27.6	27.3	0.3	1.2
15-30	27.3	27.1	0.2	0.8
30-45	27.1	26.7	0.4	1.6
45-60	26.7	26.5	0.2	0.8
60-75	26.5	26.3	0.2	0.8
75-90	26.3	26.2	0.1	0.4

Calculations:

- Overall Reduction in temperature (R) = 27.6 °C – 26.2 °C = 1.4 °C
- Cooling Capacity = 1.4 °C ÷ 1.5 hr = 0.94 °C/hr
- Efficiency = (1.4 °C ÷ 27.6 °C) x 100 = 5.07%

4.4 Eco-cooler (design 2) with water aid

A new Eco-cooler with modified design (Figure) was constructed so that to increase the amount of intake wind coming from the window. Water container was placed near the Eco-cooler mounted window as the same experimental procedure was followed. (Wind velocity at start 3.6 m/s).

Figure-3: Design 2 (A)

Figure-4: Design 2 (B)

Table-4: Observation of Cooling Capacity

Time interval (min)	Initial temp. (° C)	Final temp. (° C)	Reduction in temp. (° C)	Cooling capacity (° C/hr.)
0-15	25.6	25.3	0.3	1.2
15-30	25.3	25.0	0.2	0.8
30-45	25.0	24.6	0.4	1.6
45-60	24.6	24.1	0.2	0.8
60-75	24.1	23.9	0.2	0.8
75-90	23.9	23.8	0.1	0.4
90-105	23.8	23.7	0.1	0.4

Calculations:

- Overall Reduction in temperature (R) = 25.6 °C – 23.8 °C = 1.9 °C
- Cooling Capacity = 1.9°C ÷ 1.75 hr = 1.09 °C/hr
- Efficiency = (1.9°C ÷ 25.6 °C) x 100 = 7.42%

4.5 Result Tabulation

Table-5: Comparative analysis of Efficiencies

Design and setup	Overall Reduction in temperature	Cooling capacity overall	Efficiency
Eco-cooler-1	0.9 °C	0.72 °C/hr.	3.36%
Eco-cooler-1 with water aid	1.4 °C	0.94 °C/hr.	5.57%
Eco-cooler-2 with water aid	1.9 °C	1.09° C/hr.	7.42%

It is clear from the results that:

- More Void area/surface area gives better result and cooling capacity.
- Introducing a heat absorbent aids to the performance of the Eco-cooler.
- Wind velocity has a huge effect on the performance of the Eco-cooler, i.e., better wind flow is directly proportional to the performance of Eco-cooler.

5. Conclusions
Following are the conclusions drawn from the present research

- As the temperature around the globe has started to grow at an unprecedented rate, it has become the need of the hour to look for viable options that can help people everywhere, to help them survive the worst-case scenarios that may come in the future.
- The data that has been shown in this paper has shed a light on how the temperature variations may occur in the next few decades. Therefore, if a concrete solution to this is not developed then most of humanity will eventually have to perish.
- Apart from temperature variations, multiple other phenomena are happening that can have dire consequences upon mankind with each one of them having catastrophic effects later on.
- Multiple cooling strategies are being researched currently but most of them are at the expense of electricity thus, it is uncertain that the solution would apply to masses as the cost of the cooling device will be marginally high.
- Also, it is known that electricity is not omnipresent and because of this, not a lot of things can be used as a solution. Hence, low-cost alternatives although hard to find must be researched upon as they can prove to be a boon to people who may be economically challenged.
- In this paper, the same has been tried to establish with the help of an Eco-cooler. Temperature sensors were used to see the effect of varying parameters on the performance of the Eco-cooler inside a closed environment.
- Benefits of using an Eco-cooler are mentioned in this paper and also how it can impact the environment as well as be in harmony with the cooling capabilities. The cost of a typical Eco-cooler is roughly Rs. 200/- per square feet which is affordable for most of the Indian families.
- The construction procedure and its setup are fairly simple and hence, it can be done with simply without the need for external assistance. Apart from this, the Eco-cooler can be used in conjunction with other home appliances as such the table fan and hence, facilitating the cooling effect.
- It is also observed that a certain pre-requisite mainly, wind flow is one of the most important criteria for the working of the Eco-cooler. Hence, places where the natural wind velocity would be constant and have a higher magnitude, the Eco-cooler would perform better.
- Also, establishing the result on efficiency by the use of external aids such as a water-wall which have proven to improve the performance of the Eco-cooler, up to some extent.
- A step is taken in the direction of providing a solution which is much more economically viable and as the Air Conditioner are being used today its usage adds on to multiple emissions and although the room inside becomes cooler, it is not affordable for everyone, apart from its consequences to the environment.
- This paper aims to understand how these variables can be changed and even provide a solution that has another impact on everyone.
- At last, although the obtained results are not in accordance to the required results and the efficiency is not as much as it was predicted. Further research needs to be done in this field to improve the efficiency of the Eco-cooler and prove to be helpful.

6. Scope for future work

- Eco-coolers have a big future ahead. There are some improvements which were not included in this paper such as using phase transition materials.
- Use of phase-transition elements or heat sink can prove beneficial.
- There is a scope in using different materials for bottles and the support on which bottles are fixed.
- Different shapes of bottles can also be adopted in future to enhance the performance of the Eco-cooler.
- As the supporter or the plywood prevents the view of the outside of the window, different transparent materials can be used to overcome this problem.
- Making the design as two door panel window can be adopted.

References

[1] A F Zobaa & J S McConnach 2006 International response to climate change: an overview IEEE Power Engineering Society General Meeting Pgs 465-473.

[2] Tapan Kumar Rana, Anik Naha Biswas, Debashmita Basu, Biswarup Rana, Debaleena Mukherjee, Paboni Das and Bhaskar Kumar 2017 Impending impact of climate change on present and future generations *8th Annual Industrial Automation and Electrotechnical Engineering Conference* pgs. 22-27.

[3] Muniyandi Balasubramania, V Dhulasi Birundha Climate change and its impact on India *The IUP Journal of Environmental Sciences* vol **6** Pgs 31-46.

[4] Vidadi Samedov, Iskandarova Tunzala Hasan Kizi Study of deforestation and degraded forests as a main source of biodiversity loss and global warming *Conference: Rehabilation & Restoration of Degraded Forests* Pgs. 1-9.

[5] Himangka Kaushik, Mobi Mathew and Jami Hossain 2018 Comparative analysis of greenhouse Gas Emission from different Vehicular fuels *1st IEEE International Conference on Power Energy, Environment & Intelligent Control (*Greater Noida, Uttar Pradesh, India) Pgs. 702-706.

[6] Safiuddin Md 2001 Global ozone depletion: Cause effects and preventive measure *Borneo Science* vol. **10** Pgs 73-89.

[7] J W Jones, W F Stoecker Refrigeration and Air Condition *McGraw-Hill Edition.*

[8] J X Zhang 2017 Analysis on the effect of venturi tube structural parameters on fluid flow *AIP Advances* vol **7** Pgs 1-9.

[9] Bhanuprakash Ch, Vinod Mummina and Mahesh Chakravarthi V 2018 Performance evaluation of an eco-cooler analysed by varying the physical and flow parameters *IOP Conf. Series: Materials Science and Engineering* vol **377** Pgs. 120-124.

[10] Hanan M Taleb 2014 Using passive cooling strategies to improve thermal performance and reduce energy consumption of residential buildings in U.A.E. buildings *Frontiers of architectural research* vol. **3** issue 2 154-165.

[11] D Prakash, P Ravikumar 2019 Analysis of thermal comfort and indoor air flow characteristics for a residential building room under generalized window opening position at the adjacent walls vol **4** issue 1 Pgs. 42-57

[12] Per Heiselberg, Henrik Dam, Lars C. Sarensen, Peter V Nielsen, Kjeld Svid 2001 Characteristic of airflow from open windows *Building and Environment* vol. **36** issue 7 Pgs. 859-869.

[13] P Cosar-Jorda, R A Buswell, V A Mitchell 2018 Determining of the role of ventilation in residential energy demand reduction using a heat-balance approach *Building and Environment* vol. **144** pgs. 508-518.

[14] Chen Rena, Shi-Jie Cao 2019 Development and application of linear ventilation and temperature models for indoor environmental prediction and HVAC systems control *Sustainable Cities & Society* vol. **51** 01673.

[15] Sook-Ling Chua 2015 Sensor Selection in Smart Home *Procedia Computer Science 69* vol **23** pgs. 43-57.

[16] Alexandra Schiewecka, Erik Uhdea, Tunga Salthammer, Lea C Salthammer, Lidia Morawska, Mandana Mazaheric, Prashant Kumar 2018 Smart homes and the control of indoor air quality Vol. **94** Pages 705-718..

[17] Shanshan Tonga, Nyuk Hien Wonga, Erna Tana, Steve Kardinal Jusuf 2019 Experimental study on the impact of facade design on indoor thermal environment in tropical residential buildings *Building and Environment* vol. **166** Pgs. 1-15.

[18] Bruce Spencer, Omar Alfandi 2016 Forecasting Internal Temperature in a Home with a Sensor Network vol. **89** Pgs. 1244-1249.

[19] C Monica, S Kalaivani, Gopalakrishna Jayalalitha 2018 Forecasting the annual temperature of India by Using Exponential Smoothing *Journal of Advanced Research in Dynamical and Control Systems* **06** special issue Pgs. 972-976.

[20] Xu Hong, He Yuan, Sun Xiang, He Jia, Xu Qiongmei 2020 Prediction of thermal energy inside smart homes using IoT and classifier ensemble techniques *Computer Communication* vol **151** Pgs. 581-589.

[21] William W Kellog, Stephen H Schneider 1978 Global Air Pollution and Climate Change *IEEE transaction on Geoscience* vol. **16(1)** Pgs. 44-50.

[22] Vasily G Moshnyaga 2013 Assessment of Software Lifecycle Energy and its Contribution to greenhouse Gas Emissions *IEEE International Conference Cloud and Green Computing* vol **3rd** Pgs. 197-198.

[23] Bimal Bose 2010 Global Warming: Energy, Environmental Pollution, and the Impact of Power Electronics *IEEE Industrial Electronics Magazine* vol **4(1)** Pgs. 6-17.

[24] J Andrew Kelly, Miao Fu, J Peter Clinch 2016 Residential home heating: The potential for air source heat pump technologies as an alternative to solid and liquid fuels vol. **98** Pages 431-442.

[25] Markand Oza, Chandra Kishtawal 2015 Spatio-temporal changes in temperature over India *IEEE Current Science* vol **109** Pgs. 154-158.

[26] Shouraseni Sen Roy 2019 Spatial patterns of trends in seasonal extreme temperatures in India during 1980–2010 *Article in Weather and Climate Extremes* vol **24** Pgs. 1-7.

[27] Gulrez Azhar, Shubhayu Saha, Partha Ganguly, Dileep Mavalankar and Jaime Madrigano 2017 Heat Wave Vulnerability Mapping for India *Suistainable Environmental and Research Conf.* Vol. **14(4)** Pgs. 847-852.

[28] Fănel Dorel ȘCHEAUA 2016 Theoretical Approaches Regarding the VENTURI Effect *Conference of Hydraulics, Pneumatics, Tribology, Ecology, Sensorics, Mechatronics* issue 3 Pgs. 69-72.

[29] P Farzan, M Soheilirad, M L Othman, M Hojabri, M M Broujerdian 2013 CFD Analysis of wind blocking vs Venturi effect: A case study in tropical climate *IEEE International Conference on Renewable Energy and Sustainable Energy* (Karunya Unviversity Coimbatore, India) Pgs. 192-195.

[30] Subha Ganguly 2018 Plastic pollution and its adverse effect on Environment and Ecosystem *Internation conference on Recent Trends in Arts, Science, Engineering & Technology* (dhanalakshmi Shrinavasan Hotel, Perambalur, Tamil Nadu, India) pgs. 15-16.

[31] N Lakshmanan, S Gomathinayagam, P Harikrishna, Arthur Abraham, S Chitra Ganapathi 2009 Basic wind speed map of India with long-term hourly wind data *IEEE Current Science* vol. **96** Pgs. 911-922.

IOP Conf. Series: Materials Science and Engineering 1006 (2020) 012006 doi:10.1088/1757-899X/1006/1/012006

Case Study and Analysis of Ecological Restoration Plan of Lakes in Coimbatore

M Natarajan[1], V Johnpaul[2], R Sindhu[3]and K Jayalakshmi[4]

[1]Professor, Department of Civil Engineering, Karpagam Academy of Higher Education, Eachanari Post, Coimbatore 641021, Tamil Nadu, India

[2,3]Assistant Professor, Department of Civil Engineering, Karpagam Academy of Higher Education, Eachanari Post, Coimbatore 641021, Tamil Nadu, India

[4]PG Student, Department of Civil Engineering, Karpagam Academy of Higher Education, Eachanari Post, Coimbatore 641021, Tamil Nadu, India

Email: pmnatarajanin@gmail.com, johnpaulv2490@gmail.com,

Abstract. Lakes provide both humans and animals with great interest in water conservation because it contributes to their survival. Lakes have many applications, including leisure purposes, related to human activities. The lakes in India need to be protected and restored directly. Owing to unsustainable human activity, much of the lake landscape is a waste water collection storage area. Given this troubling failure of urban water bodies to function, it is essential that the lakes are kept free of pollution and accessible. The proposed work aims to advance the conservation of urban water systems in Coimbatore. The case study on lake rehabilitation (Periyakulam and Valankulam Lake) was examined. The approval and cooperation by Coimbatore City Corporations enhanced the hygienic condition and the ecosystem of the lakes in and around Coimbatore City. As a result, various approaches for restoring water bodies have been suggested.

1. Introduction

Lakes are characterized as the landscape that collects water, and their life is faster than the outlet, depending on the flow of inlet water. To marine life and human needs, the lake habitats are essential resources. Lakes have mild temperatures and influence the surrounding ecosystem. We regulate streamflow, restore groundwater aquifers and the frequency of droughts by storing water. Lakes provide shelter for aquatic or semi-aquatic plants and animals, which in turn provide food for many living things, increasing landscape diversity. Coimbatore District is a Tamil Nadu district and is the South Indian Manchester. The company is made up of 257.04sq. The Noyyal River is divided into the north, south, east, west and central areas, flowing through Coimbatore, forming the southern limit of the city. The town is located in the Noyyal basin and has an extensive lake system fed by the river and the runoff of rainwater. Narasampathi, Krishnampathi, Selvampathy, Kumaraswami, Selvachinthamani, Periyakulam, Valankulam,Singanallur tanks are the eight largest tank areas of Coimbatore. The lake pollution and the treatment needed to control it were carried out by Antara Bhattacharya and Anugya Shandilya (2017). Three initial evaluation stages have been established and extended to ten projects as planned for ecological conservation of the lakes. The lake was restored with rich mineral content and particular emphasis on climatic conditions. The rehabilitation of the lake with an ongoing recovery plan on the edge of the water as suggested. A researchwasconductedonthequestionofenvironmentalimpactsoftourismgrowth.Thiscasestudy review several proposals related to Periyakulam Lake and Valankulam Lake rejuvenation to compile a comprehensive report on the different rejuvenation methods, summarization of the advantages, the need to rejuvenate the lake for the people around the lake and discuss the various benefits that can be obtained following the rejuvenation. Following the effective rejuvenation of Periyakulam Lake and Valankulam Lake, it was suggested that different aspects of other famous lakes in Coimbatore, examined for implementation of a rejuvenation/improvement plan. Periyakulam Lake and Valankulam Lake profile and geographical data are shown in Table 1-4.

Table 1. Profile of Periyakulam Lake

Lake Name	Periyakulam Lake
District/Taluk/Village	Coimbatore, Tamil Nadu
Latitude and Longitude	N 10º 58' 44" E 76º 56' 43.56"
Altitude(m)	408.00MSL
Basin/Series/Sub-Series/Halla	Noyyal Basin
Lake Basin Type	Transitional
Catchment Area	8.54 sq km (as per current analysis) 10.752 sq km(as per Tank register)
Top of bund Level	97.4 m
Length of Main Bund	2880 M (as per Tank register) 2800 M (as per topographical Survey Drawing)
Water Spread area	1.245 sq km (as per topographical Survey Drawing)
Area of Lake	333 Acres 21.381 Gunta (as per topographical Survey Drawing)
Full Tank Level(FTL)	94.94 m
Gross Storage	1980763.42 Cum (as per Tank register) 1843912.0Cum (as per topographical Survey Drawing)
Length of Waste Weir, Type	28.00 m
Maximum Flood Discharge	42.62 Cum
Length of Shore Line	5240m
Mean Depth	1.5m
The average depth of water concerning FTL of the lake	2.44 m
Maximum Depth	5.87 m

Table 2. Profile of Valankulam Lake

Geographical Coordinates	N10°59.338 E 76°57.919.'
Altitude	440m above MSL
Lake Type	Natural
Area	64.75 ha
Maximum Depth	4.5m
Mean Depth	2 m
Water Spread Area	64.75 ha
Catchment Area	479.27 ha
Storage Capacity	27.88 Mcft
Inflow Water	Rainfall, Runoff, River, Drain, Wastewater Drain.
Outflow Water	Has Three outlets, surplus water is diverted directly to Sanganur Pallam drain, Maximum flood Discharge is 12.86 cumec

Table 3. Geographical data of Periyakulam lake

Catchment Area of the Lake (as per GIS Analysis)	9.014 Sq.km
Existing FTL as per Survey drawing	405.86
Proposed New FTL	405.10
Total Length of the Existing Main Bund	2800 m
Existing Surface Area of the Lake at FTL as per Survey drawing	1,249,739 Sq-m
Storage Volume of the Lake (as per survey tank register)	1,980,763m3
Storage Volume of the Lake as per Survey drawing up to existing FTL	24,40,059Cum
Estimated Flood Discharge	52,394 cumecs
Surplus Arrangements (dysfunctional)	Surplus weir-27m
Spillage Depth of Weir	0.6m Historical
Total Wastewater inflow	12.8MLD
BOD	Min:100mg/L Max:450mg/L
COD	Min:150 mg/L Max:1200 mg/L

Table 4. Geographical data of Valankulam Lake

Catchment Area of the Lake (watershed delineation using topo sheet)	11.17 Sq.Km
The current Analysis catchment area of the Lake	5.05 Sq.Km
Existing FTL as per Survey drawing	404.80
Proposed New FTL	404.50
Total Length of the Existing Main Bund	2500 m
Existing Surface Area of the Lake at FTL as per Survey drawing	4,20,528 Sq-m
Storage Volume of the Lake (as per survey tank register)	7,89,474 cu.m
Storage Volume of the Lake as per Survey drawing up to existing FTL	7,57,968 cu.m
Total Wastewater inflow	2.73MLD
BOD	Min:60 mg/L Max:160 mg/L
COD	Min:300mg/L Max:600mg/L

2. Methodology

2.1 Desilting of the lake
Desilting the lake with the max depth of 0.9m at the centre and 0.3m along the shoreline (Figure 1).

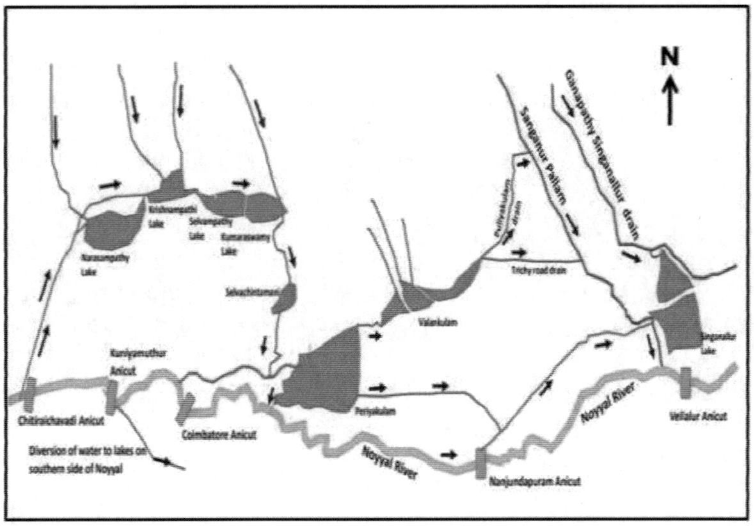

Figure 1. Lakes in Coimbatore, Tamil Nadu, India

2.2 Removal of geo fabric layer
Remove the existing geo fabric layer and provide revetment along with turfing to protect existing and proposed bund from erosion. Revetment and turfing are proposed to the water side of the bund to protect the bund from erosion. Revetments are proposed to the water side of the bund to protect the bund from erosion.

2.3 Walkway Bund profile correction and protection north western side
Revetments are proposed in the toe portion of the bund on the water side to prevent erosion of the bund.

2.4 Dismantling and reconstructing surplus arrangement
Constructing flush escape in Nala to discharge excess water directly to Noyyal River during a flood and constructing a defined drain section to accommodate flood discharge with the stone line canal.

2.5 Outnala refurbishment and refurbishment of draft channel drain
We are constructing a new box culvert or pipe culvert to accommodate the flood discharge. Construction of a defined cross-section to accommodate flood discharge with a stone-lined canal.

2.6 Construction of collection chamber near inlet
Chamber construction to link drains and connect lake via pipes. It proposed to reduce the invert level of the outlet tube to alter the depth of the water link and to alleviate floods. An irrigation sluice has become ineffective as a result of urbanization. The is proposed sluice to restored and developed to regulate water for maintenance and to act as a discharge mechanism for flooding. It proposed to provide a fixed cross-sectional line drain for 1000m for the estimated discharge of existing terrestrial drains with online channel treatment.

2.7 Island enhancement and non-motorized transport
Providing safety work for the island with a surface turfing revetment in contact with water and a non-motorized pathway is proposed.

2.8 Wastewater management
The treatment of wastewater segregated into three treatment parts, and they are pretreatment, primary and secondary treatment, and tertiary treatment.

2.9 Removal of unserviceable soil profile correction: southern side main bund
Striping the improper slope along the waterside and forming a proper bund. The gradual slope for better biodiversity integration and provided revetment on the waterside & turfing on the downstream side to protect from soil erosion. The total length for this work is 2400m.

2.10. Reformation of walkway bund and slope protection
It proposed to construct with recommended soil to maintain 1:1:5 slope. It should not be steeper than recommended. Since the northern bund does not take any water load, the slope protection works done with the formation of bund using a gentle slope of 1:3 with grass/ plants/ trees plantation to stabilize the shoreline. Provision should be made for revetment in the toe portion of the bund on the waterside to protect the soil erosion.

2.11. Diversion of storm water into the lake and outlet nala resectioning
The new surplus system has suggested for the flood discharge capacity of 35,74 m^3 / sec. For safe flood discharge passage, the outlet drains / structure near Sungam junction must be altered and adjustedaccordingly.Basedoninletlevels&drainlevels,Valankulam'sFTLproposedtoreducefrom 404.80 to404.50.

2.12. Desilting in pond A & B
Desilting recommended raising storage capacity to a maximum depth of 0.6 and 0.3 along the shoreline, as shown in drawings. When the high polluted soil with heavy metals and no sites seem to be known for the disposal of the silt, then the desilting plan can be assumed to be withdrawn. As

per the tank records, the Tank's storage capacity is 7,89,473.68 cum, and Water depth is as 4.5m as the existing surplus arrangement as FTL resulting in 7, 57,986 cum.

3. Results and discussions

The eco-restoration of lake waters is given the most critical consideration,

a. Water protection at the lake
b. Regulation of Storm water
c. Treatment of waste water

Periyakulam Lake is the largest water body located at the central place in Coimbatore; this study began with different steps to improve the environmental and pollution problems. Along with reasonable issue shielding water condition, prevention of dumping waste, the passage of contaminated water; it was decided to preserve the bunds of the lake as are creational sport for the public.

4. Conclusion

With this rejuvenation scheme, the growth of aquatic species can be increased. Water contamination can also be minimized when treated wastewater is permitted into inlets. Pollution-free water in the lake surrounding area is essential to improving the growth of aquatic animals. Therefore, the researched and proposed to investigate the remedial steps the condition of other essential lakes of Coimbatore.

References

[1] Christer N et. al. 2016 Evaluating the process of ecological restoration *Ecology and Society* **21(1)**:41 Edition- 08289-210141.

[2] Jukka H 2019 Sediment Nutrients Ecological *Status and Restoration Of Lakes, Conference Water Research* **160**: 206-208.

[3] Julita A and Grzegorz W 2016 Can we stop the degradation of lakes? *Innovative approaches in lake restoration, Ecological Engineering* **95**:714–722.

[4] Kiran et. al. 2018 Assessment of water quality and identification of pollution risk locations in Tiaoxi river (Taihu Watershed, China.)

[5] Mengmeng Z et. al. 2019 Exploring responses of lake area to river regulation and implications for the lake *Ecological Engineering* **128**:18–26.

[6] Mohan S et. al. 2018 Restoration of Mookanari lake by using Bio-Ozolyte treatment Salem corporation in Tamil Nadu, India *Inter. J. Advances, Sci.Tech* **6:1**.

[7] Sarika B and Pulluri S 2014 Sustainable approach for conserving and managing for urban water bodies *An Online International Research Journal* ISSN: 2311-3200.

[8] Sauni W and Alok R 2017 Urban development impacts on water bodies: A review in India *International Journal on Emerging Technologies* **8(1)**:363-370

[9] Wei Jetal 2018 Shear strength of purple soil bunds under different soil water contents and dry densities: A case study in the Three Gorges Reservoir Area, China, *Catena* **166**:124–133

[10] YanZhaoetal 2016 Downstream ecosystem responses to middle reach regulation of river discharge in the Heihe River Basin, China, *Hydrol. Earth Syst. Sci.* **20** 4469–4481.

[11] Yonghua Zetal 2016 Eco system restoration and conservation in the arid in land river basins of North west China: Problems and strategies *Ecological Engineering* **94**:629–637.

A Review on Correlations for Consolidation Characteristics of Various Soils

S Jayalekshmi[1] and V Elamathi[2]

[1]Professor, National Institute of Technology, Tiruchirappalli, Tamil Nadu.
[2]Research Scholar, National Institute of Technology, Tiruchirappalli, Tamil Nadu.

Email: elamathivenkat@gmail.com

Abstract. Settlement is the major problem that arises after the construction of a structure on a soil mass. Consolidation characteristics of soil such as coefficient of consolidation c_v, compression index c_c, recompression index c_r, preconsolidation pressure σ'_p play a major role in the settlement behavior of fine grained soil mass. c_v represents the rate of consolidation of soil mass. c_c and c_r are essential in calculating settlement of normally and over consolidated clays respectively. σ'_p is determined to find whether the clay is under, normally or over consolidated. These consolidation properties are obtained from graphical constructions after conducting several one dimensional consolidation-oedometer tests. Since this is time consuming, many correlations have been derived between consolidation properties and index properties of soil. Different researchers have used various soil parameters such as liquid limit (w_L), plastic limit (w_P), natural moisture content (w_n), initial insitu void ratio (e_o), dry unit weight (γ_d), plasticity index (I_P), void ratio at liquid limit (e_L) etc., as correlative parameters for deriving the correlations. Hence, it is desirable to predict the value of c_v, c_c, c_r and σ'_p from the known correlations rather than conducting several tests, to ease the procedures.

1. Introduction

The determination of settlement of soil mass is most important during the construction of structure on a soil. Settlement is due to compressibility of soil mass. The consolidation characteristics of soil like compression index c_c and coefficient of consolidation c_v play a vital role in predicting the compressibility of soil mass. Settlement is estimated from compression index c_c, which is obtained from void ratio e versus vertical effective stress σ' in the semi-logarithmic plane. Coefficient of consolidation c_v is used to predict the time required for a given amount of compression to take place i.e., rate of settlement. Casagrande's logarithm of time fitting [2] and the Taylor's square root of time fitting methods [4] are widely accepted as standard methods for determination of c_v. Other consolidation characteristics of soil such as recompression index c_r, and preconsolidation pressure σ'_p are also important in predicting the compressibility behavior of the soil mass. c_c and c_r which are slopes of virgin curve and recompression curve respectively (e versus log σ' plot) for calculating the settlement of normally and over consolidated clays respectively. In order to find whether the clay is normally or over consolidated, it is necessary to determine σ'_p. Casagrande's [3] construction is the most popular method for determining σ'_p. These consolidation characteristics are obtained from data of several one dimensional consolidation tests according to ASTM Standard [1] using oedometer apparatus for varying loading increments. The collected data are plotted as graphical construction, from which the consolidation characteristics are determined. The soil specimen in the oedometer can also be tested for various conditions like incremental loading [4,5], constant rate of strain (CRS) [6,7], controlled Gradient tests (CG Test) [8,9] and continuous loading (CL test) [10]. Thus the laboratory oedometer test takes more time to determine these consolidation characteristics. Hence equations have been developed for predicting consolidation properties, which can be used without conducting oedometer tests, thus saving considerable time. The paper reviews the correlation for c_v, c_c, c_r and σ'_p using soil properties like activity (ACT), liquidity index I_L, plasticity index I_p, shrinkage index I_s, liquid limit w_L plastic limit w_P, initial void ratio e_0, natural moisture content (w_n), initial in situ void ratio (e_o), dry unit weight (γ_d), plasticity index (I_p), and void ratio at liquid limit (e_L) as correlative parameters.

2. Prediction of coefficient of consolidation c_v from soil properties

An equation for one dimensional consolidation of cohesive soil layer under sustained loads considering the length of the drainage path given by [5] is expressed as:

$$T = \frac{c_v t}{H^2} \qquad \text{-----(1)}$$

Where t is the time for consolidation, T is the time factor and H is the length of the longest drainage path or thickness of the consolidating layer. The drainage path for double drainage is equal to H/2 and for single drainage, it is equal to H. The dimensionless factor T is derived from degree of consolidation U (%), for U<60%, T= (π/4) (U /100)^2, for U>60%, T=1.781-0.933 log (100-U) [2,12]. c_v is calculated using Casagrande's logarithm of time fitting method [3], Taylor's square root of time fitting method [12] Inflection point method [13], Analytical method [14], Velocity method [15,16], Rectangular Hyperbola Fitting Method [17] which consumes more time by conducting oedometer test for more than a week. In recent days, index properties of soil are used in correlating many engineering properties of soil.

Table 1. Correlations for determining c_v

S.No.	Equation	Applicability	source
1.	$c_v = \frac{9.09*10^{-7}(1.192+ACT^{-1})^{6.998}(4.135I_L+1)^{4.29}}{I_P(2.03I_L+1.192+ACT^{-1})^{7.998}}$ (m^2/s)	Normally consolidated clays	[18]
2.	$c_v = \frac{1+e_L(1.23-0.276\log\sigma')}{e_L}\times\frac{1}{\sigma'^{(0.353)}}\times10^{-3}$ (cm^2/s)	Normally consolidated clays with w_l 50-106% and w_p 27-47%	[21]

3.	$c_v = \frac{3}{100(I_s)^{3.54}}$ (m²/s)	Remolded soils with w_L 37% to 74%	[19]
4.	$c_v = 7.7525 I_p^{-3.1025}$ (cm²/s)	Normally consolidated clays of surat city	[22]
5.	$c_v = -4\times10^{-9} w_L + 4\times10^{-7}$ (m²/s)	Undisturbed clays of Manipur valley	[20]
6.	$c_v = 4258 w_L^{-1.75}$ (m²/yr)	middle and south Iraq with w_L 25 to 65%	[23]
7.	$c_v = 0.6155 - 0.0183 I_p$ (m²/yr)	Remolded soil with w_L 29% to 46% and I_P 8 to 18	[24]

From Table 1, It can be seen that index properties like liquid limit (w_L), liquidity index (I_L), plasticity index (I_P), shrinkage index (I_s), activity(ACT), void ratio at liquid limit (e_L) are used as a single parameter or multiple correlative parameters for predicting coefficient of consolidation c_v. Experiments were conducted on normally consolidated clay and it was observed that c_v decreases with decrease in water content and it is inversely proportional to I_p[18]. c_v has a better correlation with shrinkage index I_S than plasticity index I_P or liquid limit w_L and it was reported that ample scope exists for undisturbed and over consolidated soils [19]. Liquid limit w_L gives better correlation with c_v than shrinkage index I_s and plasticity index I_p [20].

3. Prediction of compression index c_c from soil properties

Compression index c_c is a main parameter in predicting time dependent consolidation settlement (S_c) as given in the equation 2. Hence it is necessary to find c_c in order to calculate S_c. For a layer of normally consolidated clay of thickness H, initial void ratio e_o, compression index c_c, and effective overburden pressure σ_0', the total settlement s_c under an applied load increment $\Delta\sigma'$ can be expressed as:

$$s_c = \frac{c_c}{1+e_0} H \log \frac{\sigma_0' + \Delta\sigma'}{\sigma_0'} \qquad \text{------(2)}$$

c_c is the slope of the compression curve e - log σ' plot and the expression is given as $c_c = \Delta e / \log(\sigma_2'/\sigma_1')$, where Δe is the change in void ratio for the corresponding effective vertical stress. The conventional one dimensional consolidation by oedometer test is time consuming in determination of c_c. In order to reduce the effort, many correlations exist between c_c and soil index properties like w_L, I_p, w_n, e_o, γ_d, as a single or multiple correlative parameter, as tabulated in Table 2, for different regions and applicability. Hence these empirical equations can be used directly for predicting c_c based on its applicability without undergoing conventional oedometer tests.

Table 2. Correlations for c_c

S.No.	Equation	Applicability	source
1.	$c_c = 0.007(w_L - 10)$	Remoulded clays	[11]
2.	$c_c = 0.54(e_n - 0.35)$ $c_c = 1.15(e - e_0)$	Normally consolidated clays	[34]
3.	$c_c = 0.001(w_L - 12)$	Osaka alluvial clays	[33]
4.	$c_c = 0.009(w_L - 10)$	Normally consolidated clays	[25]
5.	$c_c = 0.0083(w_L - 9)$	Various clays	[35]
6.	$c_c = 0.006(w_L - 9)$ $c_c = 0.4(e_0 - 0.25)$ $c_c = 0.01(w_n - 5)$	All clays with w_L<100%	[26]
7.	$c_c = 1.35 I_p$	Remoulded clays of Gulf of Mexico and North Sea	[36]
8.	$c_c = 0.0092(w_L - 13)$ $c_c = 0.0072(I_p + 26)$	All Clays	[37]
9.	$c_c = 0.011 w_n$	Chicago clays	[27]
10.	$c_c = 0.2343 e_L$	Remoulded normally consolidated clays	[28]
11.	$c_c = 0.329[0.027(w - w_p) + 0.0133 I_p (1.192 + ACT^{-1})]$	Normally consolidated clays	[18]
14.	$c_c = 0.256 e_L - 0{,}04$	Reconstituted clay Soil lying above A-Line	[29]
15.	$c_c = 0.009(w_L - 8)$	Osaka Bay Clay	[38]
16.	$c_c = 0.008(w_L - 12)$ $c_c = 0.007(I_s + 18)$	All Clays	[30]

17.	$c_c = 0.0012(w_L + 16.4)$ $c_c = 0.013(w_n - 3.85)$ $c_c = 0.54(e_0 - 0.37)$ $c_c = -1.6\gamma_d - 2.4$ $c_c = -0.0003w_n + 0.538e_0 + 0.002w_L - 0.3$	South coast of Korea	[39]
18	$c_c = 0.01w_n$	Undisturbed fibrous peat	[31]
19.	$c_c = 0.0061(w_L - 0.0024)$ $c_c = 0.0082I_p + 0.0915$ $c_c = 0.0091w_n + 0.0522$	Alluvial deposits, Surat, India	[40]
20.	$c_c = 0.0055(w_L - 1.8364)$ $c_c = 0.0055(I_s + 21.2364)$ $c_c = 0.0086(I_p + 24.2674)$ $c_c = 0.2875(e_0 - 0.5082)$	Highly plastic clays of Kuttanad, Kerala, India	[32]
21.	$c_c = 0.013(w_n - 0.115)$ $c_c = 0.49(e_0 - 0.11)$ $c_c = 0.014(w_L - 0.168)$ $c_c = 0.009w_n + 0.005w_L$ $c_c = 0.013w_n + 0.0w_L + 0.168$	Republic of korea	[41]
22.	$c_c = 0.0118(w_L - 20.7)$ $c_c = 0.014(w_n - 22.7)$	Irish soft soils	[42]
23.	$c_c = 0.0062w_L + 0.0165$	Remoulded soil with w_L 29% to 46% and I_P 8 to 18	[24]

Skempton[11] was the first to give empirical correlation for c_c of remoulded soils with liquid limit w_L.Colloidal clays are more compressible and possess higher void-ratios under a given pressure than the sandy clays and silts[11]. A modified equation for normally consolidated clays was given by [25]. e_0 and w_n were utilised to obtain best correlaive model for the determination of both c_c and c_r than w_L [26]. Ridge regression holds well in dealing with multicollinearity in regression models [27]. c_c decreases with decrease in the water content and it is directly proportional to the plasticity index [18]. The normally consolidated e versus log σ' curves of different fine grained soils were normalized by their respective e_L values where $e_L = w_L G_s$ and equations were developed for the stress and time dependent nature of preconsolidation pressure [28].

A normalizing parameter called void index, I_v was introduced in order to differentiate the properties of reconstituted clay (i.e., intrinsic compression characteristics of clay) from the natural clay soil [29]. It has been demonstrated that intrinsic compression line (ICL) and sedimentation compression line (SCL) is a valuable reference line in studying the compression characteristics of natural, normally and over consolidated clays. However ICL can be obtained by plotting I_v versus log σ′ for every increase in σ′. Similarly SCL can be obtained by plotting I_{v0} and log σ′. I_v and I_{v0} can be calculated from the equation 3 and equation 4.

$$I_v^* = \frac{e^* - e_{100}^*}{c_c^*} \quad \text{----(3)}$$

$$I_{v0}^* = \frac{e_0^* - e_{100}^*}{c_c^*} \quad \text{-----(4)}$$

Where I_v^* = intrinsic void index
e^* = intrinsic void ratio at various preconsolidation pressure
e_{100}^* = intrinsic void ratio at 100 kPa preconsolidation pressure
c_c^* = intrinsic compression index $= e_{100}^* - e_{1000}^*$
The normally consolidated soil lying between ICL and SCL, for a given value of I_{v0}, the effective overburden pressure taken by the natural clay is almost five times that for reconstituted clay. For overconsolidated clays, ICL and SCL are helpful in assessing the degree of overconsolidation [29].

The soil with a lower shrinkage index compresses less than a soil with a higher shrinkage index, even though their liquid limits are about the same [30]. The fibrous peat possesses very high water content of 500% to 2000% and very high permeability of 1000 times the initial permeability of soft clays. Fibrous peat deposits compresses extremely with an increase in effective vertical pressure and it exhibits highest value of secondary compression and lowest duration for primary consolidation [31]. Shrinkage index I_s is found to be the most suitable generalization parameter to characterize the compressibility behavior of highly plastic soil of Kuttanad region in the State of Kerala, India [32]. c_c in terms of w_n and w_L and normalized with e_0 was found to be more promising [24].

4. Determination of recompression index c_r

Over consolidated soils are probably encountered more often than normally consolidated soils. Moreover, if the soil is overconsolidated, it is necessary to determine settlement due to overconsolidation. The soil is overconsolidated, when the preconsolidation pressure σ'_p is greater than the existing vertical overburden pressure σ'_0. So, the first thing is to check whether the soil is preconsolidated. However, σ'_p is determined by plotting e versus log σ′ through graphical construction [3]. Equation 5 and equation 6 gives recompression index c_r for overconsolidated clays of thickness H to compute the level of settlement.

When $\sigma_0' + \Delta\sigma' \le \sigma_p'$,

$$s_c = \frac{c_r}{1+e_0} H \log \frac{\sigma_0' + \Delta\sigma'}{\sigma_0'} \quad \text{-----(5)}$$

When $\sigma_0' + \Delta\sigma' > \sigma_p'$,

$$s_c = \frac{c_r}{1+e_0} H \log \frac{\sigma_p'}{\sigma_0'} + \frac{c_c}{1+e_0} H \log \left[\frac{\sigma_0' + \Delta\sigma'}{\sigma_p'} \right] \quad \text{-----(6)}$$

Recompression index c_r is the slope of the recompression curve obtained when void ratio e is plotted against effective vertical stress σ′ in a semi logarithmic plot. Since the determination of c_r from oedometer tests is relatively time-consuming and is usually determined for a single unloading, empirical equations based on index properties can be useful for settlement estimation. Correlations have been proposed to relate the c_r of clay deposits to other soil parameters as given in Table 3. c_c and c_r are not only affected by over consolidation ratio, but also by other physical properties of soil like initial void ratio e_0. Increase in e_0 decreases c_c and c_r and increase in OCR, increases c_c and c_r [43]. Artificial neural networks (ANN) perform better in developing regression equations than the regression analysis [44].

Table 3. Correlations for detemining c_r

S.No.	Equation	Applicability	References
1.	$c_r = 0.0463e_L$	overconsolidated saturated fine grained soils	[28]
2.	$c_r = 0.14(e_0 + 0.007)$ $c_r = 0.01(w_n + 7)$ $c_r = 0.002(w_L + 9)$	All clays with w_L<100%	[26]
3.	$c_r = 0.00084(I_p - 4.6)$	remolded young natural clays	[45]
4.	$c_r = -0.025 + 0.002w_n$ $c_r = -0.024 + 0.0732e_0$ $c_r = 0.0048 + 0.001I_p$ $c_r = -0.0214 + 0.0013w_L$	Clay deposits	[46]
5.	$c_r = 0.041 - 0.0268e_0$ $c_r = 0.0131 + 0.0254c_c$ $c_r = 0.02 + 0.00052OCR$	Turkish clays	[43]
6.	$c_r = 0.0007w_L + 0.0062$ $c_r = 0.1257\gamma_d^{-2.8826}$	Mazandaran Province of Iran	[44]

5. Correlation between preconsolidation pressure σ′p and soil properties

In order to find whether the soil is preconsolidated, it is necessary to determine preconsolidation pressure σ'_p. Several procedures have been proposed to determine σ'_p. Casagrande's method [3], which is graphical construction method was widely used for determination of σ'_p. Interpretation of preconsolidation pressure from incremental loading test is more difficult, when the incremental ratio is high. Controlled strain test is the fastest method of determination of σ'_p [16]. σ'_p determined using ANN model gives more accurate result than graphical method and void ratio is the effective parameter on preconsolidation pressure [47]. Correlations for σ'_p for different applicability are given in the Table 4.

Table 4. Correlations for determining σ'_p

S.No	Equation	Applicability	References
1.	$\frac{e}{e_L} = 1.122 - 0.188 \log \sigma'_p - 0.0463 \log \sigma'$ (kN/m^2)	overconsolidated saturated fine grained soils	[28]
2.	$\sigma'_p = 137.924 - 0.179\sigma' - 30.48\left(\frac{e}{e_L}\right)$ (kN/m^2)	alluvial deposits of south Gujarat region	[40]

6. Conclusions

- It can be concluded that , the existing empirical correlations for the consolidation characteristics of soil, coefficient of consolidation c_v, compression index c_c, recompression index c_r and preconsolidation pressure σ'_p with soil parameters, as per the conditions of applicability and the methods used for deriving the correlations are reviewed extensively. The correlations are derived in terms of single or multiple soil parameters.
- Hence, by selecting the suitable equations,one can predict the values of c_v, c_c, c_r and σ'_p without conducting conventional oedometer tests.

Acknowlegement

The authors acknowledge with thanks, for the facilities offered ,by the Department of Civil Engineering, National Institute of Technology, Tiruchirappalli, India . Also,grateful to the Ministry of Education (MoE), India, for providing scholarship, to carry out the research work.

References

[1] ASTM D2435 / D2435M-11 2011 Standard Test Methods for One-Dimensional Consolidation Properties of Soils Using Incremental Loading, ASTM International.

[2] Casagrande A and Fadum R E 1940 Notes on soil testing for engineering purposes Soil Mechanics Series Harvard University Cambridge **8: 37** Pub. No. 268.

[3] Casagrande A 1936 The determination of the preconsolidation load and its practical significance In *Proceedings of 1st International Conference Soil Mechanics and Foundation Engineering* (Cambridge) **3** 60–64.

[4] Taylor D W 1942 Research on the Consolidation of Clays Massachusetts Institute of Technology Department of Civil and Sanitary Engineering Serial 82 Cambridge.

[5] Terzaghi K 1925 Settlement and consolidation of clay Engineering News-Record **95** 874-878.

[6] Hamilton J J and Crawford C B 1960 Improved determination of preconsolidation pressure of a sensitive clay ASTM STP **254** 254-271.

[7] Smith R E and Wahls H E 1969 Consolidation under constant rates of strain *Journal of the Soil Mechanics and Foundations Division* ASCE, **95(SM2)** 519-540.

[8] Gorman C T, Hopkins T C, Deen R C and Drnevich V P 1978 Constant-Rate-of-Strain and Controlled-Gradient consolidation Testing *Geotechnical Testing Journal* **1(1)** 3-15.

[9] Lowe J, Jonas E and Obrician V 1969 Controlled gradient consolidation test *Journal of the Soil Mechanics and Foundations Division American Society of Civil Engineers* **95(SM1)** 77-98.

[10] Janbu N, Tokheim O and Senneset K 1981 Consolidation tests with continuous loading Proceedings of 10th International Conference on Soil Mechanics and Foundation Engineering Stockholm **1** 645-654.
[11] Skempton A W and Jones O T 1944 Notes on compressibility of clays *Quarterly Journal of the Geological Society London* **100(2)** 119-135.
[12] Taylor D W 1948 Fundamentals of Soil Mechanics John Wiley and Sons New York.
[13] Cour F R 1971 Inflection point method for computing consolidation coefficient *Journal of the Soil Mechanics and Foundations Division* ASCE **97(SM5)** 827-831.
[14] Sivaram B and Swamee P K 1977 A computational method for consolidation coefficient *Soils and Foundations* **17(2)** 48-51.
[15] Parkin A K 1978 Coefficient of consolidation by the velocity method *Geotechnique* **28(4)** 472-474.
[16] Parkin A K 1981 Consolidation analysis by velocity method *Proceedings of 10th International Conference on Soil Mechanics and Foundation Engineering Stockholm* **1** 723-726.
[17] Sridharan A and Rao A 1981 Rectangular hyperbola fitting method for one dimensional consolidation *Geotechnical Testing Journal* ASTM **4(4)** 161-168.
[18] Carrier W D III 1985 Consolidation parameters derived from index tests *Geotechnique* **35(2)** 211– 213.
[19] Sridharan A and Nagaraj H B 2004 Coefficient of consolidation and its correlation with index properties of remolded soils *Geotechnical Testing Journal* ASTM **27(5)** 1-6.
[20] Priyadarshini Devi S, Rambha Devi K, Prasad D S V and Prasada Raju G V R 2015 Study on consolidation and correlation with index properties of different soils in Manipur valley *International Journal of Engineering Research and Development* **11(5)** 57-63.
[21] Raju Narasimha P S R, Pandian N S and Nagaraj T S 1995 Analysis and estimation of coefficient of consolidation *Geotechnical Testing Journal* **18(2)** 252– 258.
[22] Solanki C H 2011 Quick settlement computation of shallow foundation using soil index and plasticity characteristics Pan-Am *CGS Geotechnical conference* 1-5.
[23] Asma Y Al- Tae'e and Abbas F Al- Ameri 2011 Estimation of relationship between coefficient of consolidation and liquid limit of middle and south Iraqi soils *Journal of engineering* **17(3)** 430-440.
[24] Kok Shien N, Yee Ming C and Nur Izzati A L 2018 Prediction of consolidation characteristics from index properties *E3S Web of Conferences* **65** 1-5.
[25] Terzaghi K and Peck R B 1967 Soil Mechanics in Engineering Practice John Wiley and Sons New York.
[26] Azzouz A S, Krizek R J, and Corotis R B 1976 Regression analysis of soil compressibility *Soils and Foundations* **16(2)** 19–29.
[27] Koppula S D 1981 Statistical estimation of compression index *Geotechnical Testing Journal GTJODJ* ASTM **4(2)** 68-73.
[28] Nagaraj T S and Srinivasa Murthy B R 1985 Prediction of the preconsolidation pressure and recompression index of soils *Geotechnical Testing Journal* ASTM **8(4)** 199-202.
[29] Burland J B 1990 On the compressibility and shear strength of natural clays *Geotechnique* **40(3)** 329-378.
[30] Sridharan A and Nagaraj H B 2000 Compressibility behaviour of remoulded fine grained soils and correlation with index properties *Canadian Geotechnical Journal* **37** 712-722.
[31] Mesri G and Ajlouni M 2007 Engineering properties of fibrous peats *Journal of Geotechnical and Geoenvironmental Engineering* **133(7)** 850-866.
[32] Vinod P and Bindu J 2010 Compression index of highly plastic clays - An empirical correlation *Indian Geotechnical Journal* **40(3)** 174-180.
[33] Murayama S, Akai K and Ueshita K 1958 Engineering properties of diluvial clays in Osaka Soils and Foundations **6(4)** 39-47.
[34] Nishida Y 1956 A brief note on compression index of soil *Journal of the Soil Mechanics and Foundations Division ASCE* **82(3)** 1-4.
[35] Schofield A N and Wroth C P 1968 Critical-state soil mechanics (McGraw-Hill London) 134-206.
[36] Wroth C P and Wood D M 1978 The correlation of index properties with some basic engineering properties of soils *Canadian Geotechnical Journal* **15(2)** 137-145.
[37] Mayne P W 1980 Cam-clay predictions of undrained strength Journal of the Geotechnical Engineering Division *Proceedings of the American Society of Civil Engineers ASCE* **106**(GT11) 1219-1242.
[38] Tsuchida T, 1991 A new concept of e-logp relationship for clays *Proceedings of the 9th Asian regional conference on soil mechanics and foundation engineering* (Bangkok Thailand) 87-90.
[39] Gil-Lim Yoon, Byung-Tak Kim and Sang-Soo Jeon 2004 Empirical correlations of compression index for marine clay from regression analysis *Canadian Geotechnical Journal* **41** 1213–1221.
[40] Solanki C H and Desai J A 2008 Significance of Atterberg limits on compressibility parameters of alluvial deposits–new correlations *Proceedings of Indian Geotechnical Conference Bangalore India* **23** 20.
[41] Park H I and Lee S R 2011 Evaluation of the compression index of soils using an artificial neural network. *Computers and Geotechnics* **38** 472-81.

[42] Bryan A McCabe, Brian B Sheil, Michael M Long, Fintan J Buggy and Eric R Farrell 2014 Empirical correlations for the compression index of Irish soft soils *Proceedings of the Institution of Civil Engineers Geotechnical Engineering* **167(GE6)** 510–517.

[43] Gunduz Z and Arman H 2007 Possible relationships between compression and recompression indices of a low plasticity clayey soil *Arabian Journal for science and engineering* **32(2)** 179-190.

[44] Sinan N 2009 Estimation of swell index of fine grained soils using regression equations and artificial neural networks *Scientific Research and Essays* **4(10)** 1047–1056.

[45] Nakase A, Kamei T and Kusakabe O 1988 Constitute parameters estimated by plasticity index. *Journal of Geotechnical Engineering* ASCE **114(7)** 844-858.

[46] Afshin K, Farzin K, Behrouz K and Hossein M A 2015 Prediction of recompression index using GMDH-type neural network based on geotechnical soil properties *Soils and Foundations* **55(6)** 1335–1345.

[47] Celik S and Tan O 2005 Determination of preconsolidation pressure with artificial neural network *Civil Engineering and Environmental Systems* **22(4)** 217-231.

IOP Conf. Series: Materials Science and Engineering 1006 (2020) 012008 doi:10.1088/1757-899X/1006/1/012008

Road Safety Ratings for Selected Accident-Prone Sections in Urban Areas

Ujjwal Bansal[1], Sumit Kumar[2], Vilas Patil[3], Vikas K[4], L Durga Prashanth[5] and Vikas M[6]

[1,2,3,4]Student, Department of Civil Engineering, RV College of Engineering, India
[5,6]Assistant Professor, Department of Civil Engineering, RV College of Engineering, India

E-mail: vikasm@rvce.edu.in

Abstract. This study carries out a detailed analysis of Road Safety conditions at different junctions in Bengaluru. This study takes into account various factors both from Highway Engineering and Traffic Engineering. These factors govern the safety features of the roads. The factors include both the road attributes and the traffic data from various locations of which a detailed analysis is carried out.

The project proceeds with a detailed data collection and analysis of the collected data using an analytical model designed by International Road Assessment Programme (iRAP). This model uses the data collection and analysis technique to generate a Star Rating System (SRS) which is effective in keeping a track of the Road safety features. The system is suitable for all types of road users in terms of their safety and is destined to reduce road accidents. This study is very much important to know the basic flaws in the structure as well as in traffic management in order to improve the traffic facilities in India.

1. Introduction

India is the seventh largest country in terms of area. It has a huge Road network of over 5,897,671 kms. This huge road network requires a great maintenance. All the areas in India are being connected to the main districts through road networks and the government of India is spending a lot of money in the maintenance of this network as transportation plays an important role in maintaining continuity in the country. The road infrastructure in India costs over 8 Billion USD every year which is a great amount. Even after spending this much amount, the current road safety situation is not up to the mark mainly on the National Highways where the road users do not follow the guidelines issued by the Highway authorities. This is due to the negligence of the road users in India.

As per a semi structured administered questionnaire, around 75% of the road users didn't know the correct side of pedestrian walking and only around 15% could recognize all the traffic signs. It also showed that 95.5% of the total road users agreed on the safety aspects of using helmets but only 37% use helmets in day to day road commute.

There are many ways in which the accidents can be reduced. One way is to rate the existing roads on the basis of their features and characteristics. This way, the road network can be improved further in a very detailed way as per the requirements of the area and the road users and accidents can be reduced further. In this paper, we have provided the road safety ratings at selected accident-prone areas in Bangalore, Karnataka on the guidelines issued by iRAP. This rating helps us in understanding the improvements which can be adopted in the given area of interest in order to reduce the no. of road accidents and improve the road safety conditions in India.

1.1. Motivation

A country's development is based on a lot of factors in which Road safety plays an important role. As mentioned earlier, a country loses its man power as well as economy in the Road accidents.

The motivation towards choosing this as the research topic is to try finding solutions to the pre-existing major problems and take India a step forward towards development. This study involves the optimized use of both Highway Engineering and Traffic Engineering into our daily life in order to further motivate the Civil Engineers to carry more research in this field. The International Road Assessment Programme is an organisation with a group of Researchers who are motivated towards carrying research in this field and reduce road safety issues. The Road Safety Rating model is being used in many counties including India and our sole aim is to increase Road safety by using this model in and around Bangalore.

1.2.Literature review

Improvement in road safety is generally credited to introducing countermeasures related to the actions of utilities, automobiles and road users, and descriptions of such are outlined in manuals of best practice. Further rigorous analysis undertaken in particular countries have shown arithmetic correlations among accident reductions and road safety infrastructure initiatives, mandatory seatbelt use, drunken driving strategies, speed regulation etc. Some of the early technologies used in the past which were very useful in establishing a link between the road accidents and various different factors responsible for the same are discussed below:

According to Fukuda et al (2005), a new approach for improving public participation in developing a database of plant diseases across the country could be the Hiyari-Hatto method. Dynamic segmentation, geocoding of addresses and identification of intersections are the parameters on which the traffic accident situation is based. Furth (2005) used techniques, comprising simple sampling, estimated relationship beside a spread of possible auxiliary variables, representative sampling, cluster sampling, and combinations of those approaches. The later technique used on-off data from a travel sample to estimate the connection between passenger miles and potential passenger miles, a recently proposed auxiliary variable.

Chin and Quddus (2003) confirmed like others that in predicting crashes traffic volume is one of the most crucial factors to be considered. Even though these studies concluded traffic volume as the most crucial indicator variable, there are researchers who considered the effect of land use on road segment crashes, Valverde and Jovanis (2006) examined the consequences of demographic patterns and climate on shocks at the country level. Shankar et el (1996) applied a logical formulation to determine and verify the probability of accident severity conditioned on the occurrence of an accident. Their findings have the conclusion that there is a significantly higher chance of obvious injury or grievous hurt or fatality, relative to no obvious injury if the restraint system is not practiced by at least one driver at the time of occurrence of the accident.

Chand and Alex (2007) calculated the accident risk index and accident severity index for various Indian states. They are centred upon a series of accident indicators that are combined to form an index. The values of various India Stated indices are then compared with the values of these two indices. The number of accidents in a place and vehicles movements in that place are related to each other, rather than the population or the number of vehicles.

The accident rate fluctuates greatly from state to state. Different accident figures give different images. In this way, it aims to calculate the Accident Risk Index (ARI) for the states of

India. It has been calculated using a series of accident ratios combined by assigning certain weights to it. ARI is one such indicator that shows the impact of vehicle, road length, area and population on the accident number and identifies the prevailing accident rate in the district or state. A seriousness index has been found of the accident number with accidents that gives more weight to the number of fatalities than to the number of incapacitated.

Edwards (1998) [23], examined the association between weather recorded at a place and automobile accidents in England and Wales. Accident intensity for different hostile weather classifications of precipitation i.e. rain, fog and intense wind were weighed against non-hazardous weather situations. The outcomes revealed that the severity of the accident during rain decreases significantly compared to good weather conditions.

Home and Reyner (2000), conducted a study which indicates that driver exhaustion accounts for at least 20% of the total road accidents on straight roads. It urges that a lot of accidents happen on the roads which are due to excess sleep of the driver at the wheel during travel. A previous study by Home and Reyner (1995) on traffic accidents between 1987 and 1990 concluded that accidents related to driver's tiredness and sleep involved 16% of the overall traffic accidents and 23% of the accidents alone in UK.

To be precise, the main issues related to the behaviour of illegal drivers are speed, violations of priority rules, drunk driving, traffic lights, traffic lights across the road, distraction and fatigue, and illegal overtaking. It can be established from these studies that human factors are decisive in road accidents, but distraction and fatigue also appear to influence driving skills and driving behaviour (Delhomme et al 2009)

1.3. Need for star rating

There were some of the researches which focussed on reducing accidents and finding out the major reasons behind those accidents and to mitigate them. But not lot research has been done on linking the economy of a country to the road safety.

The iRAP Star Score is an unprejudiced indicator of occurrence and seriousness of an accident occurring on road. The prime focus of this study is on recognizing and documenting the characteristics of roads which, based on empirical evidence-based studies, affect the most common and serious accident types.

In this way, on a specific road section or network, the degree of road user risk can be evaluated without any detailed analysis of accident type data, that is common in low income and high accident rate countries having weak network of data availability. Evidence indicates that 1-star roads are least safe in terms of road accident and fatal injuries with 5-star (green) highways being the safest for daily travel and road usage.

2. Safety rating methodology

In order to ensure road safety, there is a need for accurate measurement of risk level that is actually considered detrimental to the safety of road section. Risk perhaps captures the combined risk arising from the interaction of road users, vehicles and road environment. This chapter aims to provide details of the procedures adopted in road safety rating system. Identifying the conditions of the road features leading to safety issues, design, plan and evaluation of the road section under consideration are explained in this section.

The process of how safe the road stretch/section is can be determined and expressed in various ways. Road safety rating through star rating for existing roads is one of many available methods which is a well-developed process through exhaustive data collection and its analysis. Since the process involves more time, most of the agencies do not follow this method and is limited to several countries. However, the outcome of this method is found beneficial in the countries wherever it is adopted. The sequential star rating process of any road section under consideration for road safety auditing is represented in Fig.1and is also listed below.

1. Data collection 2. Road coding 3. Data preparation 4. Data Analysis 5. Star Rating 6.

Investment planning

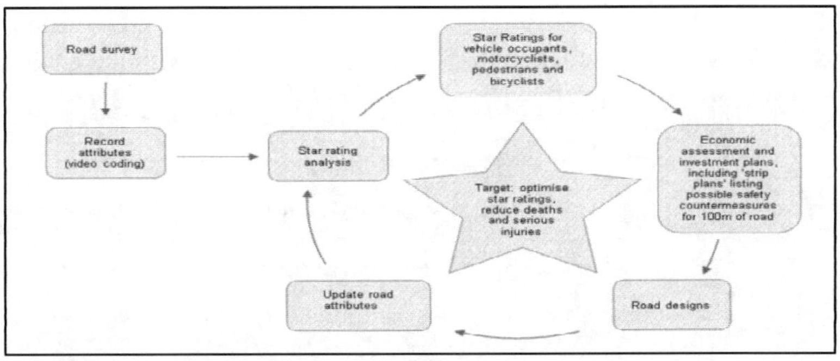

Fig. 1 Safety Rating process diagram

2.1.Data Collection

The data was collected by road survey for 4 accident prone sections of the Mysore road, namely RV College, Mylasandra Gate, Madhu Fuel Station and Kengeri Police Station.

The following data was required for the project to give star rating:

Width of lane, Road curvature, Curve Quality, Shoulder strips, Condition of Road, Skid resistance, Road side object, shoulder width etc.

2.2. Tabulation of data and calculation of star rating

All the above raw data is then tabulated and calculation is carried out. First, the star rating score is calculated. It is calculated for a 500-metre segment of the road using:

Star rating score (SRS)is given by summation of the rash type score (CTS), Where:

- SRS indicates the road user's relative risk of serious injury or death
- CTS = product of (Average speed, Flow influence, severity, Median traversability, likelihood)

Where:

- Likelihood cite the risk factors which indicate the chances for initiation of a crash
- Severity cite the risk factors that indicate the crash severity
- Operating speed cite the risk factors that relate speed and the risks associated with it
- External flow influence relates the influence of one person on another while using the road, and the risks of crash involved.
- Median traversability indicates the chances of a person to cross the median.

Star rating scores are calculated using following equation:

x – Factors multiplied (to the right)

+ - Factors added (to the right)

Factors are shown in Fig. 2. Star ratings are determined after calculation the SRS. They are calculated by matching the SRS to the range of star rating as shown:

Table 1: Star Rating Score

Star Rating	**Star Rating Score**
5	0 to < 2.5
4	2.5 to < 5
3	5 to < 12.5
2	2.5 to < 22.5
1	22.5 +

Where, 5 being the safest road and 1 being the most unsafe road.

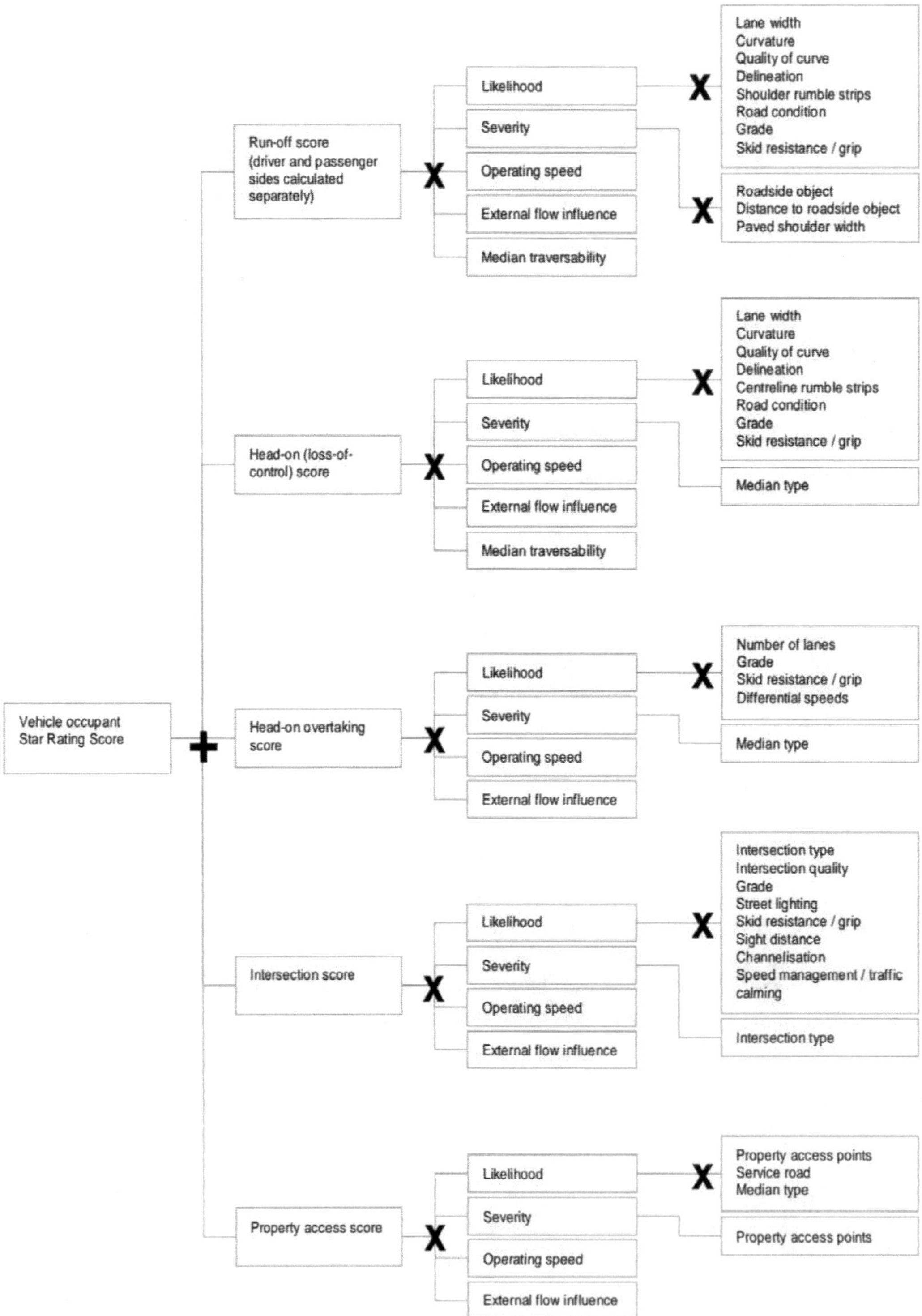

Fig 2: Calculation of Star Rating

IOP Conf. Series: Materials Science and Engineering 1006 (2020) 012008 doi:10.1088/1757-899X/1006/1/012008

3. Road safety rating

This chapter presents the details of the physical features of various road stretches under study. The existing physical conditions of each road stretch are compared with the standard features of road geometrics to check the deviations that lead to road safety risk. This information is later used for the road safety rating. The study stretch is a part of National Highway 275, commonly referred to as NH 275, that connects Bantwal town with Bangalore through Mysore and Madikeri in Karnataka state, India. It is one of the important highways in Karnataka and India.

Table 2: Analysis of traffic data collected at RV College (12.9237° N, 77.4987° E)

Risk factors	Category	Risk factor	Score
Road Attributes in likelihood			
Width of Lane	Wide	1.0	
Road Curvature	-	-	
Curve Quality	Poor	1.25	
Delineation	Poor	1.2	
Shoulder strips	Not Present	1.25	
Condition of Road	Poor	1.4	
Skid Resistance	Unsealed Adequate	3.0	
Product of the Risk factors in terms of likelihood			7.87

Road Attribute in terms of severity			
Road side Severity Distance	1-5 m	0.8	
Road side Severity Object	Tree	60	
Width of paved shoulder	1.2 m	0.83	
Product of Road (severity) risk Factors			
flow influence (External)	1500	0.50	
Traversiblity of Median	No	0	
Average Speed	37.4	0.33	
			0

Runoff speed star rating score (passenger side)			
Road Attributes in terms of likelihood			
Width of Lane	Wide	1	
Road Curvature	-	-	
Curve Quality	Poor	1.25	
Delineation	Poor	1.20	
Shoulder strips	Not present	1.25	
Condition of Road	Poor	1.4	

Skid resistance	Unsealed Adequate	3.0	
Product of the Risk factors in terms of likelihood			7.875

Road Attribute in terms of severity			
Road side Severity Distance	1-5 m	0.8	
Road side Severity Object	Tree	60	
Width of paved shoulder	1.2 m	0.83	
Product of Road (severity) risk Factors			39.84
flow influence (External)	1500	0.5	
Average Speed	37.4	0.33	
			6.57

Head on Crash (Control)			
Type of Risk factors	Category	Risk factor	Score
Width of Lane	Wide	1.0	
Road Curvature	-	-	
Curve Quality	Poor	1.25	
Delineation	Poor	1.2	
Shoulder strips	Not Present	1.25	
Condition of Road	Poor	1.4	
Skid Resistance	Unsealed Adequate	3.0	
Product of the Risk factors in terms of likelihood			7.56

Road Attribute in terms of severity			
Type of Median	Physical median 1-1.5m	80	
Product of the Risk factors in terms of likelihood			80
Flow influence (External)	15000	0.185	
Traversability of Median	No	0	
Average Speed	37.4	0.33	
Head on Crash Star Rating Score			0
Head on Crash (Overtaking)			

Skid resistance	Unsealed adequate	3.0	
Varying Speed	Not Present	1.2	

IOP Conf. Series: Materials Science and Engineering 1006 (2020) 012008 doi:10.1088/1757-899X/1006/1/012008

Lanes	3	0.01	
Product of road Attribute (Likelihood) Risk Factor			0.036
Road Attribute (severity)			
Median Type	1-5m	80	
Product of the Risk factors in terms of likelihood			80

Risk factors			
Flow influence (External)	15000	0	
Average speed	37.4	0.33	
Head on Crash (Overtaking) Star Rating Score			0

Intersection			
Type of Intersection	4-leg signalized	15	
Quality of Intersection	adequate	1	

	-	-	
Street Lighting	Present	1	
Skid Resistance	Unsealed adequate	3.0	
Sight distance for road users	Adequate	1	
Intersection characterization	Not present	1.2	
Speed Management	Not Present		
Product of Risk Factors in terms of Likelihood			54

Road Attribute in terms of severity			
Risk factor	Category	Risk factor	Score
Type of Intersection	4-leg signalized	15	
Product of Risk Factors in terms of Likelihood			15
Flow influence (External)	1000-5000 vehicle	0.125	
Average speed	37.4	0.33	
SRS (Intersection)			33.41

Access to property			
Property access point	Commercial access	2	
Service road for property access	Not present	1.5	
Product of Risk Factor in terms of Likelihood			3

Road Attribute in terms of severity			
Property access point	Commercial	50	

Product of Risk Factor in terms of Likelihood	50

Flow Influence (External)	Default	0.01	
Average Speed	37.4	0.33	
Star Rating Score (Intersection)			0.495

SRS and Vehicle occupant star rating			
Type of Crash	SRS	Star rating	
Driver side Road Runoff	0		
Passenger side Road Runoff	6.57		
Head on crash due to lost control	0		
Head on Crash due to overtaking	0		
Intersection Score	33.41		
Access to property	0.495		
Total score and Star rating	40.475	1	

4. Conclusions

After completing the data analysis at all the four locations, namely RV College, Mylasandra Gate, Madhu Fuel Station and Kengeri Police Station, the star rating was found to be 1. This shows that the road network in Bangalore is least safe for daily commute. Some points were common and can be considered as common factors responsible for the bad Star Rating Score. The common factors found were the following:

1. Absence of proper Traffic Management System at the junctions.
2. Lack of maintenance of the junctions in terms of Road attributes.
3. Traffic diversion was not available at all the junctions.
4. A facility for pedestrian traffic was absent.
5. Speed controlling measures are not taken at the junctions.

All these factors show the general trends of Indian Roads in terms of safety. Most of the Indian roads fall under the category of least safe roads and require a lot of considerations and improvements towards Road Safety.

Fig 3: Vehicles moving in wrong side

Fig 4: Positioning of Pole

Fig 5: Tree in midst of the road features

Fig 6: Damaged Traffic lights

References

[1] T Fukuda, T Ishizaka and A Fukuda 2015 Empirical study on identifying potential black spots through public participation approach: a case study of bangkok *J East Asia Soc. Transp. Stud* vol. **6** no. January pp. 3683–3696 doi: 10.11175/easts.6.3683.

[2] J Aguero-Valverde and P P Jovanis 2006 Spatial analysis of fatal and injury crashes in Pennsylvania *Accid. Anal. Prev.* vol. **38** no. 3 pp. 618–625 doi: 10.1016/j.aap.2005.12.006.

[3] V Shankar and F Mannering 1996 An exploratory multinomial logit analysis of single-vehicle motorcycle accident severity *J. Safety Res.* vol. **27** no. 3 pp. 183–194 doi: 10.1016/0022-4375(96)00010-2.

[4] V R Rengaraju and M Satyakumar 1994 Structuring category analysis using statistical technique *J. Transp. Eng.* vol. **120** no. 6 pp. 930–939 doi: 10.1061/(ASCE)0733-947X(1994)120:6(930).

[5] D Mohan and P S Bawa 1985 An analysis of road traffic fatalities in Delhi, India *Accid. Anal. Prev.* vol. **17** no. 1 pp. 33–45 doi: 10.1016/0001-4575(85)90006-5.

[6] H C Chin and M A Quddus 2003 Applying the random effect negative binomial model to examine traffic accident occurrence at signalized intersections *Accid. Anal. Prev.* vol. **35** no. 2 pp. 253–259 doi: 10.1016/S0001-4575(02)00003-9.

[7] Hashem R AI-Masaeid and Kumare C Sinha 1995 Analysis of accident reduction potentials of pavement markings J. Transp. Eng. vol. **120** no. 5 pp. 723–736.

[8] P P VALLI 2005 Road accident models for large metropolitan cities of India *IATSS Res.* vol. **29** no. 1 pp. 57–65 doi: 10.1016/s0386-1112(14)60119-9.

[9] A Mekky 1985 Effects of rapid increase in motorization levels on road fatality rates in some rich developing countries *Accid. Anal. Prev.* vol. **17** no. 2 pp. 101–109 doi: 10.1016/0001-4575(85)90013-2.

[10] J B Edwards 1998 The relationship between road accident severity and recorded weather *J. Safety Res.* vol. **29** no. 4 pp. 249–262 doi: 10.1016/S0022-4375(98)00051-6

Finite Element Analysis on Shear Strength of a Castellated Beam with Hexagonal Web Opening

R Deepha[1] **and S Jayalekshmi**[2]
[1]Research Scholar,Dept. of Civil Engineering, National Institute of Technology, Trichy. India
[2]Professor, Dept. of Civil Engineering, National Institute of Technology, Trichy. India
E-mail:rajendrandeepha@gmail.com

Abstract. A beam (flexural member) with several regular openings with different shapes is called a castellated beam. The castellated beams are well accepted in the industry, multi-story buildings, and power plants due to their increased depth of the section along with the web results in increased strength and stiffness. Though numerous simulation and experimental analyses have been made on the castellated beam, with intermediate transverse stiffener, bearing stiffener and torsion stiffener are used to increase a load-bearing capacity of the beam, but the study on the understanding of diagonal stiffener in the opening is very meager. The objective of the research is to investigate the linear behavior – ISMB150 and IC225 castellated beam of hexagonal openings with and without diagonal stiffeners and the effect of fillet radius at the corner of the openings is studied. Using finite element analysis software,Hyper works, with OptiStruct as a solver, under simply supported beam construction with four-point loading conditions was analyzed by developing an efficient 3D model for the analysis. Due to the openings in the section, the sharp corner of the opening is subjected to Vierendeel failure by the formation of plastic hinges at the corners, which makes critical hence by providing diagonal stiffener in the direction of diagonal tension and diagonal compression for the smooth flow of shear forces. The results show an improvement in load-bearing capacity of 1.51 times in the castellated beam than the parent section ISMB150. The stiffener provided further enhances the same to 1.45 times compared to the ISMB150 section. The linear results show lesser stress concentrations at the web opening corners, thus reducing the prominence of failure at the corners.

1. Introduction

Castellated beam finds its importance in engineering constructions in recent decades compared to the solid web beam. The significant reasons why the Castellated Steel Beam is gaining their importance are reduction in the total height of the construction; service let outs are easier, reduction in steel usage, reduction in overall cost incurred. The high strength of structural steel cannot be utilized to its best due to the limitations on maximum allowable deflections.Steel beams with perforated web openings are fabricated from hot rolled steel section is most famous owing to its improved strength, stiffness and ease of feasibility owing to its provision of opening.The fabrication of castellated beams includes bifurcating the web section in a crisscross pattern according to the cross-section as required, along its centerline, and joining the two halves by weld joint to such that the depth increases. The reduction in weight of the beam constitutes a larger section modulus and moment of inertia with better bending rigidity than the solid section counterpart, without an increase in weight. They are largely employed for medium and long spans constructions as roof or floor beams, joists, purlins.

Their aesthetic features attracts for numerous architectural applications as the web openings of these members serve as a passage for easy directing and installation of service utilities. The most common castellated structures constituets hexagonal web openings and circular openings in the case of cellular beams. The circular web openings are commonly referred to as cells. The recent increase in usage of castellated and cellular beams highlights the need for additional research.

Kerdal and Nethercot[1] have made significant studies on the typical failure flexural mechanism of the castellated beam. According to their findings, the castellated beam structures are more prone to failure due to web-postbuckling due to shear force; Vierendeel mechanism; Lateral torsional buckling; weld rupture. Samadhan G. Morkhade[2]extensively studied the behavior of a rectangular I beam with varying fillet radius and stiffener arrangement experimentally. The author reported that stress concentration is critical at the corners of the opening. An empirical relation is provided by Timoshenko [3] to find the lateral-torsional buckling of laterally unrestrained castellated beam given by;

$$M_{cr} = \sqrt{\frac{\pi^2 EI_y}{(kL)^2}\left(GI_t + \frac{\pi^2 EI_y}{(kL)^2}\right)} \quad (1)$$

Numerical studies on the castellated beam were carried out by a few researchers who made their significant contribution to this study field. Among the different failure mechanisms, the most important failure mechanism the beam is encountered to failure due to the Vierendeel mechanism. Very few notable experimental findings were carried out on the same. One of them is the experimental investigation conducted by FerhatErdal and Mehmet polatsaka[4] on twelve cellular beams. Different beam section was chosen in their study to identify the ultimate load-carrying capacity. Their findings reported that the beam is prone to failure by Vierendeel bending, web buckling, and lateral-torsional buckling. They also reported that all the failures are due to inadequate lateral support and normal load applied to the cellular opening. The linear torsional buckling (LTB) of the castellated beam was numerically investigated by DelphineSonck and Jan Belis[5]based on ENV3 standard – annex N. The preliminary design proposedbased on the currently existing European guidelines was given, and the empirical relationship to calculate the LTB resistance of a beam M_{Rd} can be calculated using

$$M_{Rd} = \frac{X_{LT} W_y f_y}{\gamma_{Ml}} \quad (2)$$

Where;

Reduction factor - $X_{LT} = \frac{1}{\phi + \sqrt{(\phi^2 - \bar{\lambda}_{LT}^2)}} \leq 1$ (3)

Section modulus – Wy; Nominal yield stress – fyand γ_{Ml} the partial factor- 1.0.According to the 2T design approach proposed,the critical moment for the lateral-torsional buckling is calculated by

$$M_{CR} = \frac{\pi}{L}\sqrt{\left(GI_t EI_z + \frac{\pi^2}{L^2} EI_z EI_w\right)}$$

K.F.Chung et al.[6]proposed a new design approach on Vierendeel failure mechanisms, which is critical in castellated beamsdue to the plastic hinge formation at the low and high moment sides, which reduces the load bearing capacity of the beam. The shear - moment interaction curve was identified to be applicable for the medium to the large circular opening,which serves as a conservative and effective method. As advancement in analyzing the failure mechanism, K.F.Chung et al.[7]analyzed steel beams with openings varying the shapes and sizes, the diameter of which varies between 0.5h to 0.75h. With the varying depth of opening and overall depth as parameters, shear moment interaction curves are proposed, which summarises that global shear force is a major cause for both the shear failure and Vierendeel failure. Observations also stated that the failure modes are more prominent in the castellated beam with large-sized web openings. The influence of fillet corner provision was examined by Peijun W et.al[8]numerically, focusing on the Vierendeel mechanism.

Amin Mohebkhah and HosseinShowkati[9] discussed elastic bracing requirements in the castellated beam to avoid lateral-torsional buckling of the beams. The effects of central elastic lateral restraints and large lateral deflections were studied considering the provision of elastic bracings, andit was found that the provision of central elastic restraint increases the elastic buckling strength of the beam. The increased strength not only depends on the stiffness but also majorly depends on the slenderness of the beam. EhabEllobody[10]conducted extensive research on failure modes and lateral load-bearing capacity of normal to high strength castellated steel beams. The Parametric study was carried out bythe change in geometries, strength, and buckling behavior of castellated steel beams, which summarizes a considerable decrease in the distortional buckling.

Although several reports were made on the castellated beams, very few investigations were attempted on the influence of diagonal bracing and change in the fillet radius on the load-bearing capacity of ISMB150 castellated beam. The present analysis is aimed at finite element analysis througha linear approach to identify the influence of the provision of stiffeners and change in the fillet radius of the web openings. The castellated beams were designed such that the depth is increasedup to 1.5 times the actual depth of the section with an angle of cut 56°. The shear failure and stress concentration are determined with and without providing web stiffeners on either side of the web posts.

2. Methodology

2.1 The geometry of I section and tension field action

ISMB 150 is selected for fabricating the castellated beams. The castellated beams were fabricated such that the depth of the beam is 1.5 times the original depth as IC 225 with an angle of cut 56°. Generally,the angle should be within the range of about 45° (min) and 70 °(max), where 45° and 60° are commonly used. The typical geometry of the castellated beam is shown in figure 1 below.

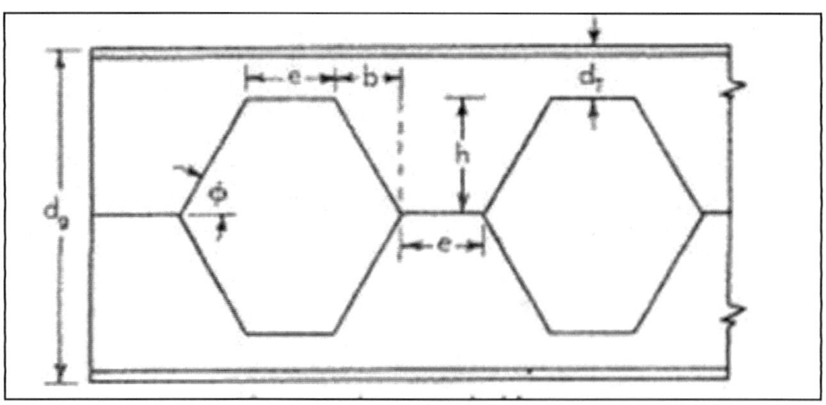

Fig. 1 Castellated beam geometry

The distribution of stress in the I section shows that flanges resist the bending moment, while web assists in resistingthe shear force.During transverse loading, normal and shearing stress are developed; hence complementary shear stress is formed to satisfy the equilibrium. This shearing stress in the beam occurs due to the internal shearing of the beam that results from shear force subjected to the beam. Consider a small element subjected to shear stress(figure 2)where, the element is subjected to principal compression in the direction AC, and the principal tension is occurring along the direction BD.As the load gradually increases and when the beam is loaded beyond the limit, first, it will deflect significantly asthe stiffness is lost further tends to buckle. The web part loses its ability to resist the diagonal compression along the direction of AC, and the beam cannot take the further load. To overcome this, diagonal stiffeners are introduced on the web opening where shear is dominating, and hence forces flow through the diagonal stiffener.

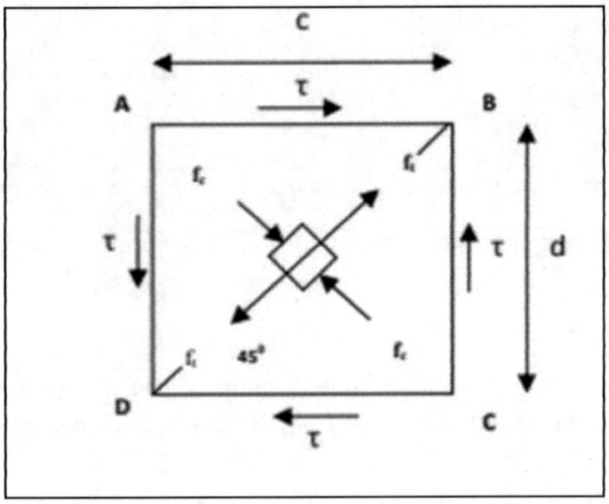

Fig. 2 Elemental representation of the shear panel

The behavior is very similar to a Pratt truss(figure 3), in which the vertical members carry compression and the tension is carried by the diagonal. The web resists only the diagonal tension and this behavior of the web is called tension field action.

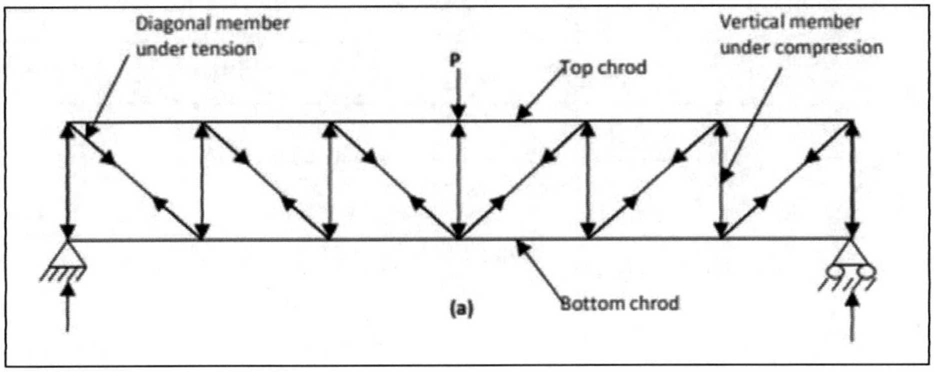

Fig. 3. Pratt Truss showing the tensile field action

2.2 Empirical equations

The following empirical relations are employed in calculating the design geometry of the castellated beam

$$Area\ of\ flange\ (A_f) = B_f\ x\ T_f \quad (4)$$

$$Area\ of\ web\ (A_w) = B_w\ x\ T_w \quad (5)$$

$$Position\ of\ centroid\ of\ castellated\ T\ section\ (\bar{Y}) = \left(\sum ay / \sum a \right) \quad (6)$$

$$MI\ of\ T\ section\ about\ neutral\ axis\ (I_T) = \left(\frac{bd^3}{12} \right) + A_f (y - \bar{y})^2 \quad (7)$$

$$Moment\ of\ resistance\ (M_R) = A\sigma_{at} d \quad (8)$$

$$Load\ applied\ over\ section\ (W) = M_R / L \quad (9)$$

$$The\ average\ shear\ stress\ at\ the\ ends\ (\tau_{va}) = P / 2d't \quad (10)$$

$$Max. mid - span\ deflection\ (\delta) = \delta_1 + \delta_2 < \left(L / 325 \right) \quad (11)$$

$$Combined\ local\ bending\ and\ direct\ stress\ (C_{st}) = P_s + S_s \quad (12)$$

$$Max. Stiffener\ width\ (W_{st}) = \left(b - t/2 \right) \quad (13)$$

2.3 Validation

The modal created in hyper works is validated against the Numerical results conducted by Wakchaure et.al[11]on the finite element analysis of the castellated beam, which shows that the error is less than 1%, and it is firm to continue. And the beam deflection value obtained from Hyperworks is also validated by a Macaulay's method theoretically, and the error is found to less than 1. The corresponding analysis results are tabulated in Tables 1 and 2, respectively.

Table 1: Numerical validation of results obtained by Hyperworks vs. Numerical results of Wakchaure et.al

S.NO	Section	Span L (m)	Deflection (mm) Numerically using Hyperworks	Deflection (mm) Numerically by Wakchaureet.all	Error (%)
1	IC 225	3	0.661	0.664	0.451%

Table 2: Theoretical validation of results obtained by Hyperworks vs. Macaulay's method

S.NO	Section	Span L (m)	Deflection(mm) Numerical usingHyperWorks	Deflection (mm) By Macaulay's method	Error (%)
1	ISMB 150	3	3.13	3.16	0.94%

2.4 Material Properties

The tensile test is carried out as per the ASTM A370:2017. Coupons drawn from flanges and web of the beam were subjected to tensile axial loading, which becomes a necessary material Properties for numerical modeling of the beams. Ultimate tensile strength, breaking strength, and maximum elongation, were recorded. From these measurements, Young's modulus, Poisson's ratio,yield strength, and ultimate strength are used in finite element modeling of the beams was also determined. The average values of yield strength and ultimate strength are determined as 361.66 MPa and 527.94MPa, respectively. The results show that the web part undergoes more stain hardening compared to the flange specimen. The specimens drawn from the flange part shows more ductility range compared to the web part. The variation in the properties can be attributed to the rolling directions of the individual members. Figure 4 shows the stress-strain graph obtained for the tensile samples drawn from the web and flange section.

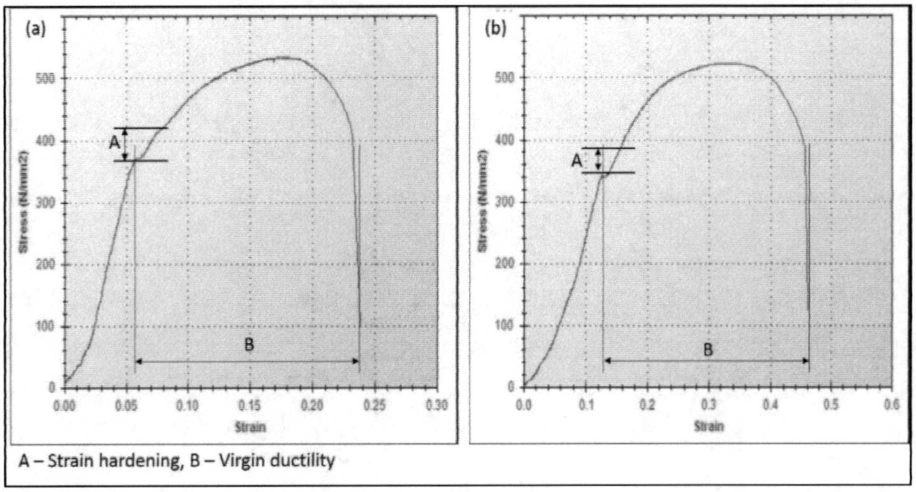

Fig. 4. Tensile stress vs. strain graph of coupons drawn from (a) Web (b) flange of ISMB150

3. Finite element analysis

The aim is to simulate the linear behavior of castellated ISMB 150 beam of I shaped cross-section with the hexagonal opening using a three-dimensional finite element model developed for determining the Y-directional deformations and vonmises failure stresses. The webs, flanges, and stiffeners were modeled by a Hexa eight noded element constituting six degrees of freedom at each node, three translations, and three rotations, which enable explicit simulation of various buckling deformations. Linear elastic material with Young's modulus $E = 2.1 \times 10^5$MPa and Poisson's ratio $\nu = 0.3$ is considered.The model was developed with simply supported beam configuration with one end hinged and roller support provided over the other end. A two-point load at l/3 and 2l/3 of the span and central 'y' deflection is calculated. At one end,the translation degree of the freedom along the three-axis was restrained at the left support to simulate the pin support while at the other end, theZ-axis degree of freedom of translation was released. The rotational degree of freedom is unrestrained at both supports, and the effects of residual stresses and welding of two parts have not been considered. The study includes three major considerations - (i) parent solid ISMB 150 compared with the castellated IC225 beam, (ii) Effects on the provision of diagonal stiffener on IC225, and (iii) Effect of provision of fillet radius on the web opening corners.

The three-dimensional models developed using CATIA was numericallyanalyzed using hyper works in the present investigation with Hypermesh as pre-processor and Opti-struct as the solver. The hyper view post-processor was employed in analyzing the results. Careful attention in mesh sizing was laid as it is the most critical factor in FEA analysis. Regardless of the competency of the software to consider the material nonlinearity, only the linear material influence was considered in the present analysis. To determine the optimum mesh size, a mesh convergence study was conducted and the results verifies that the FEA model has converged to a solution 3mm × 3mm mesh size, which provides a balance between the accuracy and computational expense. Finer meshes are generated to model areas near the openings, to improve precision and also justifies.

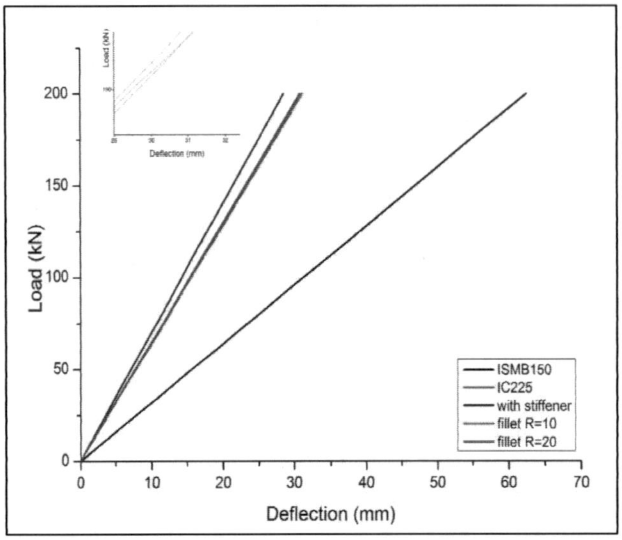

Fig. 5. Load vs. Deflection curves for different beam configurations

The deflection results arrived through the finite element analysis are represented pictorially in figure 5, which corresponds to various loading conditions for varying beam configurations. The results show that in the ISMB 150 beam, the stiffness of the beam is degressive with a continued level of increase in the load,exhibiting an increase in deflection and a reduction in the load-bearing capacity. The stiffness gets increased in the IC225 castellated beam due to the increase in the overall web height showing a considerable decrease in the deflection. The provision of diagonal stiffener arrangement has higher stiffness, which yields lesser deflection because of the truss action it takes through the diagonal stiffener.The stiffeners are provided in the direction of the diagonal tension and compression in the web for the smooth flow of shear forces. In all the cases, the FEA results show that the stiffness of the beam is decreased with anrise in the opening area and by sharp corners. The zero fillet radius has a high-stress concentration in the sharp edge of the opening,increasing overall defection. By providing a fillet radius of 10mm and 20mm at the corners of the web openings, the stiffness of the beam gets marginally increased with lesser deformation and a marginal reduction in stress concentrations.

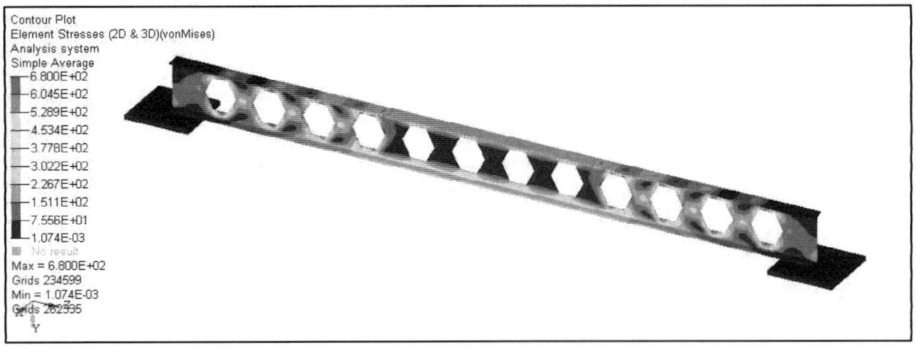

Fig. 6. Stress distribution over IC225 castellated beam @ 10kN

Figure 6 represents the critical stresses around the sharp corners of the opening in the shear zone of IC225, where shear stress is dominant. Hence, vonmisses stress is high in the shear zone at both ends and the loading points. The plastic hinge was first formed at the lower moment side following the load redistribution occurs across the web opening along the shearzone, and shearfailure is more near the opening. Subsequently,other plastic hinges were formed at different places of the opening.

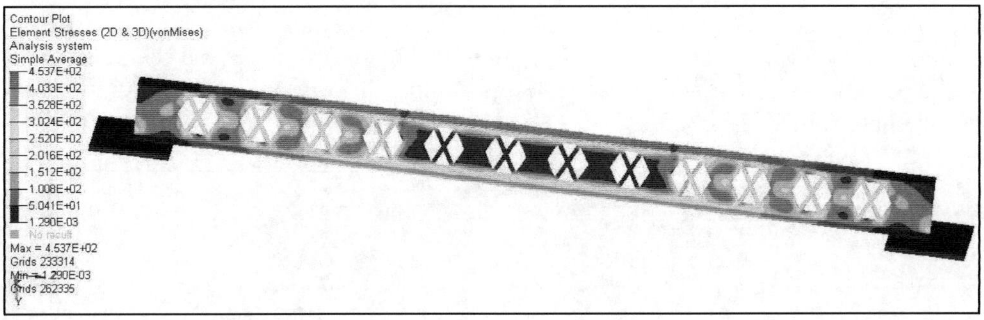

Fig. 7. Stress distribution over IC225 castellated beam with diagonal stiffeners @ 10kN

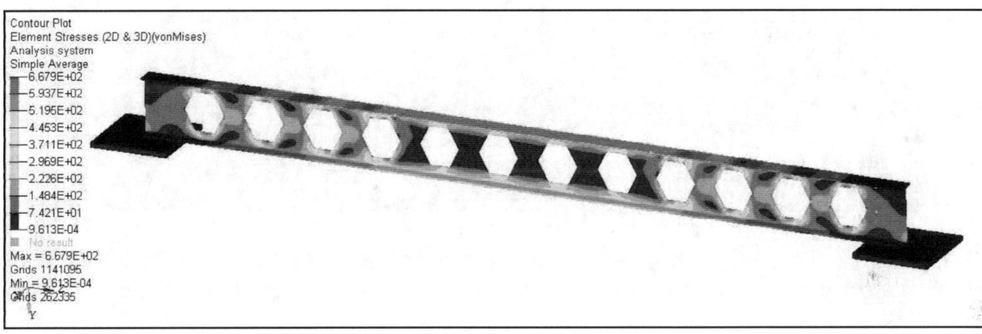

Fig. 8. Stress distribution over IC225 webopenings with fillet radius R10 @ 10kN

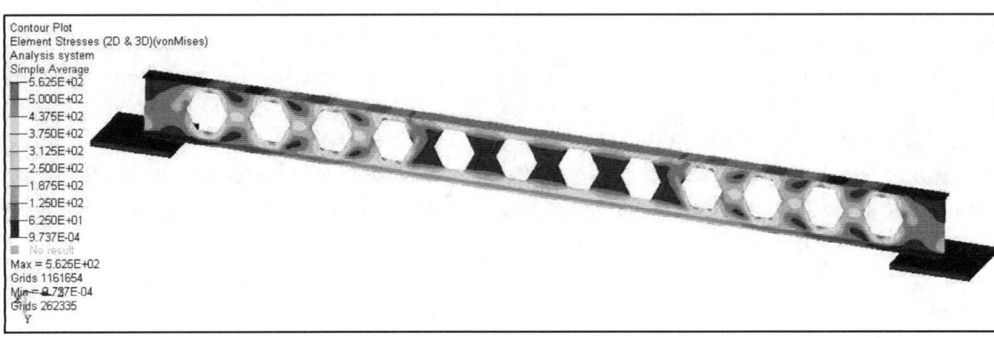

Fig. 9. Stress distribution over IC225 webopenings with fillet radius R10 @ 10kN

From the deflection results, it can be understood that the provision of diagonal stiffeners considerably increases the load-bearing capacity and decrease the overall deflection. Figure 7indicates that the diagonal stiffeners are provided in the diagonal tension, and compression directions, a truss action in the web opening of the beam occurs to ease the flow of the forces. The analysis shows that the diagonal stiffener provides effective stability than the castellated beam with rounding the corners of the opening. Also,the shear strength of the castellated beam can be improved by introducing shear stiffeners (Diagonal Stiffeners) on the web opening along the shear zone. Though the provision of stiffeners improves the overall load-bearing capacity, due to the openings, it is prone to fail by cracks in the corners. The formation of plastic hinges makes the hinges rotate till it reaches a plastic moment, further causing Vierendeel failure to occur. Figure 7,8 and 9 shows that the stress concentration at the web opening corners are lesser compared to stress concentration at the corners in case of IC225. The stress and strain distribution of different beam configurations are shown in contour plots in figure 10. The results show that the stress induced on the castellated beam with the stiffeners are lesser compared to the beam configurations provided with web opening corner fillet radius.

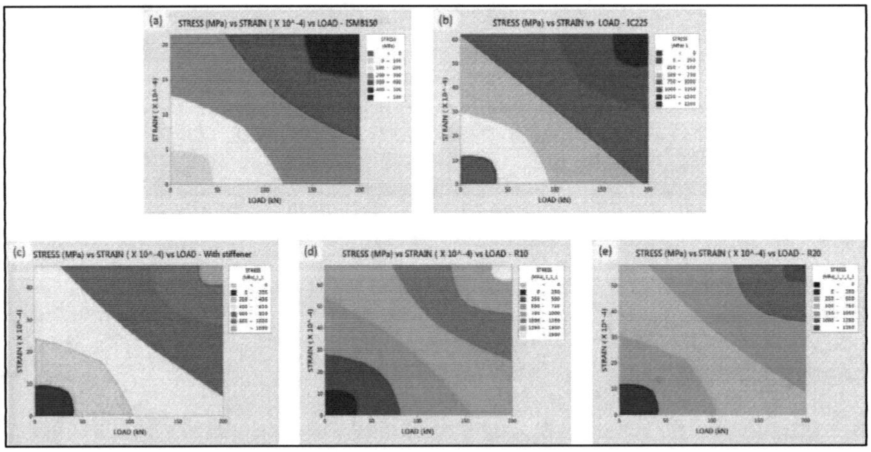

Fig. 10. Contour plots showing stress – strain distribution w.r.t load for (a) ISMB150, (b) IC225, (c) IC225 + stiffeners, (d) IC225 with fillet radius R10 and (e) IC225 with fillet radius R20

4. Conclusion

Based on the numerical simulation of the castellated beam with hexagonal web opening, the deflection and load-bearing capacity was analyzed. The results depict that the castellated beam with perforations has many local effects, influencing the load-bearing capacity. And it is found that by providing the diagonal stiffener, the stiffness of the beam is increased by a decrease in deflection. Vonmisses yield stress has been considered in the present paper, which is suitable for material like steel. The results also show that stress concentrations increase at the hole corners (Vierendeel effect) and load application point. By incorporating corrective measures like rounding the corners, providing diagonal support to increase the serviceability of the castellated beam, which paves the way for numerous structural engineering applications.

References

[1]. Kerdal D and Nethercot D.A 1984 *J. Const. Steel Res* **4** 295-315
[2]. Samadhan G M and Laxmikant M.G 2015 *Engineering Structures and Technology* **7(1)** 13-23
[3]. Timoshenko S P and Gear J M 1961 *Theory of elastic stability* (Mc. Graw – Hill)
[4]. Ferhat E and Mehmet P S 2013 *J. Const. Steel Res* **80** 335-368
[5]. Delphine S and Jan B 2015 *J. Const. Steel Res* **105** 119-128
[6]. Chung K.F, Liu T C H and Ko A.C.H 2001 *J. Const. Steel Res* **57** 467-490
[7]. ChungK.F,Liu T.C.H and Ko A.C.H 2001 *J. Const. Steel Res* **59** 1177-1200
[8]. Peijun W, Kangrui G, Mei L and Luiu Z 2016 *J. Const. Steel Res* **121** 173-184
[9]. Amin M and Hossein S 2005 *J. Const. Steel Res* **61** 1373-1386
[10]. EhabEllobody 2011 *J. Const. Steel Res*search **61** 1373-1386
[11]. Wakchaure M R and Sagade A V 2012 *Int. J Engineering and Inno.* Tech **2(1)** 368-372

An Experimental Study on Rehabilitation of Concrete Beam

Sindhu.R[1], V Johnpaul[2], Singh Sunil Gulab[3], Subash P[4] and Sohail Ahamed Z[5]

[1,2]Assistant Professor, Department of Civil Engineering, Karpagam Academy of Higher Education, Coimbatore TamilNadu, India.

[3,4,5]UG students, Department of Civil Engineering, Karpagam Academy of Higher Education, Coimbatore TamilNadu, India

Email: sindhu.r@kahedu.edu.in

Abstract. The theoretical work explains the structural behaviour of a beam reinforced with Aramid warp on repaired structures with its advantages and disadvantages. The final flexural potential of the AFRP is estimated. Experiments were performed to establish the distribution of load-deflection cracks and the width of cracks of the beams. In this work, four numbers of standard beams of the size 1500x150x100 mm are cast with the main rod diameter of 8 mm and a stirrup diameter of 8 mm with a spacing of 150 mm. After curing, the deflection is measured, and the beams under deflected condition are then wrapped in aramid fibre to improve the strength after a deflection by the epoxy resin coated on the surface. It is checked to verify the increased load capacity and is compared to the traditional beam. The result analogy is studied for the reinforcement of the Aramid fibre with the standard beam.

1. Introduction

The use of externally bonded polymer (FRP) sheets and strips was also used as a method for rehabilitating and reinforcing reinforced concrete structures. M40 mix with OPC-53-grade is used in the investigation. The nominal size of fine aggregates below 4.75 mm is mostly used worldwide, according to zone-II. The size of 2.36 mm is used, and river sand passing through a sieve of 1.18 mm is used to prepare Ferro cement. Locally available 20 mm crushed stone is used for concrete. It is used for mixing and healing concrete with a pH value of 7.2. M40 mix is built to cast 1:2.54:3.23 beam.

The effect of the AFRP confined column is studied and investigated under blast and impact load10 and investigated to enchase torsional behaviour of the RC beam using strips of aramid fibre11 and investigated the durability of concrete with aramid fibre subjected to acid attack and temperature rise12. It is observed that there is no enhancement with the increase in several layers of mesh and the betterment of rotation under the state of torsion is noticed on an under reinforced beams with ferrocement "U" wrap compared with others13. Resisting torque capacity is good in completely over reinforced beams than other states of torsion14. It is also observed that in addition to ferrocement, addition of steel angles improves the seismic performance, superior stability for stiffness degradation15. Here, concrete mixes were replaced by 40, 50 and 60% with RCA and GGBS16. It was found that 50% replacement of RCA and 40% replacement of GGBS shows adequate strength than control mix. We carried out on rectangular column Confined with Aramid Composites17. They studied the Behavior of AFRP Confined Concrete Subjected to High Strain Rate Compression18.

IOP Conf. Series: Materials Science and Engineering 1006 (2020) 012010 doi:10.1088/1757-899X/1006/1/012010

2. Material properties

2.1 Aramid fibre

Aramid fibre is high-performance fibre, with molecules connected by strong hydrogen bonds that pass mechanical stress very effectively, and distinguished by rigid polymer chains, allowing it use of low molecular weight chains. Commonly Structures are reinforced by fibre-reinforced polymer wrapping like CFRP & GFRP, whose strength and stiffness are high, is the fragile nature of failure mode. In this 1.45 bulk density, AFRP is used for reinforcement purpose.

Table: 1 Properties of Aramid Fiber

Sl. No.	Property	Aramid Fiber
1.	Laminate Strength	1430 MPa
2.	FibreStrengthG	2757MPa
3.	Strength-to-Weight	993 kNm/kg
4.	Young's Modulus	70.5-112.4GPa
5.	Specific gravity	1.44
6.	UV Degradation	degrade in Sunlight and high UV environment
7.	Abrasion Resistance	strong
8.	Tensile Modulus	131 GPa

Table: 2 Properties of Wire Mesh

Sl.No	Properties	Values
1	Average diameter	1.2mm
2	The opening size of the mesh	12.5x12.5mm

The Material Properties for various materials like Cement, Fine Aggregate, Coarse Aggregate and Aramid Fiber is shown in table 3 and 4, respectively

Table: 3 Properties of cement

Specification	Cement
Fineness	18%
Initial setting time	30 min
Soundness	0.233
Standard consistency	34%
Specific gravity	4.74

Table 4: Properties of fine aggregates

Specification	Fine aggregate (M Sand)	Coarse aggregate
Specific gravity	2.61	2.62
Fineness	1.64	5.56
Bulk density	1639 kg/m3	1653kg/m3
Water absorption	7.14%	8.66%

3. Experimental investigation

The Experimental setup and the beam specifications are shown in Figure: 1 & 2, respectively. After processing the Load deflection testing, the beam is ready to be coated with epoxy resin for aramid fibre wrapping and is tested. Ferro cement is deformed in four layers with a ratio of 0.50 w / c.

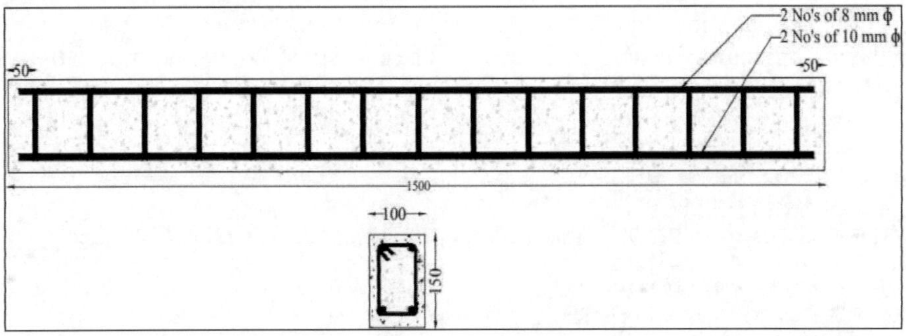

Figure 1: Beam Specification

Figure 2: Experimental Setup of beam

4. Result and discussion

Aramid Fiber- achieved better cracking performance. It is observed that the beam wrapped with 2 layers of aramid fibre gives 36.8 kN as ultimate load and beam with 1 layer gives 35.5kN. As no much difference is observed between the two, one layer can be adopted, which shows similar strength to carbon fibre. Maximum ultimate load obtained on wrapping is 36.8 kN by AF and on concrete with RCA is 36.7 kN.

Table 5: Test Result Comparisons

Sl.No	Layer of Fibre	Ultimate Load (kN)	First cracking load (kN)G	Mid Span Deflection (mm)	Stiffness (N/mm)	Flexural Strength (kN)
1	1	35.5	6.12	12.68	2.79	4.73
2	2	36.8	7.63	13.42	2.74	4.91
3	1	33.5	7.12	11.62	2.88	4.47
4	2	34.6	8.1	12.75	2.71	4.61

IOP Conf. Series: Materials Science and Engineering 1006 (2020) 012010 doi:10.1088/1757-899X/1006/1/012010

5. Conclusion

The specimens tested have achieved the strength equal to the control concrete. As the load was constantly increased, the formation of the vertical crack is visible on the surface of the beam, followed by the composition of shear cracks at 45°. Even though the cost of Aramid fibre is high compared to other materials used, the regained strength obtained is higher. Use of more layers of confinement is not beneficial in overall work. Hence it is recommended to use one layer of aramid fibre for the strengthening of deformed beams. The measured crack width is within the allowable limit as prescribed by IS 456-2000.

Reference

[1] ACI Committee 318 1983 Building code requirements for reinforced concrete (ACI 318-83). American Concrete Institute, Detroit 111.
[2] ACI-ASCE Committee 426 1973 The shear strength of reinforced concrete members. Proceedings, ASCE 99: 1091-1187.
[3]American Concrete Institute (ACI) 1995 Building code requirements for reinforced concrete ACI 318-95, Detroit.
[4] I.S.2386:1963 (Part- 3) Methods of test for an aggregate of concrete.
[5] I.S.2386:1963 (Part- 4) Methods of test for an aggregate of concrete.
[6] I.S.383:1971 Specification for coarse & fine aggregate from natural sources.
[7] IS-456, 2000 Code of Practice for plain and reinforced concrete structures, Bureau of Indian Standard, Manak Bhavan, Bahadurshah, Zafar Marg, New Delhi
[8] IS 383, 2016 Coarse and fine aggregate for concrete: Bureau of Indian Standard
[9] IS 9103, 1999 Specification for concrete admixtures: Bureau of Indian Standard
[10] Pierre Rochette and Pierre Labosiere 2014 Axial testing of rectangular column models confined with composites *Construction and Building Materials*
[11] Hui Yang, Hengwen Song and Shi Zhang 2015 Experimental investigation of the behaviour of aramid fibre reinforced polymer confined concrete subjected to high strain rate compression *Construction and Building Materials*
[12] Krishna Balakrishnan, Bhavya C 2017 Aramid fiber reinforced polymer in high strength concrete compressive structures *International Journal for Research in Applied Science & Engineering Technology (IJRASET)* Volume **5** Issue V.
[13] Sachin B Kandekar and Rajshekhar S Talikoti 2018 Torsional behaviour of reinforced concrete beam wrapped with aramid fibre *Journal of King Saud University- Engineering Sciences* 1-5
[14] Rajashekhar Siddappa Talikoti and Sachin Balkrishna Kandekar 2019 Strength and durability study of concrete structures using aramid-fiber-reinforced polymer *Fibres*.
[15] Gopal Charan Behera, T D Gunneswara Rao and C B K Rao 2016 Torsional behaviour of reinforced concrete beams with ferrocement u-jacketing experimental study *Case Studies in Construction Materials.*
[16] Ibrahim G Shaabana and Osama A Seoud 2018 Experimental behaviour of full-scale exterior beam-column space joints retrofitted by ferrocement layers under cyclic loading *Case Studies in Construction Materials* 61-78.
[17] Jayalakshmi Sasidharan Nair and Basil Johny 2016 Study of properties of concrete using gwgbs and recycled concrete aggregates *International Journal of Engineering Research & Technology* Vol. **5** Issue 09 160-166.

IOP Conf. Series: Materials Science and Engineering 1006 (2020) 012011 doi:10.1088/1757-899X/1006/1/012011

Mechanical Properties of Concrete with Bottom Ash as Partial Replacement of Fine Aggregate

S. Solai Mathi[1], V. Johnpaul[2], PR. Riyas[3]

[1,2] Assistant Professor, Department of Civil Engineering, Karpagam Academy of Higher Education, Coimbatore-21, Tamil Nadu, India.

[3] Assistant Professor, Department of Civil Engineering, Dhaanish Ahmed Institute of Technology, Tamil Nadu, India

E-Mail :solamathis@gmail.com, johnpaulv2490@gmail.com, riyascerasheed@gmail.com

Abstract. The fine aggregate is one of the important constituent in preparing concrete. Nowadays, fine aggregate is not fond in abundant quantity and so many replacements were done for river sand like crusher sand, M-sand, bottom ash etc. In this study the replacement is done partially for the fine aggregate with bottom ash which is extracted from Mettur Power Plant. The bottom ash and sand proportions is taken as 10:90, 20:80, 30:70, 40:60, 50:50 (M1,M2,M3,M4 and M5 respectively). The basic tests for all material used in concrete preparation were determined. M_{20} grade of concrete is casted and the test on fresh concrete is determined for conventional and partially replaced concrete. The specimens (beams, cylinders and cubes) were casted and kept under curing for 7 and 28 days from the cured concrete specimen's hardened concrete test were determined. The strength of concrete for ratios 10:90 and 20:80 are having nearly equal strength comparing to normal concrete. Hence ratio 20:80 is advisable by this experimental study.

1. Introduction

Concrete is a material with strength and longevity. It has emerged as the dominant construction material for the infrastructure. In addition to being durable, concrete is easily prepared and fabricated from readily available constituents like cement, sand, and coarse aggregate. It is widely used in all types of structural systems. The challenges for our civil engineering community are to develop the concept of sustainable development in construction aside and this sustainable development should be a reasonable cost. There have been some considerable developments in the research of bottom ash in concrete production. The increase in the performance of this type of concrete is only due to a delay in pozzolanic reaction. The compressive strength, flexural strength and density of bottom ash concrete decrease with an increase in the percentage of bottom ash level.

[1] Abdulhameed Umar Abubakar et all, used bottom ash and fly ash as the partial replacement for fine aggregates and cement respectively in the range of 0, 5, 10, 15 & 20% with equal percentages for M_{35} grade of concrete .from their study it is clear that increased curing is required to attain maximum compressive strength and the targeted strength was achieved at 56 days curing period. [2]. P. Aggarwal, Y. Aggarwal, s.m. Gupta, in their study they used bottom ash as a partial replacement of fine aggregate from 0% to 50% and found the strength of concrete is increased with an increase in curing time through various mechanical properties. [3]A S Cadersa

IOP Conf. Series: Materials Science and Engineering 1006 (2020) 012011 doi:10.1088/1757-899X/1006/1/012011

and I Auckburally in their study, they have replaced bottom as a partial replacement of fine aggregate in the increments of 20% up to 80% replacement and found that 20% replacement provides favorable results. [4]. Chou-Fu Liang and Hung-Yu Wang studied the moisture content of bottom ash which is used in fine and coarse aggregate and suggested the water-cement ratio of 0.45 through their test results. [5]. Dinesh Kumar G, et al, used Coal bottom ash as a partial replacement of fine aggregate for M25 grade of concrete in the increments of 10% up to 50% and found that up to 20% of fine aggregate can be replaced with bottom ash. [7]. Kandagaddala Revanth Kumar et al., used bottom ash and marine sand as partial replacement of fine aggregate and studied the strength properties of concrete and found lower compressive strength and suggested to equate to a lower grade of concrete. [8]. Kiran Kumar M S et al. used bottom ash for the partial replacement of fine aggregate up to 50% in the increments of 10% and suggested for 10% replacement of bottom ash from their results. [9]. Kula, I.et al, in their research used colemanite waste, coal bottom ash, and fly ash for the replacement of cement and found the mechanical properties of the same through the various test from their test results they have concluded the materials can be used as cementitious materials. [10] Kylasnath M and Ranjan Abraham, used bottom ash as a partial replacement in the increments of 5% up to 30 % in M60 grade of concrete, and mechanical properties were studied. Through their investigation, it is clear that the strength of concrete with bottom ash is getting decreased due to slow pozzolanic action. [11]. Ratchayut Kasemchaisiri and Somnuk Tangtermsirikul, they studied the properties of SCC incorporating Bottom ash as replacement of fine aggregate in increments of 10% from 0% to 30%. From their experimental study, they recommended a 10% replacement will give the desired strength at 56 days curing due to the pozzolanic reaction. [12]. Remya Raju, Mathews M. Paul, K. A. Aboobacker in their research they used bottom ash and replaced up to 30% for fine aggregate with M35 grade of concrete and fund there is the declination in workability and found that the tensile strength of concrete is improved with the addition of bottom ash. [13]. Sabisha Aliyar and Remya Raju used bottom ash and washed bottom ash as a partial replacement and compares the strength properties of concrete through the various test. They used the bottom ash and washed bottom ash in the increments of 5% up to 30% and found that beyond 10% replacement there are fewer changes in the strength of concrete. [14]. V. Sandeep Kumar and K.Konda Reddy used bottom ash as fine aggregate replacement in addition to steel fiber and studied the hardened properties of concrete for M25 grade of concrete. [15]. Dr. R. G. D' Souza, he studied the compressive strength of bottom ash replaced concrete for different grades of concrete and found 100% replacement of fine aggregate with bottom ash produces high compressive strength. [16]. Yogesh Aggarwal and Rafat Siddique used the by-products bottom ash and waste foundry sand in various percentages (0-60%) and studied the mechanical behavior of concrete.

From various studies undergone from the above literature, we choose to use the bottom ash (waste from Mettur Thermal power plant) as a partial replacement of fine aggregate. The replacement will be done from 0% to 50% for M20 grade of concrete. The hardened concrete test like compressive strength, Split tensile strength, and flexural strength of cured concrete was tested and results were tabulated and discussed.

2. Methodology

The amount of cement, aggregate and water required for preparing concrete is determined through mix design. The ingredients are mixed by hand mixing. The moulds are lined with oil to prevent sticking of the concrete inside the mould. The fresh concrete is then filled in layers approximately 5 cm deep and is subjected to 25 strokes with tamping rod to avoid air voids in the specimens and the surface is leveled using a trowel. The moulds with concrete are kept in room temperature for 24 hours, the marked specimens are then removed from the moulds and submerged in water for curing. The curing water is to be renewed for every seven days and the testing specimens are to be kept in contact with water till it is taken for testing. Before testing a specimen, the surfaces of the specimens and testing machine are wiped clean and placed in the machine to apply the load.

3. Preliminary tests conducted

The initial and final setting time for cement is determined to be 30 minutes and 6 hours 30 minutes respectively. The specific gravity test of fine aggregate and coarse aggregate was determined and found to be 1.51 and 2.58 respectively. In coarse aggregate other test like impact value, crushing value tests are conducted and the following results are observed respectively 10.43% and 25.17%. The slump value of fresh concrete is found as 30mm.

4. Mechanical properties of concrete

4.1 Cube's compressive strength

Table 1. Cube's Compressive Strength

S. No	Mix ratio	Compressive value of 7 days (N/mm^2)	Compressive value of 28 days (N/mm^2)
1	N.C	19.20	25.5
2	M1	21.00	27.6
3	M2	19.25	24.8
4	M3	12.40	18.9
5	M4	8.00	16.4
6	M5	9.70	13.8

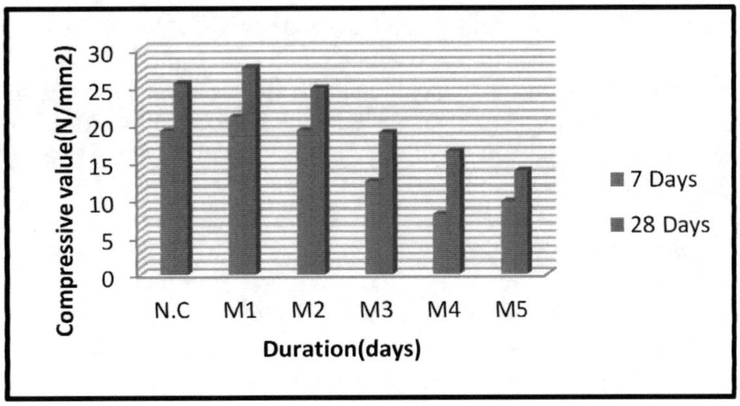

Figure 1. Compressive Strength of Cubes

Cube's Compressive Strength after 7 and 28 days curing is listed in the Table 1 and respective bar chart is given in Fig 1. This clearly shows that, the strength of ratio M2(20:80) gives nearly equal strength as conventional concrete for curing periods of 7 days and 28 days.

4.2 Cylinder's compressive strength

Table 2. Compressive Strength of Cylinders

S. No	Mix ratio	Compressive value of 7 days (N/mm^2)	Compressive value of 28 days (N/mm^2)
1	NC	3.5	6.7
2	M1	4.1	7.5
3	M2	3.7	6.6
4	M3	3.43	5.36
5	M4	3.12	4.56
6	M5	2.08	3.37

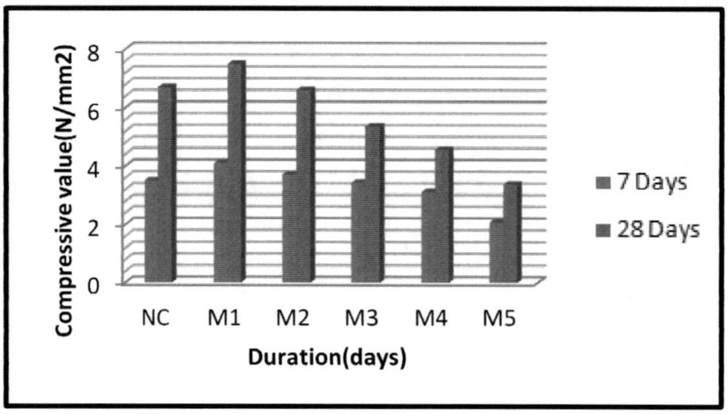

Figure 2. Compressive Strength of Cylinders

Cylinder's Compressive strength after 7 days and 28 days curing is listed in theTable 2 and respective bar chart is given in Fig 2. This clearly shows that, the ratio M3(30:70) gives nearly equal strength as conventional concrete for 7 days curing and strength of ratio M2 (20:80) gives nearly equal strength as normal concrete for 28 days curing.

4.3 Cylinder's split tensile strength

Table 3. Cylinder's Split Tensile Strength

S. No	Mix ratio	Split Tensile strength of 7 days (N/mm^2)	Split Tensile strength of 28 days (N/mm^2)
1	NC	2.8	3.5
2	M1	2.5	2.08
3	M2	2.3	3.98
4	M3	1.74	2.71
5	M4	1.56	1.53
6	M5	1.29	1.35

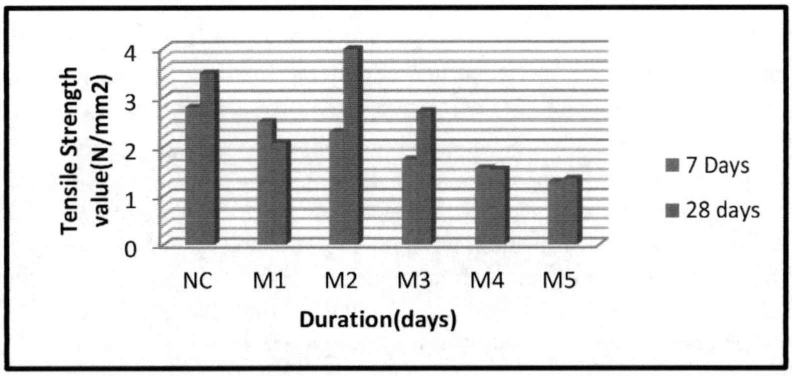

Figure 3. Split Tensile Strength of Cylinders

Cylinders Split tensile strength after 7 days and 28 days curing is listed in the Table 3 and respective bar chart is given in Fig 3. This clearly shows that, the ratio M1(10:90) gives nearly equal strength as conventional concrete for 7 days curing and strength of ratio M2 (20:80) gives nearly equal strength as normal concrete for 28 days curing.

4.4 Flexural Strength of beam

Table 4. Flexural Strength of Beams

S. No	Mix ratio	Flexural Strength of 7 days (N/mm^2)	Flexural Strength of 28 days (N/mm^2)
1	N.C	11	10.2
2	M1	10.125	9.83
3	M2	10.75	12.81
4	M3	8.04	9.33
5	M4	8.5	7.6
6	M5	7.55	6.4

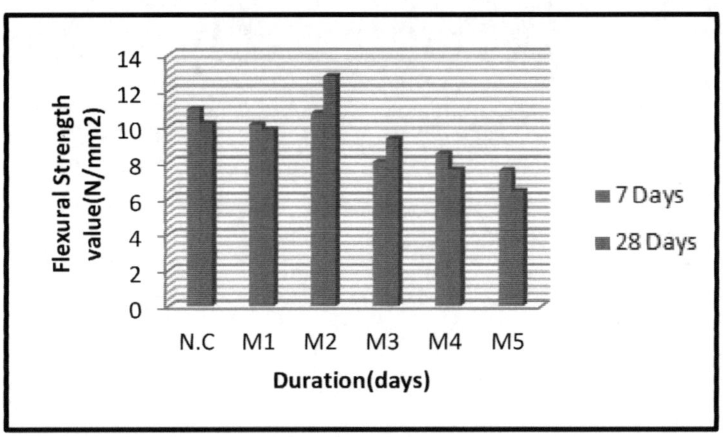

Figure 4. Flexural Strength of Beams

Flexural strength of concrete beams after 7 and 28 days curing is listed in the Table 4 and respective bar chart is given in Fig 4. This clearly shows that, the ratio M2(20:80) gives nearly equal strength as conventional concrete after 7 and 28 days curing period.

5. Conclusion

Density of concrete will decrease with increase in bottom ash content due to the low specific gravity of bottom ash. Bottom ash absorbs more water hence the workability of concrete is decreased. Bottom ash can be used with admixtures. The strength of concrete for ratios 10:90 and 20:80 are having nearly equal strength comparing to normal concrete. Hence ratio 20:80 is recommended. The replacement of Bottom ash enables the maximum waste utilization from the thermal power plants and can avoid dumping of the same in land.

References

[1]. Abdulhameed Umar Abubakar, Khairul Salleh Baharudin Properties of concrete using tanjung bin power plant coal bottom ash and fly ash *International Journal of sustainable Construction Engineering & Technology,* ***3(2)*** 56-69.

[2]. P Aggarwal, Y Aggarwal, S M Gupta Effect of bottom ash as replacement of fine aggregates in concrete *Asian Journal of Engineering (Building and Housing)* **8(1)** 49-62.

[3]. A S Cadersa and I Auckburally Use of unprocessed coal bottom ash as partial fine aggregate replacement in concrete *University Of Mauritius Research Journal.* **20** 62-84.

[4]. Chou-Fu Liang and Hung-Yu Wang The effects of moisture content of recycled incinerator bottom ash on permeable mrc mixing.

[5]. *G* Dinesh Kumar , Mohd Younus Mohiuddin, Ma Haleem An experimental study on partial replacement of fine aggregate with coal bottom ash in concrete *International Journal of Research Sciences and Advanced Engineering* **2(15)** 39-49.

[6]. Ghafoori N and Bucholc J Investigation of lignite-based bottom ash for structural concrete *Journal of Materials in Civil Engineering* **8 (3)** 128–137.

[7]. Kandagaddala Revanth Kumar, Gunupudi Yalamesh, Ramakrishnan K, Study of strength of concrete by partially replacing fine aggregate with bottom ash and marine sand *International Journal of ChemTech Research* **8(8)** 161-166.

[8]. Kiran Kumar M S et al. Experimental Study on partial replacement of fine aggregate by bottom ash in cement concrete *International Research Journal of Engineering and Technology (IRJET)* **5(5)** 1505-1508.

[9]. Kula I et al. Effects of colemanite waste, coal bottom ash and fly ash on the properties of cement *Cement and Concrete Research.* **31** 491-494.

[10]. Kylasnath M and Ranjan Abraham Performance of high strength concrete prepared by partially replacing fine aggregate with bottom ash *International Research Journal of Advanced Engineering and Science* **2(2)** 170-174.

[11]. Ratchayut Kasemchaisiri and Somnuk Tangtermsirikul, Properties of Self-compacting concrete in corporating bottom ash as a partial replacement of fine aggregate *ScienceAsi* **(34)** 87-95.

[12]. Remya Raju, Mathews M Paul, K A Aboobacker Strength performance of concrete using bottom ash as fine aggregate *International Journal of Research in Engineering & Technology* **2(9)** 111-122.

[13]. Sabisha Aliyar and Remya Raju A comparative study on the strength properties of concrete using bottom ash and washed bottom ash as partial replacement of fine aggregate *International Journal of Engineering Research & Technology (IJERT)* **5(9)** 132-135

[14]. V Sandeep Kumar, K Konda Reddy Partial replacement of fine aggregate with bottom ash & with addition of steel fibres *International journal for innovative Engineering and Management Research* **8(9)** 64-73.

[15]. Dr. R G D' Souza Replacement of fine aggregate with bottom ash in concrete and investigation on compressive strength *International Journal of Engineering Research & Technology (IJERT)* **6(8)** 172-175.

[16]. Yogesh Aggarwal and Rafat Siddique Microstructure and properties of concrete using bottom ash and waste foundry sand as partial replacement of fine aggregates *Construction and Building Materials* **54** 21-223.

A Comparative Study on Various Stages of Level of Details in Advanced 3D Building Construction Using BIM Tools

V Tejaswini[1], P Kesava Rao[2], G Jaya Shankar[3], R Nagaraja[4], S K Sinha[5]

[1]Junior Research Fellow, CGARD, NIRDPR, Hyderabad, India
[2]Associate Professor and Head, CGARD, NIRDPR, Hyderabad, India
[3]Professor, Department of Geo Engineering, Andhra University, Visakhapatnam, India
[4]Geospatial Chair Professor, CGARD, NIRDPR, Hyderabad, India
[5] Deputy Survey General, IISM, Survey of India, Hyderabad, India

E-mail: venkumahantitejaswini9@gmail.com

Abstract. Building Information Modelling is one of the most emerging technologies that provide productivity, efficiency and compliance in the construction industry. There are various platforms on which the 3D buildings can be created with large viewing capabilities. The current paper concentrates not only on the construction of 3D buildings with extrusions but also compares the various levels of details in two different BIM based 3D softwares namely ESRI City Engine and Revit Architecture. A 3D building object can be represented basically with four LOD levels at each stage of the building construction. The focal point of the current study also reflects the creation of building interiors and their corresponding attribute information to each element of the building. In City Engine software developed by ESRI, CGA codes are modified and customized to enhance the advancement in the building construction such as generation of building mass, segments, Roofs, Facade Creation, and other Architectural elements. These basic blocks of the building are produced by using Rule based pseudo codes in City Engine. The study illustrates a sample building and incorporates these codes to generate the 3D Building and the interiors are designed with more detailed components in Revit software. Hence, BIM is a path that provides effective monitoring, coordination and collaboration in construction field.

1. Introduction

Building Information modelling (BIM) is a boon to the construction industry. There are many milestones involved while planning and constructing a building from base level to the roof level. “Plan before construct” is achieved more easily with the BIM technology. There are various stages of construction in a building development since beginning. The floor plan design, development and construction stages of the building domain are specified using the BIM models. Numerous softwares providing great access to build such models efficiently. Virtual building models are the scope for the future construction industry. It performs the task of a civil engineer, Architect, Electrical engineer, planning industry and provides an important contribution to the builders in order to make the analysis, estimate the cost and other decision making time frame. BIM is a cloud computing which is accompanied with integration and management of the Building life cycle related project planning, geometrical distribution, design, construction and maintenance [1]. When it comes to the real existence of the building model, the BIM Tools help to attain the original planning scenario and develops a technology revolution in the construction field. Hence, BIM technology can be simply understood as a source and sink of large amount of information regarding the building models.

2. Objectives

The objectives of the current study include comparison of level of details involved in the building lifecycle that enables reliable construction and to generate building architecture using rule based pseudo code and obtains a sceneric view of the topology by taking two sample buildings as inputs.

3. Methodology

The methodology and procedure involves the identification and flow of work by considering two sample buildings in two softwares ESRI City Engine and Autodesk Revit Architecture. Each stage of the building construction are analysed at different level of details from LOD - 0 to LOD - 4 as shown in the below methodology flowchart.

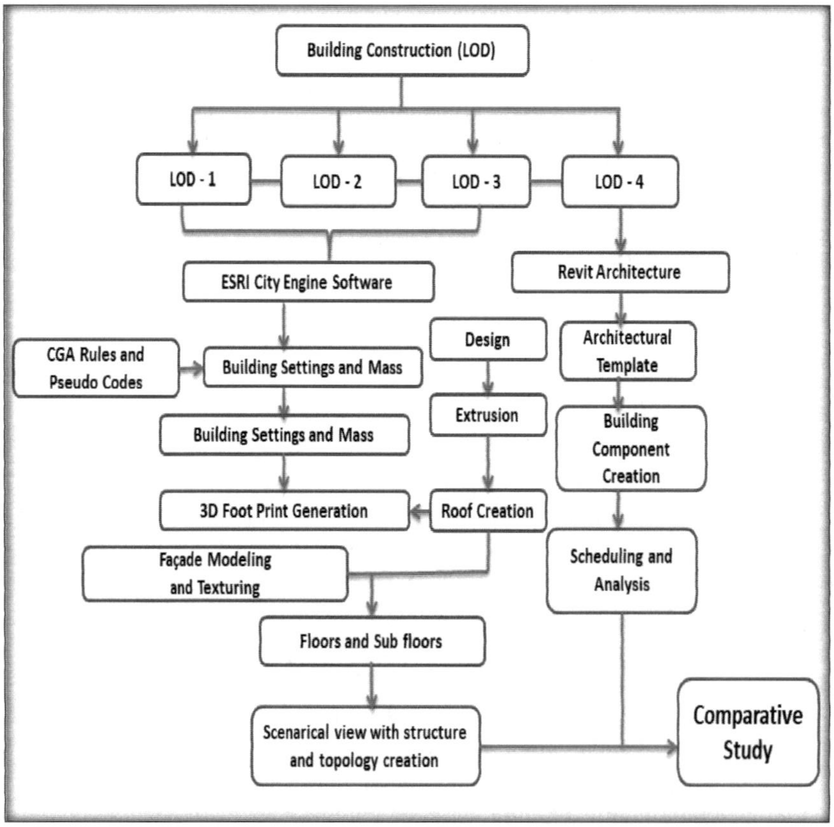

Figure 1: Methodology flow diagram of the current project

4. Building Creation in CITY ENGINE Software

4.1. Building block from foot print

In order to construct a building the basic block is the generation of the foot print. There are various sources from which a foot print can be delineated. These sources may include open source base maps, Toposheets, High Resolutions Satellite Images, Cadastral Maps (Aerial Survey). The current study incorporates the footprint from the ESRI base map. In city Engine software, CGA Pseudo code has already been inbuilt which enables the user to modify the codes according to the design of the construction.

4.2. Building attributes

Each CGA Code in the City Engine software should begin with the defining of building attributes. If they are initiated in the middle of the rule, user understanding may be an issue. Once, these attributes are defined in the rule set, based on the call out functions, they are automatically appeared in the inspector tab so that CGA Grammar editor is enabled for the customization of the geometry and other asset management.

4.3. Asset creation and management

The assets required for the creation of the current model are well defined in the CGA Code itself. As soon as they are defined, the corresponding assets are called out from the asset management folder of the City Engine. The created assets in this building are shown in the below code.The advantage of this kind of CGA pseudo coding is that one need not write the code from beginning instead the software itself generates it at the background section **[2]**. But user should need to customize the code as per the requirements of the building model.

```
version "2018.1"

attr height                  = 15
attr roofht                  = 3
attr groundfloor_height         = 8
attr floor_height            = 5
attr tile_width              = 3.1
attr ledge_width             = 0.5
attr wallColor               = "#ffffff"
//attr door_width            = 2.1
attr window_width            = 2.3
attr window_height_1            = 3
```

```
const window_asset       = "assets/window/sash_window.obj"
const round_wintop_asset     = "facades/round_windowtop.obj"
const tri_wintop_asset       = "facades/triangle_windowtop.obj"
const halfarc_asset          = "facades/arc_thin.obj"
const ledge_asset        =
                            "facades/textures/mg block_ledges.JPG"
const modillion_asset        =
              "facades/ledge_modillion.03.for_cornice_ledge_closed.lod0.obj"
const wall_inset = 0.4
```

Figure 2: CGA Code in City Engine defining the asset creation

4.4. Declaration of the texture and colouring

It is observed that the assets are defined in the early stages of the CGA Code flow. Similarly, textures of the walls, windows, roof etc. are also declared. After the declaration, the software itself performs the tasks of loading the texture from the corresponding in built folder.

```
const wall_tex               = "facades/textures/brickwall.jpg"
const wall2_tex              = "facades/textures/brickwall_bright.jpg"
const dirt_tex               = "facades/textures/dirtmap.15.tif"
const doortop_tex            = "facades/textures/doortoptex.jpg"
const roof_asset             = "assets/solarPanel_1.obj"

randomWindowTex = fileRandom("assets/window/sash_window_7.jpg")
randomDoorTex = fileRandom("*facades/textures/doortex.*.jpg")
Lot --> extrude(height) Building
Building --> comp(f) { front : FrontfacadeTex | side : Sidefacade |top : roof}

TopLedge -->

split(y){  ~0.1 : Wall(0) | ~1 : s('1,'2.5,0.9) LedgeAsset}

LedgeAsset -->
i(ledge_asset) ledgetext
ledgetext -->
    setupProjection(0,scope.xy, '2, '1)
    projectUV(0)
    texture("facades/textures/mg block_ledges.JPG")
```

Figure 3: CGA Code in City Engine declaring the texture and colouring of Facades

IOP Conf. Series: Materials Science and Engineering 1006 (2020) 012012 doi:10.1088/1757-899X/1006/1/012012

4.5. Creation of the building mass

The creation of the building mass starts with the LOT Rule and extrude command is used to elevate the building to a specific height. As a result of the command "LOT → Extrude (height) building", the building block is created. The LOT Rule is followed by Façade modelling code. To this particular building model, there exist two types of façade models namely front façade and side facade. This is driven by using the component split command. The shape is split by the code of the building component split which is used to divide the mass models into faces providing front facade (Front shape of the building and several side shapes along with the roof shape **[3]**. The below commands explain the front and side facade shape of the actual building photograph and the building is constructed based on this CGA coding. The horizontal and vertical ledges are also clearly mentioned in the rule set defining the dimensions and the offset values.

```
FrontfacadeTex -->
setupProjection(0, scope.xy, 2.25, 1.5, 1)
setupProjection(2, scope.xy, '1, '1)
Frontfacade


Frontfacade -->
setupProjection(0, scope.xy, scope.sx, scope.sy)
split(y){ groundfloor_height    : Floor(split.index)     // Groundfloor
    | floor_height : Floor(split.index)            // First Floor
    | floor_height : Floor(split.index)            // Second Floor
    | {~floor_height : Floor(split.index)}*        // Mid Floors
    | floor_height : Floor(999)           // Top Floor, indexed
                                                    // with 999

    | 0.5 : s('1,'1,0.3) LedgeAsset}          // The top ledge just
                                                // below the roof
```

```
split(x) {15 : structure1  }
structure1-->
setupProjection(0, scope.xy, scope.sx, scope.sy)
split(x) {2.5 : a | 10 : b | 2.5 : c }
a -->
setupProjection(0, scope.xy, scope.sx, scope.sy)
b -->
setupProjection(0, scope.xy, scope.sx, scope.sy)

c -->
setupProjection(0, scope.xy, scope.sx, scope.sy)


 Sidefacade -->
      split(y){ groundfloor_height    : Floor(split.index)    // Groundfloor
        | floor_height : Floor(split.index)          // First Floor
        | floor_height : Floor(split.index)          // Second Floor
        | {~floor_height : Floor(split.index)}*      // Mid Floors
        | floor_height : Floor(999)          // Top Floor, indexed
                                                     // with 999
        | 0.5 : s('1,'1,0.3) LedgeAsset}         // The top ledge just

                                         // below the roof
```

Figure 4: CGA Code for the Creation of current Building Mass

4.6. Creation of Building Components (Windows, Walls, Door and Roof)

The architectural modelling of the building is achieved by making a few customized and modified City Engine CGA coding for roof, walls, doors, window assets and by dividing the floors using the pseudo code.

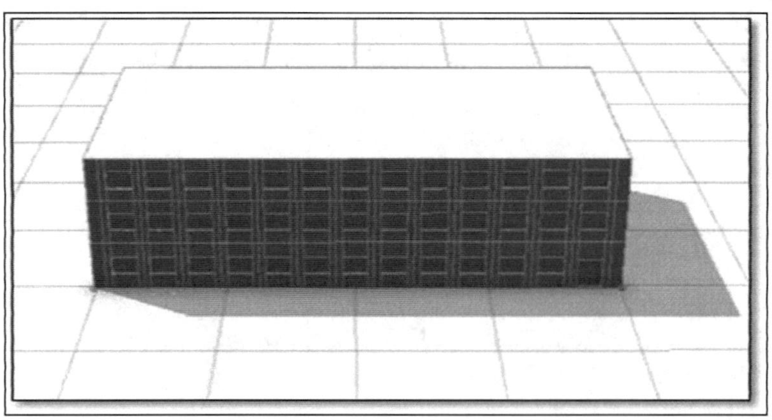

Figure 5: Skeleton of Building block (LOD – 2) and a part of LOD -3 (City Engine)

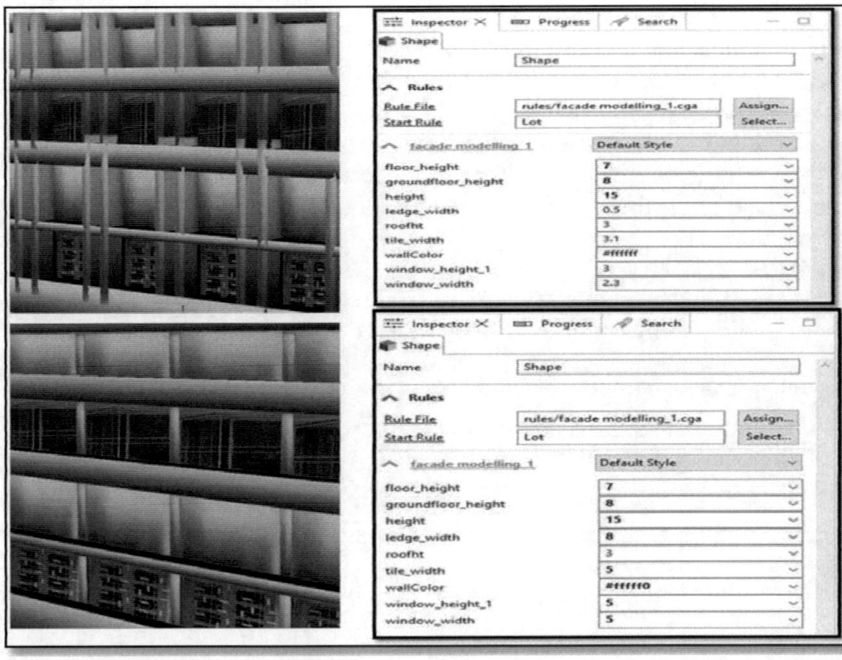

Figure 6: Facade creation and modelling with different constants in inspector mode

4.7. Results in City Engine

The obtained building after the creation of the facades based on the complete CGA pseudo code modified based on the building design such as ledges, window unit dimensions, floors, front and side façade modelling is shown below.

Figure 7: Overall Building Model created with CGA Code (sceneric view along with topology)

5. Creating LOD-4 with Revit Architecture

Autodesk Revit Architecture Software is used to develop Building Information Modelling (BIM) which retains data and consequently adapts to the related changes. Revit is helpful for extracting various drawings from the each model that is possible for architecture, mechanical, plumbing, electrical, electronics and communications [4].

5.1. Floor plan preparation
Another sample building is constructed in the Revit Architecture with tentative measurements of the floor plan. The total length of the building is taken as 200 meters which contains two blocks and three floors ground floor, first floor and second floor) for illustration purpose.

5.2. Wall extends
The settings of the wall in the Revit software can be directly drawn in the architectural template at different levels extending each wall element with respect to the base and top constrains.

5.3. Creation of floor and ceiling
Creating the floors and ceiling in Revit Architecture is as easy as digitization procedures. The levels to which floor/ceiling is planned need to be created and the walls are picked up for the rectangular stretch.

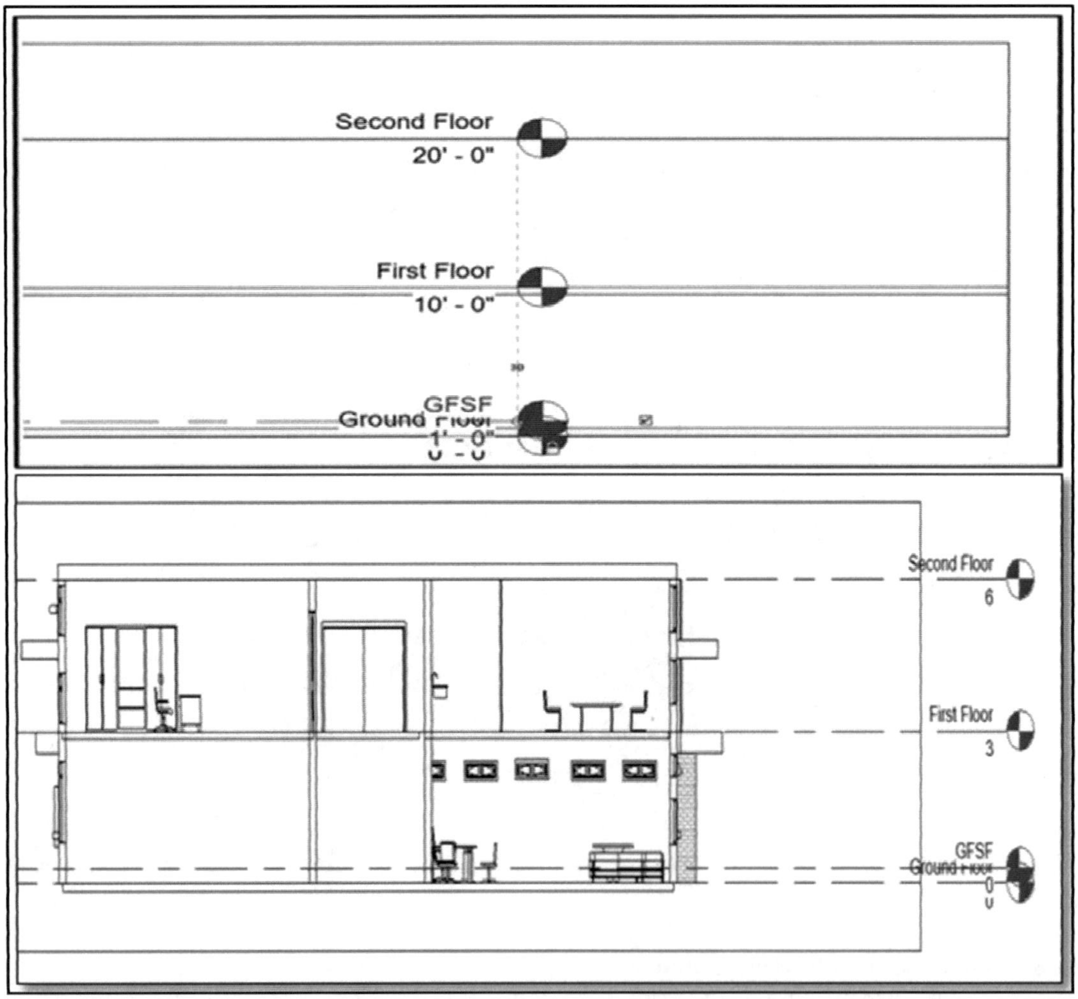

Figure 8: Creation of Levels in Revit Architecture with elevation gaps (Section View)

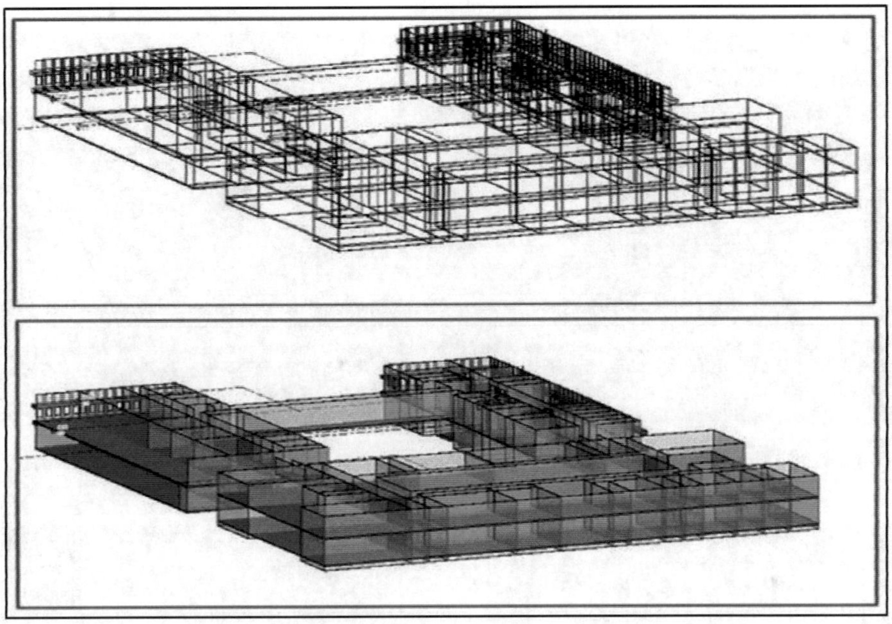

Figure 9: Wireframe and Realistic view of sample building created in Revit Architecture

5.4. Staircase construction and revit families

It is essential to identify that level of detail 4 (interior construction) contains all the building elements including staircases and interior room partitions and other furniture components. These building elements such as furniture components are represented in the form of Revit families. The creation of stair case contains Runs, landings, supporters and railings which necessitate the individual creation. The stairs connecting the level 1 to level 2 and so are designed as below figure. An integrated management of the building information and life cycle is the key point in enhancing and improving the efficiency of BIM Models [5]. Analysis on BIM also explores the benefits of cross section of the construction industry [6].

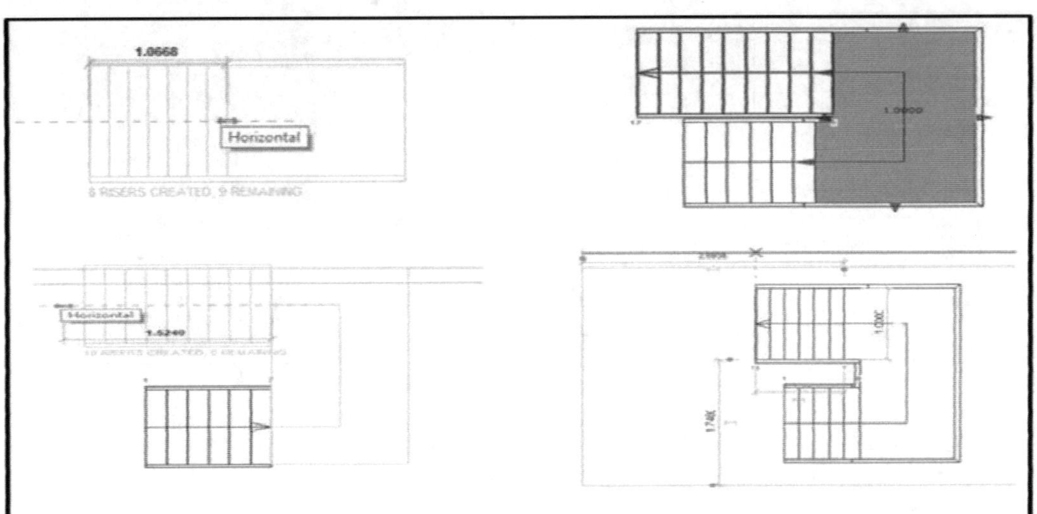

Figure 10: Skeleton construction of staircase with Runs and Landing declarations

Figure 11: Realistic view of stairs created in Revit cutting down the floor to reach next level

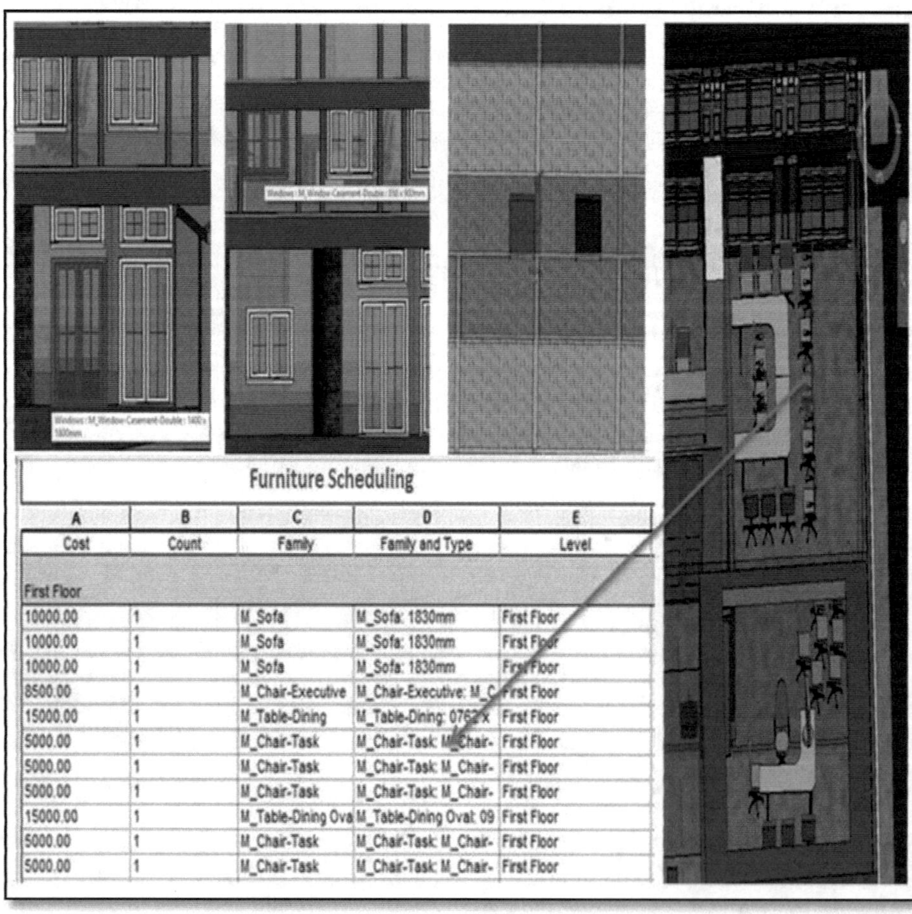

Furniture Scheduling

A	B	C	D	E
Cost	Count	Family	Family and Type	Level
First Floor				
10000.00	1	M_Sofa	M_Sofa: 1830mm	First Floor
10000.00	1	M_Sofa	M_Sofa: 1830mm	First Floor
10000.00	1	M_Sofa	M_Sofa: 1830mm	First Floor
8500.00	1	M_Chair-Executive	M_Chair-Executive: M_C	First Floor
15000.00	1	M_Table-Dining	M_Table-Dining: 0762 x	First Floor
5000.00	1	M_Chair-Task	M_Chair-Task: M_Chair-	First Floor
5000.00	1	M_Chair-Task	M_Chair-Task: M_Chair-	First Floor
5000.00	1	M_Chair-Task	M_Chair-Task: M_Chair-	First Floor
15000.00	1	M_Table-Dining Ova	M_Table-Dining Oval: 09	First Floor
5000.00	1	M_Chair-Task	M_Chair-Task: M_Chair-	First Floor
5000.00	1	M_Chair-Task	M_Chair-Task: M_Chair-	First Floor

Figure 12: LOD – 4 building interiors Doors, Windows, Rooms and Furniture components

6. Conclusion

Building Construction involves stage by stage design and development. Virtual Building Models can also be constructed as realistic as possible by using the BIM Software tools. In the present analysis, the two sample building models are compared based on the level of details. It is observed that this kind of analysis provides a revolution in the building planning and the implementation of workflow because it involves detailed attribute domains and data sharing among the various stakeholders. Though 2D building plans are beneficial to some extent, 3 Dimensional virtual models provide more insight and effective visualization for the user understanding. The comparison of the four levels in the study states that LOD 0 is merely building foot print that can be generated by using toposheets, satellite data or Aerial Photographs etc. LOD 1, the next phase to LOD 0 includes the z factor that is building elevation. An extruded building along with the roof shape which comes under LOD 2 and an added building model such as Facade details, Structure and Architecture are obtained by using City Engine Pseudo code generation. LOD 4 is the complete finishing of the building along with the cumulative level of details (LOD 0, LOD 1, LOD 2, and LOD 3). Revit Architecture has provided efficient details and modelling in this study to bring out the interior structure of the building including attribute data. Computer Generated Architecture (CGA) rule based pseudo codes inbuilt in the software contributing more access to modify and customize the BIM Models based on its structure and Architecture whereas Revit Architecture is used to bring out level of detail 4 (interior construction) reflecting the actual façade models , walls, windows, doors and other components. The difficulty in assigning the non-spatial data information detailing the building at every phase of BIM Development. The analysis provides more scope for the construction and planning to demonstrate their 3D Building models with more accurate results and productive representation of the Building Information Models.

References

[1] Tianqi Y and Lihui L 2016 Research on building information model technology *World Construction* vol **5** p1 DOI 10.18686/wcj.v5i1.1.

[2] Koehl M and Roussel F 2015 Procedural modelling for reconstruction of historic monuments *ISPRS Annals of Photogrammetry, Remote Sensing and Spatial Information Sciences* vol **2-5/W3** pp 137–144.

[3] ESRI 2020 *City Engine Tutorial* **6** Basic Shape Grammar retrieved from https://doc.arcgis.com/en/cityengine/latest/tutorials/tutorial-6-basic-shape-grammar.htm

[4] Onur Zeynep A and Fatemeh N 2019 BIM software in architectural modelling *International Journal of Innovative Technology and Exploring Engineering* vol **8** DOI 10.35940/ijitee.K1968.0981119

[5] Qing L, Tao G and Jianping W 2011 Research on application of BIM technology in construction project *International Conference on Computer Science and Service System (CSSS*)(Nanjing: China) pp. 2463–67 DOI 10.1109/CSSS.2011.5974366.

[6] Bryde D, Broquetas M, and Volm J M 2013 The project benefits of building information modelling (BIM) *International Journal of Project Management*, **31(7)** pp. 971– 980.

High Strength Lightweight Foam Concrete

V Johnpaul[1], R Abiraami[2], R Sindhu[3], N Balasundaram[4] and S. Solai Mathi[5]

[1,3,5]Assistant Professor, Department of Civil Engineering, Karpagam Academy of Higher Education, Coimbatore-21, Tamil Nadu, India.
[2]Assistant Professor, Department of Civil Engineering, Christian College of Engineering and Technology, Oddanchatram, Tamil Nadu,India
[4]Professor, Department of Civil Engineering, Karpagam Academy of Higher Education, Coimbatore-21, Tamil Nadu, India.

E-Mail: johnpaulv2490@gmail.com

Abstract. This research is to develop high strength lightweight foam concrete; Foam concrete is also called as lightweight concrete. Light Weight foam concrete enables in reducing the dead weight of the total Structure. The foaming agent used in this research is an Animal base, to create foam for foam concrete. The consistency of foam concrete should be equal to 1000kg/m^3. The compressive strength of the Light Weight Concrete varies from 6Mpa - 14Mpa. The main scope of the research is to increase the strength of the foam concrete by easily available sources. Foam concrete is an inventive applied science for viable construction in civil engineering applications. The property of foam concrete is to develop air bubbles with greater strength, stable during its manufacturing process by well mixing of ingredients, weathering condition on the site at the time of placing, and drying of concrete. The entire ratio is carried out only as trial mixes. The cube has been prepared for 7, 14, 28 days. Due to the presence of foam the cube should not be immersed in the water, only dry curing has been done. According to compressive strength result, the ratio which had the highest compressive strength can be used for both filling concrete and it can be used to create (Cellular Light Weight Concrete) CLC Blocks.

1. Introduction

Foam concrete is a combination of cementitious material with fly ash, water, foaming agent, gypsum, GGBS, quarry dust, and lime powder which is used as a foaming agent. It enables in heavy, reducing weight in the concrete of equally spread air bubbles at a constant rate. The unit weight of foam concrete is decided by its key factor ingredients. Foam concrete is developed by the homogeneous circulation of air bubbles entrapped in the whole density of concrete. Bubbles create a barrier inside the concrete which ensures a durable mixing, transmitting, draining, and deposition of a raw concrete mixture. Distinct particles of air bubbles range from 0.1 - 1mm. Foam concrete is a highly watery fluid and hence compaction need not be necessary. It is both fire and water-resistant. [1]experimented with foam concrete mixes with and without the addition of fine aggregate to confirm a higher density mix among them and compared with the normal weight concrete. [2]proposed a study on the incorporation of fly ash in various percentages and a different cement content weighed using expanded clay aggregate to achieve a high strength lightweight concrete. [3]recommended the usage of cellular lightweight concrete blocks in the construction industry. [4]discussed the transformation of aerated lightweight concrete into foamed and autoclaved concrete. [5]investigated the experimental work carried out on recycling the rice husk ash and fly ash in the concrete mixes. The results viewed to be worked better in physical and mechanical properties when compared to conventional concrete. [6]explained the practical job which was done by comparing the conventional clay bricks with cellular lightweight concrete blocks for computation of compressive strength between them at certain days of curing. [7] proposed a lightweight concrete using locally available lightweight materials and conducted an assessment in fresh and hardened nature. [8] experimented with the partial replacement of coarse aggregate with pumice and various cementitious materials were added in the mix to achieve a lightweight aggregate concrete and identified a strength increment with a reduction in weight at different trials. [9] described the evaluation of strength attainment was considerably greater in partial replacement of coarse aggregate using pumice stone when compared to conventional concrete. [10] carried out the quarry dust based foam concrete study on finding out the optimum dosage of foam thus minimizing the concrete weight. [11] described a lightweight aggregate concrete and limestone aggregate concrete were subjected to uniaxial tension through nitrogen gas pressure. Straining nature of lightweight concrete was moderately better than that of limestone aggregate concrete. [12] conducted an experimental investigation on foam polyethylene waste as proper utilization in lightweight concrete construction to prevent the coefficient of thermal expansion in concrete. [13] added an expanded polystyrene aggregate partially by the coarser one to ensure a lightweight concrete and further, these were tested to the basic properties study with suitable concrete density making lesser weight. [14]conducted a comparative survey on indoor radon concentration rate exhalation of lightweight concrete wall with a normal concrete wall and the results showed to be a less radon rate exhalation for the lightweight concrete wall when compared to normal concrete wall.

In this research, the effects of different mix proportions have been studied in detail on the compressive behavior of concrete using mineral admixtures at five mix ratios and comparative study has been made.

2. Experimental details

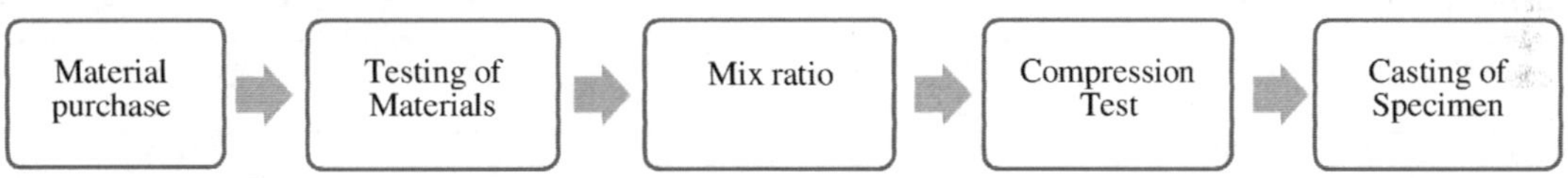

Figure 1: Methodology of Proposed study

2.1 Materials

Foam concrete is in the form of a slurry of cementitious material, fine aggregate, water, fly ash, other materials, and foam generated by foam generating machine. Cement OPC 53, Lime Stone Powder, Fly Ash, Quarry Dust, Gypsum Powder, Foaming Agent, GGBS.

Cement OPC vanishes crack propagation and offer less heat of hydration thereby surface cracks get eliminated. PPC gives a uniform gradation in the concrete mix. It attains greater strength even after reaching 28 days. Better surface finish due to higher fineness. Ultra Tech Cement is used. It was purchased in SNA Steels, Malumichampatti. Limestone Powder which is fine is used. It is originated from sedimentary rock and composed of calcium carbonate-rich minerals including calcite and aragonite. It is purchased from Sethiya Solvents & Chemicals, Sukrawarpet, Coimbatore. Fly ash, also known as fuel debris which is meant to be a by-product of coal-generated from the thermal power station, mettur. Fuel debris production has been increased due to landfill cost and feasible utilization in the construction industry. It is a product obtained from the blasting of rocks. During this process, rocks get split into various sizes which are collected in the form of dust. In building construction, quarry dust can be used essentially as a source of fine aggregate. This promotes the economy in construction. It is bought from Ultra CLC Blocks, Veerapandi, Tirupur.It's a good binding material which will help to bind all the material together and it will help the concrete to attain more strength by keeping all the materials bonded. The foaming agent is derived from the animal protein. Foam is obtained by adding 1 liter of a foaming agent to the 40 liters of water. It is the formation of foam as a blowing agent. It increases its colloidal activity by hinder the fusion of bubbles and makes it stable. Foaming Agent is bought from CHRYSO. Ground Granulate Blast Furnace Slag is obtained by dampening fused thermal slag from a blast furnace in water or steam. Its obtained from the Thermal power station or steel manufacturing companies. It is bought from Mark Associates,Tiruppur.

2.2 Material test

The chemical composition (EDOX) test of various mineral admixtures like Gypsum, Dust, Lime, Flyash, are done in karunya institute of science and technology, Coimbatore, compared with cement and tabulated from Table 1 to 5below.

Table 1. EDOX test on Gypsum

Spectrum: GYPSUM					
Element	**Series**	**Unn.C [wt. %]**	**Norm.C [wt.%]**	**Atom. C**	**Error (3 Sigma) [wt.%]**
Carbon	K-series	5.09	10.04	20.25	3.75
Oxygen		12.03	23.74	35.95	6.51
Aluminum		0.12	0.23	0.21	0.11
Sulfur		12.46	24.58	18.57	1.43
Calcium		20.97	41.40	25.02	1.93
	Total	50.66	100	100	

Table 2. EDOX test on Dust

Spectrum: Dust					
Element	**Series**	**Unn.C [wt. %]**	**Norm.C [wt.%]**	**Atom. C**	**Error (3 Sigma) [wt.%]**
Carbon	k- series	3.12	15.83	25.85	1.86
Oxygen		7.26	36.79	45.09	3.07
sodium		0.34	1.73	1.48	0.15
Aluminium		1.49	7.57	5.50	0.30
Silicon		4.23	21.42	14.96	0.62
Potassium		1.61	8.16	4.09	0.23
Calcium		0.07	0.36	0.17	0.09
Iron		1.61	8.14	2.86	0.22
	Total	19.73	100	100	

Table 3. EDOX test on Lime

Spectrum: LIME					
Element	**Series**	**Unn.C [wt. %]**	**Norm.C [wt.%]**	**Atom. C**	**Error (3 Sigma) [wt.%]**
Carbon	k- series	8.67	11.12	20.20	4.21
Oxygen		29.86	38.28	52.19	13.16
Magnesium		0.30	0.39	0.35	0.14
Silicon		0.05	0.06	0.05	0,09
Calcium		38.93	49.89	27.16	3.50
Niobium		0.20	0.26	0.06	0.11
	Total	78.01	100	100	

Table 4. EDOX test on Quarry

Spectrum: FLYASH					
Element	**Series**	**Unn.C [wt. %]**	**Norm.C [wt.%]**	**Atom. C**	**Error (3 Sigma) [wt.%]**
Carbon	K-series	3.88	8.79	14.74	3.19
Oxygen		16.32	36.96	46.51	6.99
Sodium		0.10	0.24	0.21	0.11
Magnesium		0.09	0.21	0.17	0.10
Aluminium		11.77	26.66	19.90	1.78
Silicon		10.31	23.34	16.74	1.42
Potassium		0.31	0.71	0.36	0.12
Calcium		0.80	1.82	0.92	0.17
Iron		0.56	1.27	0.46	0.17
	Total	44.15	100	100	

Table 5. Cement Properties Test

Sl.No	Characteristics	Requirements as per IS	Test results
3.	Magnesium oxide (% MgO)	10	3.44
4.	Sulphur trioxide (% SO3)	3	1.32
5.	Sulphide Sulphur (%S)	1.5	0.20
6.	Loss on ignition (%)	5	0.50
7.	Insoluble Residue(%)	4	0.32
8.	Chlorides (%)	0.1	0.010
9.	Finess (M^2/kg) (Min)	225	379
10.	Setting time Initial Final	 30 600	 180 255
11.	Soundness Le chatelier (mm) Autoclave (%)	 10 0.8	 1.0 0.03
12.	Compressive Strength (Mpa)	33	38

2.3 Mix proportions and preparation of samples

Mix design is done as per the, trial and error method by taking the properties of mineral admixture, fine aggregate, and cement. A total of five mixes has been prepared out of 5 ratios 4th one is the nominal mix, we usually do it in conventional concrete. The specimens have been cast of size 150*150*150mm.

Table 6. Different Mix Ratio

Materials	Ratio 1	Ratio 2	Ratio 3	Ratio 4	Ratio 5
Cement	1	1	1	1	1
Sand	1.5	1.5	1.5	1.5	1.5
Fly Ash	15	10	20	30	25
Quarry dust	30	25	35	30	25
Lime Stone	30	25	35	-	25
Gypsum	25	20	30	-	25

3. Casting of Cubes

3.1 Material fineness
All the materials should be mixed evenly and no material should be in the size of the granular it will result in the differential settlement which will result in the strength of the concrete. A machine like a ball milling can be used to break all the granular particles. All the materials will be feed into the machine and the iron ball will be added equal to the weight to grind the materials into microparticles.

3.2 Mixing of concrete
All the materials are evenly mixed using concrete mixer and Foam is created separately with the pressurized air and it is mixed with materials after it has mixed. Mixing of concrete should be even and gentle vigorous mixing will result in breakage of bubbles.

3.3 Placement
As the workability of concrete is high there is no need for compaction, the flow of concrete is high so that the concrete should be placed before it starts to settle.

3.4 Curing
As the foam is presented in the concrete it should not be placed or immersed inside the water which will result in the weakening of concrete. The curing has been done for 7, 14, 28 days. The curing that has been followed is dry Curing as show in figure 2.

Figure 2. Picture showing casted specimen for curing

4. Result and discussions

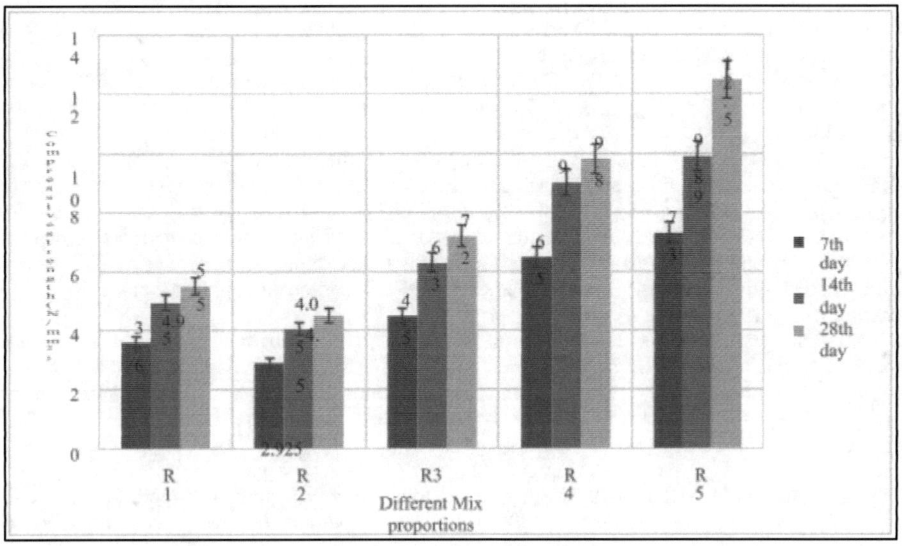

Figure 3. Shows the comparative Compressive strength results on lightweight concrete

With the idea of trial and error method, 5 Different mixes have been taken by referring different journals, in that the Mix 5 has obtained more strength while comparing all other Mix. The compressive strength of lightweight concrete differs from 3Mpa to 17Mpa. Strength attained is 12Mpa, on further development, it can be used in any type of structure. As the core idea has no reference we have only taken the Compressive Strength, on development all other types of tests can be taken like tensile strength, Young's Modulus, etc.... This mix ratio without any further development can be used in the partition walls in the framed structure and it can be used to manufacture (Cellular Light Concrete) CLC Blocks. A comparison is shown in figure 3.

5. Conclusion

- Mix 5 has obtained more Compressive strength than all other Mix as clearing mentioned in figure 3.
- Foam (Lightweight) concrete tends to obtain strength even after curing and 28days when compared to conventional concrete.
- On developing the process this type of concrete can be used in the massive structures which reduces the overall self-weight of the structure.
- Obtained results are drawn in terms of graphical form and presented for better understanding and further scope of experimentation.

References

[1] G D U Maheswari, N Sakthieswaran, G S Brintha and O G Babu 2016 Experimental study on high strength concrete vol. **4** no. V pp. 426–428.
[2] S Subaşi 2009 The effects of using fly ash on high strength light weight concrete produced with expanded clay aggregate *Sci. Res. Essays* vol. **4** no. 4 pp. 275–288.
[3] N Mustapure and H Eramma 2014 Experimental investigation on cellular light weight concrete blocks for varying grades of density *Int. J. Adv. Technol. Eng. Sci.* www.ijates.com, no. 08 online pp. 2348–7550 [Online]. Available: http://ijates.com/images/short_pdf/1407002401_P10-18.pdf.

[4] A J Hamad 2014 Materials, production, properties and application of aerated light weight concrete: Review," *Int. J. Mater. Sci. Eng*. vol. **2** no. 2 pp. 152–157 doi: 10.12720/ijmse.2.2.152-157.

[5] M Amin and B A Abdelsalam 2019 Efficiency of rice husk ash and fly ash as reactivity materials in sustainable concrete *Sustain. Environ. Res*. vol. **1** no. 1 pp. 1–10 doi: 10.1186/s42834-019-0035-2.

[6] T M S and P H Man 2015 Experimental work on cellular light-weight concrete *Int. J. Adv. Eng. Res. Dev*. vol.**2** no. 03 pp. 313–319 doi: 10.21090/ijaerd.020352.

[7] M A M Abdeen and H Hodhod 2010 Experimental investigation and development of artificial neural network model for the properties of locally produced light weight aggregate concrete *Engineering* vol. **02** no. 06 pp. 408–419 doi:10.4236/eng.2010.26054.

[8] L K Minapu, M V Ratnam, U Rangaraju and P G 2007 Student Experimental study on light weight aggregate concrete with pumice stone, silica fume and fly ash as a partial replacement of coarse aggregate *Int. J. Innov. Res. Sci. Eng. Technol.* (An ISO Certif. Organ. vol. **3297** no. 12 pp. 18130–18138 doi: 10.15680/IJIRSET.2014.0312051.

[9] S Rajeswari and S George 2016 Experimental study of light weight concrete by partial replacement of coarse aggregate using pumice aggregate *Int. J. Sci. Eng. Res* vol. **4** no. 5 pp. 50–53.

[10] S Taufique, S Shakil, S P Bhise and P V Malkapur 2019 Experimental study of compressive strength on foam concrete with quarry dust and fly ash.

[11] F D Lydon and R V Balendran 1980 Some properties of higher strength lightweight concrete under short-term tensile stress *Int. J. Light. Concr.* vol. **2** no. 3 pp. 125–139 doi:10.1016/0262-5075(80)90015-9.

[12] T N Kumar, V Johnpaul and N Balasundaram 2015 Experimental study of polypropylene fibre concrete Int. *J. Eng.Technol. Sci. – IJETS* vol. **2** no. 5 pp. 104–107 doi: 10.4010/2016.1544.

[13] A Mandlik, T Sarthak Sood, S Karade, S Naik and S Phule 2013 Lightweight concrete using EPS *Int. J. Sci. Res.* vol. **4** no. 3 pp. 2319–7064 (Pune University) [Online]. Available: www.ijsr.net.

[14] K N Yu, T Cheung and S Y Koo 1999 Effects of using light-weight concrete on indoor radon concentration in high-rise buildings *Radiat. Prot. Dosimetry* vol. **86** no. 2 pp. 147–152 doi: 10.1093/oxford journals.rpd.a032934.

An Experimental Study on Partial Replacement of Cement and Fine Aggregate with Industrial Waste in Concrete Paver Blocks

Dr. J Venkateswara Rao[1], Mr. S Rajesh[2], Mr. Gopi Shankar[3]

[1]Professor, Civil Engineering, Lakireddy Bali Reddy College of Engineering, A.P, India.
[2]Site supervisor, Grasim industries Ltd, BB puram, A.P, India.
[3]Associate Professor, Godavari Inst. of Engineering and Tech., A.P, India.

E-mail:jampanivenkat8@gmail.com

Abstract. The present experimental investigation reports the relative results of flexural and compressive strength of conventional uni-paver concrete blocks of M-40 grade with the paver blocks produced by the combined partial replacement of the cement and fine aggregate with fly ash and pond ash respectively. Incorporation of these industrial products resulted in the saving of cement subsequently an overall saving of about 10% in the economy per each paver block production, while not compromising with its performance in respect to its strength aspect. The paver block considered in this experimental work is of zig- zag shape specimen of size 201 mm × 100 mm × 80 mm. Initially, fine aggregate and cement were replaced in percentages of 5, 10, 15 and 20 separately with pond ash and fly ash respectively. Later on, both fine aggregate and cement were replaced and the performance of paver block is evaluated in terms of its mechanical properties.

1. Introduction

Due to the special benefits associated with the interlocking paver blocks from the point of view of operational and environmental constraints they are being extensively used all over the world especially in the locations of footpaths, parking areas, airport runways, entrance to offices, driveways ,parkways, small and medium market roads, container yards etc. [1-2]. To make the production of such concrete paver blocks more economical utilization of industrial waste as a partial replacement of the fine aggregate and cement is a good alternative. This also simultaneously solves the problem of disposing these wastes. In recent times, many researches are attempting to evaluate the effective use of these industrial wastes in terms of their modified mechanical properties and the feasibility of their economical production. Effective use of waste glass and copper slag as a partial replacement of fine aggregate in the concrete paver blocks was examined in the recent investigations. Results indicated that up to 30% these industrial wastes can be used without affecting the mechanical properties of paver blocks [3-4]. Through the experimental investigations, it was concluded that crushed stone dust can be successfully utilized as a partial replacement for sand in the preparation of concrete paver blocks [5-6]. Previous researchers evaluated the feasibility of utilization of waste tire crumb rubber and PVC plastic material as partial replacement of fine aggregate in precast concrete paver blocks. Results showed that up to 15-20% these wastes can be employed without any loss in strength properties [7-8]. Earlier investigations also reported the feasibility of partial replacement of cement by various mineral admixtures namely run-of-station ash (60%), plasterboard gypsum (5%), ground granulated blast-furnace slag (55%), and cement bypass dust (25%) without significant influence on the strength and durability properties [9].

2. Experimental analysis

2.1. Materials characterization

In this experimental study 53 grade Ordinary Portland Cement confirming to IS: 12269-1999 was used for casting of concrete paver blocks. For the cement the physical properties are provided in Table.1. Class F Fly ash and pond ash are collected from the nearby source NTPC Visakhapatnam, India. The Pond ash particle size is varying between 1.18 mm and 150 μ. Specific gravity is found to be 2.25. The fineness and specific gravity of class F fly ash sample was determined as 361 m^2 / kg and 2.05 respectively. Chemical composition of fly ash sample is presented in Table.2. For fine aggregate locally available sand which confirms to Zone III, having a specific gravity of 2.64 is used. Particle size of fine aggregates lies in between 4.75 mm and 150 μ. Coarse aggregate of maximum nominal size 12 mm with a specific gravity and water absorption as 2.72 and 0.45 % is used. A water-reducing admixture, Sikament FF, having a relative density of 1.25 and pH range of 8-12 is also used in the production of paver blocks.

Table 1. Physical properties of OPC-53 grade ((IS 12269-1999)

Specific gravity	Fineness	Consistency	Initial setting time	Final setting time
3.15	7%	31%	140 min	31 min

Table 2. Chemical composition of Fly ash

Chemical composition (%)	SiO_2	Al_2O_3	Fe_2O_3	Na_2O	MgO	CaO	SO_3
Fly ash	61.24	25	8.71	0.09	0.09	4.42	0.49

2.2. Mixture design

Trial tests are performed in achieving a concrete mix of strength corresponding to M-40 grade. According to the guidelines of IRC 44 and IS 15658:2006 the mixture proportions details are finalized. The mixture proportions of the M-40 grade and the mix designations of all remaining mixes, prepared by partially replacing fine aggregate and cement are represented in Table.3 and Table.4. Fine aggregate (F_{Agg}) and Cement (OPC) in the conventional concrete paver block mixes are partially replaced by Pond ash (PA) and Fly ash (FA) respectively in percentages of 5, 10, 15 and 20 %.

Table 3. Mix proportions (kg/m^3)

Mix ID	Cement	Fine aggregate (F_{agg})	Coarse aggregate	Water	Chemical admixture	Water/ Cement
M_{0-0}	443	822	1051	161.78	2.21	0.365

Table 4. Mix Designations

Mix ID	F_{Agg} : PA	OPC : FA	Mix ID	F_{Agg} : PA	OPC : FA
M_{5-0}	95:5	100:0	M_{10-15}	90:10	85:15
M_{10-0}	90:10	100:0	M_{10-20}	90:10	80:20
M_{15-0}	85:15	100:0	M_{5-5}	95:5	95:05
M_{20-0}	80:20	100:0	M_{5-10}	95:5	90:10
M_{10-5}	90:10	95:5	M_{5-15}	95:5	85:15
M_{10-10}	90:10	90:10	M_{5-20}	95:5	80:20

2.3. Specimen preparation and test procedures

All the raw materials which are stored separately are weighed for accuracy and transferred into the mix in the sequence of: coarse aggregate, fine aggregate, cement, and pigment of the desired color.

Figure 1. Compression and flexural strength test setup for Paver Block

First these materials are mixed for about three to four minutes. Later, additional water is added along with the admixture and again mixed again for about three to four minutes. After measuring the slump it was transferred to moulds of zig- zag shape. In order to maintain identical slump the admixture content is slightly adjusted. The size of moulds used is 201 mm × 100 mm × 80 mm. The moulds later vibrated for about 15 seconds and then they are kept undisturbed for 24 hours. After this period specimens are de-moulded and kept in water for the curing process to take place until the time of testing. Both the flexural and compressive strength tests were performed in accordance with IS 15658:2006. The specimens during the testing are arranged as shown in Figure 1 and the average value of compressive and flexural strengths for four paver blocks was evaluated.

3. Results and discussions

The compressive strength and flexural strengths of concrete paver blocks are evaluated at 3, 7 and 28 days are evaluated and the average of four specimens is reported as the respective strength. The compressive and flexural strengths of all mixes are discussed below.Figure 2 delineates the comparison between average 3, 7 and 28 days compressive strength of concrete paver blocks with 5, 10, 15 and 20% replacement of aggregate fines by pond ash and the average compressive strength of conventional concrete paver blocks.

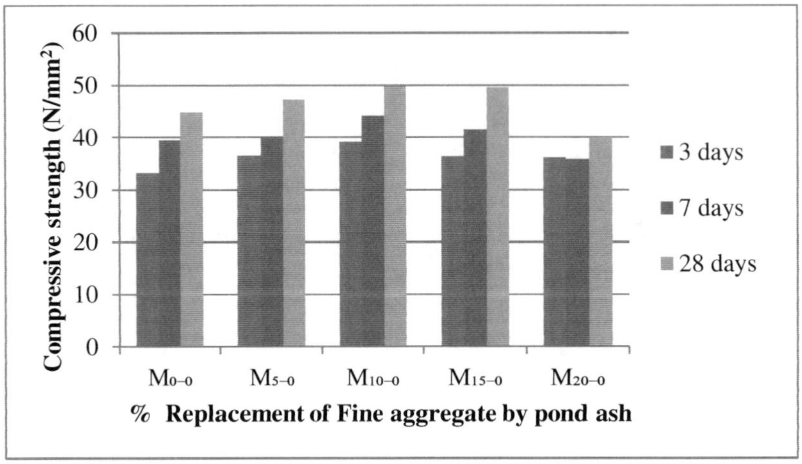

Figure 2. Compressive strength of Concrete Paver blocks with the partial replacement of FA

It can be observed that up to 15% replacement of fine aggregate by pond ash there is an increment in the compressive strength of paver blocks at 3, 7 and 28 days when compared to the conventional concrete paver blocks. The maximum gain in the compressive strength is found to be at 10% replacement of aggregate fines by pond ash. The rate of increment in the average 3, 7 and 28 days compressive strength of paver blocks is found to be 18, 12 and 11% respectively at 10% of replacement of fine aggregate. At 20% replacement there is a decline in the average 28 days compressive strength and the rate of decrement is found to be 10% when compared to the conventional concrete paver blocks. Interestingly, except for 20% replacement by pond ash the average compressive strength of paver blocks at 3,7 and 28 days is greater than conventional concrete paver blocks wherein the sand alone is used as fine aggregate. Figure 3 represents the flexural strength variation of concrete paver blocks when the fine aggregate is replaced by the pond ash in percentages of 5, 10, 15 and 20%. It can be noted that at 5% replacement of pond ash the flexural strength is improved compared to the paver blocks made from conventional concrete.

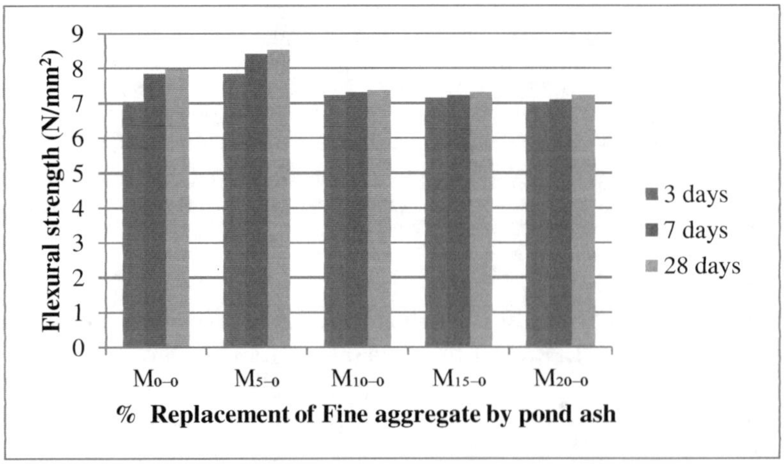

Figure 3. Flexural strength of Concrete Paver blocks with the partial replacement of FA

There is a slight decrement in 7 and 28 days flexural strength of concrete at 10, 15, 20% replacement with pond ash compared to the conventional paver blocks although marginally higher flexural strengths are attained at 3 days curing period. Compared to the conventional blocks the percentage decrements in the average flexural strengths at 28 days are found to be 7.3, 8.2, 8.9 at 10,15 and 20 % replacement of aggregate fines with the pond ash. On the other hand, percentage increments in the average 3, 7, 28 days flexural strengths at 5% replacement are found to be 11, 7.3 and 7.2 % respectively. Figure 4 illustrates the influence of fly ash replacement in the concrete paver blocks wherein 10% of the fine aggregate was replaced with the pond ash. From figure 4, it is clear that as the fly ash replacement increases, there is an increase in the average 28 days compressive strength of concrete paver blocks up to 15%, although at 3 and 7 days, the compressive strengths are marginally low. This may be due to the low heat of hydration arising with the presence of fly ash.

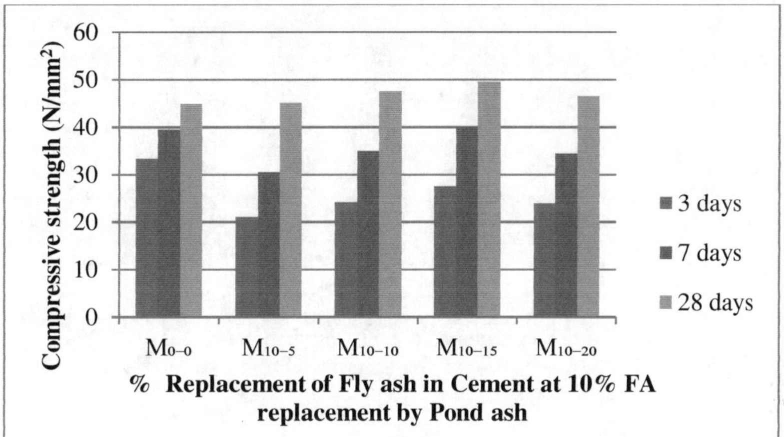

Figure 4. Compressive strength of Concrete Paver blocks with the partial replacement of FA (10%) and cement

At 20% replacement of cement by fly ash, there is a marginal reduction in the compressive strength of paver blocks compared to 15% replacement. The percentage gain in 28 days compressive strength of 10, 15, and 20% replacement is found to be 6, 11 and 4 % when compared to conventional paver blocks. At 3- days curing period, when 15% by weight of cement is replaced with fly ash a decrement of 17% in average compressive strength of paver blocks It can be noticed while at 7 days a marginal increment in the compressive strength was observed. Hence, the maximum gain in compressive strength occurred at 10% replacement of aggregate fines by pond ash and 15% replacement of cement by fly ash.

Figure 5 depicts the flexural strength variation of paver blocks when 10% of the fine aggregate is replaced by pond ash and cement is replaced by fly ash in percentages varying from 5 to 20% in increments of 5%. From Figure 5 it is evident that the flexural strength of the paver blocks with partial replacement of fine aggregate are marginally lower than conventional concrete paver blocks except at 15% replacement of cement by fly ash when the curing period of paver blocks attained 28 days. At this curing period, the highest gain in the flexural strength was found to be 3.5% when compared to the flexural strength of conventional concrete paver blocks. Therefore, similar to the trend of compressive strength, when cement and fine aggregate are replaced by 15% of fly ash and 10% pond ash, respectively better flexural strengths of paver blocks are achieved compared to the conventional blocks. Figure 6 shows the compressive strength development of paver blocks at the age of 3, 7 and 28 days when fine aggregate and cement are partially replaced with pond ash and fly ash respectively. From Figure 6, it is observed that as the in fly ash content increases the compressive strength of paver blocks is improving, but beyond 15% replacement of cement by fly ash, the compressive strength is dropped down even below the strengths of conventional paver blocks. Addition of fly ash resulted in lesser compressive strengths at 3 and 7 days curing whereas significant strength developments at 28 days are observed. The maximum compressive strength gain occurred when 15% of the cement is replaced by fly ash and 5% of fine aggregate was replaced by pond ash.

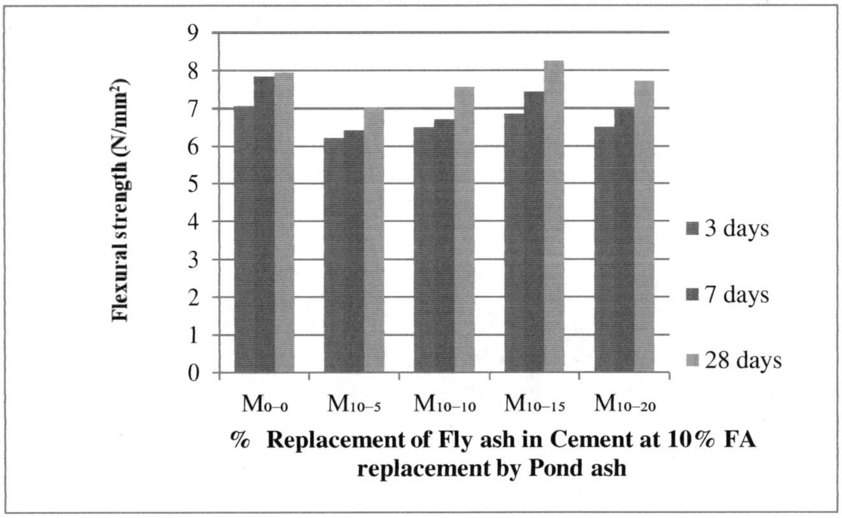

Figure 5. Flexural strength of Concrete Paver blocks with the partial replacement of FA (10%) and cement

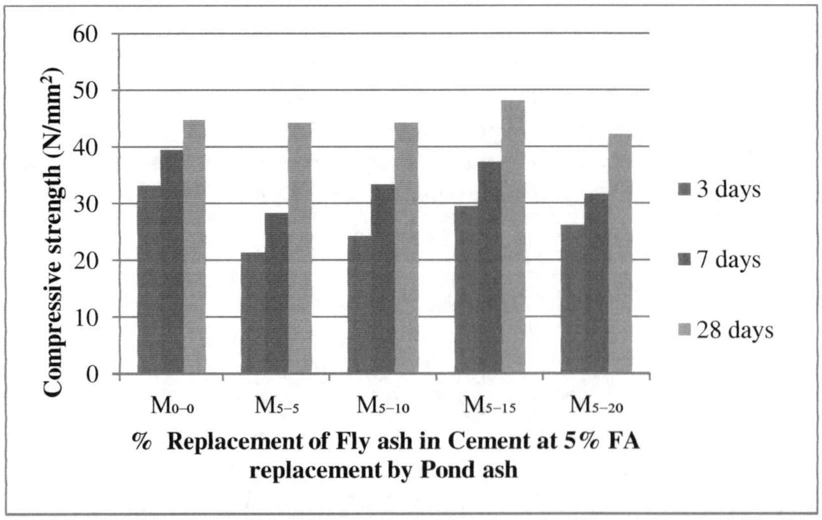

Figure 6. Compressive strength of Concrete Paver blocks with the partial replacement of FA (5%) and cement

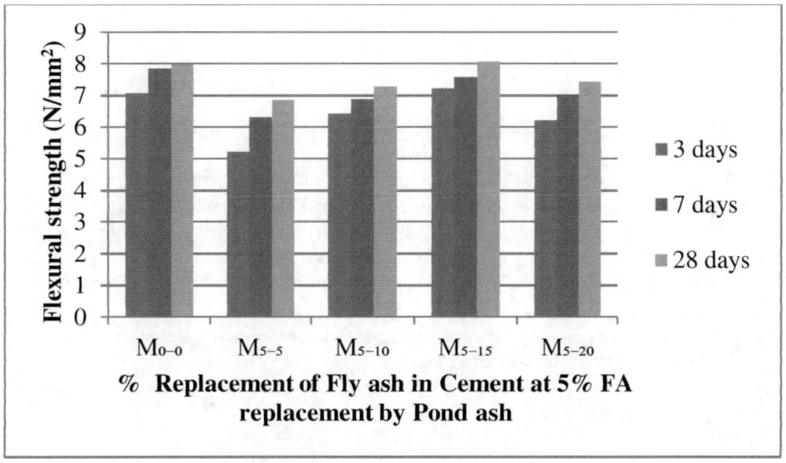

Figure 7. Flexural strength of Concrete Paver blocks with the partial replacement of FA (5%) and cement

The maximum strength gain in this case is found to be 7.3% when compared to conventional paver blocks. This gain in the strength is relatively lower than the strength gain occurred when 15% of the cement and 10% of the aggregate fines are replaced with fly ash and pond ash respectively. Figure 7 describes the variation in the flexural strength development of paver blocks when 5% of the fine aggregate is replaced by pond ash and fly ash replaces cement in percentages of 5, 10, 15 and 20%. It can be noted from Figure 7 that it AT early ages the flexural strength development is low when compared to the conventional paver blocks. A slight increment in the flexible strength is found at 28 days when 15% of cement and 5% of the fine aggregate are replaced with fly ash and pond ash respectively. The maximum flexural strength gain is found to be 1.3% compared to the conventional paver blocks. It can also be noted that this strength is less than the strength attained by paver blocks when 15% of the cement and 10% of the aggregate fines is replaced by fly ash and pond ash respectively.

4. Conclusions

From the experimental investigation conducted on the effective use of industrial wastes i.e. pond ash and fly ash as an alternative source material for fine aggregate and cement, these following conclusions can be made.:

- The compressive strength of conventional paver block can be improved by 11% by replacing 10% of aggregate fines by weight with pond ash. The maximum gain in the flexural strength was obtained when 5% of aggregate fines was replaced by pond ash
- Industrial wastes namely pond ash and fly ash can be effectively used to achieve economy in the production of concrete paver blocks simultaneously improving their mechanical properties.
- Use of 10% pond ash and 15% fly ash as partial replacement of the fine aggregate and cement resulted in an increase in the compressive strength by 11%.
- The maximum gain in the flexible strength of paver blocks was attained when 10% of flash and 15% of cement are replaced by pond ash and fly ash respectively. A maximum increment in the flexural strength was observed to be 3.5% compared to the conventional concrete paver blocks.

- Replacement of 10% of fine aggregate and 15% cement with pond ash and fly ash by weight results in economy savings up to 9.8% for single concrete paper block.

References

[1] Navya G and Rao J V 2014 Experimental investigation on properties concrete paver block with the inclusion of natural fibers *Int. Jour. of Eng. Res. and App.* **4** pp 34-38

[2] Kumar B R and Rao and J V 2015 Effect of inclusion of glass fibers and fly ash in concrete paver blocks *International Journal for Research in Applied Science & Engineering Technology* **3** pp 437-443

[3] Nishikant K, Nachiket A, Avadhut I and Sangar A 2016 Manufacturing of concrete paving block by using waste glass material *International Journal of Scientific and Research Publications* **6** pp 61-77

[4] Kalaiselvi S, Prabhakaran S, Jagadeesan K 2015 An experimental investigation on partial replacement of copper slag as fine aggregate in paver block *Int. Jour. of Chem.Tech. Res.* **10** pp 306-310

[5] Nanda R P, Das A K and Moharana N C 2010 Stone crusher dust as a fine aggregate in concrete for paving blocks *International Journal of Civil & Structural Engineering* **1** pp 613-620

[6] Gupta A and Tiwari A 2016 Effect on mechanical properties of paver block consist crusher stone dust as fine aggregate with inclusion of steel fiber *International Journal for Research in Applied Science & Engineering Technology* **4** pp.529-535

[7] Rethinavelsamy B M and Chidambarathanu N 2016 Investigation on precast concrete paver block with waste tyre crumb rubber *Road Materials and Pavement Design* **17** pp719-736

[8] Mahadevi R, Abirami S, Jananipriya P, Karunya J and Sakthipriya M 2018 An experimental investigation on concrete paver block by using PVC plastic material *Int. jour. of mod. Tr. in eng. and res.* **5** pp 112-115

[9] Ganjian EJalull G and Sadeghi-Pouya H 2015 Reducing cement contents of paving blocks by using mineral waste and by-product materials *Journal of Materials in Civil Engineering* **27** p 04014106

Agro-Climatic Zonation for Tamil Nadu Using GIS and AHP Techniques

Rajkumar R[1], Magesh Kumaran M[2], Mathan Prasath R K[3] and Senthilshanmugavelayutham S P[4]

[1]Assistant Professor, Department of Civil Engineering, PSG Institute of Technology and Applied Research, Coimbatore, Tamil Nadu, India.
[2,3,4]UG Student, Department of Civil Engineering, PSG Institute of Technology and Applied Research, Coimbatore, Tamil Nadu, India.

E-Mail id:rajkumar@psgitech.ac.in

Abstract. Agro-climatic zonation is a classification technique that uses several meteorological and topographical data of our targeted region for proper zonation with higher priority towards crop suitability factors. This study deals with creating such zonation using 7 major factors for the entire state of Tamil Nadu using GIS and AHP. The Agro-climatic variables include precipitation, evapotranspiration, land surface temperature, land use and land cover data, soil moisture, soil order and ground water level. On incorporating the data in GIS with attribute tables, thematic maps were created through inverse distance weighting interpolation. Such maps were over-laid so that zoning using all the defined factors could be made feasible. Weightage given for each factor was determined based on Analytical Hierarchy Process, with precipitation and soil order with the highest priority. A final map was obtained with Agro-climatic zones defined based on the degree of suitability for agriculture from highest to lowest. This classification can further help to define the suitability of various crops on a region and can help in planning an appropriate crop pattern for that region.

1. Introduction

Productivity of crops in an area mainly depends on some major topographical and meteorological factors. Such factors are to be considered before deciding the crop to be cultivated. So, based on such factors, distinguishing regions of different nature helps in making utmost yield of crops that are more suited to the conditions. Agro-climatic zone classification system enables taking up the desired meteorological factors into account and distinguish the target area into separate regions based on the factors for the main aspect of taking developmental actions accordingly in each of the zones[1,5]. The aim of this zonal classification is to aid in crop suitability issues in a region [2,9]. In this study, the target area is taken as the entire state of Tamil Nadu. The major factors of consideration are precipitation, soil order, land surface temperature, land use and land cover, evapotranspiration groundwater level and soil moisture. Mean values of the data are taken into consideration and incorporated in ArcGIS as separate data maps [3]. Analytical hierarchy process is used for arriving appropriate weightage of each factor over the other based on the desired priority. On taking such weightage into account, weighted overlay could be done quite precisely. Regions with similar range of data are taken as unique zone and crop suitability analysis is made to grow such crops throughout the zone.

2. Materials and methods

2.1 Study area

The state of Tamil Nadu is one of the 28 states of India. Its capital and largest city is Chennai. It is located in the south-eastern part of Indian peninsula between 8° 5' and 13° 35' North latitude and 76° 15' and 80° 20' East longitude and is bordered by the union territory of Puducherry and the South Indian states of Kerala, Karnataka, and Andhra Pradesh.The total population of the study area as per 2011 census is 72,147,030. The state covers 1,30,058 km^2 and is the 10th largest state in India.The state is significantly known for its agriculture and is the second largest producer of rice in the country [4].The study area is depicted in figure 1.

Figure 1. Study area map

2.2 Meteorological and topographical data

Meteorological data obtained for a period were averaged and the mean value was considered for further analysis [5]. The meteorological and topographical data taken for analysis are listed below.

2.2.1.Precipitation

Moist air floats up higher into the sky which is colder than the land surface and condenses into tiny liquid droplets of water forming clouds. Several cloud droplets fuse to form heavier clouds which makes it unable to float and in turn it starts to precipitate in any form such as rain, snow, hail etc. The most predominant form of precipitation in our study area is rainfall. We have collected rainfall data across Tamil Nadu over the last five years for the analysis from India-WRIS [6].The rainfall data are integrated into Arc map Software and the layout for the precipitation is prepared and shown in Figure2.

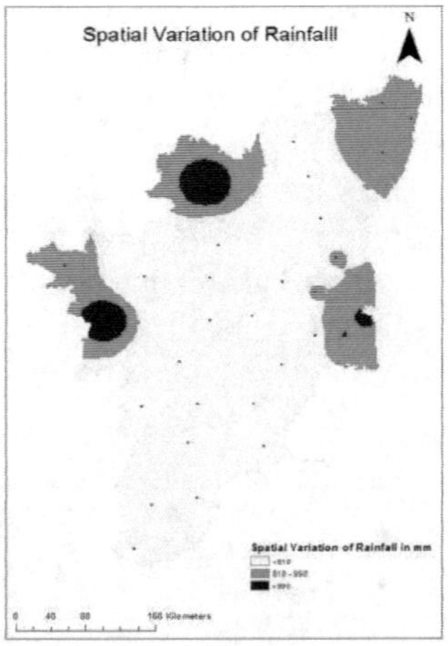

Figure 2: Spatial Variation of rainfall

2.2.2. Soil order

Soil order or soil taxonomy is a soil classification system based on one or two dominant physical, chemical, and biological properties. It is generally classified into 12 categories or orders with each of them having distinct characteristics and ecological significance. In our study area, 7 orders are predominant namely Alfisols, Entisols, Inceptisols, Mollisols, Ultisols, Vertisols and other miscellaneous .These 7 orders are considered for our analysis and it is shown in figure 3. We have acquired a soil order map of Tamil Nadu from the Department of Remote Sensing and GIS, Tamil Nadu Agricultural University.

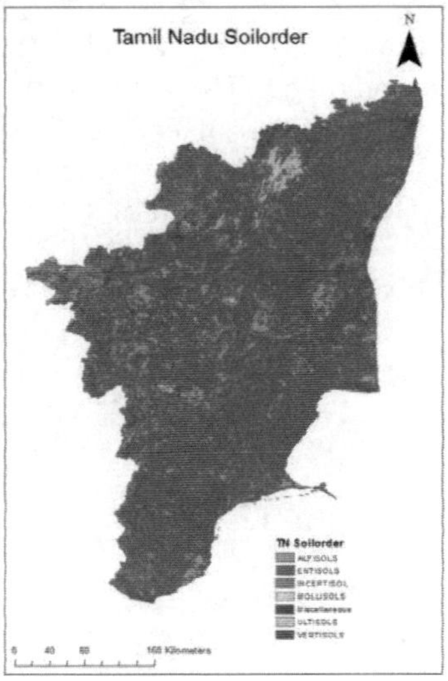

Figure 3: Soil order of Tamil Nadu

2.2.3. Land surface temperature

Land surface temperature (LST) is the radiative temperature of the Earth's surface due to solar radiation. In general, it varies from place to place depending upon the climate and weather, the type

IOP Conf. Series: Materials Science and Engineering 1006 (2020) 012015 doi:10.1088/1757-899X/1006/1/012015

of surface and its properties. We have acquired the average LST over the last 5 years through LANDSAT-8. It is the satellite that is predominantly used to determine LST across the world [7].

2.2.4. *Land use and land cover*

Land use and land cover (LULC) denotes the physical land type of a region and the type of usage of the land cover by people. It is displayed with help of maps over a topographical area. LULC maps are created using of satellite or aerial imagery with the help of remote sensing and geographical information system. We have used the current LULC map of Tamil Nadu for our analysis shown in figure 4.

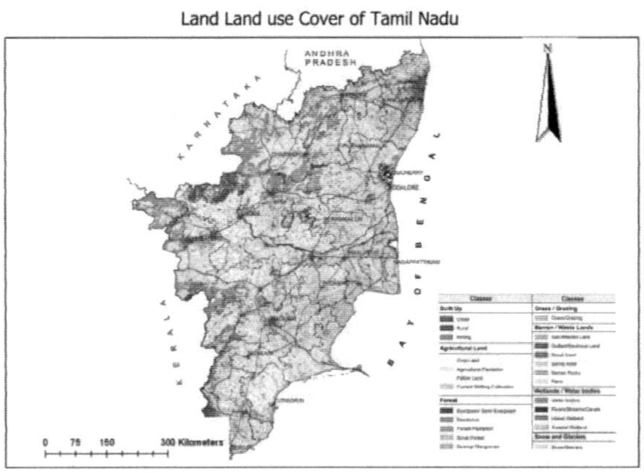

Figure 4: LULC Pattern of Tamil Nadu

2.2.5. *Evapotranspiration*

Evapotranspiration is the loss of water from the surface of the Earth to the atmosphere through evaporation and plant transpiration combined. The amount of evapotranspiration loss depends on the temperature and pressure at the point on the surface. It also accounts for the movement of water through air from variable sources. We have collected evapotranspiration data across Tamil Nadu over the last five years for the analysis from India-WRIS.

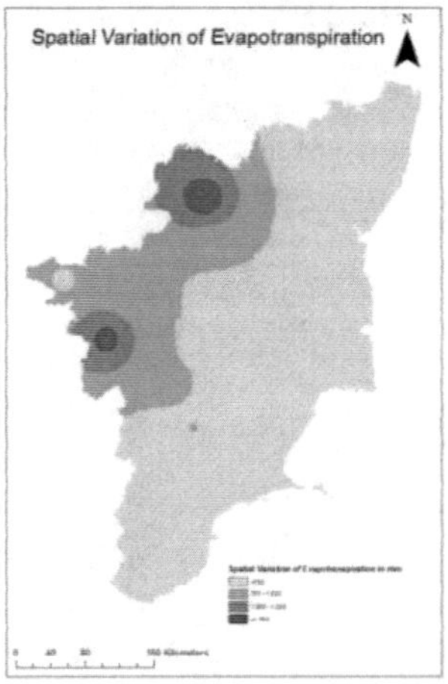

Figure 5: Layout of Evapotranspiration

IOP Conf. Series: Materials Science and Engineering 1006 (2020) 012015 doi:10.1088/1757-899X/1006/1/012015

2.2.6. Ground water level

Ground water level is an important parameter that defines the property and the applications of a particular cover of land. We have collected ground water level data across Tamil Nadu over the last five years for the analysis from India-WRIS. The layout of groundwater level is shown in figure 6.

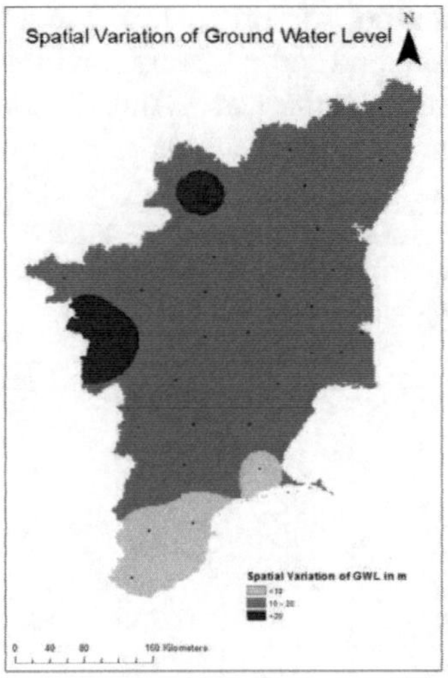

Figure 6: Variation of groundwater level

2.2.7. Soil moisture

Soil moisture is the amount of water (in all forms) held between the soil particles on the land surface. Soil moisture affects the crop growth and sustainability on a piece of land. We have collected soil moisture data across Tamil Nadu over the last five years for the analysis from India-WRIS shown in figure 7.

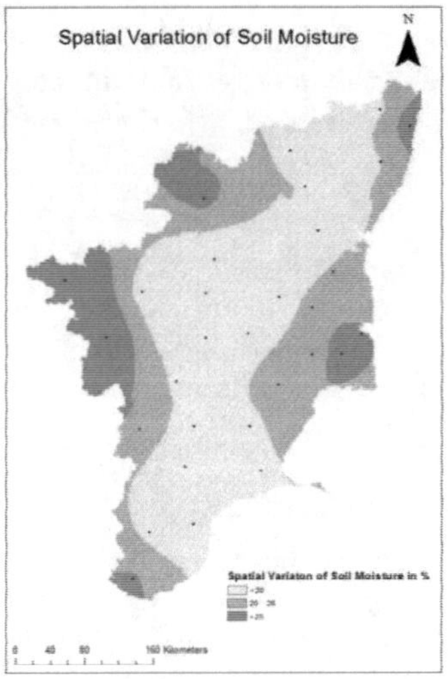

Figure 7: Variation of Soil moisture

2.3 AHP analysis

Analytical Hierarchy Process (AHP) is a structured technique used for organising and analysing complex decisions. It is powerful tool in Multi Criteria Decision Making (MCDM) technique. The parameters considered are compared pair wise with the help of a matrix in Table 1 in which each of them is given a relative importance over another (shown using numbers 1 to 9) based on priorities and their weights are determinedby the evaluation scale developed by Saaty [8].The parameters that are considered here include precipitation, soil order, land use and land cover, surface temperature, groundwater level, evapotranspiration, and soil moisture. Precipitation and soil order are given with higher priorities while soil moisture with the least. And finally, a consistency checks in done to ensure whether the original preference ratings were consistent or not. The Consistency ratio for this process is about 0.036 which is lesser than the maximum acceptable value of 0.1.

Table 1. Criteria AHP pair-wise comparison matrix

Criteria	Prep	Soil Order	LST	LULC	EVP	GWL	Soil Moisture	AHP Weights
Prep	1	1	3	5	7	7	9	0.32
Soil Order	1	1	3	5	7	7	9	0.32
LST	1/3	1/3	1	3	5	5	7	0.17
LULC	1/5	1/5	1/3	1	3	3	5	0.09
EVP	1/7	1/7	1/5	1/3	1	1	3	0.04
GWL	1/7	1/7	1/5	1/3	1	1	3	0.04
Soil Moisture	1/9	1/9	1/7	1/5	1/3	1/3	1	0.02
Total Weight								1.00

2.4 GIS analysis

GIS analysis was carried out in ArcGIS. The point shape file, integrated with the study area shape file, for each parameter across the study area was created by using the spatial data consisting of the latitude and longitude of each district and the attribute data consisting of the corresponding mean value calculated and individual thematic maps for the parameters were developed using the Inverse Distance Weighted (IDW) interpolation method [5]. Based on the criteria weights determined by AHP analysis, raster weighted overlay analysis technique was used to prepare the final agro-climatic zonation map.

3. Results and discussion

The weight of criteria resulting from AHP is presented in Table 1. The results indicate that the value of the consistency ratio (Cr) is 0.1. This indicates that the results are valid in accordance with the threshold recommended by Saaty (2008).The weighting results showed that the soil criterion has the highest weight. This result indicates that soil order is very crucial in the successful cultivation of vegetables crops. The next weights are precipitation, followed by land use land cover and temperature. The weights calculated are assigned to the thematic layers created in Arc map. The weighted overlay analysis is performed in GIS tool and the final result for the agro-climatic zonation is obtained. The agro-climatic zonation map is shown in the figure 8.

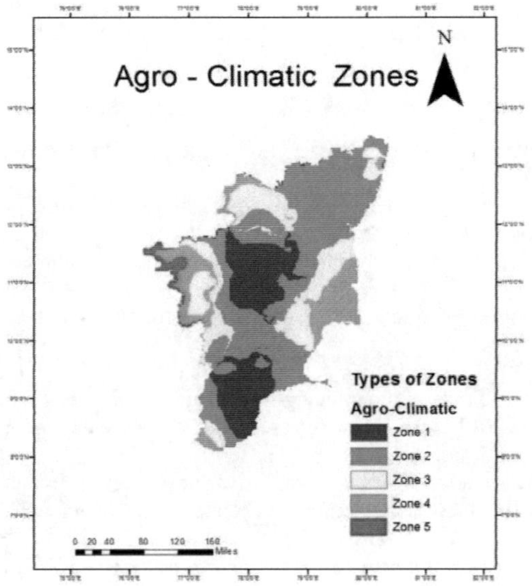

Figure 8: Agro-climatic Zonation of Tamil Nadu

The study region is divided into five different zones as per the results obtained in the weighted overlay analysis performed in Arc GIS. Zone 1 covers southern part of Tamil Nadu and a part central belt in which rainfall is relatively high and soil is more fertile. The crops such as coconut, sugarcane are feasible for the zone 1. Similarly five zones are divided and the feasible crop recommendations are given for each zone and are indicated in the table 2.

Table 2: Zonation of recommended crop

S.NO	Zone	Recommended Crop
1.	Zone 1	Sugarcane, Pulses, Cereals and Millets
2.	Zone 2	Groundnut, Gingelly, sunflower, castor and cotton
3.	Zone 3	Coriander and Chillies
4.	Zone 4	Paddy, Cholam and ragi
5.	Zone 5	Gingelly and Chillies

4. Conclusions

Climatic variability affects agriculture production and these affects are severe in mountain regions as these regions more prone to climate change. The agro climatic suitability assessment needs to be given greater thrust for meeting future food demand and ensuring food security. By using geospatial techniques and GIS based models this research performs suitability assessment of crops to be cultivated in Tamil Nadu. Application of GIS-based multi-criteria analysis proved to be useful to assess climatic (temperature and precipitation), soil and topographic suitability for crops in Tamil Nadu region. The agro climatic suitability assessment need to be given greater thrust for meeting future food demand and ensuring food security.

References

[1] Showkat A, GanaieM, Sultan Bhat et al 2014 Delineation of micro agro-climatic zones of Jammu and Kashmir *Int. J. Agricult. Stat. Sc* **10** 219-225.
[2] Widiatmaka 2016 Integrated use of GIS, AHP and remote sensing in land use planning for tropical high altitude vegetable crops *Journal of Applied Horticulture* **18(2)** 87-99.
[3] Wart, Justin & Bussel, et al 2013 Use of agro-climatic zones to upscale simulated crop yield potential *Field Crops Research* **143** 44–55.
[4] Anand A &Mankad et al 2009 Agro-climatic zonation of Maharashtra State using GIS *Transactions of the Institute of Indian Geographers* **31** 25-36.
[5] Dr. N R Patel, Jyoti Singh 2015 Geospatial assessment of shift in agroclimatic suitability of food grains and plantation crops in Himachal Pradesh under changing climate *Indian space research organisation*(ISRO)
[6] Bisht, Himani & Nain, Ajeetet al 2013 Agro-climatic zonation of Uttarakhand using remote sensing and GIS *Journal of Agro meteorology* **15** 30-35.
[7] Government of India 2020 Department of Water resources, river development & Ganga rejuvenation *India-WRIS*.
[8] Geletič Jan & Lehnert Michal 2016 GIS-based delineation of local climate zones: The case of medium-sized Central European cities *Moravian Geographical Reports* retrieved from 242-12.
[9] Saaty T L 1980 *the Analytic Hierarchy Process* (New York: McGraw Hill)
[10] Román-Figueroa, Celián & Herrera et al 2019 Methodology for the estimation of land suitability for Atriplex cultivation in arid and semi-arid regions *Arid Land Research and Management* **33** 412-426.
[11] https://indiawris.gov.in/wris Ministry of Jal Shakti Government of India.

IOP Conf. Series: Materials Science and Engineering 1006 (2020) 012016 doi:10.1088/1757-899X/1006/1/012016

Theoretical Investigation on Pervious Concrete Pavement (PCP) as Sustainable Pavement Technology (SPT)

Pala Gireesh Kumar[1],Abhirami Priyanka Pathivada[2], Velugoti Sasi Jyothishma Himaja[3], Korukonda Rachana Tejaswi[4], Garapati Keerthi Liniya[5]

[1]Associate Professor and Head, Department of Civil Engineering, Shri Vishnu Engineering College For Women, Bhimavaram, West Godavari District- 534202, India.

[2,3,4,5]U.G Student, Department of Civil Engineering, Shri Vishnu Engineering College For Women, Bhimavaram, West Godavari District- 534202, India

E-mail: palagireeshk@gmail.com, abhiramipriyanka4@gmail.com

Abstract. A pavement has always been in trend with lots of amelioration and furtherance. From gravel and cobblestone pavements to asphalt and concrete pavements, the world has seen a lot of new design criteria with the theoretical, field, and laboratory test analysis. From ages, the concept of Pavement has been holding on to improvising its esteemed characteristics such as compressive strength, durability, and life span. With varying mix designs and new methodologies of pavement evaluation and improvisation, the concept of sustainable pavements has gained enough priority. Better and durable quality with economical benefit and ecological balance is a much-needed pavement as per the present scenario of Mother-Earth. For a tidier society, the need for a new advancement in pavements has become crucial. With the rising revolution, Pervious Concrete Pavement (PCP) has paved its way straight to the top with an exceptional working mechanism in the category of Sustainable Pavement Technology (SPT). PCP has gained its empowered usage with beneficiary factors such as controlling storm-water runoff, reduction in pavement noise, providing vehicle-friendly pavements with a good riding platform, and being budget-friendly. With various conceptual studies and from several literature reviews, it is proven that PCP has become the future of pavement engineering with more add-ons and advantages. Therefore, this review paper, in particular, deals with the advancements of pavements, the necessity of sustainable pavements and, the transmogrification of Pervious Concrete Pavement. Following the theoretical concepts, the reviewed articles and proven test results will be scrutinized. This study concludes that available design mixes will be concentrated and accentuated for further analysis thus, proving PCP to be more efficacious in SPT.

1. Introduction

Transportation has been a major area of business economy for every country. The major import and export business deals with the mode of transport. Therefore, a country's growth rate is closely influenced by the advanced transportation facilities that a country holds. Among the various transportation sectors, roadways are designated to be the most economical and easy of all. In this regard, the major concern is emphasized on the working ability of roads and as a part of it, various studies and experimental works are carried out upon the road and pavement quarter. The pavement plays a key role in giving cut-rate and comfortable transport. Adding to this context, the Indian government scrutinized the need for a pavement to be sustainable. Many pieces of researches have been put forth and several proposals were made in the field of pavements. There has been a revolution of add-ons to the pavements and also amendments were suggested in the mix designs. As known already, pavements are categorized into flexible and rigid pavements with each one shouldering several assets and stags. Of the two, flexible pavements have proven to be more in use especially in the areas of major highways. According to research, seventy-eight percent of rural and eighty-six percent of urban highways in the United States are of flexible surfacing. It is experimentally suggested that the type of pavement is concerned with ground conditions and traffic flow. However, the carbon footprint analysis and budget assessment revealed that rigid pavements are a step more advanced than flexible pavement with lesser emission of energy and cut-rate[1]. In addition to the various considerations, there are several new reforms in sustainable pavement designs like pavements of hot mix asphalt, coarse grade asphalt, stone mastic, composite, etc., Unanimously pervious concrete has substantiated to be the most eco friendly with a lot more precedence. Pervious concrete pavement can be quoted as one of the emerging trends in sustainable pavement technology. Due to an increase in terms of energy emissions of conventional concrete along with a soaring rate of construction, there has been a desideratum for innovation. In addition to the levy and outpouring nature of concrete, excess precipitation crises have made a hike in the problem statement criterion. To resolve the nodus, the commencement of pervious concrete was made. Unlike the conventional concrete, pervious concrete authorizes the swirl of water into the ground and thus, gives the advantage of further utilization of the collected water. This paper entitles a crystal clear idea on the design criteria of pervious concrete pavement with the inclusion of various theories proposed by numerous authors accounting for the replacement of conventional concrete with pervious concrete.

1.1. Chronicle history
As per various citations available and information congregated, it is discernible that the first usage of pervious concrete is in the form of a pavement surfacing done in Europe in the early 1800s. The substantial usage is due to the prior requirement to cut down the content of cement used in the construction or paving of roads[2]. Later in England and Scotland, in the 1920s, pervious concrete was utilized in constructing two-storey buildings. But it became sought after the second world war where the availability of cement has fallen to the ground. In late 2000, India saw the rise of pervious concrete as an empowering guard. Well in advance with its evolving aids, pervious concrete has set a mark in the pavement area as the most satisfying paving surface that holds tremendous pros.

1.2. Materials and mix proportioning of PC
Pervious concrete(PC) works on the principle of permeability. In other words, the mechanism of Pervious Concrete Pavement (PCP) entails the porosity and thereby, diverting the action of porosity into a fringed aid of collecting the water. The PC comprises aggregates coarser than 9.5 mm mixed with normal portland cement along with water to attain a concrete mix with desirable voids [3]. The batching of materials is highly dependent on the requirement in the working area. Apart from the basic mix of pervious concrete, a little amount of fine aggregate can also be added just in case of enhancing the strength properties [4]. But for its porous nature, the PCP has attained many merits and has been a constant work field in the Sustainable Pavement Technology(SPT)[5]. The mix design of pervious concrete is likely to vary but a standardized ratio of aggregate to cement is found to be 1:3 for concrete of M20 grade [6]. However, it is notified that the strength of the PC can be escalated with the addition of plasticizers and superplasticizers which reduces the water content and thereby, increases the strength properties. The minimum percentage of voids required for an easy permeability of water is stated to be 20% which would accommodate the strength desired along with an acceptable rate of infiltration. The predominant factor affecting the properties of PCP is the water-cement(w/cm) ratio. The value of w/cm would be so beneficial if it ranges from 0.26 to 0.4 [7].

1.3. Design considerations and construction of PCP

The design of pervious concrete pavement predominantly intensifies the mechanical properties like strength and stiffness characteristics. Also, PCP design thickness is expected to rely on hydraulic properties such as the rate of infiltration and the void percentage in the mix[8]. Depending upon hydrological cycle, the environmental aspects, and the rate of traffic flow, the PCP needs to be fashioned in the above-cited properties and needs to be paved with a thickness of 200 mm to 300mm to sustain the heavy loads and enormous traffic flow(in relation with the passenger car unit theory). Also, care should be taken while compacting the soil of the subgrade. It is known that if a soil is more compacted less will be the voids and as a consequence, permeability decreases resulting in stagnation of water on the pavement. The next major consideration in PCP usage is the rate of infiltration. For a PCP to be paved over soil, the minimal infiltration rate expected is 2.5 mm per hour. This context has cleared many norms that PCP can be used for any type of soil provided the minimal infiltration rate meets the design infiltration rate and in case of infiltration rate lesser than 2.5 mm per hour, pivot tubes are inserted in the subgrade to regulate and divert water to constrained channels [9]. It is foremost to have a proper subgrade to avoid failures. Well experienced and efficient workmanship is a must to have a PCP laid. In rare cases where freezing and thawing are neglected, PCP can be directly placed over the subgrade with no requirement of subbase [10]. This needs to be determined only after a complete analysis of the site conditions and field tests of soil by the project manager. Furthermore, the slackening of water content creates a raveling effect on the PCP. To avoid the occurrence of raveling, the PC mix should be placed swifty and curing needs to be done instantly to maintain the water content in the mix. This helps in the reduction of raveling nature and adds an aesthetic metallic finish to the PCP. Appropriate curing needs to be done to ensure a good pavement design with a longer life[11].

1.4. Engineering properties of PC

The various engineering properties of pervious concrete are stated in Table 1.

Table 1: The engineering properties of pervious concrete[12]

ENGINEERING PROPERTIES	PERVIOUS CONCRETE
Stiffness	Stiffer than conventional concrete pavements
Placing	1 hr to 1.5 hr (can be altered with addition of admixtures)
Slump	20mm
Density	1600 to 2000 kg/m^3
Void Content	15% to 25%
Permeability	80 % more than Traditional concrete
Compressive Strength	3.5 MPa to 20 MPa
Flexural Strength	1MPa to 3.8MPa
Shrinkage(First 10 days of placement)	50% to 80%
Abrasion Resistance	Raveling and Abrasion is common
Sulphate Resistance	More prone to get affected due to open pores

1.5. Maintenance of PCP

The predominant excellence of PCP is the low service requirement. The cost of maintenance of pervious concrete pavement is comparatively lower than the conventional concrete pavement. Sometimes, it is observed that good pervious concrete pavement often results in zero maintenance for long years. However, in the case of the pores getting clogged on the top surface, removal of the debris and the waste material is recommended to revive the lost infiltration rate. Prior to the design of PCP, several factors like the elevation of surrounding drainages, presence of rocks, soil, residues, etc., should be cross-checked to avoid obturation of porosity. vacuuming is one of the best courses of action to maintain the premiums of the PCP. Furthermore, pressure washing, sweeping, rinsing as well as power blowing can also be done to restore the permeability of the PCP. To avail of the methods of cleaning the PCP, the duration of decrease of permeability should be known. However, there are no accurate and definite methods of precise information. Rather, indirect methods are preferred. One such method is the Standard Test Method (ASTM C1701) where the PCP is tested for its infiltration rate at randomly noted points on the surface soon after the pavement is laid and ready for use. After a week, the procedure is repeated and the infiltration rate is compared to analyze the rate of dwindling of permeability. Based on the results, maintenance is often planned and executed[13].

1.6. Advantages of PCP

There are numerous aids and merits of PCP as a sustainable pavement technology. Table 2 stipulates the traits of pervious concrete pavement.

Table 2: Merits of PCP[14]

GRADE	ASSETS
Appearance	1) More attractive than asphalt pavements 2) metallic finish 3) Rice-Krispie texture
Cut-Rates	1) Best surrogate to management of stormwater. 2) Low maintenance. 3) Budget friendly with low construction cost. 4) Economy on Rain Water Harvesting can be cut short. 5) Life-cycle cost is low
Environmental Aspect	1) Eco Friendly in nature with lesser emission of heat 2) Hydrological cycle is more effectively utilized. 3) No stagnation of precipitation over the road thereby, reducing runoff. 4) Enhances the groundwater table. 5) Rate of pollution is dwindled
Structural Aspect	1) Prevents hydroplaning 2) Provides a smoother travel in cases of rain and snow 3) Reduction in the glare during night

2. Literature Reviews

Several pieces of research have been made and numerous reports were submitted in favor of replacing conventional concrete pavements with pervious concrete pavements. Adding further, the crucial anchor is categorizing pervious concrete pavement as one of the sustainable pavement technology. In this aspect, many theoretical, laboratory, and field tests were conducted and several proposals are depicted in favor of PCP. Few such literature reviews concerning the PCP are as follows.

John T. Kevern et. al.(2012) carried an experimental analysis on the working procedure of previous concrete pavement concerning the Urban Heat Island Effect (UHI). In his research, he clearly stated the advantage of replacing conventional pavements with pervious concrete pavement for better sustainable pavement technology. Sensors that detect temporal variations are taken as the key materials in the test procedure and they are installed in both conventional and pervious concrete pavement at varying depths. After a five-day analysis, the storage of heat for each day is calculated for both conventional and PC pavements. From the results, it is experimentally proven that PCP has stored less energy in comparison to conventional type pavement. Consecutively, the UHI effect can be dwindled in PCP due to low heat stored property. The predominant inducement leading to low heat storage is the presence of voids in the PCP, which acts as a heat cutter for the pavements[15].

Hilal El-Hassan et. al (2018) stated that the weighty succor of a pervious concrete pavement is to reduce the rate of runoff. He instigated the inclusion of ground granulated blast furnace slag commonly known as GGBS to the PCP. With the additive incorporation of polypropylene fibers, it is observed that the PCP has exhibited high porosity with low compressive and tensile strength. Experimental analysis was done on a PCP with GGBS of 50%, aggregates of 10 mm, and 20 mm. The results inferred that the PCP has exhibited a better performance in the name of low UHI effect, low carbon emissions (reduced by 54% in comparison to conventional concrete pavement), and of course a less cost of construction and maintenance. It is quoted that a 40-year life can be achieved with the removal of dust and debris at regular intervals. Furthermore, he proposed an inverse relation of permeability with aggregate size and direct relation with porosity[16].

Youngmin Joung et. al (2008) emphasized greater concern on the strength characteristics of pervious concrete pavement. Research work was directed in the urban stream to optimize the drawbacks concerning the strength characteristics. As known already, urban areas are more prone to stagnation of rainwater due to improper drainage facilities. To counteract this trend, PCP is regulated to have more permeability. But permeability being inversely proportional to strength parameters, an increase in infiltration rate decreases the strength of the pavement. To provide a balance, the void ratio to be maintained without disturbing both strength and permeability characteristics is found to equalize 33%. Future research to compensate both parameters is also on the verge of completion in this regard[17].

Dale P. Bentz (2008) has given a virtual approach in association with pervious concrete pavements. The outcome of his analytical research profounds the virtual representation of the voids in the PCP. A 3D visualization was set up using the data from past researches, records, and chronicle data. He evaluated the three-dimensional structure of the pervious concrete pavement by considering a small section of it. A correlation between the permeability characteristics following percolation and transport of percolated water is formed. Initially, the 2D hybrid HCSS working model was used in developing the images of the void structure of PCP in 2D. Later, after obtaining all the data relevant to the development of 3D models, the venture of percolating void structure has come into the picture. In addition to the data represented, the author has involved the virtual grant of PCP in other aspects like clogging, freezing, and thawing resistance as well [18].

Malhotra. V. M (1976) stated the various properties dealt with pervious concrete pavement as a sustainable concrete pavement. It was experimentally found that the engineering properties of PC are more likely to differ from the conventional type of concrete. The density of pervious concrete was found to be 70% of the traditional concrete with the same constituents. He also instigated that the use of rammers or vibrators in compacting the PCP would result in more compaction thereby lowering the voids and as a result, decreasing the permeable nature of PCP. He also added that properties like skid resistance are comparatively more for PCP than conventional concrete pavements. This has provided a clear way of making PCP more worthy of increased positive outcomes [19].

John Tristan Kevern (2008) quoted the need to look into the effect of freezing and thawing in the case of PCP. The amount of cement paste is much less in PCP as compared to conventional type pavement making it vulnerable to the freeze-thaw effect. There are tremendous chances of the coarse aggregate getting exposed to the open atmosphere due to the lack of intense coating of cement paste around them. In this aspect, the durability criteria were mentioned as one of the supreme contexts. There are direct test procedures in measuring the workability of conventional concrete but there is no such comfort for pervious concrete to date. Indirect means were focused to know the standards of pervious concrete. The ease to lay PCP and the density is considered in explaining the workability of pervious concrete as a pavement material. Also, the less water content ratio of pervious concrete strengthens the process of non-uniform curing thereby, resulting in the raveling of the pavement surface. A long term investigation and future scope on the strengthening of PCP was being carried out by them with a preliminary focus on pavement aging and distress problems to be minimized [20].

Rohini lomte (2018) illustrated the advancement of PCP in accordance with mechanical and hydraulic properties. She stated the dependency of strength on the constituents of PCP. From the study conducted, it is scrutinized that no coarse aggregate of size less than 10mm is to be used and also amplified the need to avoid the flaky and elongated type of aggregates. From the research done, the nominal air content for an optimistic PCP is given as 13% to 28%, aggregate to cement ratio as 4:1. In furtherance to the study, admixtures like crushed granite, VMA, HRWR, superplasticizers were also recommended [21].

Suraj F. Valvi et. al (2017) demonstrated the study of pervious concrete pavement concerning factors affecting the sustainability of pavement. From the test analysis, it is evinced that the mechanism of PCP needs to alter to suit the ground and climatic conditions especially in rural areas that come under the zone of heavy rainfall. Stormwater runoff is the most hectic factor that needs a large scale focus and with the replacement of conventional concrete with PCP, the majority of the snag is sorted. With a well planned and sophisticated construction of PCP, the hitch of overflow is controlled and also a phase of water table recuperation is observed. The results from the experiment stated that with an increase in the duration of the PCP construction, the permeability rate also increases with durability and strength remaining unchanged. Also, an inverse relation is quoted between water absorption and durability of concrete. Apart from all the known data, a keynote was added explaining the strength variation with change in grade of cement. According to this research paper, OPC 53 grade cement with an aggregate of 9.5 mm in a mix of 1:6 sounds to have higher compressive strength than any other mix. But the standard mix proportion preferred in PCP is 1:3 mix [22].

Darshan S. Shah et. al (2013) elucidated a coherent explanation of the cost-effectiveness of PCP in rural areas where land sustainability is at high risk. This research paper depicted the need for replacing traditional pavements with PCP to reduce the runoff of stormwater, escape the high budget construction and maintenance of pavements, enhance the level of the water table, effective use of land, and reduction in pollution. Following the design considerations, a PCP should be constructed based on the traffic flow and effective road utilization. It is preferably used in areas of low traffic and just in case of heavy flow, additives are suggested to enhance the strength of pavement [23].

Mary Vancura et. al (2011) posed a straight clear-cut exegesis of the structural dependency of PCP. Quotidianly, PCP is designed for serving the purpose of transportation as well as a stormwater management system. PCP plays a dual role when it comes to the working procedure. Thus, it is mandatory to heighten the strength parameter as well as the infiltration rate. As compared to conventional concrete pavement, PCP lags in the value of elasticity modulus, compressive, tensile, and flexural strengths and also in subgrade modulus. This paper has provided a solution to maintain the desirable patterns without lowering the permeability of the PCP. After an evidential study and fieldwork, with proper thickness of the pavement and base course, an acceptable bearing capacity can be achieved. Reverse calculation of deflections to load, thereby comparing the deflection from falling weight deflectometer and simulations of ISLAB2005 using the Westergaard model has given a conclusion that the bearing capacity of both conventional and pervious concrete pavement can be equalized. Furthermore, the increase in modulus of rupture along with thickness is directly proportional to the fatigue life of PCP. Apart from the structural point of view, external factors like maintenance and removal of clogged materials also add up to the successful working of PCP with low cost of maintenance as a sustainable pavement technology [24].

3. Conclusion

From the performed theoretical study, it is perspicuous that the pervious concrete pavements have been emerging in the trend of sustainable pavement technology with new alterations in the aspect of design and materials. After a long appraisal of various considerations, PCP has ratified several advantages over conventional pavements in terms of better infiltration rate, reasonably economical in both construction and maintenance, environmentally sound with non-polluting and energy-efficient aspects. In addition to the excellence of PCP, it has been authenticated to be a quintessential alternative to counteract the Urban Island Effect. With the increasing demand for land sustainability, the pervious concrete pavement has given a good source of replacement thereby utilizing the infiltration rate for recuperating the dwindling groundwater table and also been a source to take the edge off for stormwater management. Further researches are being made in the flow to find solutions to some mechanical problems like stiffness and strength. But a reduced pavement thickness and less content of cement usage have made pervious concrete pavement to stand as one of the best replacements for conventional pavements. In accordance with the traffic flow and bearing capacity, experiments are still in progress. From all the analysis performed various characteristics of PCP, it is absolutely and utterly clear that Pervious Concrete Pavement(PCP) can be termed as a Sustainable Pavement Technology(SPT)

References

[1] Prof Satish Chandra, Ph. D Director, CSIR-Central Road Research Institute Delhi, Flexible Pavement versus Rigid Pavement.
[2] Report on Pervious Concrete 2010 *American Concrete Institute* ISBN 9780870313646
[3] S Rajesh Kumar 2015 Characteristic Study on Pervious Concrete *International Journal of Civil Engineering and Technology (IJCIET)* Volume **6** Issue 6 Pp. 165-176 Article Id: 20320150606017.
[4] John T Kevern, Vernon R. Schaefer & Kejin Wang 2011 Mixture Proportion Development and Performance Evaluation of Pervious Concrete for Overlay Applications *Materials Journal* American Concrete Institute. **108** (4): 439–448
[5] Salunkhepatil S S , Patil O D , Dhemare A M , Jadhav S A, Bhirud SV 2018 A Review On Mix Design For Pervious Concrete *8th National Conference On Emerging Trends In Engineering And Technology,NCETET* ISBN: 978-93-87793-03-3
[6] M Harshavarthana Balaji , M R Amarnaath, R A Kavin, S Jaya Pradeep 2015 Design of Eco Friendly Pervious Concrete *International Journal of Civil Engineering and Technology (IJCIET)* Volume **6** Issue 2
[7] Dang Hanh Nguyen, Nassim Sebaibi, Mohamed Boutouil , Lydia Leleyter, Fabienne Baraud 2014 A Modified Method For The Design Of Pervious Concrete Mix *Construction and Building Materials* **73** 271–282
[8] Wanielista M, Chopra M, Spence J and Ballock C Hydraulic Performance Assessment of Pervious Concrete Pavements for Stormwater Management Credit Storm water Management Academy University of Central Florida **81** pp
[9] Leming M L, Malcolm H R and Tennis P D 2007 Hydrologic Design of Pervious Concrete EB **303** Skokie Illinois
[10] NRMCA 2004 Freeze Thaw Resistance of Pervious Concrete *Silver Spring* MD **17** pp.
[11] Report on Pervious Concrete 522R-10 American Concrete Institute Committee **522** Farmington Hills MI USA
[12] Karthik H Obla 2010 Pervious Concrete – An Overview *The Indian Concrete Journal*
[13] Vimy Henderson and Susan L Tighe 2011 Evaluation Of Pervious Concrete Pavement Permeability Renewal Maintenance Methods At Field Sites In Canada *Canadian Journal of Civil Engineering* DOI: 10.1139/l11-105

[14] Mohammadali Pishdadakhgari 2017 Pervious Concrete: Potential Advantages in Regions with High Precipitation *International Journal of Innovations in Engineering and Science*, Vol. **2** No.11

[15] John T Kevern, Liv Haselbach, Vernon R Schaefer 2012 Hot Weather Comparative Heat Balances in Pervious Concrete and Impervious Concrete Pavement Systems *Journal of Heat Island Institute International* Vol. **7-2**

[16] Hilal El-Hassan & Peiman Kianmehr 2018 Pervious Concrete Pavement Incorporating GGBS to Alleviate Pavement Runoff And Improve Urban Sustainability Road Materials and Pavement Design **19:1** **167** 181 DOI:10.1080/14680629.2016.1251957

[17] Youngmin Joung, Zachary C Grasley 2008 Evaluation and Optimization of Durable Pervious Concrete for Use in Urban Areas Research Report 08/167163-1 Southwest Region University Transportation Center Texas Transportation Institute

[18] Dale P Bentz 2008 Virtual Pervious Concrete: Microstructure, Percolation, and Permeability *ACI Materials Journal* Title no **105**-M35

[19] Malhotra V M 1976 Permeable Concrete – Its Properties and Applications *Journal of the American Concrete Institute* Vol. **73**. No.11. pp 628 – 644

[20] Kevern John Tristan 2008 Advancements In Pervious Concrete Technology *Retrospective Theses and Dissertations* **93**

[21] Rohini lomte 2018 A Review on Study and Analysis of Strength, Permeability and Void Ratio of Pervious Concrete *International Journal for Research in Applied Science & Engineering Technology (IJRASET)* ISSN: 2321-9653 IC Value: 45.98 SJ Impact Factor :6.887 Volume **6** Issue I

[22] Suraj F Valvi, Anil P Thoke, Abhijit A Gawande, Manoj B Godse, Prof D D Shelke 2017 Use Of Pervious Concrete In Road Pavement *International Journal of Advance Research in Science and Engineering* Vol. No.**6** Issue No. 03

[23] Darshan S Shah, Prof Jayeshkumar Pitroda, Prof J J Bhavsar 2013 Pervious Concrete: New Era For Rural Road Pavement *International Journal of Engineering Trends and Technology (IJETT)* – Volume **4** Issue 8

[24] Mary Vancura, Kevin MacDonald and Lev Khazanovich 2011 Structural Analysis of Pervious Concrete Pavement Transportation Research Record *Journal of the Transportation Research Board* DOI:10.3141/2226-02

A Study on the Consolidation and Permeability Behaviour of Soils of State of Tamil Nadu, India

S Jayalekshmi[1] and V Elamathi[2]

[1]Professor, National Institute of Technology, Tiruchirappalli, Tamil Nadu.
[2]Research Scholar, National Institute of Technology, Tiruchirappalli, Tamil Nadu.

Email: elamathivenkat@gmail.com

Abstract. In order to study the consolidation behavior, clay soil samples from 20 various locations of Tamil Nadu, India were collected and subjected to one dimensional consolidation tests. Coefficient of consolidation c_v, coefficient of volume change m_v, and coefficient of permeability k were plotted against consolidation pressure σ for all the soils. The results showed that the variation of c_v with pressure σ is not constant for all the soils. It was observed that various soils have shown various trends. Hence, the trends obtained in the study were compared with the literature. From the study, it was found that the variations in the trends were due to mineralogical effects. The consolidation behavior of soil is found to be influenced by clay mineralogy. Similarly k-σ, m_v-σ trends were also compared. The study shows that clay mineralogy has reasonable effects on the consolidation and permeability behavior of soil.

1. Introduction

Settlement occurs when a foundation is laid over the clay soil and leads to severe damages to the superstructure. Settlement is due to consolidation of soil beneath the foundation. Hence it is necessary to find the parameters causing the consolidation of soil. Consolidation is controlled by coefficient of consolidation c_v [4] in which c_v gives the rate of settlement of the soil. It is necessary to know the rate of settlement of clay for design purposes. The time required for a given amount of settlement is predicted using c_v. Since consolidation test is a time consuming process, many authors attempted to correlate c_v with index properties like activity, liquidity index I_L and plasticity index I_P [5], shrinkage index Is [6], liquid limit w_L [7,9], and Plasticity index I_P[8,9]. Coefficient of consolidation c_v is a function of coefficient of permeability k and coefficient of volume change m_v [4] which is given as:

$$c_v = k / \gamma m_v \quad (1)$$

As the consolidation pressure increases, the soil shrinks in volume, because of expulsion of water from the soil pores. The rate of expulsion of water depends on the permeability of soil k. Hence, the consolidation and permeability of soil is studied using the parameters coefficient of consolidation c_v, coefficient of volume change m_v, and coefficient of permeability k for the applied consolidation pressure σ. It is also observed that c_v-σ trends for all soils were not same. c_v of some soils were increasing and some were decreasing. Analysing the variations in the c_v-σ trends obtained in the laboratory for the clay minerals kaolinite, illite and montmorillonite and powdered quartz with water and CCl_4 as pore fluids, it was found that, the c_v increased with consolidation pressure σ for kaolinite and illite whereas it decreased for montmorillonite [11]. Variations in the c_v-σ trend from various literatures have been tabulated in Table 1. According toTable1, c_v obtained in the present study is strongly influenced by clay mineralogy. The soils with increasing trends in the c_v- σ plot may contain kaolinite and illite as dominant mineral and those with decreasing trend may have montmorillonite in greater amount.

Coefficient of volume change m_v and coefficient of permeability k decrease with increase in consolidation pressure σ [11]. Higher range of values for m_v was observed for montmorillonite (1x10-^{2} to 9x10^{-4} m^2/kN), whereas, illite (1.5x10^{-3} to 1.5x10^{-4} m^2/kN) and kaolinite (2x10^{-3} to 1x10^{-4} m^2/kN) showed lower range of values. Kaolinite (7x10^{-7}cm/s to 1x10^{-7} cm/s) showed higher range of values of k whereas montmorillonite (1x10^{-8} to 1x10^{-10} cm/s) showed lower range of values, illite (7x10-^{7} to 1x10-^{7} cm/s) lies between the two[11]. The present study has also shown decreasing trends for both m_v and k similar to [11].Volume change behavior of soils is strongly influenced by mineral composition [17].The influence of volume change behavior on c_v increases with increase in consolidation pressure σ [17].

Table 1. Variation in the c_v-σ trends from literature

Soil Type	**w_L**	**Dominant mineral**	**c_v-σ trend**	**Reference**
Kaolinite	49.0	Kaolinite	Increases	[15]
BC soil-1 Illitic soil	73.5 73.4	Montmorillonite Illite	Decreases Increases	[16]
Montmorillonite Illite Kaolinite	321 131 53	Montmorillonite Illite Kaolinite	Decreases Increases Increases	[11]
Group M-I Group I-Q Group K-I	M-486 I-72 K-77	Montmorillonite Illite Kaolinite	Decreases Increases Increases	[13, 14]
Bisaccia clay Marino clay	110 50	Smectite Kaolinite	Decreases Increases	[17]

The coefficients of permeability k of clays are controlled by both mechanical and physico-chemical factors [12]. The compressibility behavior of kaolinite, illite and powdered quartz is governed by mechanical factors and the compression behavior of montmorillonite is governed by physicochemical factors [11]. Mechanical factors denote size, shape, and the geometrical arrangement of the clay particles whereas physiochemical factors denotes i.e. diffused double layer [12]. Hence clay minerals like montmorillonite, kaolinite and illite plays a major role in consolidation behavior of clay. The individual effects of montmorillonite, kaolinite and illite on the consolidation behavior of soil were studied [11]. 40 different mixtures of montmorillonite, Kaolinite, Illite, and Quartz of various percentages were examined. The variation of compressibility and permeability of soil with the proportion of minerals and it was observed that compressibility of soil is heavily dependent on the mineral composition of soil specimen [13,14]. However, natural soil contains clay minerals in different proportions which are to be quantified in order to find the amount of dominant minerals in the soil.

Behavior of soil depends on the percentage of dominant mineral present in the soil. Each of the constituent clay mineral contributes to the macroscopic behavior. The proportions of the minerals in a soil sample are primarily responsible for the variations in the index properties, coefficient of consolidation, permeability, and the compression index [13].

2. Materials and experiments

Clay soil samples from 20 different locations of various districts of state of Tamil Nadu, India were collected in order to study the consolidation behavior of soil. Test pit method is used for collecting samples. Test pits are made at depths of 1 to 1.5 m and bulk, disturbed samples are collected from the pit. Then the soils are air dried, hand crushed and sieved in 425μ sieve and subjected to experimental tests in accordance with ASTM standards [1,2,3].

Soil samples are remolded at 1.1 times the liquid limit and compacted in three layers in the consolidation ring of 60 mm diameter and 20 mm height. The soil samples are subjected to one dimensional consolidation for the consolidation pressure range σ of 25 kPa, 50 kPa, 100 kPa, 200 kPa, 400 kPa, 800 kPa, 1600 kPa.

3. Results and discussion

All the soils have been tested for index properties [1,2] and consolidation properties [3] . Liquid limit of the soils tested ranges from 25 to 61% and the clay fraction varies from 19 to 64 %. Values of consolidation characteristics of c_v and m_v for all the soils are obtained from one dimensional consolidation tests. Coefficient of permeability k is determined from equation 1.

When coefficient of consolidation c_v is plotted against consolidation pressure σ, variations were observed in the trends of c_v-σ plot. Table 2 lists c_v-σ response and the dominant mineral in the soils of the present study predicted based on the findings in Table1. From Table 2, it can be seen that, liquid limit ranges from 25-30% for kaolinite, 29-40% for illite and 46-62% for montmorillonite dominated soils. Coefficient of volume m_v and coefficient of permeability k decreases with increase in consolidation pressure σ for all the soils tested.

Table 2. Predicted dominant mineral based on literature review for all the soil samples

Sample No.	W_L	USCS classification	c_v-σ response	Dominant mineral predicted based on literature review
S1	30	CL	Increases	Kaolinite
S2	62	CH	Decreases	Montmorillonite
S3	49	CH	Decreases	Montmorillonite
S4	25	ML	Increases	Kaolinite
S5	61	CH	Decreases	Montmorillonite
S6	60	CH	Decreases	Montmorillonite
S7	44	CH	Decreases	Montmorillonite
S8	46	CH	Decreases	Montmorillonite
S9	42	CH	Decreases	Montmorillonite
S10	29	CL	Increases	Illite
S11	37	CL	Increases	Illite
S12	46	CH	Decreases	Montmorillonite

S13	40	CL	Increases	Illite
S14	30	ML	Increases	Kaolinite
S15	32	CL	Increases	Illite
S16	40	CL	Increases	Illite
S17	32	CL	Increases	Kaolinite
S18	50	CH	Decreases	Montmorillonite
S19	58	CH	Decreases	Montmorillonite
S20	55	CH	Decreases	Montmorillonite

In order to study the variations clearly, three samples S4, S16 and S18 with predicted dominant mineral as kaolinite, illite and montmorillonite respectively are considered and analysed using c_v-σ, m_v-σ and k-σ plots. Figure 1 gives log of time versus compression curve for the consolidation pressure of 25 kPa for the considered soil samples S4, S16 and S18.

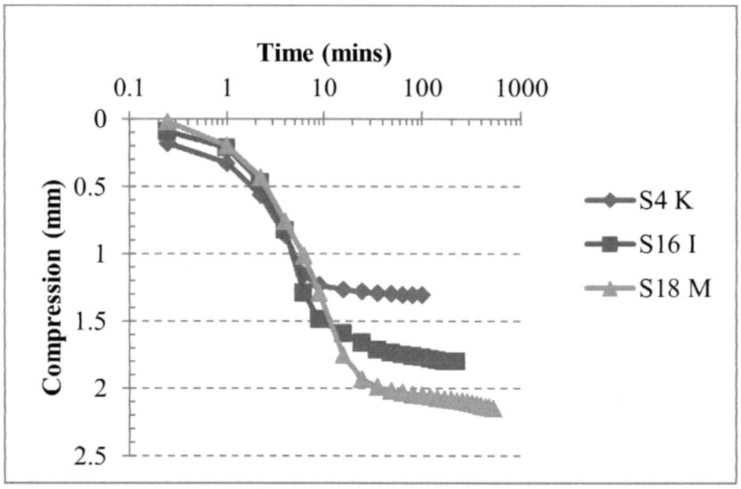

Figure 1. Time – compression curve for the consolidation pressure of 25 kPa for the soil samples S4, S16, S18 where K, I, M indicates Kaolinite, illite, montmorillonite dominated soil.

3.1. Factor affecting the variations in the c_v-σ plot

Soils S4,S14,S17,S1,S15,S10,S16,S11,S13,S7 have shown increasing trends whereas soils S9,S8,S20,S19,S5,S12,S2,S3,S18,S6 have shown decreasing trends. Table 1 show that the variation in the c_v-σ plot is due to the presence of dominant mineral in the soil. However soils will not contain only one mineral. But the soil exhibits the behavior of dominant mineral as shown in Table 1. The consolidation and swelling behavior of kaolinite and illite dominated soils were similar [13]. This is evident from figure 2 in which S4 and S16 has shown increasing trend and the dominant mineral is predicted as kaolinite and illite for S4 and S16 respectively. S18 with montmorillonite as dominant mineral has shown decreasing trend.

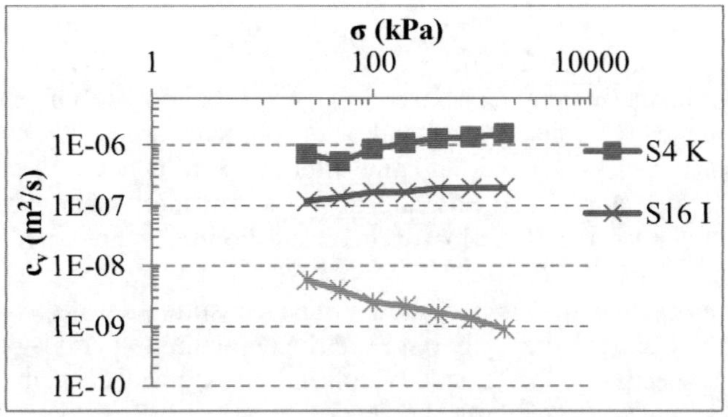

Figure 2. Variation of c_v with σ for the soil samples S4, S16, S18 where K, I, M indicates kaolinite, illite, montmorillonite dominated soil.

3.2. Volume change behavior of soil samples tested

As the consolidation pressure increases, water diffuses away from the soil, thus soil gradually takes up the pressure and shrinks in volume. Volume change is influenced by stress level [17]. The volume change is governed by two mechanisms. For kaolinite clay shearing resistance at inter particle level predominate and for montmorillonite clay, the volume change is caused by the diffuse double layer repulsive forces [19]. The shear resistance of kaolinite is a function of clay fabric and surface friction [20]. Figure 3 shows the response of m_v with σ for the soil samples S4, S16 and S18 for which dominant mineral is predicted as kaolinite, illite, and montmorillonite respectively. The plot obtained in the present study is similar to that obtained for kaolinite, illite and montmorillonite in [11]. Thus the predicted dominant mineral matches.

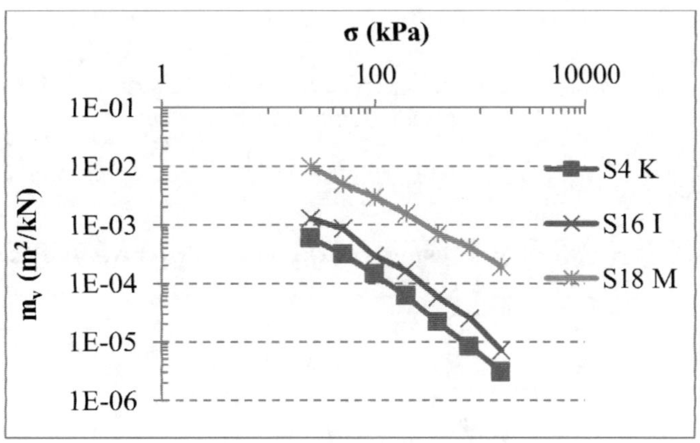

Figure 3. Variation of m_v with σ for the soil samples S4, S16, S18 where K, I, M indicates kaolinite, illite, montmorillonite dominated soil

3.3. Permeability behavior of soil samples tested

k-σ plot is obtained for soil samples S4, S16 and S18 as shown in figure 4. Coefficient of permeability k showed higher range of values for kaolinite, lower range of values for montmorillonite and illite lies between the two [11]. From figure 4, it is seen that higher range of values of k were observed for S4 and S16 with kaolinite and illite as dominant mineral and lower range of values were observed for S18 with montmorillonite as dominant mineral. The trend obtained in the present study is similar to [11]. Permeability of soil k depends on mechanical and physico chemical i.e., kaolinite and illite are greatly influenced by mechanical factors whereas montmorillonite is influenced by physico chemical factors [18].Thus, the behavior of kaolinite and illite are governed by short-range particle interactions due to grain size and shape whereas the behavior of montmorillonite is governed by long-range particle interactions due to diffused double layer [11].

Kaolinite and illite dominated soils exhibited similar consolidation behavior. This is evident from Fig. 4 in which S4 and S16 show increasing trend and the dominant mineral is predicted as kaolinite and illite for S4 and S16 respectively. The trends of S4 and S16 for all c_v-σ, m_v-σ, k-σ plots lies closer whereas more gap exists for S18. This indicates that kaolinite and illite dominated soils exhibit similar behavior (short range particle interaction), but montmorillonite exhibits physico chemical behavior (long range particle interaction).

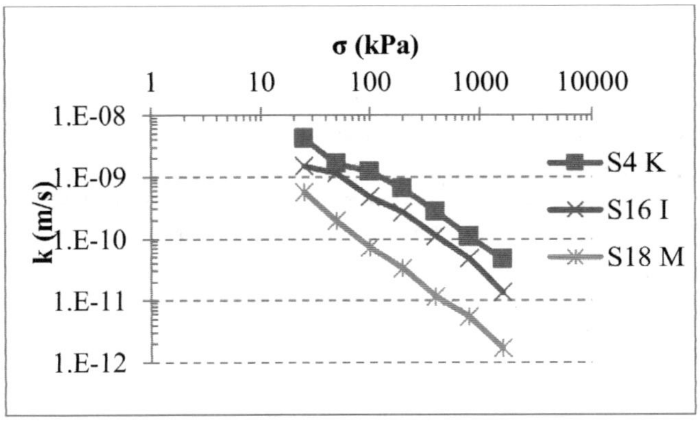

Figure 4. Variation of k with σ for the soil samples S4, S16, S18 where K, I, M indicates kaolinite, illite, montmorillonite dominated soil

3.4. Relationship between c_v and Index properties

Table 3. Correlations for c_v using index properties from various literatures

S.No.	Equation	R^2	Source
1.	$c_v = \frac{3}{100(I_s)^{3.54}}$ (m^2/s)	0.94	[16]
2.	$c_v = -4\times10^{-9} w_L + 4\times10^{-7}$ (m^2/s)	0.8298	[10]
3.	$c_v = 0.6155 - 0.0183 I_p$ (m^2/yr)	0.994	[8]

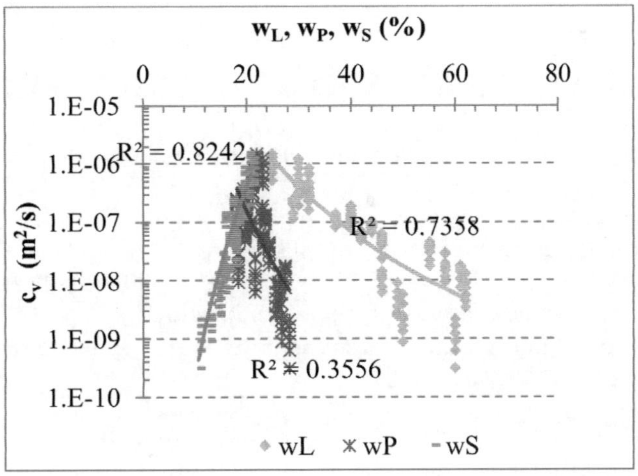

Figure 5. Variation of c_v with w_L, w_p, w_S for all pressure ranges.

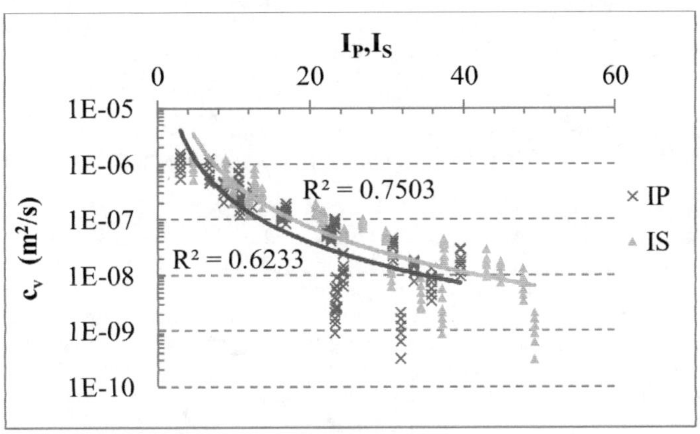

Figure 6. Variation of c_v with I_p, I_S for all pressure ranges.

In order to calculate c_v, correlations between c_v and index properties given by many literature are listed in Table 3. Index properties w_L, w_p, w_S, I_P and I_S are plotted against c_v for all consolidation pressure range σ of 25 kPa, 50 kPa, 100 kPa, 200 kPa, 400 kPa, 800 kPa, 1600 kPa as shown in figure 5 and figure 6. For the soil samples tested, shrinkage limit correlates better with c_v than any other parameter, and with R^2=0.824 for which correlation is as follows:

$$c_v = 3E^{-21} I_s^{10.86} \quad (2)$$

4. Conclusion

- The present study shows that shrinkage limit has a better correlation with coefficient of consolidation than the liquid limit, plastic limit and shrinkage index.
- Variation in c_v-σ, m_v-σ, k-σ plots were observed for all the soils and it was found that the dominant clay mineral present in the soil sample is responsible for the variation.

- The dominant clay mineral is predicted for the soil samples of the present study on the basis of literature review (Table 1).
- The trends obtained for c_v-σ, m_v-σ, k-σ with kaolinite, illite and montmorillonite dominated soils matches well with [11].
- The response of c_v-σ, m_v-σ, k-σ for kaolinite and illite dominated soils were similar as stated by [13].
- Works were carried out on pure minerals and mineral mixtures (controlled condition) [11,14]. But the present study involves testing of natural soil samples.
- The proportion of each mineral present in the soil varies for all the soil. Each mineral amount has their individual effect on the consolidation behavior of soil.
- Hence quantification of minerals is necessary in order to study effect of individual mineral on the consolidation behavior of the soil.
- Here in this study only three clay minerals were considered. But the presence of other clay minerals wills also have their significant effect on the behavior of soil.
- In summary, it is concluded that when combinations of mineral exists in sampled soil, especially containing clay, consolidation behavior and permeability behavior at various consolidation pressure, can be indicative of the dominant mineral present.

Acknowledgement
The authors acknowledge with thanks, for the facilities offered, by the Department of Civil Engineering, National Institute of Technology, Tiruchirappalli, India. Also, grateful to the Ministry of Education (MoE), India, for providing scholarship, to carry out the research work.

References
[1]ASTM D4318 - 17e1 2017 - Standard test methods for liquid limit, plastic limit, and plasticity index of soils.
[2] ASTM Standard D427-04 2007 Determination of shrinkage limit using mercury.
[3] ASTM D 2435-96 1996 Standard test method for one-dimensional consolidation properties of soils.
[4] Terzaghi K and Peck R B 1967 Soil mechanics in engineering practice John Wiley and Sons (New York).
[5] Carrier W D III 1985 Consolidation parameters derived from index tests *Geotechnique* **35(2)** 211– 213.
[6] Sridharan A and Nagaraj H B 2004 Coefficient of consolidation and its correlation with index properties of remolded soils *Geotechnical Testing Journal* ASTM **27(5)** 1-6.
[7] Asma Y, Al Tae'e, and Abbas F Al- Ameri 2011 Estimation of relationship between coefficient of consolidation and liquid limit of middle and south Iraqi soils *Journal of engineering* **17(3)** 430-440.
[8] Kok Shien N, Yee Ming C and Nur Izzati A L 2018 Prediction of consolidation characteristics from index properties E3S *Web of Conferences* **65** 1-5.
[9] Solanki C H 2011 Quick settlement computation of shallow foundation using soil index and plasticity characteristics *Pan-Am CGS Geotechnical conference* 1-5.
[10] Priyadarshini Devi S, Rambha Devi K, Prasad D S V and Prasada Raju G V R 2015 Study on consolidation and correlation with index properties of different soils in Manipur valley *International Journal of Engineering Research and Development* **11(5)** 57-63.
[11] Robinson G R and Allam M M 1998 Effect of clay mineralogy on coefficient of consolidation *Clays and Clay Minerals* **46(5)** 596-600.
[12] Olson R E and Mesri G 1970 Mechanisms controlling the compressibility of clays *Journal of the Soil Mechanics and Foundations Division* **96 (6)** 1863-1878.
[13] Tiwari B and Ajmera B 2011 Consolidation and swelling behavior of major clay minerals and their mixtures *Applied Clay Science* **54** 264–273.
[14] Tiwari B and Ajmera B 2011 Consolidation and permeability of clay minerals - Expansive to Non-expansive Geo-Frontiers ASCE 2414-2423.
[15] Sridharan A, Sivapulliah P V and Stalin V K 1994 Effect of short duration of load increment on the compressibility of soils *Geotechnical Testing Journal* ASTM **17** 488-496.

[16] Sridharan A and Nagaraj H B 2004 Coefficient of consolidation and its correlation with index properties of remolded soils *Geotechnical Testing Journal* **27(5).**
[17] Di Maio C, Santoli L and Schiavone P 2004 Volume change behaviour of clays: the influence of mineral composition, pore fluid composition and stress state *Mechanics of Materials* **36** 435–451.
[18] Mesri G and Olson R E 1971 Mechanisms controlling the permeability of clays *Clays and Clay minerals* **19** 151-158.
[19] Sridharan A and Venkatappa Rao G 1973 Mechanisms controlling volume change of saturated clays and the role of the effective stress concept *Geotechnique* **23 (3)** 359-382.
[20] Sudhakar M Rao and Sridharan A 1985 Mechanism controlling the volume change behavior of kaolinite *Clays and Clay Minerals* **33(4)** 323-328.

IOP Conf. Series: Materials Science and Engineering 1006 (2020) 012018 doi:10.1088/1757-899X/1006/1/012018

Level of Service of Urban and Rural Roads- *A Case study in Bhimavaram*

Pala Gireesh Kumar[1], Sathya Ranjan Samal[2], L. Prasanthi[3], V. Bhavitha[4], J. Mounika Devi[5]

[1]Associate Professor, Department of Civil Engineering, Shri Vishnu Engineering College for Women (Autonomous), Vishnupur, Bhimavaram-534202, West Godavari Dist., India
[2]Associate Professor, School of Civil Engineering, KIIT Deemed to be University, Bhuvaneswar, India
[3,4,5] U.G Students, Department of Civil Engineering, Shri Vishnu Engineering College for Women (Autonomous),Vishnupur, Bhimavaram-534202, West Godavari Dist., India

E-mail: bhavithavetukuri@gmail.com; palagireeshk@gmail.com; prasanthilokam277@gmail.com

Abstract. With rapid growth of population in urban and rural areas, purchase of motorized vehicles is also grown and the demand for proper transportation network which can give good compatibility to the road users has become a major challenge in India. Level of Service (LOS) is one such parameter in terms of compatibility that gives a quality measure for the operational conditions within a traffic stream, i.e., generally in terms of service that is provided by the road to the user. This paper investigates on the existing level of service models for urban and rural roads given by researchers globally. To determine the Level of service of a particular road section, there are few commonly used methods like, Cluster Analysis, Genetic Algorithm, Fuzzy Set Theory and Neural Network etc., which are used, were discussed and reported in this paper. A new approach for finding level of service was introduced, i.e., using volume to capacity ratio (v/c), average speed of the vehicles, percentage speed reduction is also discussed and presented. This paper brings out the collective information on LOS and its findings using analytical models which can estimate/predict compatibility levels of various road users in heterogeneous traffic conditions in urban environment. The objective of this project is to study Los for various selected roads were obtained and discussed. Research work was carried in Bhimavaram (India) town, Ramayanpuram, Jaganadhapuram villages. Both study area-1,3 was observed to have LOS-D grade whereas study area-2,4 was observed to have LOS-E grade. Some of the remedial measures were given in order to increase its Los grades towards better satisfaction of traffic conditions.

1. Introduction

Over the past decade, use of motorized vehicles has been significantly increased with rapid growth of urbanization in India. This robust increase in population has given a challenge for the development of modern society which has turned into a major concern in metropolitan cities in developing countries like India. At present, the world population crosses seven billion and population residing at urban areas are nearly half of seven billion counts which is expected to raise up to 60 % over worlds entire population by the end of 2025 [5]. Indian cities are well connected with urban roads and rural roads everywhere but, there exists poor transportation network/connectivity with negligible intra and intercity facilities [5]. Hence, traffic flow in heterogeneous conditions is highly complex and is difficult to predict the flow behaviour on urban roads. Driver's comfort, convenience, traffic volume, lane width, grade type, geometric design, travel delay and safety are the major concerns on Indian urban roads which are to be taken care. Monitoring traffic volume and level of service which represents quality transport has become indispensable [13].

In recent years, traffic congestion and decreasing level of service has become major issues in most of the metropolitan cities and results its impact on urban economy and its environment [1], [9], [11]. Good transportation facilities are needed to be provided as it plays a major role in growing India's economy [4], [13] [14], [15]. From past studies, it was found that, obtained traffic volume plays a vital role in determining the existing conditions and even possible to predict the future traffic volume conditions [9], [11]. There are certain factors which are responsible for decreasing level of service and traffic congestion i.e., speed and travel time, traffic interruptions and restrictions. This downfall of level of service has become a serious threat as it creates difficulty for vehicle manoeuvring. Similarly, traffic congestion also creates air pollution and noise pollution which is caused by the delays in traffic jams. This traffic congestion and poor level of service on Indian roads have huge impact on its economic growth [1], [9]. There are few commonly used methods like, Cluster Analysis, Genetic Algorithm, Fuzzy Set Theory and Neural Network etc., which are used to determine the Level of service of a particular road section.

This paper aims to develop sound understanding of Level of Service (LOS) on urban and rural roads in Indian conditions from the previous studies conducted. The detailed information on various parameters that are used to define LOS are also reported and presented. The classifications of LOS grades along with its operating characteristics are also provided.

2. State of the art and methodology

This section gives sound knowledge on level of service concept under heterogeneous traffic conditions on Indian urban roads. A detail literature study was carried out and presented an outline of various researchers' findings.

2.1. Literature review

B Robin et al (2016) carried at research work on finding level of service for urban areas under free flow conditions, load factor, peak hour factor and volume to capacity ratio respectively. The classification of LOS and its operating characteristics were presented in this paper which is mostly based on run time and delay time. Concludes that, LOS for urban roads are unlike similar to that of rural roads [5].

S Biswas et al (2016) field studies were carried out using video graphic technique to observe and collect traffic flow characteristics on urban arterials in Kolkata metropolises city. Companion has been made among the collected data like, mean speed, standard deviation and coefficient of variation with standard values. Kolmogorov -Simonov (K-S) test was performed to check compatibility of the modelled distribution curves. Clustering algorithm was used to classify PSR into number of sub groups. Study concludes that, the metropolitan's study operational conditions on roads falls under LOS F; which indicates necessity of improvement on LOS with immediate effect [16].

F M John and Al Werner (1990) an attempt hasbeen made to investigate the traffic performs slow moving vehicles rather concentrating on speed to measure the traffic performances. Supply and demand were considerate to determine overtaking ratios which is a measure of LOS anduse to compare it with 1985 (HCM). Concludes that, effect of passing lanes on the level of service can be measured with the help of over-ratio concept [12].

C R Patel and G J Joshi (2012)this paper aims to study the behaviour of heterogeneous traffic speed and flow rate at signalized intersections in urban article roads of Gujarat state. Data was collected manually and using video graphic technique for six lane divided road segments in Surat, Gujarat state of India, multi regime speed flow relation was developed with the data from field survey and obtained LOS based on v/c ratio on cluster analysis approach. Concludes that, road way section capacity would be calculated based on traffic, control and road way conditions. LOS with respect to capacity and volume measures the traffic quality which can help further to improve long and short-term infrastructure facilities for the existing scenario by conditions [8].

B Someshwar Rao et al (2017)focused on estimating the critical gap using various methods like Wu harder, Raff's and sum of the proposed methods (by IIT Rookie). The study area compresses of three places in Vishakhapatnam and peak hour traffic data, traffic volume vehicle entry time, vehicle exit time, was collected with video graphic technique. Conclude that harder method gives higher critical gap values when compared with the other method adopted [6].

S Chatterjee et al (2017) platoon parameters such as follower rate; follower density, percentage follower and platoon rate were used to study the level of service for multi lane highway in India. K-means clustering technique is used to classify traffic flow, i.e. free flow, stable flow and constrain flow. Stream parameters like flow rate, density, v/c ratio were examined the all six sections which was considered for the study taking time head way into count as a parameter to define platoon criteria. A new approach has been adopted which is based on platoon parameters in order to predict LOS further in the research [17].

B Prasanta Kumar Bhuyan and K V Krishna Rao (2011) conducted survey corridors of Kolkata city and data like average travel speed was collected using GPS were fitted on mobile velocity. The probe vehicle was fitted with tremble Geo-xt GPS receiver, capable of collecting speed data information continuously using HIC methodology [2]. Concludes speed ranges on urban streets in Indian contexts are different from corresponding values mentioned in HCM (2000).

C. K. Haritha et al (2017) a relation between flow and speed have been developed and it can be used to estimate capacity of road way. Video graphic method is adopted to collect the data such as density and speed averages of vehicles which are used to measure LOS. Results concludes that, the obtain capacity value by static PLU method is 11% higher than the recommended value by IRC and decrease in carriageway width results increase in level of service [7].

D V Jitendra Bhai et al (2016) spot speed and traffic volume studies were carried out traffic data and they're by providing recommendation towards congestion mitigation that ultimately results in improving existing level of service [10].

B R Marwah and Bhuvanesh Sigh (2000) conducted studies using simulation models for urban roads under heterogeneous traffic conditions in Kanpur metropolitan city. Calibration and validation havebeen carried out for traffic system. Road section lengths and roadway characteristics were used to design simulation experiments. Results conclude that, level of service is classified into four groups which can be used to identify deficiencies and helps to plan for alternative improvement measures [3].

A Patel and K Bhatt (2017) level of service on different road segments in urban areas are well investigated and presented a critical review from the previous studies. This paper highlights the basis for LOS and different categories which are given by HCM. The various definitions, congestion parameters, factors affecting LOS and factors affecting capacity and congestion affecting LOS are well addressed. The study concludes that, level of service various from country to country and have differentmethodologies to measure it [1].

S Raji and A Jagannatham (2017) correlation between level of service and capacity is well addressed in this paper. Level of service concept and its classification from 'A' to 'F' has been explained along with graphical representation. Some of the inputs like evolution in level of service, LOS and safety, road traffic influences, heterogeneous traffic conditions, LOS and road user's data were used [18].

D B Qadr et al (2019) conducted video graphic surveys on 60-M ring-road in Erbil city, Iraq to collect data for traffic volume and spot speed of vehicles. Results shows that, average vehicle speed is about 19 km/hr which is less than the designed speed of 60 km/hr that leads LOS for that particular road segment is at 'F' category. Peak hour factor is also calculated and obtained 0.97 which exceeds the range of typical values that are given for urban areas. Concludes PHF can be improved with LOS and by solving traffic congestion issue which would be a result of design and construction of bridge ramps [9].

E N Joseph and M S Nagakumar (2014) different parameter types like LOS, capacity, vehicleto capacity ratio, average delay time, peak hour, average journey time were studied and are well addressed in this paper. Reconnaissance survey is carried out on road segments along with field surveys for traffic volume and for spot speed of vehicles [11].

Sanjay Kumar Singh (2012) mainly focused on urban transportation problems like pollution, traffic, accidents and congestion etc. the challenges in India over urban transportation such as transport availability, vehicular growth in metropolitan cities, number of restricted motor vehicles and private vehicles in India has been briefly reported. Information about vehicular emission, road safety and its effect on health as well as policy measures for urban transportation were provided [19].

2.2. Study corridors and Objectives

Four different study corridors in and around Bhimavaram, Andhra Pradesh, India were chosen to define LOS for different types of roads. In which two corridors come under urban roads categories (Refer Fig. 1, Fig. 2) and two corridors comes under Rural roads categories (Refer Fig. 3, Fig. 4). Bhimavaram is densely populated town with different type of transport network. After the result is obtained it is easy to know how LOS will be vary for urban and rural roads. Section length of 21 meters was considered.

The objectives of this project are to find volume of vehicle in each corridor, to find difference in composition in peak hour and non-peak hour, to determine Average speed of vehicles in each corridor, to find difference in LOS grades for urban and Rural areas.

Table: 1 Road Characteristics

S.No	Name of Road	Class of Road	Lane Characteristics	Length of Road (m)
1.	Vishnu college road	Sub Arterial	2 Lanes (Two Lane)	2900
2.	Undi Bypass road	Sub Arterial	2 Lanes (Two Lane)	2200
3.	Ramayanapura m	Sub Arterial	2 Lanes (Two Lane)	1300
4.	Jaganadhapuram	Sub Arterial	2 Lanes (Two Lane)	1500

Fig. 1 Vishnu College Road (Corridor-1)

Fig. 2 Undi Bypass Road (Corridor- 2)

Fig. 3 Ramayanapuram Village Road (Corridor-3) **Fig. 4** Jagannadapuram Village (Corridor-4)

2.3. Methodology

The vehicles were classified into four categories: motorized two wheelers, cars/Autos, heavy vehicles, pedestrian. In this project LOS grades were defined mainly by using V/C ratio and Average speed method which were possible in selected corridors. There are many methods like cluster analysis, Percentage speed reduction, Fuzzy set theory and many more to define LOS of roads but these two methods were common and best methods to define LOS in suburban and rural areas. All four corridors were free from pedestrian activities, bus stop, vehicles parking.

3. Level of service (LOS) concept

The term level-of-service has been introduced by Highway capacity Manual (HCM) which represents the level of facility a user can experience from a road under various operating characteristics and traffic volumes. The Highway capacity manual is a publication National Academics of Science which contains concepts, guidelines and procedures for knowing capacity and quality of service of highway facilities. Capacity gives a quantitative measure of traffic, whereas Level of service gives a Qualitative measure. Capacity of the road could be constant whereas the actual flow rate will be different for different days, times in a day itself. Depending upon v/c ratio and travel speed, HCM has defined six Level of Services i.e. A to F (Refer from Fig. 5 to Fig. 10). A represents the best operating conditions and F represents worst condition.

i) LOS A: This grade represents best operating conditions for a user with free flow of traffic and with an average speed of vehicles obtaining 90% of its free flow speed. Volume to capacity ratio for this grade is less than 0.125.

ii) LOS B: This grade indicates stable flow of traffic and users can go in their desired speed. The volume to capacity ratio for this grade lies between 0.125 and 0.276 and the average speed is 70 % to 80 % of free flow speed.

iii) LOS C: This grade also indicates stable flow of traffic. The average speed varies from 50 % to 60 % leads to gradual decrease in level of comfort. The volume to capacity ratio for this grade is between 0.276 and 0.479.

iv) LOS D: This grade indicates nearly an unstable flow of traffic with high density of vehicles. The average speed of vehicles is about 40 % 50 % and the drivers are restricted to go in their desired speed. The volume to capacity ratio for this grade lies between 0.479 and 0.715 that indicates poor level of comfort.

v) LOS E: This grade cannot allow drivers to go with their desired speed as it indicates high density of vehicles with average speed of vehicle dropped to 30 % to 40 %. The volume to capacity ratio for this grade lies between 0.715 and 1.0 which gives extremely poor comfort and convenience to passengers.

vi) LOS F: This grade indicates the amount of traffic is very high and the average speed is dropped to 25 percent to 35 percent indicating the driver's comfort, patience and convenience as very poor. The volume to capacity ratio is greater than 1.000 for this grade.

Fig. 5 Level of Service Grade A

Fig. 6 Level of Service Grade B

Fig. 7 Level of service Grade C

Fig. 8 Level of Service Grade D

Fig. 9 Level of service Grade E

Fig. 10 Level of Service Grade F

3.1 Level of service based on volume to capacity (v/c) ratio and passenger speed reduction (PSR)
This study emphasis on the level of service for urban and rural roads under heterogeneous traffic conditions in India. Critical findings of Level of service based on Characteristics for urban and suburban arterials, Average speed, volume to capacity ratio and percentage speed reduction are well addressed and the various ranges and their LOS are given (Refer Table 2 to Table 5) respectively. Methods to find parameters required for LOS determination are also discussed.

Table 2 LOS Characteristics for urban and suburban Arterials

S. No.	LOS	Operating Characteristics
1	A	✓ Average overall travel speed is 50 kmph or more ✓ Free flow ✓ v/c ratio is 0.6 or less and factor at intersections is 0.0 ✓ Peak Hour Factor is 0.7 or less.
2	B	✓ Average overall travel speed is 40 kmph or more ✓ Stable flow but negligible delays ✓ v/c ratio is 0.7 or less and load factor at intersections is 0.11 or less ✓ Peak Hour Factor is 0.8 or less.
3	C	✓ Average overall travel speed is 30 kmph or more ✓ Stable flow and acceptable delays ✓ v/c ratio is 0.8 or less and load factor at intersections is 0.3 or less ✓ Peak Hour Factor is 0.85 or less.
4	D	✓ Average overall travel speed is 25 kmph or more ✓ Unstable flow and tolerable delays ✓ v/c ratio is 0.9 or less and load factor at intersections is 0.7 or less ✓ Peak Hour Factor is 0.9 or less.
5	E	✓ Average overall travel speed is 25 kmph or more ✓ Unstable flow & intolerable delays ✓ Congestion and load factor at intersections is 1.0 or less ✓ Peak Hour Factor is 0.95 or less.
6	F	✓ Average overall travel speed is 15 kmph or less ✓ Forced flow & Jammed conditions ✓ v/c ratio exceeds 1.0 and overloaded intersections

Table 3: LOS based on Average travel speed of vehicles on urban roads

S. No.	Urban street class	1	2	3	4
1	LOS A	>72	>59	>50	>41
2	LOS B	>56-72	>46-59	>39-50	>32-41
3	LOS C	>40-56	>33-46	>28-39	>23-32
4	LOS D	>32-40	>26-33	>22-28	>18-23
5	LOS E	>26-32	>21-26	>17-22	>14-18
6	LOS F	<26	<21	<17	<14

IOP Conf. Series: Materials Science and Engineering 1006 (2020) 012018 doi:10.1088/1757-899X/1006/1/012018

Table 4: LOS based on PSR

S. No.	Level of service	Percentage Speed Reduction (PSR)
1	LOS A	≤ 13
2	LOS B	>13-24
3	LOS C	>24-33
4	LOS D	>33-41
5	LOS E	>41-50
6	LOS F	>50

Table 5: LOS based on V/C Ratio

S. No.	LOS	v/c ratio
1	A	<0.125
2	B	0.125-0.276
3	C	0.276-0.479
4	D	0.479-0.715
5	E	0.715-1.000
6	F	>1.000

Table 6: Methods to find parameters required that defines LOS

S. No	Author	Model	Purpose
1	Abdullah Ahmad (IIT Rookie)	Critical Gap method	To calculate critical gap
2	Mentioned in HCM 2000	Capacity of specific movement method	To calculate capacity of a specific movement at intersections.
3	Subhadip Biswas, Bhupendra Singh, ArpitaSaha	Probability Density method	To calculate probability density function
4	Rousseau (1987)	Silhouette method	Silhouette index for i^{th}member

3.2 Theoretical capacity

Capacity is maximum no. Of vehicles that can pass a point in lane in one hour. The formula for theoretical capacity is given below,

Theoretical capacity(C)=1760*V/I

Where V=constant vehicular speed expressed in miles per hour

I=intra vehicular lead expressed in yards

4. Results and discussions

Field studies were carried out manually i.e. for volumetric survey no.of vehicles passed through that corridor in specific time was counted and noted.

4.1. Volume study in peak hour and non-peak hour

Both peak hour and non-peak hour traffic composition for selected four corridors were shown below in the form pie charts. Traffic composition was taken by manual method. In all four corridor's maximum vehicle composition is occupied by Bikes above 50% whereas non-motorized vehicles such as bicycles and pedestrian occupy less composition below 7% (Refer Fig. 11 to Fig. 14).

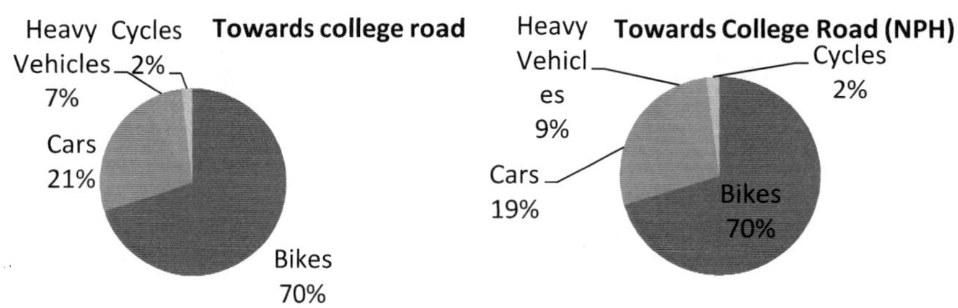

Fig. 11 (a) During peak hour Fig. 11 (b) During Non-peak hour

Fig: 11 Traffic volume at Vishnu college road

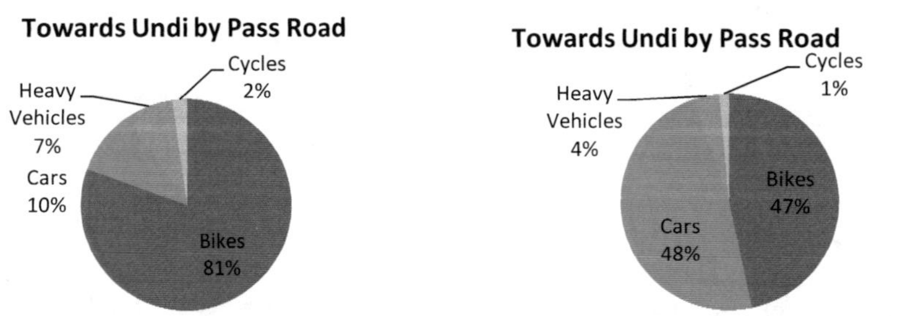

Fig. 12 (a) During peak hour Fig. 12 (b) During Non- peak hour

Fig: 12 Traffic volume at Undi Bypass road

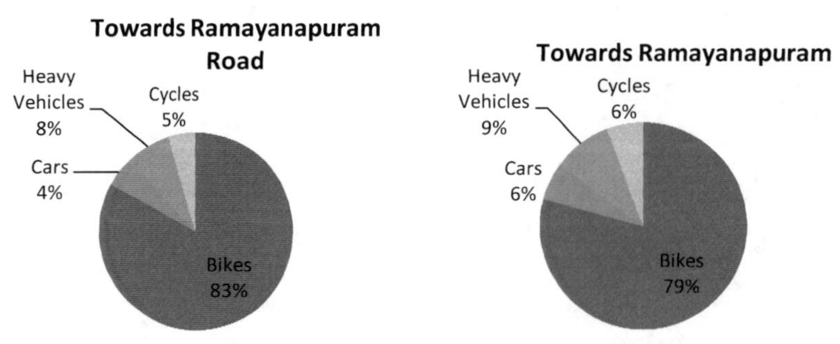

Fig. 13 (a) During peak hour Fig. 13 (b) During Non-peak hour

Fig: 13 Traffic volume at Ramayanapuram road

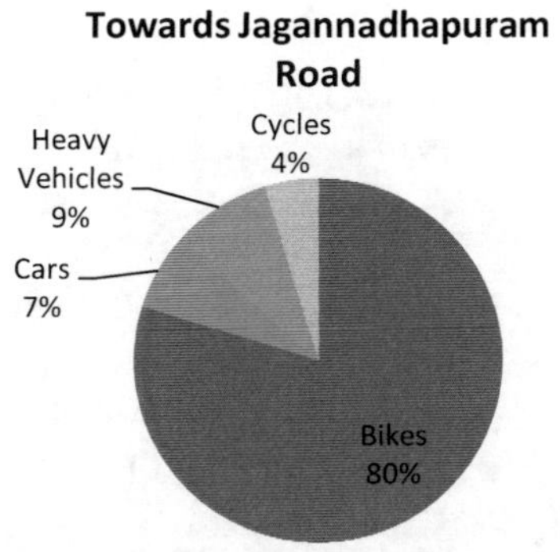

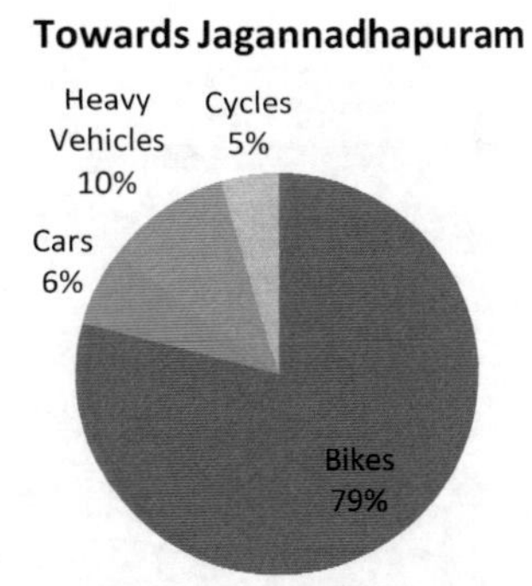

Fig. 14 (a) During peak hour Fig. 14 (b) During Non-peak hour

Fig. 14 Traffic volume at Jaganadhapuram road

4.2. Average speed studies

Average speed studies were carried out along a selected study Corridor by manual method i.e. using stop watch method. These studies were conducted in both peak and non-peak hour. Average speeds for all selected four corridors were obtained from speed distribution curves (Speed vs Frequency). In Corridor-1 maximum no. of vehicles travelled at the speed of 45-50 kmph (Refer Fig. 15). In Corridor-2,3,4 the maximum vehicles travelled at a speed of 25-30 kmph (Refer Fig. 16 to Fig. 18).

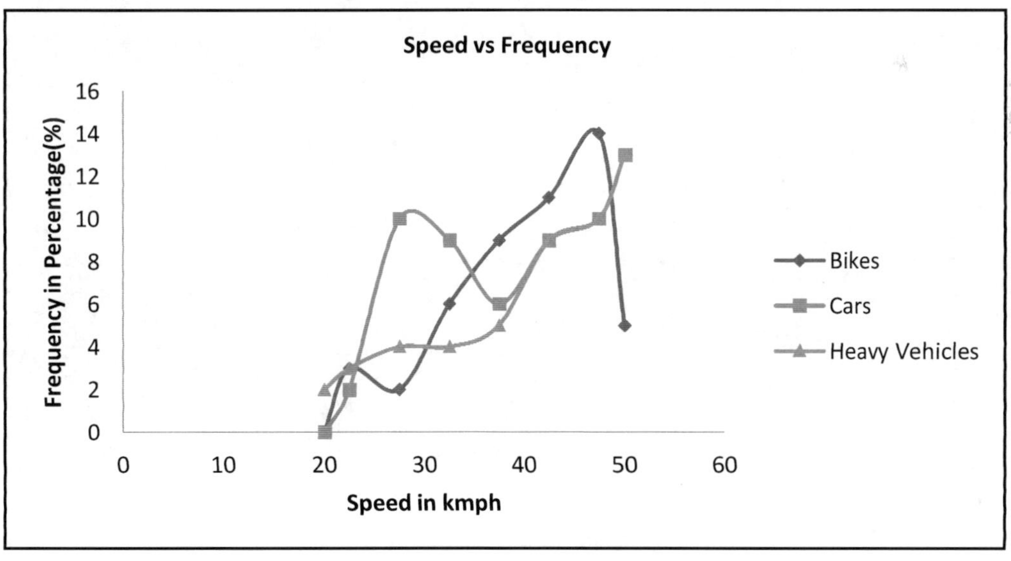

Fig. 15 (a) Average speed of vehicles at corridor-1 in peak hour

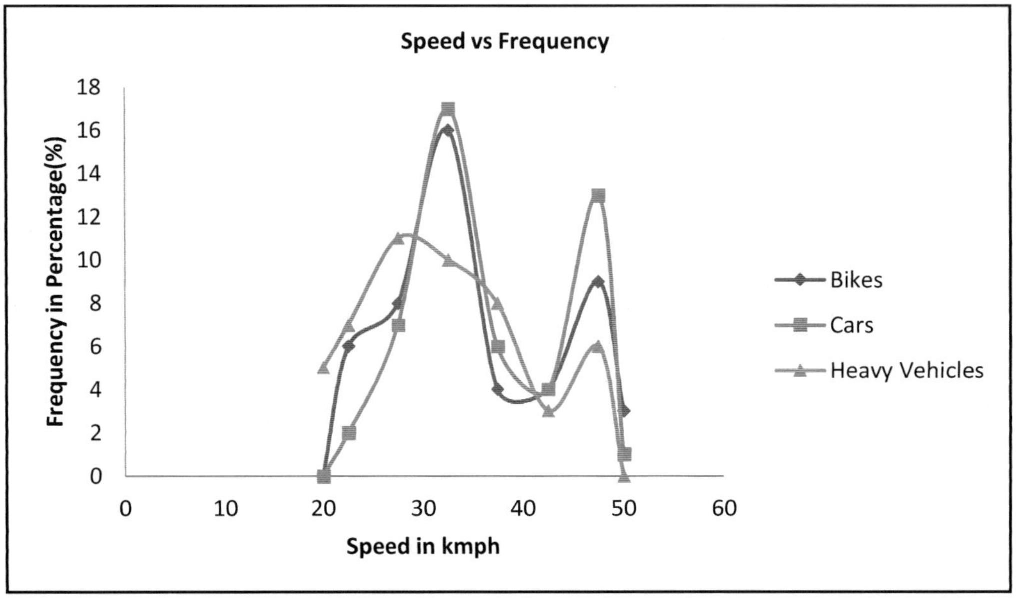

Fig. 15 (b) Average speed of vehicles at corridor-1 in non-peak hour

Average speed of vehicles travelled in corridor-1 was 31.65 kmph. Speed of vehicles travelled in Corridor-1 was higher than remaining selected study corridors.

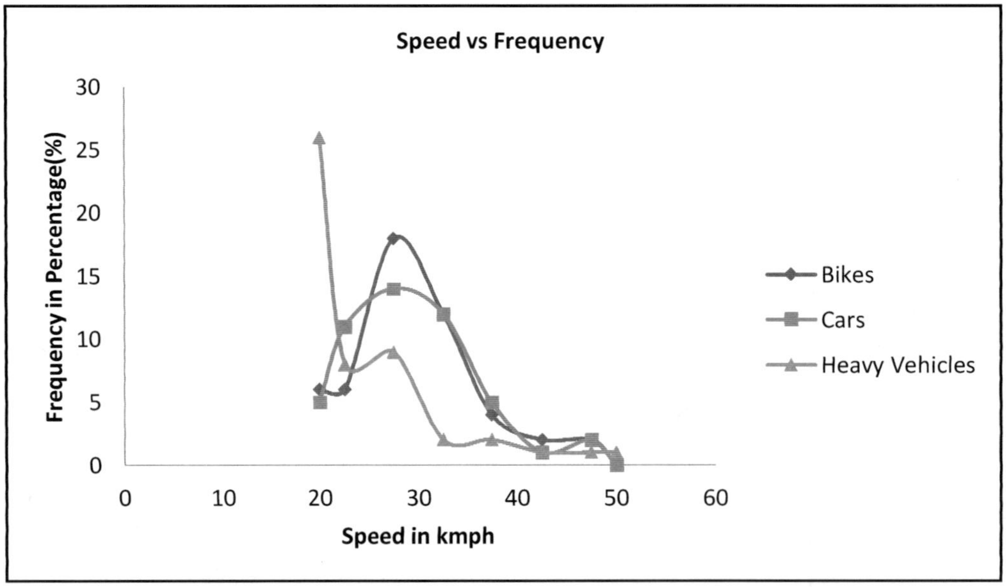

Fig. 16 (a) Average speed of vehicles at corridor-2 in Peak hour

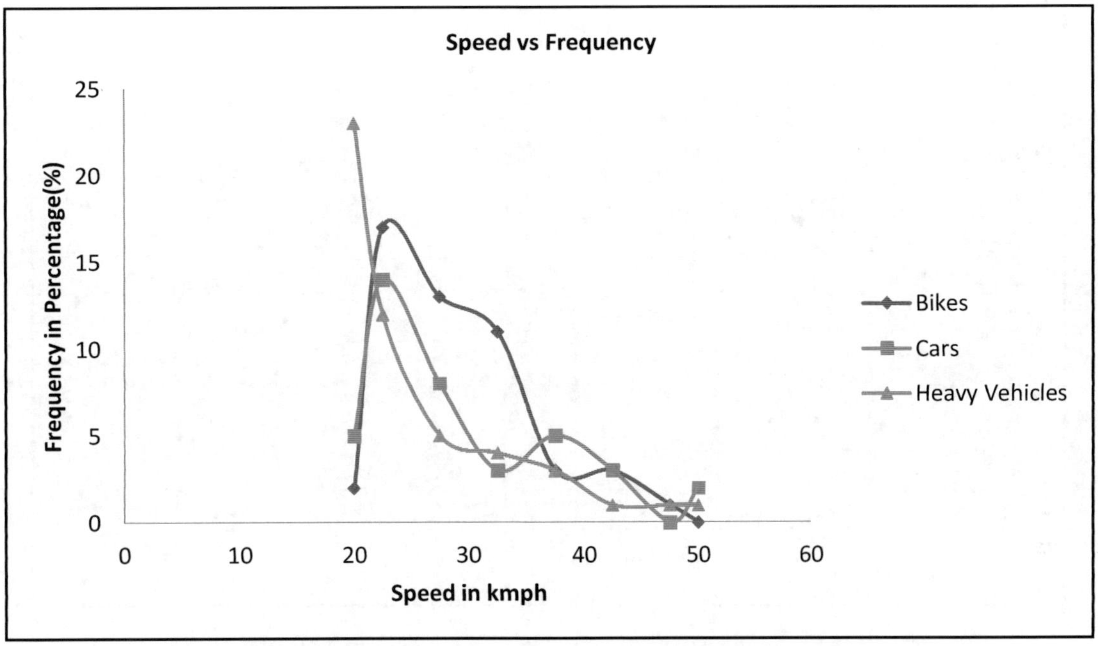

Fig. 16 (b) Average speed of vehicles at corridor-2 in Non-Peak hour

Average speed of vehicles travelled in corridor-2 was 24.46 kmph. The Average speed of vehicles travelled in this Corridor was less than remaining selected corridors.

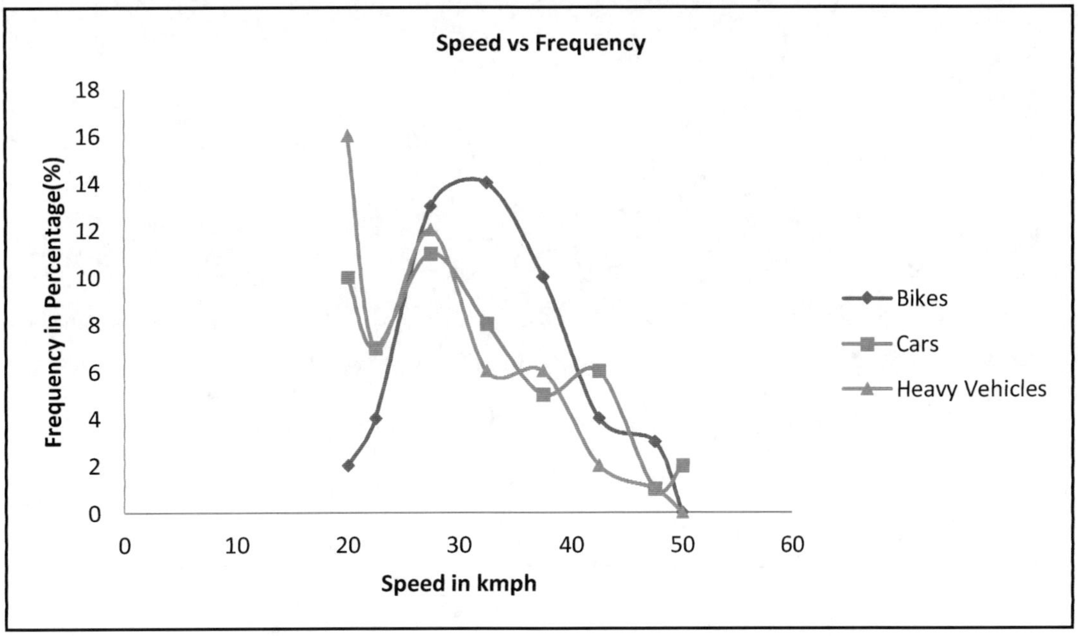

Fig. 17 (a) Average speed of vehicles at corridor-3 in Peak hour

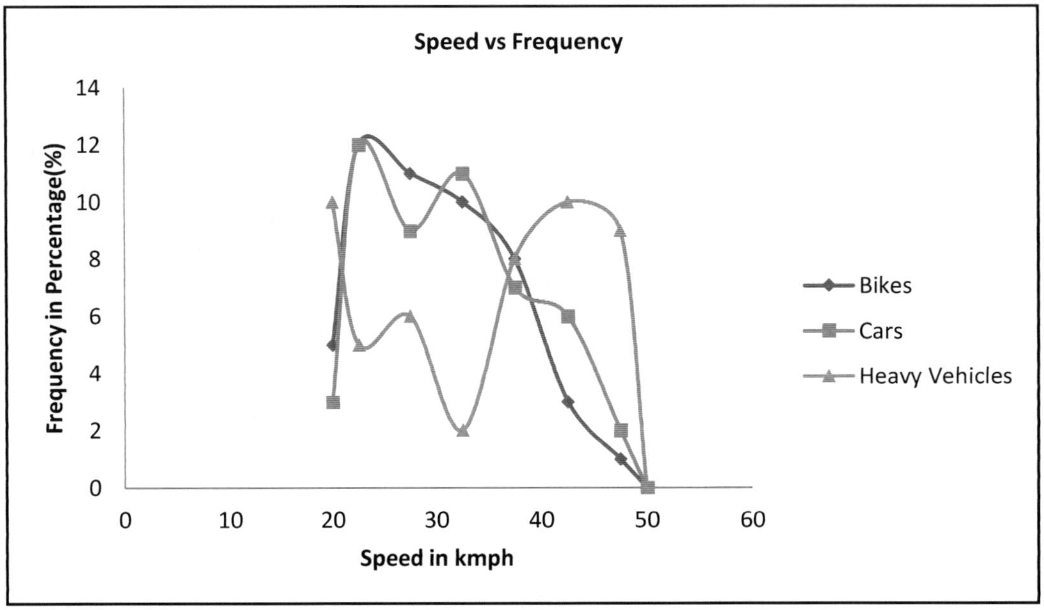

Fig. 17 (b) Average speed of vehicles at corridor-3 in Non- Peak hour

Average speed of vehicles travelled in corridor-3 was 28.51kmph.

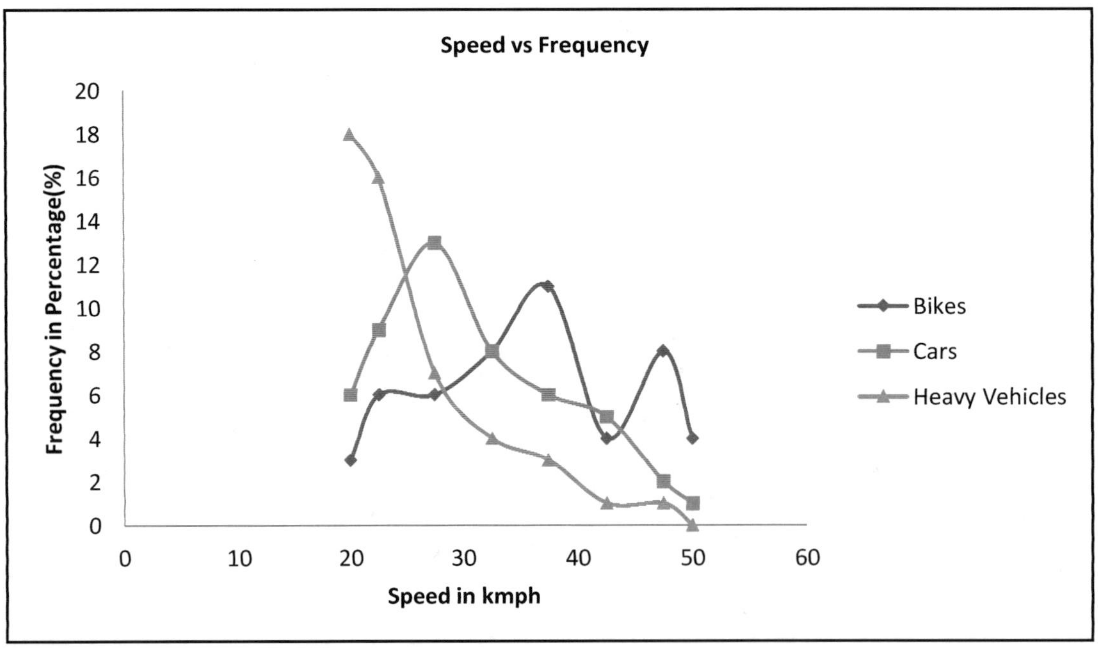

Fig. 18 (a) Average speed of vehicles at corridor-4 in Peak hour

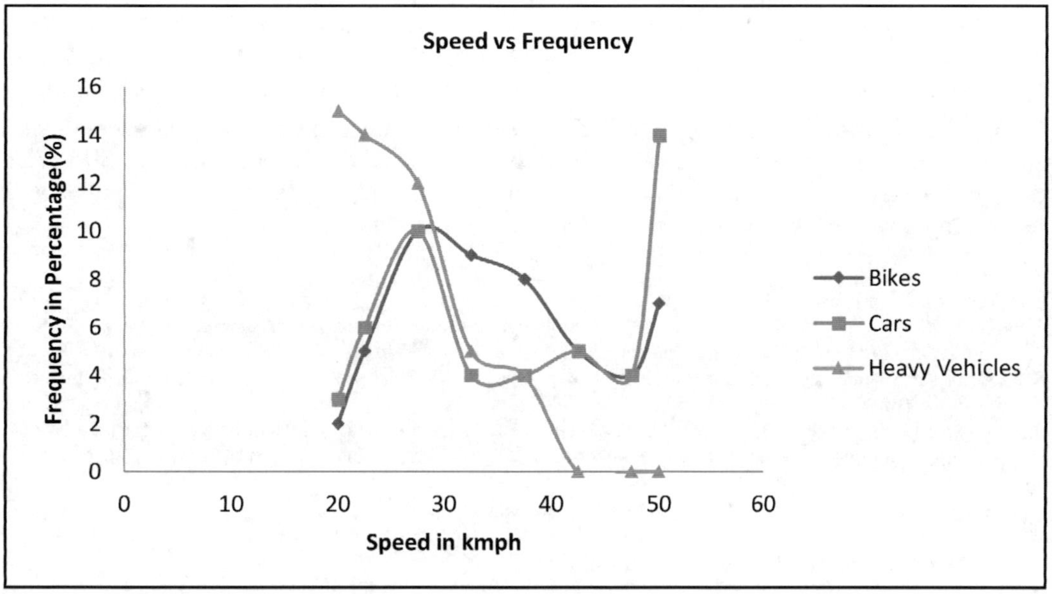

Fig. 18 (b) Average speed of vehicles at corridor-4 in Non-peak hour

Average speed of vehicles travelled in corridor-4 was 25.81kmph.

4.3 LOS based on Average speed and V/C ratio

Above observations were calibrated and values were compared to standard values which is given in HCM 2000 (Refer Table-2, Table-4). The final LOS grades of selected four corridors were given in below table.

Table:7 LOS of selected corridors based on V/C ratio and Average speed

S.No	Speed (kmph)	V/C Ratio	LOS Grades
Corridor-1	31.65	0.48	D
Corridor-2	24.46	0.96	E
Corridor-3	28.51	0.48	D
Corridor-4	25.84	0.73	E

5. Conclusion

LOS criterion based on various factors are discussed and presented in detail. A keen insight into level of service concept for urban and rural roads was given and methods to define compatibility level of urban roads under heterogeneous traffic were addressed based on the literature survey study. Volume to capacity ratio is found to be the easiest method to determine LOS from the various methods which are available. Level of service characteristics for urban and sub-urban arterials roads were also summarized for 6 categories (Refer Table 1). LOS based on average travel speed, v/c ratio and percentage speed reduction (PSR) are discussed and presented (Refer Table 2 to Table 5). It is observed that LOS grades for Vishnu college road (corridor-1) and Ramayanapuram village (corridor-3) was LOS-D and Undi bypass road (corridor-2) and Jaganadhapuram village (corridor-4) were LOS-E grade. Corridors-2, Corridor-4 was observed to have poor LOS grades because the road conditions were not up to the mark. The limitation of our study is that we need to consider more no. of vehicles in order to obtain better results.

Acknowledgement
We thank our mentor and guide Dr Pala Gireesh Kumar for the kind support and encouragement given for doing this present work. We would like to thank the management of Shri Vishnu Engineering College for Women (Autonomous) for their acceptance to support to implement the project in Bhimavaram, West Godavari District, Andhra Pradesh.

References

[1] A Patel and K Bhatt 2017 Level of service on different segment road: Review *Int. Jr. Scientific Research in Science, Engineering and Technology* vol. **3** pp. 451-454.
[2] B Prasanta Kumar and K V Krishna Rao 2011 Defining level of service criteria of urban streets in Indian context *European Transport* vol. **49** pp. 38-52.
[3] B R Marwah and Bhuvanesh Singh 2000 Level of service classification for Urban Heterogeneous traffic a case study of Kanpur Metropolis *Transportation Research Circular E-C018*: 4th Int. Symp. on Highway Capacity pp. 271-286.
[4] B Rabeya et. al. 2014 Changing modes of transportation: A case study of Rajshahi city corporation.
[5] B Robin, S Viranta and K D Ajay 2016 Level of service concept in urban roads *Int. Jr. of Engineering Science Innovation research and Development* vol. **3** pp. 1-5.
[6] B Someswara Rao, T Rambabu and G Venkata Rao 2017 Analysis of capacity and level of service at uncontrolled intersections under heterogeneous traffic conditions *Int. Jr. of Civil Engineering and Technology* vol. **8** pp. 181-190.
[7] C K Haritha, N P Bindiya and V S Ravindra 2017 Estimation of capacity and LOS for urban arterial road – A case study of Rajkot city *Int. Jr. of Innovative Research in Science* vol. **6** pp. 9377 – 9383.
[8] C R Patel and G J Joshi 2012 Capacity and LOS for urban arterial road in Indian mixed traffic conditions *Procedia – Social and Behavioral Sciences* vol. **48** pp. 527-534.
[9] D B Qadr et. al. 2019 Improving the level of service of a portion of 60-M ring road in Erbil city *Cihan University-Erbil Scientific Journal* vol. **3** pp. 12-17.
[10] D V Jitendra Bhai et. al. 2016 Estimation of LOS through congestion on urban road - A case study of Verundavan cross road *Int. Re. Jr. of Engineering and Technology* vol. **3** pp. 317-320.
[11] E N Joseph and M S Nagakumar 2014 Evaluation of capacity and level of service of urban roads *Int. Jr. Emerging Tech. and Eng.* pp. 1-7.
[12] F M John and Al Werner 1990 Measuring level of service of two-lane highways by overtaking *Transportation Research Board* pp. 62-69.
[13] K Abdulla-Al et. al. 2018 Estimating traffic volume to identify the level of service in manajor intersections of Rajshahi, Bangladesh *Trends in Civil Engineering and its Architecture* vol. **2** pp. 292-309.
[14] M A A Imran, B Ahmed and M K Bhuiyan 2014 Investigation of highway geometric problems and remedial measures in Rajshahi City Corporation area, Bangladesh *International Journal of Advanced Structures and Geotechnical Engineering* vol**.** **3** pp. 126-133.
[15] S Ashish Kumer, B A Moiur Rahaman and T Tasnim Nahar 2013 Analysis of traffic congestion and remedial measures at traffic mor in Pabna City, Bangladesh *International journal of recent development in engineering and technology* vol. **1** pp. 23-26.
[16] S Biswas, Bhupendra Singh and S Arpita 2016 Assessment of LOS on urban arterials case study in Kolkata metropolis *Int. Jr. for Traffic and Transport Engineering* vol. **6** pp. 303-312.
[17] S Chatterjee et. al. 2017 Level of service criteria on Indian multilane highways based on platoon characteristics *Transportation Research Board* pp. 1-3.
[18] S Raji and A Jagannatham 2017 Evolution and application of level of service concept – A literature study *Int. Jr. of Informative and Futuristic Research* vol. **4** pp. 8450-8474.
[19] Sanjay Kumar Singh 2012 Urban transport in India: Issues, challenges, and the way forward *European Transport* vol. **5** pp. 1-26.

IOP Conf. Series: Materials Science and Engineering 1006 (2020) 012019 doi:10.1088/1757-899X/1006/1/012019

A Study on Effect of Steel Tube in Axial Behavior of CFST Stubs

S.Jayaprakash Narayana[1] and Kolan Manvi Rao[2]

[1]Assistant Professor, Mahatma Gandhi Institute of Technology, Gandipet, Hyderabad-500075, Telangana, India

[2]Assistant Professor, Mahatma Gandhi Institute of Technology, Gandipet, Hyderabad-500075, Telangana, India

E-mail: jp@mgit.ac.in, manviraokolan@gmail.com

Abstract. Concrete filled steel tubular (CFST) columns are becoming popular due to its advantages from both steel and concrete in resisting loads. CFST members are made up of steel hollow section of rectangular or circular sections filled with plain or reinforced concrete. CFST columns can be used effectively in high raised structures and in bridges. The reduced cross sections of columns can be achieved by the composite action of steel and concrete. The steel tube provides confinement to the concrete and helps to improve load carrying capacity. The concrete core offers higher resistance to axial compression. The inward buckling of steel tube will be avoided due to concrete and the out buckling resistance may be improved due to higher stiffness of the column due to higher gross moment inertia with combined effect of steel and concrete. This experimental work made an attempt to understand the effect of thickness of steel tube in behavior of CFST stubs.

In the present experimental study, a total of 24 cylindrical concrete filled steel tube specimens were prepared to evaluate the uniaxial compressive behavior and to understand the effect of thickness and aspect ratio (h/D) on the compressive strength. Two different thickness of steel tubes 3 mm and 5 mm and four aspect ratios (h/D) 1, 2, 3 and 4 are considered in the present study. M30 grade of concrete is used. From this study, as the aspect ratio increases the buckling loads decreases. The decrement is up to 37% in aspect ratio 4 when compared to aspect ratio1.

1. Introduction

Concrete-Filled Steel Tubes (CFST) members comprises of a steel tube in filled with concrete. It has the advantage of both concrete and steel. The steel tube provides confinement to the concrete core and is capable of enhancing the core strength and ductility [2]. Inward buckling of tube is restrained by the concrete, while the steel tube is provided as tensile reinforcement for the concrete [5].

The steel at the outer perimeter is most effective in resisting bending stresses. Typically, the stiffness of the CFST is improved because of steel. Since steel has a greater elastic modulus than the concrete. Steel also contributes more to the moment of inertia as it is situated farthest from the centroid.

Therefore the stiffness of the member also greatly improved. The concrete core helps in delaying and often avoids inward buckling of steel tube and sustains the compressive loading in typical applications. Local buckling takes place at higher loads than the usual. CFST columns have exceptional structural benefits for earthquake resistance of structures due to its large energy absorption capacity and high ductility [4]. With the elimination of permanent form work construction time can be considerably reduced. Steel tube of CFST provides both longitudinal and transverse reinforcement and also acts as a confinement to concrete core, which favours the introduction of tri axial stress state to core of the concrete. The concrete core stiffens steel tube. The mechanical material properties like yield strength of the steel and compressive of the concrete determines the resulting ultimate strength of CFST columns
CFST columns are used mainly in bridges, and high rise buildings. Thesecomposite columns however can require additional protection, in structures under fire conditions. The cross section of CFST columns is smaller in comparison to RCC columns.

1.1. Advantages of CFST columns
CFST columns have many advantages compared with RC system or steel construction. The important advantages are as follows.

1.1.1. Composite behaviour
The occurrence of inward buckling of the steel tube is avoided and outward buckling is delayed. The residual strength after the local buckling is not very severe due to the resisting effect of concrete. The axial capacity of concrete is increased because of the confinement from the steel tube. The spalling of concrete is also prevented by the steel tube.

1.1.2. Sectional properties
In the RC sections and encased column sections the ratio of reinforcement is much lesser than those in the CFST cross section. In CFST sections steel will be plasticized under bending since the steel location is outside of the section.

1.1.3. Construction efficiency
Construction efficiency will improved with time, manpower and cost as form work is not required. Concreting can be done by pumping method and reinforcement can be avoided depending upon the requirement. The CFST columns are even suitable to precast construction.

1.1.4. Fire resistance
The performance of CFST columns under the fire can improved by use of fire resistance coating material since steel is exposed to fire.

1.1.5. Cost
The cost of the structure may bit higher than RC but comparatively overall cost of the structures will be reduced due the advantages listed above.

2. Experimental investigation
In the present study, an experimental program carried out to understand the behavior of CFST short

columns with different aspect ratios (h/D) and different thicknesses of steel tubes. As load resisting capacity of concrete short columns increases in the presence of steel tubes compared to RCC columns due to confinement of concrete and yielding of steel. A total 24 specimens are prepared for the present study. The steel tubes are of two different thicknesses of 3 mm and 5 mm and outer diameter is 100 mm. The steel tubes were cut to desired dimensions with aspect ratios of 1, 2, 3, and 4. Each aspect ratio contains three columns of each thickness. The steel tubes used are of mild steel and the grade of concrete used was M30. The grade of cement used was OPC 53 grade. The mix design was done as per IS 10269: 2009. The mix proportions of different ingredients used per m3 are shown in table 1. Trial mixes were made and cube specimens were prepared.28days of curing was done by watering. The characteristic compressive strength was found to be 32 N/mm^2.

Table 1: Ingredients of concrete per m^3

Cement (kg)	**Fine Aggregate (kg)**	**Coarse Aggregate(kg)**	**Water (Liter)**	**Chemical admixture(kg)**
335	888	1095	143	6.68

For the specimen preparation, steel tubes which were cut to the desired shape by taking proper care to have plane surfaces at top and bottom. The concrete was mixed to designed proportion in pan mixer and filled in the circular cylindrical steel tubes in 4 layers. Each layer is compacted manually with a compacting rod for 25 blows as per IS 516:1959. Before placing concrete in steel tubes were cleaned off the dust and applied with oil. The care is taken for avoiding segregation and bleeding. Then the concrete filled steel tubes are finished on the top with cement paste. All prepared samples are shown in Fig.1. To these columns curing compound is applied to retain the moisture in the concrete. The prepared specimens are allowed to cure for 28 days. The cured specimens were tested for axial compression in compression testing machine of loading capacity 3000kN. A dial gauge with maximum capacity of 25mm was used to measure the longitudinal deformations of the CFST columns. The loading set up is shown in Fig.2.

Fig 1: CFST stub specimens

Compression testing is setup of axially loaded CFST segments. These sections are pin ended at the closures and the load applied exactly on the longitudinal axis of the column. A dial gauge was mounted on platen to record the amount of axial shortening for the given load at regular intervals of loading. Proper care was taken and observes the alignment of the test set up periodically to avoid the eccentricity of the loading which may lead to damage of alignment of platens of compression testing machine. Continuous monitoring was done to note the buckling load and its corresponding shortening and test was continued till the maximum piston displacement or failure of the specimen whichever is earlier. The buckling patterns and the failure patterns are also observed during the compression test.

Fig 2: Compression testing machine and test setup

3. Results and discussions

3.1. Load carrying capacity

The experiments are conducted on 3000 kN capacity semiautomatic CTM. A rate of loading 140 kg/cm^2/min was maintained and the deformations were recorded using dial gauge. Continuously monitored during the testing and buckling loads and axial deformations are noted. For each thickness and aspect ratio, three specimens were tested and averages of the results were considered in results. The specimens which were failed early due to eccentricity of loading due to initial imperfection were avoided in considering results. Due to the limitation in equipment the maximum deformation measured only 25mm. The load vs axial deformation curves for 3mm thick and 5 mm thick samples are presented in Fig 3 and Fig 4.

From Fig 3 and Fig 4, It is observed that all the specimens are have initial portion ascending curve merely linear and then yielding and strain hardening portion after certain loading. This behavior indicates clearly that the concrete contributes to taking load up to certain loading and once it crushes steel start taking load. Once steel reaches to its yield strength in compression and it start buckle. In the specimens of 3 mm and 5 mm thick CFST stubs, the aspect ratio 1 i.e. 3CF100 and 5CF100 are resisting higher loads than other aspect ratio specimens. The behavior of aspect ratio 3 and 4 specimens are almost close and it indicates the higher than aspect ratio 2, the load carrying capacity in compression is not increasing and the specimens have long column effect. Since complete failure of specimens are not observed, the loads at 15 mm axial deformations are compared for understanding capacities. At 15 mm axial deformation the load capacities in 3CF100, 3CF200, 3CF200 and 3CF400 are 1800kN, 1380kN,1200kN and 1168kN respectively. As the aspect ratio is increasing from 1 to 2 the capacity is decreasing 23%, 1 to 3 the capacity is decreasing by 33% and from 1 to 4 the capacity is decreasing by 35%.

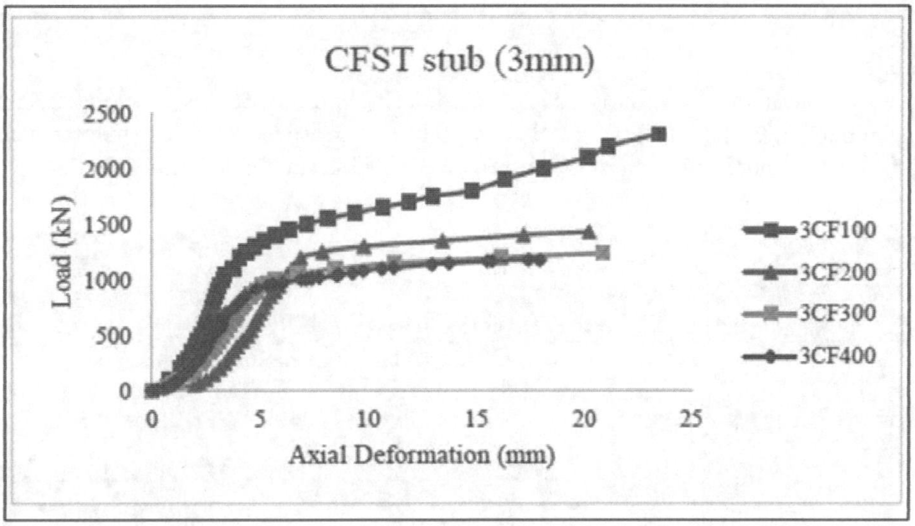

Fig 3. Load Vs Axial deformation for 3 mm thick specimens

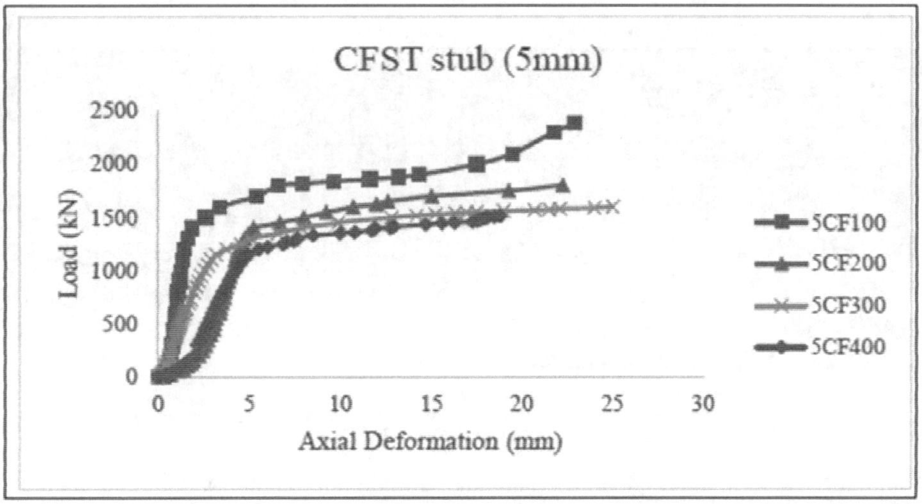

Fig 4. Load Vs Axial deformation for 5 mm thick specimens

At 15 mm axial deformation the load capacities in 5CF100, 5CF200, 5CF200 and 5CF400 are 1950kN, 1710kN, 1530kN and 1480kN respectively. As the aspect ratio is increasing from 1 to 2 the capacity is decreasing 12% , 1 to 3 the capacity is decreasing by 21% and from 1 to 4 the capacity is decreasing by 24%. On comparing capacities between thickness of CFST stubs, as the thickness increases the capacity of specimen increases. the capacity is increasing by 8% in aspect ratio 1, 23% in aspect ratio 2, 27% in aspect ratio 3 and 26% in aspect ratio 4. The aspect ratio 1 specimen behaves more stiffer than other aspect ratios. It can be observed that as the thickness increases the stiffness of the member also increasing.

3.2. Buckling behavior

The failure specimens are shown in Fig 5. The mode of failure of CFST is concrete crushing followed by steel tube buckling. The buckling effect is more in higher aspect ratio specimens which buckles at lower loads due to slender effect. Some specimens were buckled at end locations.

Fig 5. Tested specimens- failure modes

The buckling load results are presented in Fig 6. It is clearly shows that as the h/D ratio increases the buckling load decreases. For a particular h/D ratio, as the thickness increases the buckling load increases. In 3 mm thick specimens, the buckling loads in aspect ratio 1,2,3 and 4 are 1624 kN,1223kN,

1122 kN and 1062 kN respectively. As the aspect ratio is increasing from 1 to 2 the capacity is decreasing 24% , 1 to 3 the capacity is decreasing by 30% and from 1 to 4 the capacity is decreasing by 34%. In 5 mm thick specimens, the buckling loads in aspect ratio 1,2,3 and 4 are 1802 kN, 1590 kN, 1500 kN and 1133 kN respectively. As the aspect ratio is increasing from 1 to 2 the capacity is decreasing 11% , 1 to 3 the capacity is decreasing by 16% and from 1 to 4 the capacity is decreasing by 37%.

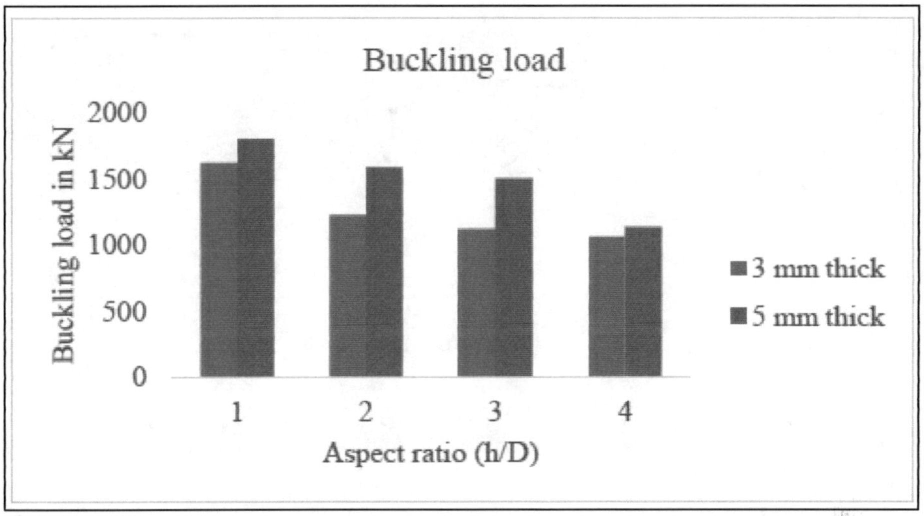

Fig 6. Buckling load Vs Aspect ratio

4. Conclusions

- As the aspect ratio increases the buckling load capacity is decreasing. In 3 mm thick CFST short columns of aspect ratio 2, the buckling load capacity decreased by 24%, for aspect ratio 3 , is decreased by 30% and for aspect ratio 4 , is decreased by 34% when compared to aspect ratio 1.
- In 5 mm thick CFST short columns for aspect ratio 2, the buckling load capacity decreased by 11% and for aspect ratio, is decreased by 16% and for aspect ratio 4 , is decreased by 37% when compared to aspect ratio 1.
- The aspect ratio 4 and more the percentage decrement in buckling load is almost is same irrespective of thickness of the CFST column.
- The aspect ratio 1 specimens behave more stiffer than other aspect ratios. It can be observed that as the thickness increases the stiffness of the member also increasing.

References

[1] Abdelmaseeh Bakos Keryou and Gailan Jibrael Ibrahim 2014 Axial load carrying capacity of thin walled HSS stub columns filled with waste glass concrete *International Journal of Enchanced Research* vol **3** issue 5.

[2] An HeOu Zha 2019 Experimental and numerical investigations of concrete-filled stain less steel tube stub columns under axial partial compression *Journal of Constructional Steel Research* **158** 405-416.

[3] Arivalagan and Kandasamy 2009 Energy absorption capacity of composite beams *Journal of Engineering and Science and Technology* vol **2** issue 1.

[4] D R Panchal, V P Sheta 2016 Experimental study on circular and square concrete filled steel tube columns subjected to axial compression loads vol **5** issue 5.

[5] Fa-xing Ding et.al. 2015 Mechanical behavior of circular and square concrete filled steel tube stub columns under local compression *Thin Walled Structures* **94** 155-166.
[6] Farid Abed, Mohammad AlHamaydeh, Suliman Abdalla 2013 Experimental and numerical investigations of the compressive behavior of concrete filled steel tubes (CFSTs) *Journal of Constructional Steel Research* **80**
[7] IS 516:1959 Indian standard Methods of Tests for Strength of Concrete *Bureau of Indian standards* (New Delhi).
[8] IS 10262:2009 Guidelines for concrete mix design proportioning *Bureau of Indian standards* (New Delhi) 110020.

IOP Conf. Series: Materials Science and Engineering 1006 (2020) 012020 doi:10.1088/1757-899X/1006/1/012020

Flexural Behavior of Quaternary Blended Fiber Reinforced Self Compacting Concrete Slabs Using Mineral Admixtures

G Sree Lakshmi Devi[1]and Dr P Srinivasa Rao[2]

[1]Associate Professor, Department of Civil Engineering, Vignana Bharathi Institute of Technology, Aushapur(V), Ghatkesar (M)India.
[2]Professor in Civil Engineering &Vice Principal, JNTUH College of engineering Hyderabad 500085. India .

E-mail: sreelakshmidevi159@gmail.com, srinivasa.pt@gmail.com

Abstract. Greatly used material in the construction industry is concrete in the world environmental concern. The use of SCM's (Supplementary Cementitious Materials) in concrete is popular in these days and enlarged along with the amassment of concrete industry, due to the discussion of problems associated with the manufacture of cement like taking out of raw material, co_2 emission etc. These effects brought pressures to minimize cement consumption by utilization supplemental cementitious materials are Fly ash, Micro Silica, Rice Husk Ash (RHA), GGBS, etc., use of these byproducts facilitates sustainable development at present world scenario. Developing of self-compacting concrete has brought a positive Perception due to several advantages as reduction in pollution of noise due to vibration, decreases in construction time and power consumption through the utilization of supplemental materials. If binder content in concrete is more it leads to environmental impacts. By incorporating maximum cement content with mineral admixtures it improvises properties along with self-compacting concrete. In this Study an experiment made to study the behavior of QBSCC slabs such as Flexural strength, load Vs deflection curves with addition of crimped steel fibers 0%, 0.5%, 1.0%, and 1.5%.

1. Introduction

This paper deals with the supplemental cementitious materials Fly ash, GGBS and micro silica as in self compacting concrete. The studies were conducted in this quaternary blended self-compacting (QBSCC) concrete with 10% micro silica 25% Fly ash and 25% GGBS. Partial replace of Fly ash, GGBS and micro silica leads to a good environmental and produces eco-friendly concrete. previous researches show that the Micro Silica ameliorate the early life performance of concrete GGBS reduces permeability and Fly ash continuously refines the properties of hardened concrete as it matures[4-11]. Previous studies of various authors show that to resist cracking and crack propagation in reinforced concrete with help of fibers. As a result of this ability to arrest cracks, at first crack and at failure fibers in reinforced concrete can possess increased extensibility and tensile strength particularly under flexural loading [1-2].

2. Experimental study

An attempt has been made in this study to investigate the behavior of The flexural behavior of steel fiber reinforced Quaternary blended self – compacting concrete slabs made with w/b ratio of 0.30 is studied in terms of load-deflections characteristics. The parameters to be observed are first crack load and ultimate load carrying capacity of Quaternary blended self – compacting concrete slabs. Loads and deflections are plotted for the reference SCC and fiber reinforced QBSCC mixes. The flexural behavior of reference SCC beams (without admixtures and 0% fibers) and QBSCC slabs made with addition of steel fiber varying 0 to 1.5 % by the volume of concrete are studied. The reinforced concrete slabs specimens of size 1200mm X 1200mm X 100mm is designed as a under reinforced section. Reinforcement details are 8mm ø @ 150mm c/c in both directions and crimped steel fibers of length 30 mm and diameter 0.5mm (aspect ratio 60) are used.

Table 1: Materials required for 1 m^3 of concrete for 0.3 w/b ratio for 100 % cement -SCC & Quaternary Blended SCC

% of Admixtures	W/B ratio	Cement (kgs)	Fly Ash (kgs)	GGBS (kgs)	Micro Silica (kgs)	F.A (kgs)	C.A (kgs)	Water (L)	SP (L)	VMA (L)
100% Cement	0.3	614	-	-	-	884	749	177	7.36	0.61
40% OPC + 25% Fly Ash+ 25% GGBS+ 10% Micro Silica	0.3	246	153	153	62	884	749	177	7.36	0.61

Table 2 : The fresh properties of SCC and QBSCC according to EFNARC guidelines.

S.No.	W/B ratio	Type of mix	Slump flow	T50 cm slump flow	L-box	U-box
Limits as per EFNARC specifications – 2002			**650-800 mm**	**2-5 (sec)**	**0.8 -1.0**	**0 -30mm**
1	0.30	SCC 2	675	3.6	0.84	27
2	0.30	QBSCC 2	715	3.2	0.87	24

3. Testing procedure

In this present study quaternary blended fiber reinforced concrete slabs were casted with various % of steel fibers and tested @ 28 days shown in figure 1&2.A hydraulic jack of 500KN is used in this experimental work. A dial gauge with least count of the deflection at 0.01mm is used for measuring deflection at the center. The readings of deflections measured at 20KN intervals throughout the investigation, loading at first crack is also noted. 10 slabs were tested, load deflection test results were averaged and tabulated in table 3. Load deflection characteristics of slab with w/b ratio 0.3 for quaternary blended concrete with 0%, 0.5%, 1.0 % and 1.5 % fibers and control self-compacting concrete.

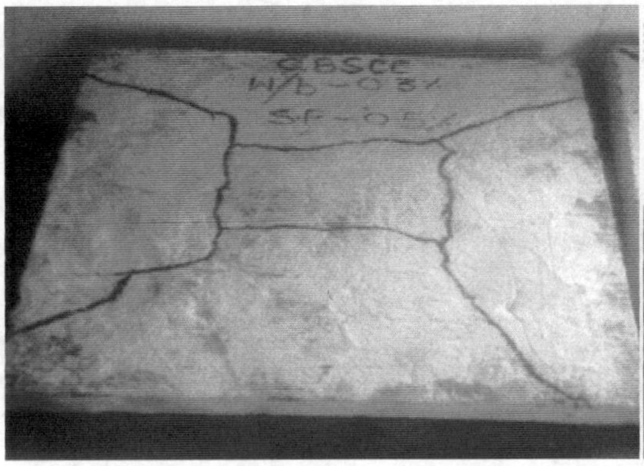

Figure 1 Slab specimens

Figure 2. Testing of slab at 28 days

Table3: Load –Deflection characteristics of slabs with W/B Ratio 0.3 at 28 days

S.No	SCC		QBSCC-0 %		QBSCC -0.5 %		QBSCC-1.0 %		QBSCC-1.5 %	
	Load (kN)	Deflection (mm)	Load (kN)	Deflection (mm)	Load (kN)	Deflection (mm)	Load (kN)	Deflection (mm)	Load (kN)	Deflection (mm)
1	20	0.309	20	0.308	20	0.303	20	0.29	20	0.25
2	40	0.87	40	0.91	40	0.84	40	0.75	40	0.65
3	60	1.25	60	1.28	60	1.15	60	0.98	60	0.85
4	80	2.2	80	2.27	80	2.05	80	1.85	80	1.65
5	100	3.35	100	3.45	100	3.25	100	2.95	100	2.84
6	120	4.85	120	4.82	120	4.77	120	4.32	120	3.94
7	140	5.98	140	5.91	140	5.64	140	5.21	140	4.65
8	160	7.68	160	7.61	160	7.45	160	7.02	160	6.25
9	180	8.54	180	8.59	180	8.32	180	8.11	180	7.54
10					200	9.25	200	8.89	200	8.24
11									220	8.85

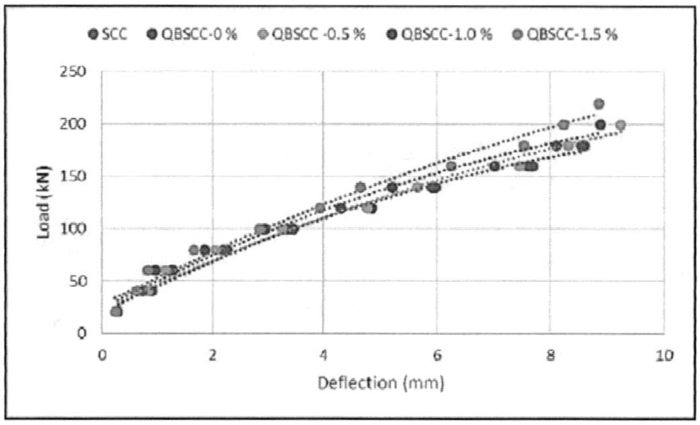

Figure 3 : Load –Deflection characteristics of slabs with W/B Ratio 0.30

4. Conclusions

From the experimental study results the following conclusions can be made.

- The ultimate flexural strength or flexural load bearing capacity of plain quaternary blended SCC slabs made with w/b ratio 0.30 is almost similar to that of reference SCC mixes made with w/b ratio 0.30 but QBSCC slabs with & without fibre deflected more than plain SCC slabs. QBSCC slabs with 1.5% fibre has improved deflection capabilities than other percentage fibre reinforced QBSCC slabs.
- Quaternary Blended Self – Compacting Concrete slabs made with fibre content of 1.5% steel fibres showed better flexural strength at w/b ratios 0.30.

References

[1] Amir Hossein Jodeiri, Ronaldo J Quitalig 2012 Effect of Wirand FS7-II Steel Wire Fiber on Flexural Capacity of Reinforced Concrete Beam *Journal of Civil Engineering Research* **2(6)** 100-107.

[2] C B Kukreja and S Chawla 1989 Flexural Characteristics of Steel fiber reinforced concrete *The Indian concrete journal* Page 154.

[3] Nan Su, Kung-Chung Hsu, His-wen chai 2009 A simple mix design method for self- compacting concrete *Cement and Concrete Research* **31** (2001) 1799-1807

[4] Eva Vejmelková, Martin Keppert, Stefania Grzeszczyk, Bartłomiej Skaliński, Robert Černý 2011 Properties of self-compacting concrete mixtures containing metakaolin and blast furnace slag *Construction and Building Materials* Volume **25** Issue 3 Pages 1325-1331 ISSN 0950-0618,

[5] Fereshteh Alsadat Sabet, Nicolas Ali Libre, Mohammad Shekarchi 2013 Mechanical and durability properties of self consolidating high performance concrete incorporating natural zeolite, silica fume and fly ash *Construction and Building Materials* Volume **44** Pages 175-184 ISSN 0950-0618,

[6] Sameer Hamoush, Miguel Picournell-Dardner 2011 Freezing and Thawing Durability of Very High strength concrete *American Journal of Engineering and Applied sciences* **4(1)** 42-51

Dynamic Analysis of Vertical Irregular Tall Structural System

HB Akhilesh[1] and BO Naveen[2]

[1] Postgraduate Student, Department of Civil Engineering, The National Institute of Engineering, Mysuru, India.
[2] Assistant Professor, Department of Civil Engineering, The National Institute of Engineering, Mysuru, India.

E-mail: hbakhilesh111@gmail.com, naveenbo@nie.ac.in

Abstract. The behaviour of any structuredepends upon the structural element present in it. The critical aspect on which the structural configuration depends are shape, length and geometry of the building. In the present study the performance of vertical geometric irregular tall structures under the wind load is evaluated as per IS 16700: 2017 provisions. Three type of vertical geometric irregular tall reinforced concrete buildings having different plan dimensions with cantilevered offset length varying from 11% to 15% of the building lateral dimension with increment of 0.5% are modelled. Modal analysis and wind analysis are performed using structural analysis software ETABS to assess the behaviour of vertical geometric irregular tall structural system with cantilevered offset. Parameter such as time period, frequency, storey displacement and storey drift are studied and compared with the regular building. Multiple linear regression analysis is carried out to formulate a general expression for the responses of the vertical irregular tall buildings.

1. Introduction

The population growth of large cities is increasing due to relocation of individuals from rural to urban areas for employment and to raise their living standards. The limited construction area available in the heart of the city leads to the design and construction of tall buildings. In many cases the design of tall building is governed by the stiffness requirement rather than strength demand. The strength requirement can be easily achieved by modern methods, but these buildings are mainly affected by the lateral loads so there should be sufficient lateral stiffness of the building. The strength requirement can be easily achieved by modern methods, but these buildings are mainly affected by the lateral loads so there should be sufficient lateral stiffness of the building. The stiffness to the lateral load can be achieved by selecting the proper structural system and proper structure design to the lateral loads. Structural engineer should choose the structural system wisely to design the building economically. There are mainly two types of lateral load, wind loads and seismic loads. Behavior of the structure to lateral loads can be studied by numerical analysis as well as from experiments. Wind tunnel test is carried out to study the behavior of structures to wind loads and shaking table test is conducted to study the response of the structure to ground motions. Static and dynamic analysis is carried out in numerical analysis, these methods includes linear and nonlinear analysis. Tall buildings situated in high wind speed location are more affected by the wind forces, so the tall structures need to be analyzed by selecting the proper analysis method for these forces.

There are many factors which affect the behavior of the high-rise building such as height of the building, plan aspect ratio, slenderness ratio, building geometry, damping of the building etc. Proper analysis method should be adopted for analysis purpose according to building type, environmental conditions and all the parameters which will affect the behavior of high-rise building should be considered while designing.

In many cases, it is very difficult to estimate wind load and their associated structural behavior by analytical methods. For example, uncommon aerodynamic shape of the building or the building is flexible so that motion of the building affects the forces acting on it. Wind tunnel test is carried out in those cases to find the wind loads and study the behavior of the structures to wind action. The tall buildings having square geometric configuration are studied for dynamic responses for interference excitation [1] and stated that the kinetic energy is the main factor observed to be in upstream buildings which will alter the response induced along the wind and cross wind for the interference excitation. Wind tunnel test has been conducted to identify the source of error which arises when the mode shape of the buildings is non-ideal and methods of applying correction which include 2 adjustment factors. The first adjustment factors is related to static load and second one is relate to displacements and accelerations [2]. Shielding and channeling effects between two or more buildings are studied [3]. Upstream interfering buildings will decrease the mean wind load on downstream building by providing certain shielding effect. The vertical geometric irregularity structures are studied for their performances [4]. The parameter studied are as base shear, time period, inter storey drift ratio and maximum displacement. The results of regular structures are then compared to the results of the irregular structure. Nonlinear static approach is carried out with used defined hinges for different setback ratios. From the study they have conclude that performance of vertical irregular structure when designed conferring to IS code provisions are inferior related to regular structure. Five different structural systems of tall buildings are compared, and optimal structural system was determined for a certain building height [6]. The ratio of building height to the minimum base width is called as slenderness ratio. The maximum slenderness ratio of the building for different structural system and seismic zones are given in code [5]. Building with different slenderness ratio from 2 to 8 has been studied [7] and the design governing factor for all the configuration has been specified. High strength RC buildings performance and their cost utilization are studied [8]. The parameters considered for the cost analysis are utilization of steel, concrete and formwork. The static analysis and dynamic analysis of tall buildings are studied. From the study they have concluded that steel is the main parameter which contributes to the increase the cost. Member size can be reduced by adopting high strength concrete, the behavior of high strength concrete structure is same as that of equivalent normal strength concrete with respect to top displacement, base shear and maximum inter storey drift ratio. The damped outrigger concept for high rise buildings to increase damping of the structure are proposed [9]. In this concept lateral design force will be reduced so there will be a reduction in the member size of the structure and cost of construction can be saved. Damped outriggers are introduced to reducing dynamic wind effects. Performance based design using nonlinear pushover analysis is carried out [10]. Steel reinforcement is used to control the lateral drift beyond the limit, and it provides ductility to the RC building. Pushover analysis is carried out using “SAP2000” software. Principle of virtual work and Taylor series approximation are used for the inelastic performance-based design.

The objectives of this study are to evaluate the performance of vertical irregular tall structures under wind loads as per IS 16700: 2017 provisions, to assess behaviour of vertical geometric irregular tall structural system with cantilevered offset to wind loads and development of correlation for responses using regression analysis.

2. IS 16700: 2017

In 2017 bureau of Indian standards has published a new code IS 16700: 2017 "Criteria for structural safety of tall concrete buildings" which address the issues associated with the reinforced concrete (RC) tall buildings design. According to this code buildings which are having height between 50 m to 250 m are considered as tall buildings. The following are the salient features addressed by this standard:

a) Different structural systems that can be adopted
b) General requirement such as maximum height of the building that can be adopted for different structural systems and for different seismic zone, plan aspect ratio, lateral drift, storey stiffness and strength, vertical acceleration, floor systems, material properties and progressive collapse mechanism
c) Loads and load combinations
d) Structural design parameters
e) Foundation design parameters

Buildings exceeding the specifications provided by this standard (for example, buildings height greater than 250 m, buildings height exceeding the height specified for particular structural system, or using of other structural systems which are not specified by this standards) and buildings which are not covered in this standards are called as code exceeding tall building. Such building requires an approach of performance-based design to meet defined performances of the structure. All code exceeding high rise buildings should go through two processes, review by structural design reviewer and review by an expert review panel.

3. Vertical geometric irregularity

Now a days, buildings shapes are becoming more complex because of its aesthetic architecture and function. Vertical geometric irregularity is the common type of irregularity in the buildings. In the code [11] they have given the definition of irregular buildings, where vertical geometric irregularity is a type in the vertical irregularities. According to the code this irregularity should be considered when the lateral dimension of lateral force resisting system at any storey is greater than 125 percent of the storey below. Different type of vertical geometric irregularity are given in the code are as shown in figure 1.

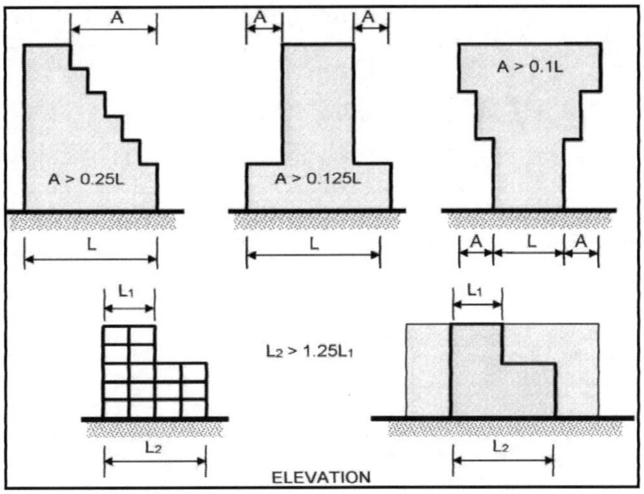

Figure 1. Vertical geometric irregularities

4. Structural modeling

Three type of symmetric building Type A, Type B and Type C having 3, 4 and 5 bays respectively are considered for the present study where cantilever projection is provided in the plan at 18th, 19th, 20th and 21st stories and the projection length is doubled for the 22nd, 23rd, 24th and 25th stories. In each type of building one reference building without vertical geometric irregularity (without any cantilevered offset) and nine vertical irregular buildings are modelled. The reference models are considered as base models to study and compare the responses of vertical geometric irregular buildings. Similar member properties are considered for both regular and irregular buildings. The building data considered for the present study are as shown in Table 1.

As per IS 1893:2016, the building having lateral dimension of the lateral force resisting system in any storey greater than 125 percent of the storey below are considered as vertical geometric irregular building. For the buildings having cantilevered offset in the top stories, lateral dimension in any storey is more than 120 percent of the storey below then they are considered as vertical geometric irregular buildings. The length of the cantilevered offset in each side of the plan is varied from 11 percent to 15 percent of the building horizontal length with the increment of 0.5 percent. Cantilevered offset length considered for Type A, Type B and Type C buildings are as shown in Table 2. Moment frame with structural wall system is selected as the structural system for modelling the tall RC buildings and L shaped structural wall is positioned at the opposite corners of the building so that the location of the shear wall can be maintained same for all models.

Table 1. Building data

Type of Building	Components	Details
Type A	Plan dimension	9 m × 9 m
	Storey height	3 m
	Building height	75 m (25 storey)
	Column spacing	3 m
	Number of bays	3
Type B	Plan dimension	12 m × 12 m
	Storey height	3 m
	Building height	75 m (25 storey)
	Column spacing	3 m
	Number of bays	4
Type C	Plan dimension	15 m × 15 m
	Storey height	3 m
	Building height	75 m (25 storey)
	Column spacing	3 m
	Number of bays	5

Table 2. Cantilevered offset length

Type A			Type B			Type C		
Model no.	**Cantilevered offset length – A (m)**		**Model no.**	**Cantilevered offset length – A (m)**		**Model no.**	**Cantilevered offset length – A (m)**	
M-1	Regular		M-11	Regular		M-21	Regular	
M-2	0.110 × L	0.990	M-12	0.110 × L	1.320	M-22	0.110 × L	1.650
M-3	0.115 × L	1.035	M-13	0.115 × L	1.380	M-23	0.115 × L	1.725
M-4	0.120 × L	1.080	M-14	0.120 × L	1.440	M-24	0.120 × L	1.80
M-5	0.125 × L	1.125	M-15	0.125 × L	1.500	M-25	0.125 × L	1.875
M-6	0.130 × L	1.170	M-16	0.130 × L	1.560	M-26	0.130 × L	1.950
M-7	0.135 × L	1.215	M-17	0.135 × L	1.620	M-27	0.135 × L	2.025
M-8	0.140 × L	1.260	M-18	0.140 × L	1.680	M-28	0.140 × L	2.100
M-9	0.145 × L	1.305	M-19	0.145 × L	1.740	M-29	0.145 × L	2.175
M-10	0.150 × L	1.350	M-20	0.150 × L	1.800	M-30	0.150 × L	2.250

ETABS 16.0.0 software is used for modelling of the vertical irregular tall RC buildings with different cantilevered offset and plan size of the building. The grid system are used for the modelling. Figure 2 represents the typical plan and 3D–Render view of the models with the building data as shown in Table 1. The modal analysis and wind analysis is performed using ETABS software to study the dynamic characteristics of the building such as time period, frequency, displacements, drift and base shear for 3 regular and 27 vertical geometric irregular tall RC buildings.

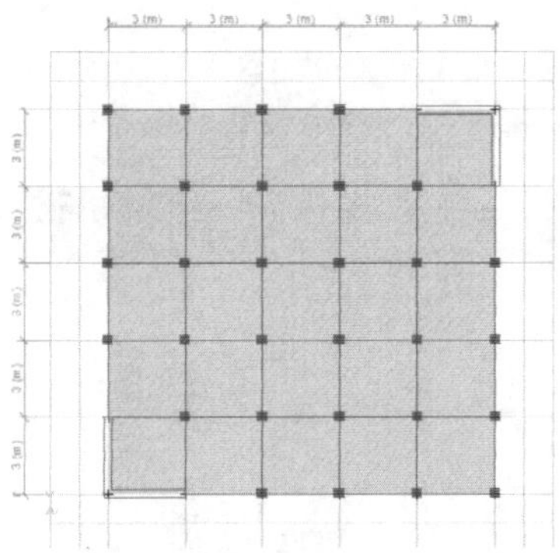

(a) Plan of bottom stories

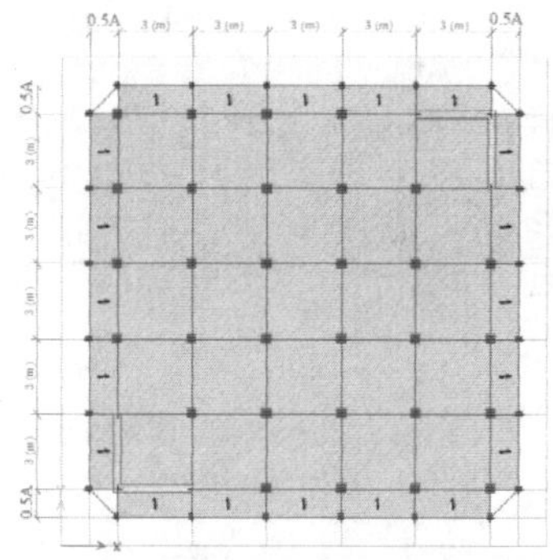

(b) Plan of stories from 18 to 21

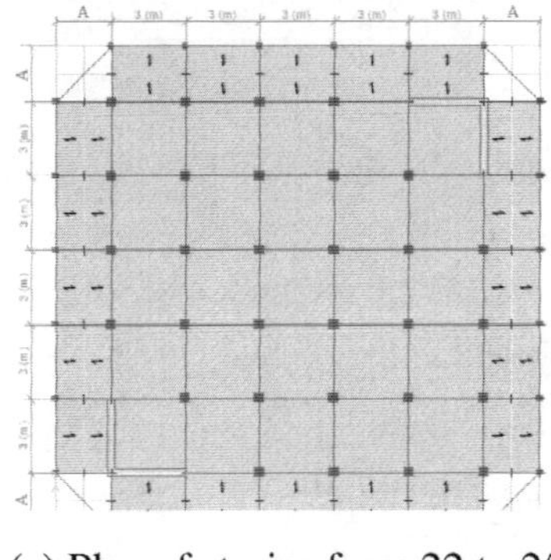

(c) Plan of stories from 22 to 25

(d) 3D - Render view

Figure 2. Typical view of the Buildings

The material properties and section properties considered for the study are shown in Table 3 and Table 4 respectively.

Table 3. Materials data

Components	Materials
Concrete	M 40 (Beams & Columns) M 25 (Slabs)
Rebar's	HYSD 500 (Longitudinal bars) HYSD 415 (Ties)

Table 4. Sectional properties of the buildings

Components	Dimensions
Columns	400 × 400 mm
Floating Columns	230 × 300 mm
Beams	230 × 400 mm
Periphery Beams in projection area	230 × 300 mm
Shear wall Thickness	300 mm
Slab Thickness	150 mm
Slab Thickness in projected area	125 mm

Dead loads, super dead loads, live loads and wind loads are applied to the structure. The super dead loads as per IS 875 (Part 1): 1987, live loads as per IS 875 (Part 2): 1987 and wind load factors as per IS 875 (Part 3): 2015 are considered. The loads and wind load factors consider are shown in Table 5 and Table 6 respectively for the city Darbhanga. The p delta effect of the building is considered as per IS 16700: 2017.

Table 5. Loads considered

Components	Details
Live load	4 kN/m^2
Roof live load	1.5 kN/m^2
Super dead load	1 kN/m^2
Glazing	1 kN/m

Table 6. Wind load factors

Components	Details
Basic wind speed (V_b)	55 m/s
Terrain category	3
Structural class	C
Risk coefficient factor (K_1)	1.08
Topography factor (K_2)	1

5. Results and discussion

5.1. Modal analysis results – time period

Time period is an inherent property of the building which is controlled by its mass and stiffness. Safety of structure is governed by natural time period and amount of damping in each mode of vibration. Fundamental natural time period is an important parameter to understand the dynamic behaviour of building. Estimation of approximate translational natural time period is given in IS 1893 (Part 1): 2016, clause 7.6.2 which depends on base dimension, height of the building and type of the building. Time period is defined as the time taken by the building to complete one cycle of motion during vibration.

According to IS 16700: 2016 natural period of fundamental torsional mode of vibration shall not exceed 0.9 times the smaller of the natural period of the fundamental translation mode of vibration in each of the orthogonal directions in plan. As the modes increases the time period will get decreases. It is observed that in 12 modes of vibrations there will be a 9-translation mode of vibration and 3 torsional mode of vibration at mode 3, mode 6 and mode 10 for all type of models. Time period of torsional mode of vibration is less than 0.9 times of the smaller of the time period of translation mode of vibrations for all type of models. Total modal mass participation of the buildings in the considered modes are above 90 percent for all buildings. From the results it is observed that increase of time period with respect to increase of cantilevered offset is less in Type C buildings than Type B buildings than Type A buildings.

The time period increases with increasing the cantilevered offset from 0.99 m to 1.35 m. There is a 2.33% increase in the time period for increase of 4% cantilevered offset for 3 bays buildings. The time period for vertical irregular buildings are increases from 11.26% to 15.30% as compared to base model. The graphical representation of time period with respect to modes from modal analysis are as shown below in figure 3 for Type A models.

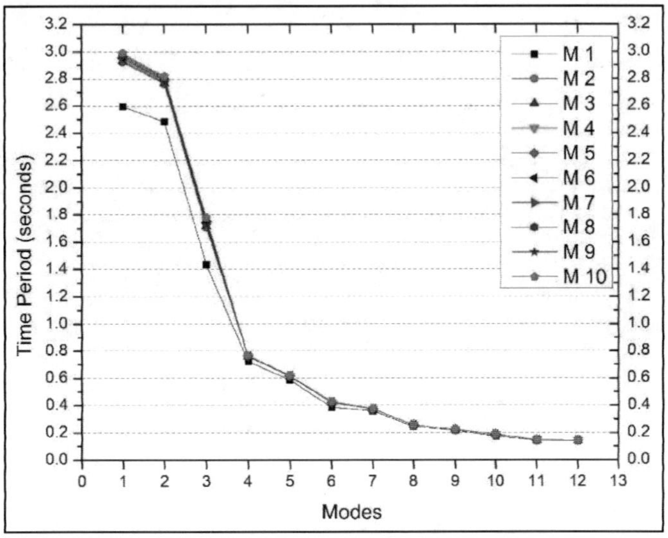

Figure 3. Mode Vs Time period – Type A

The time period increases with increasing the cantilevered offset from 1.32 m to 1.8 m. There is a 2.47% increase in the time period for increase of 4% cantilevered offset for 4 bays buildings. The time period for vertical irregular buildings are increases from 11.95% to 14.72% as compared to base model. The graphical representation of time period with respect to modes from modal analysis are as shown below in figure 4 for Type B models.

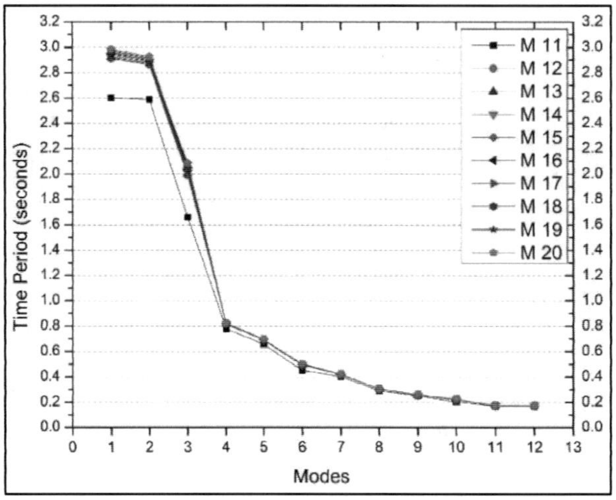

Figure 4. Mode Vs Time period – Type B

The time period increases with increasing the cantilevered offset from 1.65 m to 2.25 m. There is a 2.28% increase in the time period for increase of 4% cantilevered offset for 5 bays buildings. The time period for vertical irregular buildings are increases from 10.24% to 12.75% as compared to base model. The graphical representation of time period with respect to modes from modal analysis are as shown below in figure 5 for Type C models.

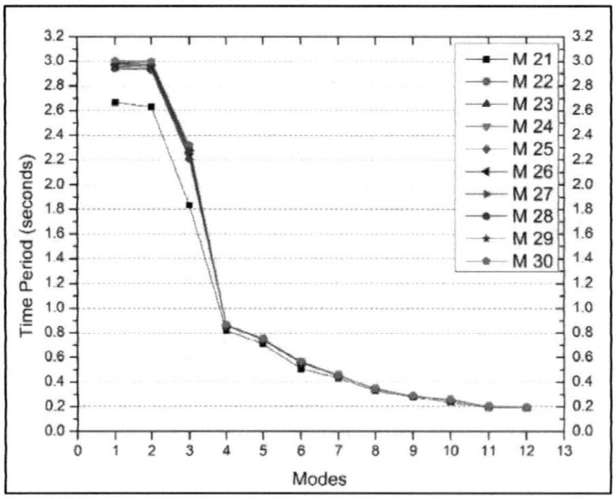

Figure 5. Mode Vs Time period – Type C

5.2. Modal analysis results – Frequency

It is the number of times the building sways and comes to its original position in one second after its excitation. It also depends on the mass and stiffness of the building. It is the inverse of the time period of the building. From the results it is observed that decrease of frequency with respect to increase of cantilevered offset is less in Type C buildings, Type A and Type B buildings frequency are almost equally decreased.

The frequency decreases with increasing the cantilevered offset from 0.99 m to 1.35 m. There is a 2.34% decrease in the frequency for increase of 4% cantilevered offset for 3 bays buildings. The frequency for vertical irregular buildings decreased from 11.4% to 13.47% as compared to base model. The graphical representation of frequency with respect to modes from modal analysis are as shown below in figure 6 for Type A models.

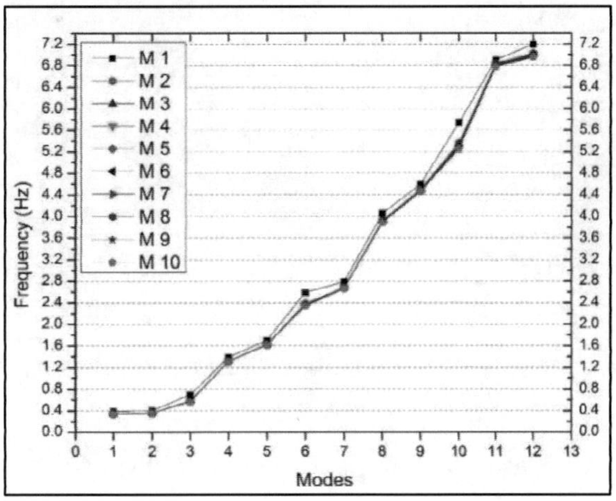

Figure 6. Mode Vs Frequency – Type A

The frequency decreases with increasing the cantilevered offset from 1.32 m to 1.8 m. There is a 2.33% decrease in the frequency for increase of 4% cantilevered offset for 4 bays buildings. The frequency for vertical irregular buildings decreased from 10.68% to 12.76% as compared to base model. The graphical representation of frequency with respect to modes from modal analysis are as shown below in figure 7 for Type B models.

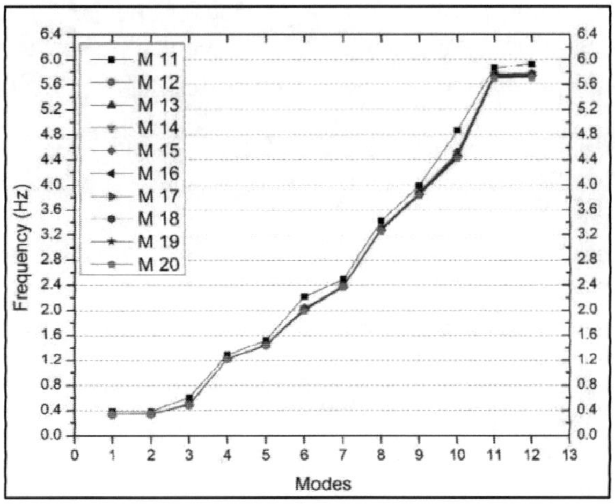

Figure 7. Mode Vs Frequency – Type B

The frequency decreases with increasing the cantilevered offset from 1.65 m to 2.25m. There is a 2.06 % decrease in the frequency for increase of 4 % cantilevered offset for 5 bays buildings. The frequency for vertical irregular buildings decreased from 9.33 % to 11.20 % as compared to base model. The graphical representation of frequency with respect to modes from modal analysis are as shown below in figure 8 for Type C models.

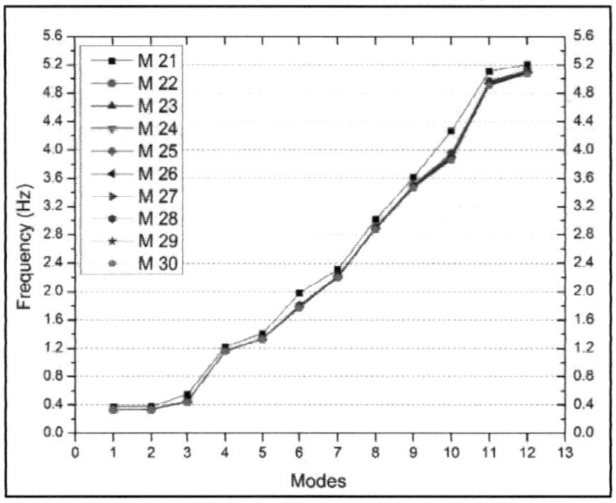

Figure 8. Mode Vs Frequency – Type C

5.3. *Wind analysis results – Storey Displacement*

Wind force acting on a building will cause displacement of the building. Storey displacement is defined as displacement of a storey with respect to base of the structure. The tall buildings and buildings which are located in the cities where wind speed is high are more affected by the wind force. From the results it is observed that increase of displacement with respect to increase of cantilevered offset is more in Type A buildings than Type B buildings than Type C buildings.

The storey displacement increases with increasing the cantilevered offset from 0.99 m to 1.35 m. There is a 3.61% increase in the displacement for increase of 4% cantilevered offset for 3 bays buildings. The maximum storey displacements for vertical irregular buildings are increases from 10.9% to 14.9% as compared to base model. The graphical representation of displacement with respect to stories from wind analysis are as shown below in figure 9 for Type A models.

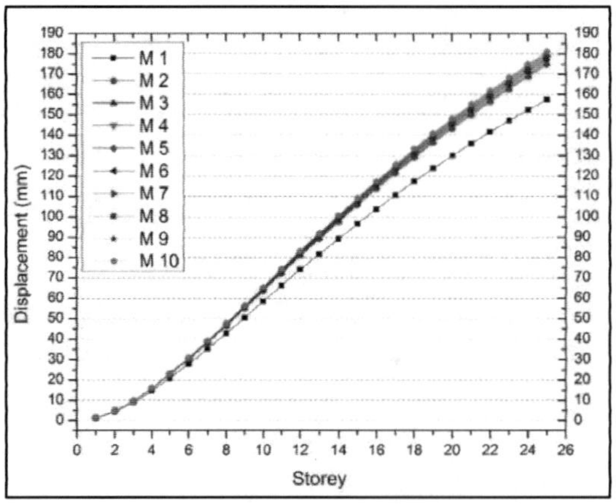

Figure 9. Storey Vs Displacement – Type A

The storey displacement increases with increasing the cantilevered offset from 1.32 m to 1.8 m. There is a 3.58 % increase in the displacement for increase of 4% cantilevered offset for 4 bays buildings. The maximum storey displacements for vertical irregular buildings are increases from 10.76% to 14.72% as compared to base model. The graphical representation of displacement with respect to stories from wind analysis are as shown below in figure 10 for Type B models.

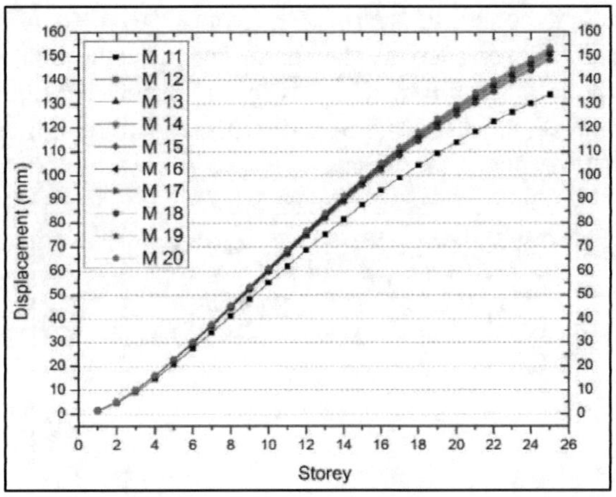

Figure 10. Storey Vs Displacement – Type B

The storey displacement increases with increasing the cantilevered offset from 1.65 m to 2.25 m. There is a 3.50% increase in the displacement for increase of 4% cantilevered offset for 5 bays buildings. The maximum storey displacements for vertical irregular buildings are increases from 10.78% to 14.65% as compared to base model. The graphical representation of displacement with respect to stories from wind analysis are as shown below in figure 11 for Type C models.

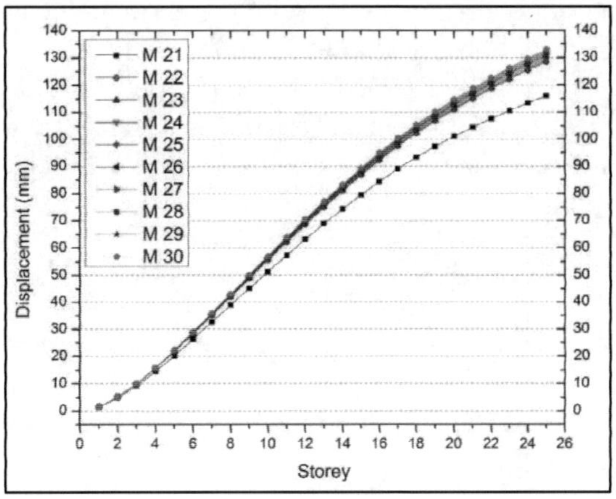

Figure 11. Storey Vs Displacement – Type C

5.4. Wind analysis results – Storey Drift

Under the actions of lateral force induced due to wind action, building tends to displace which affects the lateral stability of the structure. Therefore, storey drift is one of the important factors to check the lateral stability of the structure which is defined as difference of displacements between two consecutive stories of the building. According to IS 16700: 2017 lateral drift ratio should be limited to H/500 for the wind loads where H is total height of the building. Since all the models are having the same height 75 m, lateral drift ratio is limited to 0.15, so the maximum allowable storey drift for the present study is 11.25 mm. The storey drift and storey levels at which there will be maximum storey drift was observed to be decreasing as the plan of the building increases for vertical geometric irregular tall RC buildings.

The top storey drift is increasing for every 1.5 % increase in the cantilevered offset. The storey drift increases with increasing the cantilevered offset from 0.99 m to 1.35 m and maximum drift was observed to be in storey 11 for each model including the base model. The maximum storey drift observed in Type A models is 8.9 mm which is less than the maximum allowable storey drift 11.25 mm. Thegraphical representation of lateral drift with respect to stories from wind analysis are as shown below in figure 12 for Type A models.

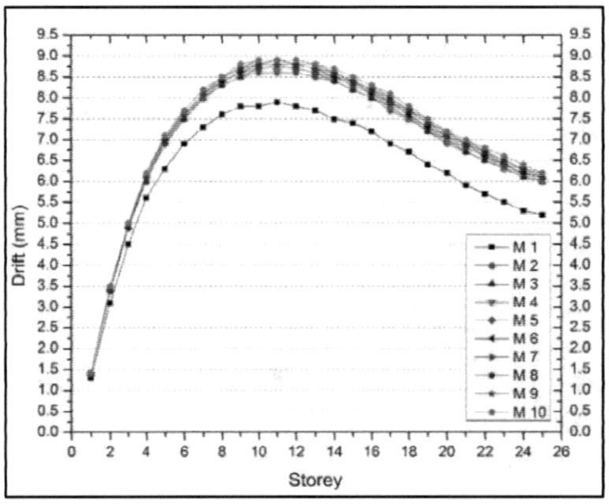

Figure 12. Storey Vs Drift – Type A

The storey drift obtained from the wind analysis for critical stories of Type B models are shown below in the Table 4.11. The storey drift increases with increasing the cantilevered offset from 1.32 m to 1.8 m and maximum drift was observed to be in storey 9 and 10 for each model including the base model. The maximum storey drift observed in Type B models is 7.9 mm which is less than the maximum allowable storey drift 11.25 mm. The graphical representation of lateral drift with respect to stories from wind analysis are as shown below in figure 13 for Type B models.

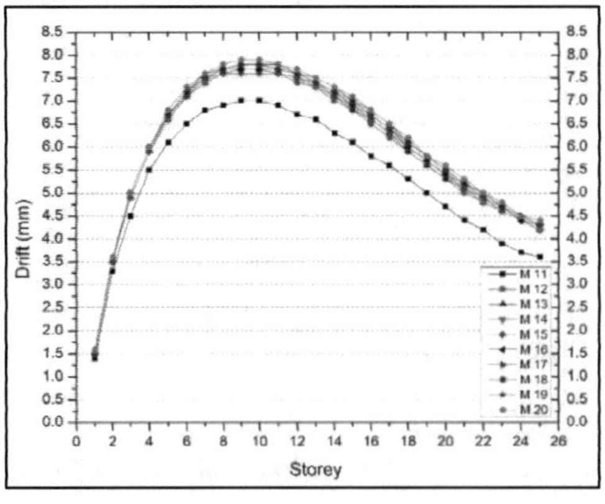

Figure 13. Storey Vs Drift – Type B

The storey drift obtained from the wind analysis for critical stories of Type C models are shown below in the Table 4.12. The storey drift increases with increasing the cantilevered offset from 1.65 m to 2.25 m and maximum drift was observed to be in storey 8 and 9 for each model including the base model. The maximum storey drift observed in Type C models is 7 mm which is less than the maximum allowable storey drift 11.25 mm. The graphical representation of lateral drift with respect to stories from wind analysis are as shown below in figure 14 for Type C models.

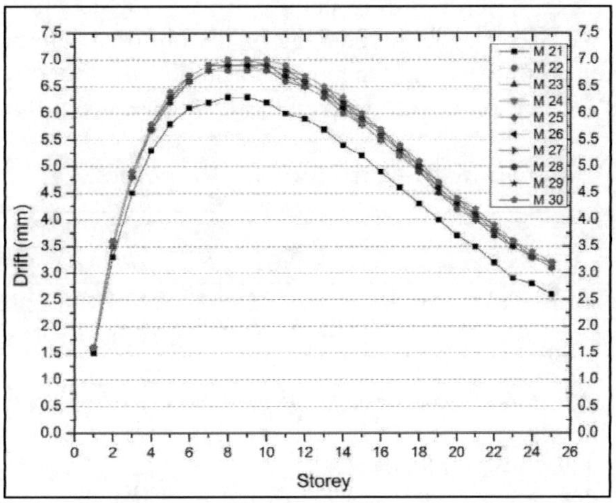

Figure 14. Storey Vs Drift – Type C

5.5. Wind analysis results – Base Shear

Base shear is termed as estimation of maximum expected lateral force on the base of the structure. According to IS 1893 (Part 1): 2016 design base shear is calculated for the complete building as a whole. The design base shear is distributed to the complete building at a various floor levels at the respective centre of the mass. Base shear of all the models from wind analysis are determined. The graphical representation of base shear with respect to models for wind analysis are as shown below in figure 15. It is observed that there is an equal increase of 2% base shear as the cantilevered offset is increased from 11% to 15% for all type of buildings.

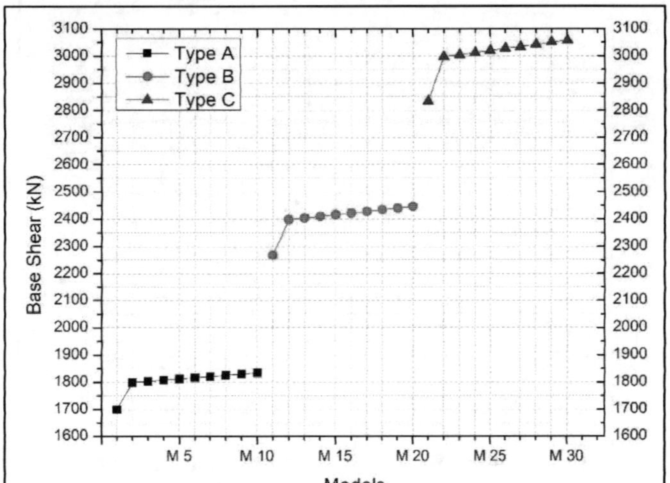

Figure 15. Models Vs Base Shear

5.6. *Correlation for responses*

In this section a correlation for responses are presented in order to find the response characteristics of building for different overall dimension and cantilevered offset. This correlation is well suited for the symmetric buildings having plan dimension between 9 m × 9 m to 15 m × 15 m and cantilevered offset up to 15% of the building lateral dimension.

Multiple linear regression analysis is carried out to formulate an equation to find the maximum displacement and base shear of the building caused by wind load and time period from modal analysis. Cantilevered offset in terms of percentage of the building lateral dimension and building lateral dimension is taken as the two independent variable and the responses are taken as one dependent variable. The correlation developed for the responses from this study are shown below in Table 7.

Table 7. Correlation for Responses

Responses	**Correlation**	**R^2**
Time Period	$y = 0.0253A + 0.0033L + 2.5898$	0.980
Maximum Displacement	$y = 1.337A - 7.768L + 228.977$	0.992
Base Shear	$y = 11.914A + 200.566L - 139.842$	0.999

Where, y = Responses, A = Cantilevered offset in terms of percentage of the building lateral dimension

L = Building lateral dimension at bottom

The coefficient of determination R^2 is a statistical measure that measures the proportion of variance in the dependent variable (responses) that is predictable from independent variables (% of cantilevered offset and horizontal length of building). R^2 value will tell how well the regression predication approximate the real data points. The range of R^2 in the multiple linear regression is 0 to 1. Residual is the difference between the actual values and the values obtained from the equation (predicted values). Residual represent how much the predicted value is deviated from the actual value. The positive residual represent that the predicted value is less than the actual value and negative residual represent that the predicted value is more than the actual value. The graphical representation of residual Vs predicted values from the formulated equations are shown below in figure 16 to figure 18.

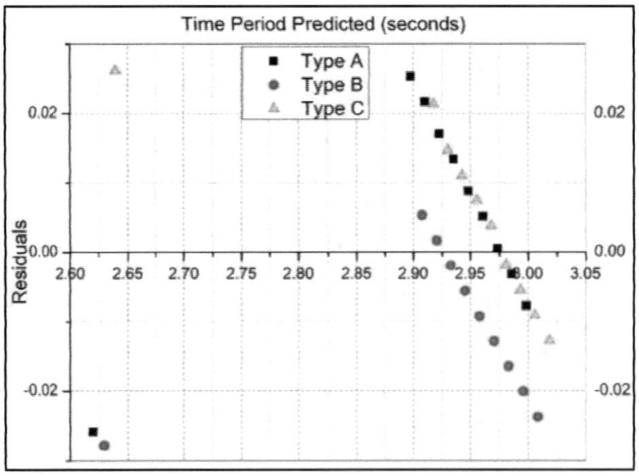

Figure 16. Time Period Predicted Vs Residuals

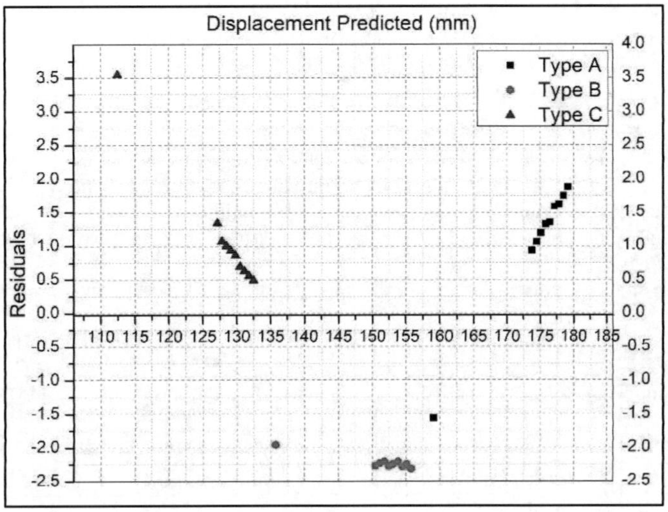

Figure 17. Wind Analysis – Displacement Predicted Vs Residuals

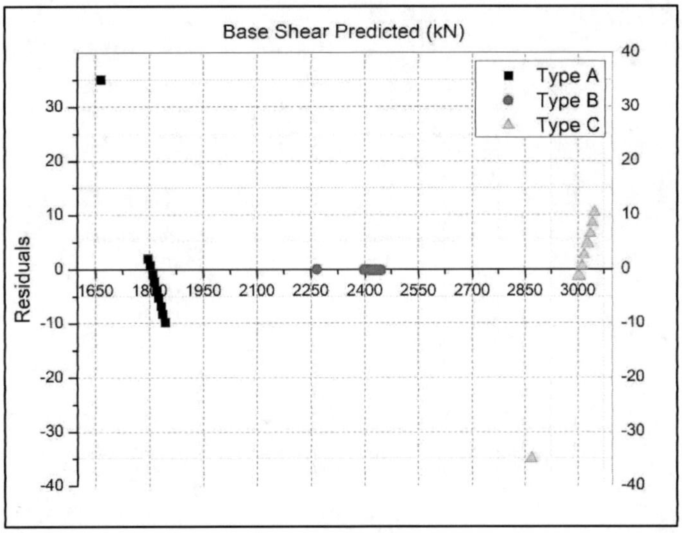

Figure 18. Wind Analysis – Base Shear Predicted Vs Residuals

6. Conclusions

The analytical study on vertical geometric irregular tall structural system was carried out in this study. From the results and comparative study, following conclusions are drawn:

- From the modal analysis it can be concluded that, the increase of time period with respect to increase of cantilevered offset is less in buildings having larger plan dimension. Time period of torsional mode of vibration is less than 0.9 times of the smaller of the time period of translation mode of vibrations for all type of buildings as per IS 16700: 2017.
- The increase of lateral displacement of the buildings with increasing the cantilevered offset is more for buildings having smaller plan dimension for the action of wind load.
- The storey drift and storey levels at which there will be maximum storey drift was found to be decreasing as the plan of the building increases for vertical geometric irregular tall RC

buildings. The storey drifts are within the limits provided by the IS 16700: 2017.

- There is an equal increase of 2% base shear as the cantilevered offset is increased from 11% to 15% for all type of buildings.

References

[1] Xie Z N and Gu M 2004 Mean interference effects among tall buildings *Eng. Struct.* **26** 1173
[2] Bailey P A and Kwok K C S 1985 Interference excitation of twin tall buildings *J. Wind Eng. Ind. Aerod.* **21** 323
[3] Boggs D W and Peterka J A 1989 Aerodynamic model tests of tall buildings *J. Eng. Mech.* **115** 618
[4] Rahul L, Saraswathy B and Udaya L 2014 Effect of vertical irregularity on performance of reinforced concrete framed buildings *Proc. of the Second Intl. Conf.* **52**
[5] IS 16700. 2017 Indian standard criteria for structural safety of tall concrete buildings. New Delhi: Bureau of Indian Standards.
[6] El-Leithy N F, Hussein M M and Attia W A 2011 Comparative study of structural systems for tall buildings *J. American Sci.* **7** 707
[7] Mele E, Imbimbo M and Tomei V 2019 The Effect of slenderness on the design of diagrid structures *International J. High-Rise Build.* **8** 83
[8] Laogan B T and Elnashai A S 1999 Structural performance and economics of tall high strength RC buildings in seismic regions *The Structural Design of Tall Buildings*. **8** 171
[9] Smith R J and Willford M R 2007 The damped outrigger concept for tall buildings *The structural design of tall and special buildings*. **16** 501
[10] Zou X K and Chan C M 2005 Optimal seismic performance-based design of reinforced concrete buildings using nonlinear pushover analysis *Eng. Struct.* **27** 1289
[11] IS 1893 part 1. Indian standard criteria for earthquake resistant design of structures. New Delhi: Bureau of Indian Standards; 2016.

IOP Conf. Series: Materials Science and Engineering 1006 (2020) 012022 doi:10.1088/1757-899X/1006/1/012022

The Comparative Study of Treated Sewage Water on Mechanical Properties of High Performance Concrete M40 with Conventional Concrete

Lakshmi Lavanya Damaraju[1] , Jonna Bharath Kumar [2], B Prathyusha Yadav[3], MD Basid[4]

[1]Assistant Professor, Vignana Bharathi Institute of Technology
[2,3,4]Student, Vignana Bharathi Institute of Technology

E-mail: lavanya.damaraju@vbithyd.ac.in[1], lldamaraju@gmail.com[1], bharathkumarj1807@gmail.com[2], payalyadavboddu@gmail.com[3], basidcivil@gmail.com[4]

Abstract. Concrete, the widely used building material in construction industry.Cement is the primary constituent of concrete, during its production releases the large amount of carbon dioxide in the atmosphere creating the serious environmental threats like air pollution . The paper discuss the ways to reduce cement composition as building material, so that the construction industry will be more environment friendly by replacing the cement with fly ash to an extent . Fly ash is waste product in thermal power plants creatingdisposal problems on landfills.Treated water effluent from STPs are used in the preparation of concrete makes benefits to the environment .The study investigates the water quality tests like Ph, dissolved oxygen , turbidity , COD,BOD, TSS .The Mix proportions was obtained for the preparation of M40 grade concrete and the Destructive (compressive strength)tests and non-Destructive Tests (rebound hammer , ultrasonic pulse velocity test)was conducted and analysed and compared with conventional M40 grade concrete test values for different ages of concrete 7,14,28 days
.

1. Introduction

Concrete a most important Building material causes a large cement production in the world.Cement produces 2.5 billion tons of CO_2 Causing a huge greenhouse effects over worldwide.Fly-ash is a by-product of thermal power plants which presently creating a serious waste Disposing Problem .This study highlights the application of percentage of fly-ash content in mixing the concrete, which slowlydecreases the production of cement to some extent .Fly-ash itself acts as a mineral admixtures and increases the life of concrete, plays a major role in reduction of Energyuse, decreases the amount of waste reaching to the landfills. Thus, usingfly-ash in the concrete is a eco-friendly. Constructionindustry, utilizes more amount of water in preparing the concrete, curing, of concrete blocks, curing of walls, using of water to clean the concrete mixer after utilization etc. In present scenario the increasing trend of water scarcity and water deficit in areas over the world leads to the conservation of water. The present study was done in order to reduce the usage of tap –water in construction industry, and to reuse the treated Sewage water, the treated water is used in concrete mix. Water, the key ingredient forms a paste when mixes with cement and forms a chemical bond through Hydration Process. The quality of water and amount of water used in concrete preparation plays a major role in the fresh and Hardened Properties

including Workability, Compressive strength,durability, Destructive Properties etc.The water parameters variations like chlorides , iron ,alkalinity inorganic salts shows their effects on initial strength , strength achievement ,bond between cement paste and aggregate ,property of concrete .
To improve the properties Admixtures are added to concrete during freshly mixing and to achieve certain effectively. Water consumption for the Building construction of 1000 sq.ft

- 20000sq.m X 1000L/sq.= 20,000,000 L or 20000m^3

The demand or Availability of water varies with respect to season, with respect to source (either Ground water or tap water).In present Scenario the eventualities from the tanker costs Rs. 300 /m^3

- 300/m^3 X 20000m^3 = 60.00 lakhs

2. Methodology

In present study,Concrete cubes wasprepared withcement,sand, 20mm aggregates and collected treated water samples. Superplasticizer FOSROC SP 430 to maintain the workability .In the study M 40 grade concrete was selected.

Table 2.1: Specific gravity tests was conducted on the materials

S.No	Material	Specific Gravity
1	Cement	3.15
2	Fine aggregate	2.7
3	Coarse aggregate	2.7

According to the IS 456:2000(CODE OF PRATICE AND REINFORCED CONCRETE), Materials used in the preparation of concrete,

- Cement : 53 grade
- Flyash : IS 3812
- Coarse aggreagate: 20mm passed aggregate IS:383
- Super Plasticizer : FOSROC SP 430

Table 2.2: Mix proportioning of the concrete was tabulated

Materials used	Volume in Kg/m^3	Proportion
Cement + flyash	465	1
water	186	0.4
Fine aggregate	684	1.47
Coarse Aggregate	1160	2.494

Fig. 2.1 Workability test

IOP Conf. Series: Materials Science and Engineering 1006 (2020) 012022 doi:10.1088/1757-899X/1006/1/012022

The treated water is collected from STP ,PATEL CHERVU , NACHARAM, HYDERABAD, Sewage treatment plant (STP) is 25 MLDPLANT
The sewage treatment involves mainly the following steps

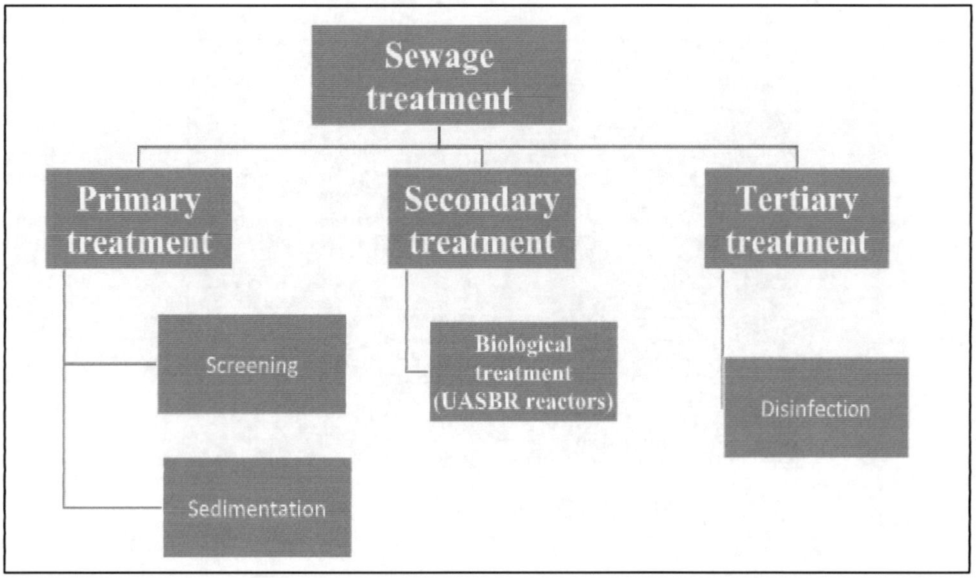

Fig. 2.2: Flowchart of Sewage treatment

2.1 Ultra sonic pulse velocity
It is an non-destructive insitu test , to check the quality of the concrete, the strength and quality of concrete is assumed by measuring the velocity of an ultrasonic pulse passing through a concrete structure. From the pulse velocity travelling through the concrete, the length of the cube density of concrete Dynamic modulus of elasticity values can be calculated

Fig 2.1.1 Ultrasonic Pulse Velocity Test

2.2 Rebound hammer
Also known as a Swiss hammer or a rebound hammer or concrete hammer test , the elastic properties or strength of concrete can be measured by this hammer test

Fig 2.2.1 Rebound Hammer Test

2.3 Comprressive strength
This gives the capacity of the concrete to withstand loads. it gives all the characteristics of concrete , under the action of compressive forces , the resistance offered by the concrete can be measured

Fig 2.3.1 Compressive strength test

2.4 Water quality parameters
The treated water is analysed in the laboratory to determine the various parameters like Alkalinity,Ph,chloride, calciumchloride,iron,sulphates, Total dissolved and suspended salts The water quality for construction plays a major role in the initial strength of concrete, workability of

IOP Conf. Series: Materials Science and Engineering 1006 (2020) 012022 doi:10.1088/1757-899X/1006/1/012022

the concrete under drying conditions, the salts can rise upon the surface by evaporation process and in extreme conditions causes cracking and scaling. Suspended solids presence creates the turbidity, which reduces the bonding nature between cement paste and aggregate further reduces the strength of concrete considerably. The presence of chloride slats in the mixing shows it effect in long-term on reinforcement of the slab.

Fig 2.4.1 Water sample

3. Results & discussions:

The concrete cubes of sizes 150X 150X150 $mm^3$was prepared by the treated water and as well as with normal water has been tested for UPV, Rebound hammer, Compressive Strength and Dynamic Modulus of elasticity.

Table 3.1: comparison of test results of normal water cubes and treated water cubes

S No	Age of Concrete in Days	UPV in km/sec		Rebound hammer result in N/mm2		Compressive strength in N/mm2		Dynamic modulus of elasticity in GPa
		Normal water	Treated water	Normal water	Treated water	Normal water	Treated water	Treated water
1	7	4.65	4.5	38.96	33.86	33.7	30	46.3
2	14	4.7	4.6	40	31	36.3	32.88	47.4
3	28	4.8	4.7	41.25	35.66	42.77	40.77	49.57

Some physical and chemical properties of treated water are evaluated in the laboratory from the tests namely Ph ,Alkalinity, chemical oxygen demand , biological oxygen demand , dissolved oxygen , turbidity

Table3.2 : Results of physical and chemical properties of treated water

S.no	Parameter	Unit	Test value
1	pH	-	7.99
2	Dissolved oxygen	mg/L	4.5
3	Turbidity	NTU	3.00
4	COD	mg/L	30
5	BOD	mg/L	7
6	chlorides	mg/L	280

4. Conclusions

a. Based on the test results, the slight variations has been observed in the mechanical properties of concrete , when comparison was done between normal concrete and treated water concrete .
b. The usage of treated water decreases the cost of construction to an extent in a place where tanker water are used in construction .
c. Both reuse of treated water and utilization of flyash in construction is a eco-friendly method avoiding the dumping problem of waste in landfills and also conserving of fresh water .
d. The variations was studied for 28 days test results having UPV 2.08%, Rebound hammer test 13.55% and compressive strength 4.67% .
e. Rebound hammer test value indicates the long-term effects might show on the concrete prepared with treated water .
f. The presence of chlorides in treated water shows its hardened effect causing the variation in compressive strength percentage

Acknowledgement

Authors would like to acknowledge the computer centre funded by DST-FIST at Vignana Bharathi Institute of Technology.

References

[1] Kucche K J, Jamber S S, Dr Sadgir P A 2015 Quality of water for making concrete-A Review of literature *International Journal of Scientific and Research Publication,*Volume **5** ,Issue 1
[2] Jabri K S AL, Saidy A H, Taha R and Kemyani A J AL 2011 Effect of using Wastewater on the Properties of High Strength Concrete *International Journal of Scientific and Research* Volume **7**
[3] Gadzama E W, Ekele O J, Anamtemfiok V E and Abudakar A U 2015 Effect of Sugar Factory Wastewater as Mixing Water on the Properties of Normal Strength Concrete *International Journal of Science Environment and Technology* Volume **4** No.3
[4] Nirmalkumari K and Shivkumari V 2008 A Study on the durability impact of concrete by using recycled waste water *Industrial Pollution Control* **24(1)**
[5] IS 3025: 2009 Method of Sampling and Test (Physical and Chemical) for Water and Wastewater [12] IS 456:2000
[6] Plain and Reinforced Concrete *Code of Practice.*
[7] Rickert J and Grube H 2000 Analysis of recycled water component *VDZ, Concrete Technology Reports* (Düsseldorf, Germany).

An Experimental Investigation on Soil Stabilization using Terrazyme

Pala Gireesh kumar[1], Nagubandi Madhunika[2], Yakkala Sindhura[3], Saladi Dwarakamai[4], Addanki Praneetha[5], Talapanti Harika Sravani[6] and Puchakayala Sridevi[7]

[1]Associate Professor and Head, Department of Civil Engineering, Shri Vishnu Engineering College For Women, Bhimavaram, West Godavari District- 534202, India.

[2,3,4,5,6]U.G Student, Department of Civil Engineering, Shri Vishnu Engineering College For Women, Bhimavaram, West Godavari District- 534202, India

E-mail: palagireeshk@gmail.com; nmadhunika468@gmail.com; sindhuray111@gmail.com

Abstract. Soil is defined as minerals, gases, organic matter, liquid and organisms which together supports the life. Black cotton soil being clayey, deep, and impermeable, expands and becomes sticky during rainy season and contracts during the dry season engendering deep cracks. Black cotton soil being more expansive in nature, the shear strength is very less and thereby, lowering the bearing capacity too. To overcome the problem of low strength and bearing capacity of locally available soil and to stabilize soil for a longer life of buildings, improvisations have become a necessity. In this present study an effective technique is provided for ground improvement for soil stabilization using bio enzyme. Bio-enzyme namely Terrazyme is added to improvise Unconfined Compressive Strength (UCS) and California Bearing Ratio (CBR) value of soil, and these are further compared with virgin soil's CBR and UCS values. The laboratory tests were conducted like California Bearing Ratio Test, Atterberg Consistency Limits and Compaction test on soil samples with differential Terrazyme. Percentages for redefining the properties. From the results, it is evident that Terrazyme has proven to be a most satisfying add-on that helps in improvising the supreme characteristics of the soil with the usage up to 0.2 - 0.4 ml. Being economical and eco-friendly makes it to stand out of all and has made it more preferable than other enzymes. The study concludes, the soil has become more stable to support a greater life span and more strengthened in the presence of Terrazyme. The experimental investigations and further discussions concerning Terrazyme usage can be more precisely contemplated in the paper.

1. Introduction

Soil stabilization is defined as alteration of one or more properties of soil, by chemical or mechanical means, to create improved material of soil which possesses the engineering desired properties. Soil is stabilized to prevent erosion and dust generation or to increase strength and durability or Irrespective to the purpose for stabilization, the desired result is the formation of a soil material or soil system that will remain in position under the conditions of design use the life of the project. There exist various methods aimed at stabilization of soil like mechanical, chemical, cement, lime, bituminous, thermal, electrical stabilization, stabilization by grouting etc.

1.1 Basic principles of soil stabilization

a. Properties of stabilizing soil will be evaluated.
b. Deciding the lack of property of soil and choose economical and effective method of stabilization the soil.
c. Designing the stabilized when values are durability.

1.2 Need for soil stabilization

a. To provide a complete road network system to build in conventional method, Limited financial resources are used.
b. Use of locally available soil and appropriate stabilizing agent effectively.
c. Encouraging the use of wastages of industries in building low cost construction of roads.

Enzymes remain adsorbed by clay lattice, and then releases upon exchange through the metals cations. These makes a significant effect on clay lattice, which causes them to expand initially and then becomes tighten. Enzymes can also be captivated by colloids enabling them to be transported through the soil electrolyte media. Enzymes also helps release hydrogen ions from the soil bacteria, which results in pH gradients at the surfaces of particles of clay, which supports in breaking the structure of the clay. By definition, enzyme is an organic catalyst which helps in speeding up the chemical reaction or else it happen at the slower rate, and it doesn't become the part of end product. To form an intermediate reactant, enzyme combine with the large organic molecule, which exchanges ions with the structure of clay, causing the cover-up effect and breaking down the lattice, which prevents the further loss of density and absorption of water.

The thickness of electrical double layer reduced by organic cation which allows enzyme treated soils to be more compacted and more tightly together. Enzyme supports in the development of cementitious compounds using the following reaction:

$$H_20 + \text{clay} \xrightarrow{\text{Enzyme}} \text{CSH}.$$

Terrazyme is ancation-reactive synthetic compound which forms a protective coating on oily layers of clay and particles of soil. Terrazyme reduces ion exchange and ion mobility. Simultaneously by eliminating the absorption of water the material becomes hydrophobic. The result obtained is a soil material which is less sensitive to moisture and more workable. And it can be compacted for a better particle-interlock state by traffic forces and equipment. Improved bearing capacity and higher internal friction leads to better particle interlock. It also means penetration of water is less and greater density. The reagent which tends to be active is always bonded to material particles and if reagent is in excess, additional water will facilitate deeper penetration into the soil horizon till the entire reagent has be absorbed. The basic effect of enzyme action in the structure of the soil can be summarized as, initially, the film of absorbed water is reduced or broken entirely. By purely mechanical method extension and shrinkage of colloidal soil constituents cannot be eliminated completely. However, by means of addition or removal of water, temperature effects with mechanical pressure, it is possible to vary the amount of water held in this manner. Such variations are attended by shrinkage or swelling. This provides an ideal point of operation for the enzyme.

In this Present study Black Cotton Soil (BCS) is considered with high plastic and high shrinkage and swelling properties. BCS is the crucial segment of this nature and road improvement industry knows the essential of it for asphalt work. So as to get the specified limits for plasticity index and liquid limit and other strength characteristics for airfields and roads as sub-base course, soil stabilization was tried using liquid stabilizer Terrazyme. By using of these materials economizes the overall cost and improves stability of sub-base for pavements. BCS were stabilized by using different Enzyme dosages and stabilized soil strength has been evaluated after curing period of 0 day. The tests are carried out to determine consistency limits, CBR test of the soil specimens without and with stabilizer Bio-enzyme.

In present investigation attempt is made to stabilize BCS with Enzyme (Terrazyme). The consistency limit test, compaction and CBR tests are carried out in the laboratory for different mix proportions of Bio-enzyme with BCS.

2. Literature review

Faisal Ali et.al(2012) This research focused on three natural residual soil's to improvement these engineering properties and liquid chemicals are mixed with different proportions. Unconfined compressive strength (UCS), chemical performances as soil stabilizing agents and evaluate the effectiveness in moisture-density relationship (compaction).And can reduce plasticity and shrinkage by addition of liquid stabilizer by water molecules re-absorption elimination. On surface of clay platelets by ionizing and exchanging the water molecules it reduces optimum moisture content. Re-arranging the clay platelets by neutralizing it, maximum dry density increases and by increasing inter particles bonding increase compressive

C. Venkata Subramanian et.al(2011) Strength parameters were studied after the application of enzyme of four different dosages for period of two to three weeks with three different soils.

Penget al.(2011) On three soil's unconfined compression tests is conducted: soil l ,soil ll ,soil lll are named as fine-grained, slit loam and coarse grained textures respectively. With quick lime and an enzyme three soils were stabilized. In two different conditions upto 60 days the samples were cured; The two containers are air-dry and in sealed container. At room temperature samples were allowed to dry in air - dry curing where as in samples during moisture was pressured in curing time in sealed containers. In air - dry curing the enzyme are more effective for soil l and soil ll than quicklime. For soil lll air - dry curing is not effective and sealed curing for three soils too. The quick lime was more effective in sealed containers than enzyme in specimens which are not allowed to evaporate quick lime further hydration is promoted.

Shukla.M et al (2010) In order to assess its suitability we use organic, eco-friendly bio enzyme, non- toxic stabilizers that are treated with expansive soils are made experiments .when bio enzyme stabilizer is used in experiment it is effective and on wet side of OMC expansive soil of swelling decreases.

M B Mgangira et al (2009) Effects of enzyme with liquid chemicals as soil stabilizer. Plasticity index(I_p) of 35 soil and 7 PI of other. Tests- atterberg limits standard Proctor and unconfined compressive strength

- decrease in PI of both soil treatment are with enzyme based products
- Two products of soils are treated with enzyme based chemicals on UCS showed a mixed effect. UCS improvement is not consistence significant could be attributed for treatment.

A.U.Ravishankar et. .al(2009) Terrazyme soil stabilizers product with lateritic soil conducted a comprehension study in dakshinakannada and udupi districts are used in pavements as base course doesn't satisfy the requirements it satisfies the atterberg limits until unless soil is blended with the sand at various proportion proportions to improve its properties. In unconfined compressive strength (UCC) soil and blended soil are effected of enzyme

R.A.Velasquez et.al (2006) In this study two types of soils are investigated and effectiveness of stabilization and mechanisms of enzyme products are also investigated using chemical analysis and resilient modulus testing.

Marasteanu et al. (2005)In this study soils are stabilized using two enzymes and conducted two- axial and resilient modulus tests. Chemical analysis included for the enzyme are PH and metal concentrations.

Miburn & parsons(2004) Different tests conducted on soils which are stabilized with cement, lime, terrazyme and class c fly ash. Tests conducted are leach test, atterberg limits, freeze-thaw, wet-dry and the strength tests.

Alan F. Rauch et.al (2002) In this study, measure changes in the engineering properties standard soil tests are conducted. In this study three reference clays are used. And three representative liquid stabilizers. Each soil is characterized in terms of the compacted unit weight, Atterberg limits, undrained triaxial shear strength and 1D free swell potential.

In the present investigation attempt is made to stabilize BCS with enzyme (Terrazyme). Various experiments were carried out in the laboratory for finding the index and engineering properties of the different materials used during the study. California Bearing ratio (CBR), Consistency limits and compaction tests were conducted by using various percentages of terrazyme mixed with expansive soil for finding optimum percentage of terrazyme. The details of the tests are given below

3. Laboratory tests

3.1 Atterberg's limit

Liquid limit (L.L) is defined as the water content, where soil tends to behave as liquid. Plastic limit (P.L) can be expressed as at which point soil ceases to be plastic in nature. Plasticity index is the measure of soil plasticity. It can be also expressed as difference between liquid and plastic limits.

Fig.1 Liquid Limit test

Fig.2 Plastic Limit test

Table 1 Influence of terrazyme on plasticity characteristics

S.No	Dosage of Terrazyme (ml)	Liquid Limit (%)	Plastic Limit (%)	Plasticity Index (I_p) I_p= L.L-P.L
1.	0	68.00	25.00	43.00
2.	0.2	67.71	25.50	42.21
3.	0.4	67.52	26.22	41.30
4.	0.6	66.93	26.81	40.12
5.	0.8	66.26	27.15	39.11
6.	1	65.81	27.35	38.46
7.	1.2	65.45	27.56	37.89
8.	1.4	64.84	27.89	36.95
9.	1.6	64.22	28.12	36.10
10.	1.8	63.91	28.59	35.32
11.	2	63.15	28.96	34.19

3.2 : Differential free swell index

Known as the increase in the volume of the soil which doesn't consists any external constraints, on submergence in the water.

Table 2 Variation of Differential free swell for soil samples treated with the various percentages with terrazyme

S.No	Dosage of Terrazyme(ml)	Differential Free Swell Test (%)
1.	0	140
2.	0.2	136
3.	0.4	127
4.	0.6	118
5.	0.8	107
6.	1	98
7.	1.2	86
8.	1.4	79
9.	1.6	66
10.	1.8	62
11.	2	57

3.3 Compaction test

It is conducted to reduce the space of voids in the soil sample and also to understand the different compaction characteristics for various types of soils.

Fig.3 Compaction Test

Table 3 Variation of compaction parameters for BCS treated with various percent of Terrazyme

S.No	Dosage of Terrazyme(ml)	Dry Density (kN/m^3)
1.	0	15.82
2.	0.2	18.1
3.	0.4	20.4
4.	0.6	22.7
5.	0.	22.9
6.	1	21.1
7.	1.2	20.3
8.	1.4	17.6
9.	1.6	16.9
10.	1.8	15.0
11.	2	14.2

3.4 California Bearing Ratio Test
Also known as CBR. defined as ratio of the force per unit area.

Fig.4 California Bearing Ratio Test

4 Results and discussions
Atterberg limits like liquid and plastic limits were found in the laboratory tests for the expansive Black cotton soil and conducted tests on BCS with addition of terrazyme, i.e; 0.2%, 0.4%, 0.6%, 0.8%, 1%, 1.2%, 1.4%, 1.6%, 1.8%, 2% by adding the percentages of terrazyme.

Water content significantly affects the designing properties of soil. From the present study the maximum water content occurred at the BCS is doesn't reacted with terrazyme. By changing the percentages of Terrzyme the water content is gradually decreases. In the present study, at 25 blows the water content is 63% as per Fig.5.

IOP Conf. Series: Materials Science and Engineering 1006 (2020) 012023 doi:10.1088/1757-899X/1006/1/012023

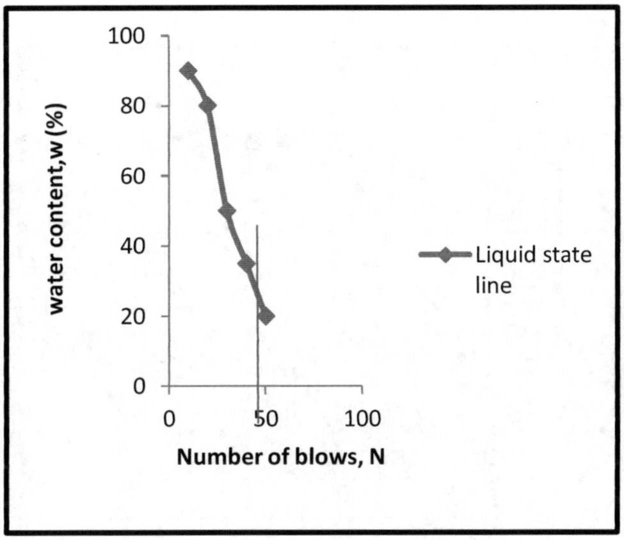

Fig .5 -Liquid limit for various percentage of terrazyme

In present study, content of terrazyme increases water content where plastic limit is increased gradually. The water content occured at BCS is not treated with the terrazyme.

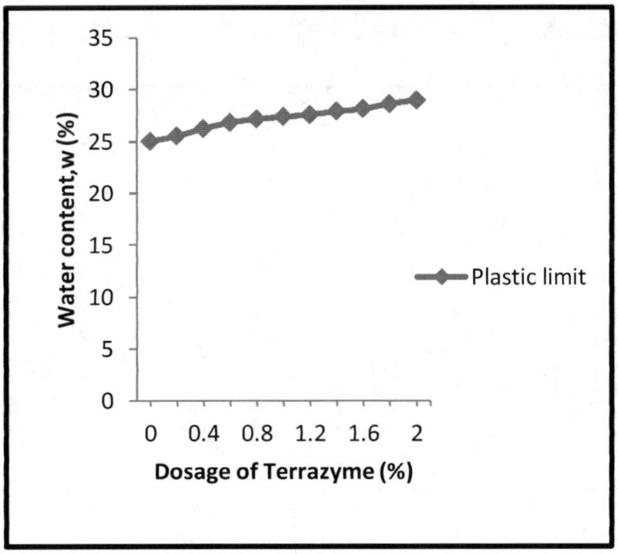

Fig .6 Plastic limit for various percentage of terrazyme

In the present study, minimum plasticity is occured when the BCS is not reacted with the terrazyme. Defined as range of the content of water at which soil tends to behaves plastically. By adding terrazyme more in percentage, plasticity index increases gradually as shown in the figure 7.

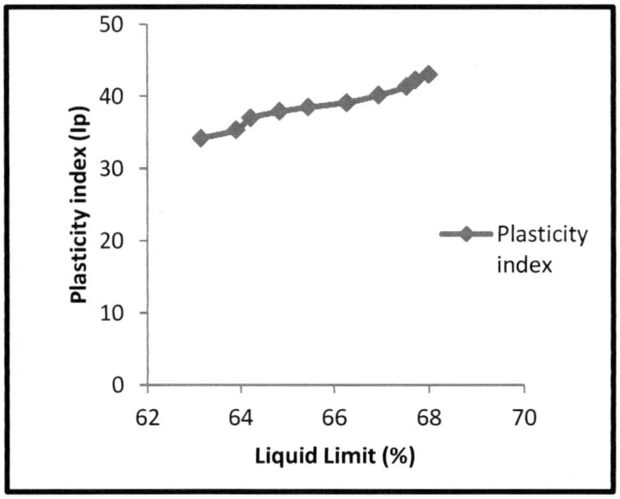

Fig .7 Plasticity index vs liquid limit fir various percentage of terrazyme

Variation in the free swell index at various percentages of terrazyme as represented in figure 8. In this figure we can observe that free swelling index is 140% of parent soil without adding terrazyme. After the addition of terrazyme from 0.2% - 2%, free swell index decreases gradually.

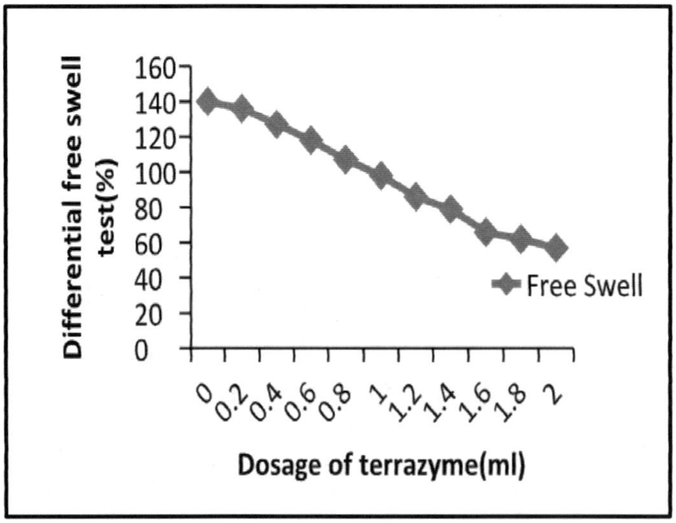

Fig.8 Differential free swell test for various percentages of terrazyme

After completion of compaction test on BCS with and without addition of terrazyme in various percentages, we can observe the changes in Maximum Dry Density (MDD). MDD for parent soil is 18.2KN/m^3 respectively. But increase in percentage terrazyme, MDD changes gradually.

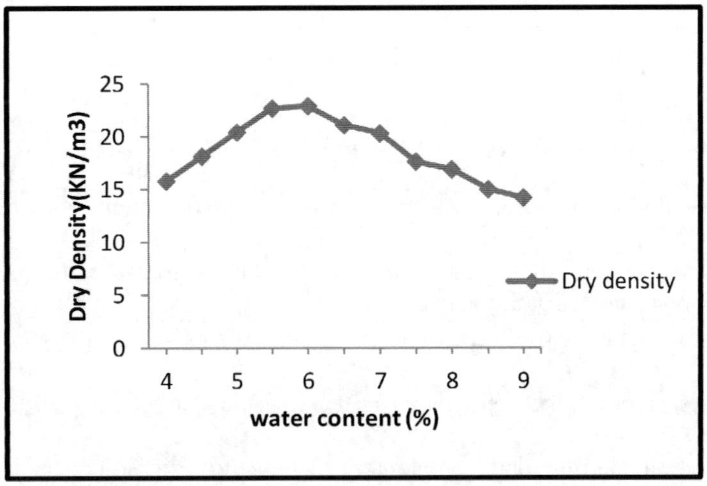

Fig .9 –MDD for various percentage of terrazyme

Samples are prepared and tested as per Is code procedure to find out the CBR value, We can observe that compressive strength of the samples increases gradually. At 2ml of terrazyme sample showing a maximum value. The curve from the Fig.10 showing the various occurs. The CBR values are gradually increases when BCS is treated with terrazyme.

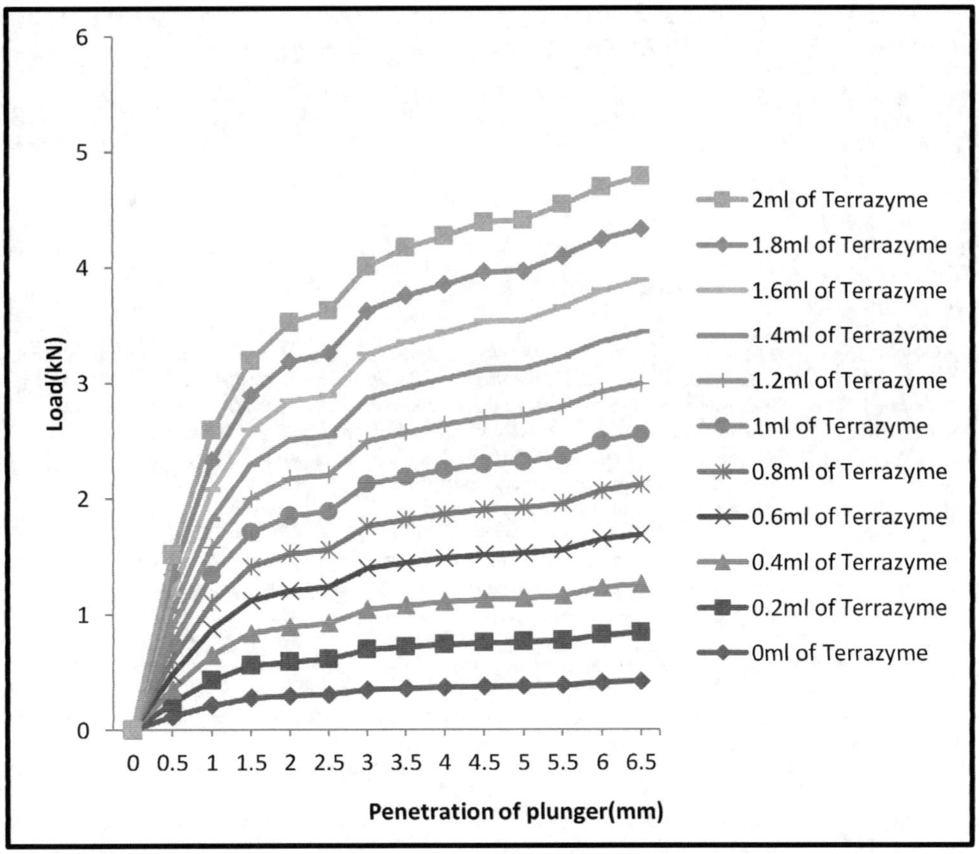

Fig.10 Penetration of plunger vs load for various percentage of terrazyme

5. Conclusions

- Addition of terrazyme in the soil increases the pecertange of coarser particles and reduce in the clay content and also reduces the liquid and plastic limit of the undefined soil. This liquid and plastic limit decreases which are irrespective to percentage of terrazyme used in the soil.
- Terrazyme is a non-corrosive, natural, non-toxic and biodegradable liquid. It is eco-friendly and does not pose any harm to the user.
- Usage of terrazyme increases strength of soil which is prooved by the increase in CBR and UCS values.
- Terrazyme also reduces the void spaces in the soil and thus, increase in the copaction and the density soil.
- Consistency limits and the optimum content of moisture soil is decreased due to action of terrazyme.
- By decreasing the soil permeability terrazyme makes soil water resistance.

References

[1] Faisal Ali 2012 Stabilization of residual soils using liquid chemical *Electronic Journal of Geotechnical Engineering (EJGE)* Vol **7** 116-117.
[2] Akshaya Kumar 2012 A study on some geotechnical properties of lime stabilized expansive soil –Quarry dust mixes *International Journal of Emerging trends in Engineering and Development (IJEIT)* Vol **4** 193-196 ISSN 2249-6149
[3] Venkatasubramanian C & Dhinakaran G. 2011 Effect of bio-enzymatic soil stabilization on unconfined compressive strength and California bearing ratio *Journal of Engineering and Applied Sciences* **6(5)** 295-298.
[4] Peng H T, Su H T, Zhang X P & Wang J 2011 An experimental comparison of compressive strengths of soils stabilized with enzyme and ground quicklime *Advance Material Research* **280** 9-12.
[5] Shukla M, Bose S & Sikdar P 2010 Enzyme for stabilization of soil in road construction a cost effective approach *IRC Seminar in Integrated development of rural and arterial road networks for socio-economic development* (New Delhi) Vol **4** 454-459.
[6] Ali F H, Wong L S and Hashim R 2010 Engineering properties of improved fibrous peat *Scientific Research and Essay* Vol **5.** 154-169
[7] M B Mgangira 2010 Evaluation of the effects of enzyme-based liquid chemical stabilizers on sub-grade soils *Southern African Transport Conference* (Pretoria).
[8] Shankar A U, Rai H K & Mithanthaya R 2009 Enzyme stabilized Lateritic Soil as a Highway material *Indian Roads Congress Journal.*
[9] R A Velasquez, M O Marasteanu & R M Hozalski 2006 Investigation of the effectiveness and mechanisms of enzyme products for sub-grade stabilization *International Journal of Pavement Engineering* **7:3** 213-220.
[10] Marasteanu M O, Hozalski R, Clyne T R & Velasquez R 2005 Preliminary laboratory investigation of enzyme solutions as a soil stabilizer *Minnesota Department of Transportation, Research Services*
[11] Mihai O M, Ray H, Timothy R C and Raul V 2005 Preliminary laboratory investigation of enzyme solutions as a soil stabilizer MN/RC – 2005-25 Report *Minnesota Department of Transportation.*
[12] Alan F Rauch, Jacqueline S Harmon, Lynn E Katz and Howard M Liljestr Measured effects of liquid soil stabilizers on engineering properties of clay *Transportation Research Record* 1787 Paper No. 02-3244.
[13] Tentative guidelines for the design of flexible pavements. IRC-37 2012 1-108.
[14] The Carbon Group, L.Perma-Zyme-11x soil stabilization for road construction and natural liners.
[15] Indian Standard - IS: 2720 (Part 2) 1980 Methods of test for soils determination of moisture content (New Delhi, India).
[16] IS-2720 part 3 1985 Reaffirmed 1995 Indian standard method of test for soils (Specific gravity) Bureau of Indian standards, New Delhi.
[17] IS-2720 part IV 1985 Reaffirmed 1995 Indian standard method of test for soils (Grain size analysis) Bureau of Indian standards, New Delhi.
[18] IS-2720 part V 1985 Reaffirmed 1995 Indian standard method of test for soils (Determination of liquid and plastic limit) Bureau of Indian standards, New Delhi.
[19] IS-2720 part VI 1972 Reaffirmed 1995 Indian standard method of test for soils (Determination of Shrinkage factor) Bureau of Indian standards, New Delhi.
[20]IS-2720 part VII 1980 Reaffirmed 1999 Indian standard method of test for soils (Determination of water content-dry density relation using light compaction) Bureau of Indian Standards, New Delhi.

[21] IS – 2720 part 10 1987 Indian standard method of test for soils (Laboratory Determination of UCS) Bureau of Indian standards, New Delhi.
[22] IS – 2720 part 16 1987 Indian standard method of test for soils (Laboratory Determination of CBR) Bureau of Indian standards, New Delhi.
[23] IS-2720 part 40 1985 Reaffirmed 1995 Indian standard method of test for soils (Free swelling index) Bureau of Indian standards, New Delhi.
[24] IS-2720 part 41 1985 Reaffirmed 1995 Indian standard method of test for soils (Swelling pressure) Bureau of Indian standards, New Delhi.

Dynamic Analysis of Multi-Layered Grid Space Structures

K M Mahanthesh[1] and B O Naveen[2]

[1]Post Graduate Student, The National Institute of Engineering, Department of Civil Engineering, Mysuru, India
[2]Assistant Professor, The National Institute of Engineering, Department of Civil Engineering, Mysuru,India

E-mail:[1]kori.mahanth@gmail.com,[2]naveenbo@nie.ac.in

Abstract. In the present study, an attempt has been made to understand the structural behavior of a two-way lattice grid space structure. Two-way lattice grid space structures are analyzed for dynamic earthquake loads such as time-history analysis by STAAD. Pro V8i software, using structural steel hollow tubular sections for a span of 30 m for 4.0 m height column and 3 storeys were considered for a total height of 12 m. Further the dynamic analysis such as displacements and modal analysis to find natural frequency and time period, were carried out for the eight different configurations and three different grid sizes of space truss structure.

1. Introduction

Space structures refers to a structure made of assemblage of linear members interconnected to each other in space. The self-weight of these roofs is less than conventional type of roofs. It is the new structural form which accommodate large unobstructed areas such as sports arenas, canopies, stadiums, exhibition pavilions, reservoir covers, airplane hangars, transportation terminals, workshops, warehouses, etc.

1.1 Structural loads considered

The structural loads considering in structures is discussed in this section. Loads basically divide into vertical loads and horizontal loads. Low and medium rise building designs generally affected by vertical loads like dead and live loads. High rise building design affected by both vertical and horizontal (wind and seismic) loads.

1.2 Time history analysis

This is one of the method0of dynamic analysis, this method overcomes the disadvantages of response spectrum analysis. This method requires greater efforts for computations for calculate response of building at smaller intervals. In this method base of the structure is exposed to selected earthquake motion. For the full period of the earthquake, immediate stress throughout the structure is evaluated at smaller intervals. From the output records we can obtain maximum stress in any member. This method is not generally used in seismic analysis because of its lengthy computer running time.

1.3 Modelling of the structure in STAAD.Pro

The analysis is carried out using STAAD.Pro V8i software. In this work the total roof to be covered is 33 m × 33 m is considered. The double layer of grids is provided by connecting the two layers by inclined members. The dimension of bottom layer is 30 m × 30 m. The interior supports were given are 8 numbers in which 4 columns are provided at the edges and another 4 are provided at corners. The spacing provides between the columns is 12 m. Three stories of each 4 m height are considered for the analysis. At each story levels the lateral beams are connected on all the sides of the structure.

Table 1. Structure details

Components	Details
Plan dimension	33 m × 33 m
Number of grid layers	2
Grid spacing	2 m, 3 m and 4 m
Grid depth	2 m
Story height	4 m
No. of story	3 story
Total height	14 m
Column spacing	12 m
Foundation depth	1.5 m

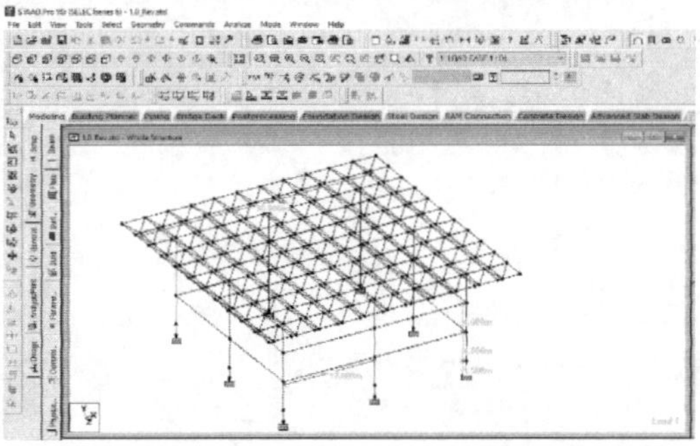

Figure 1. Isometric view of structure

1.4 Properties of the model.

The structural steel tubes were used for the members. Section PIP2445H is used for the columns and the beams which are connected laterally. Section PIP1524H is used for members of grid roof. The sectional properties of the members are discussed in table 2.

Table 2. Section properties from IS 1161- 1998.

Sections	PIP1524H	PIP2445H
Class	Heavy	Heavy
Thickness (mm)	5.4	5.9
Weight (kg/m)	19.6	37.4
Area of cross section (mm^2)	25.0	44.2
Moment of Inertia (mm^4)	674.5	3149

Modulus of section (mm^3)	88.5	258
Radius of gyration (mm)	5.20	8.44

2. Dynamic Analysis of Structure

To study the dynamic characteristics of the structure like horizontal and vertical displacement for the model which has got the optimum depth is chosen which is 2.0 m. The time-history analysis is done for Bhuj earthquake data and Equivalent static method for zone 2 and zone 5. The seismic analysis is also done for 8 different space structure configurations and for module sizes of 2m, 3m and 4m which are given in the table below.

Table 3. Studied Grid Space Structure Cases

S.no.	Configuration	Modul e sizes	Type of analysis	Total Cases
1	Two-way double layered grids	2 m	Time-history method	24
2	Three-way double layered grids			
3	Four-way double layered grids	3 m		
4	Three-way on Periphery			
5	Four-way on Periphery			
6	Two-way three-layered grids	4 m		
7	Three-way three-layered grids			
8	Four-way three-layered grids			

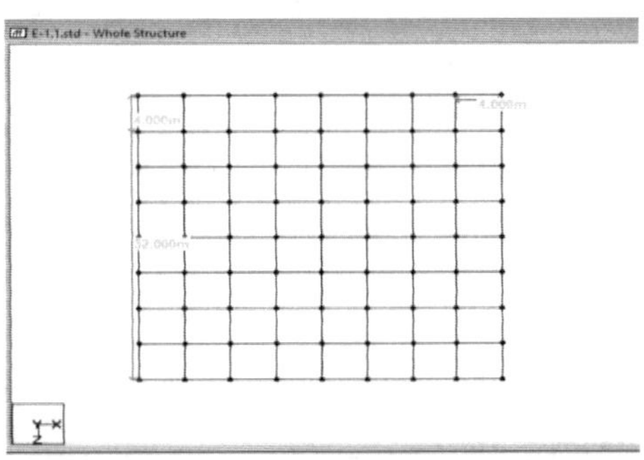

Figure 2. Two-way double layered grids

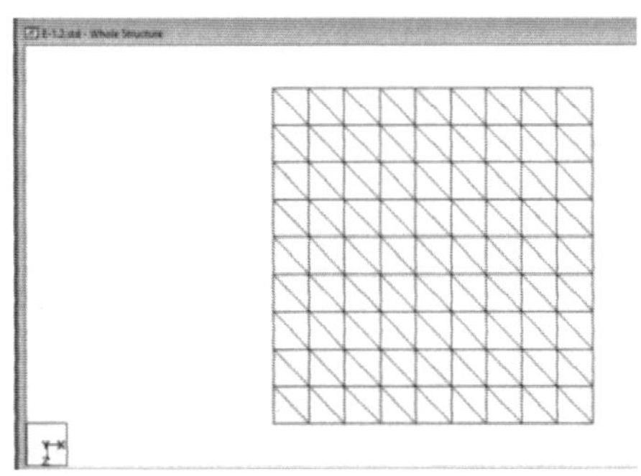

Figure 3. Three-way double layered grids

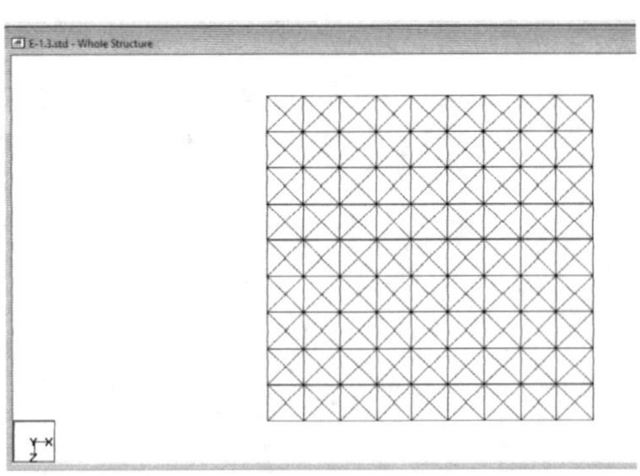

Figure 4. Four-way double layered grids

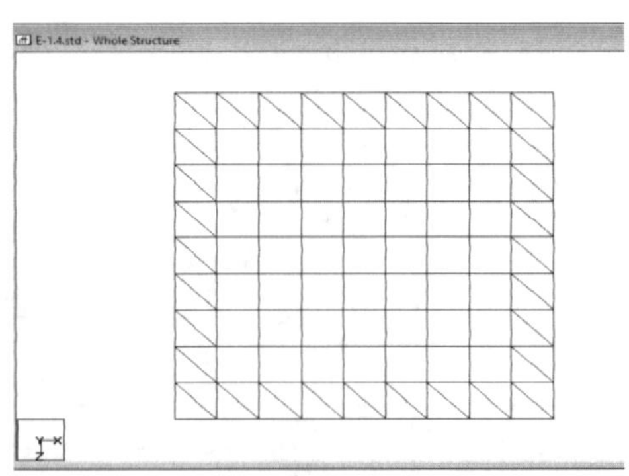

Figure 5. Three-way on Periphery

IOP Conf. Series: Materials Science and Engineering 1006 (2020) 012024 doi:10.1088/1757-899X/1006/1/012024

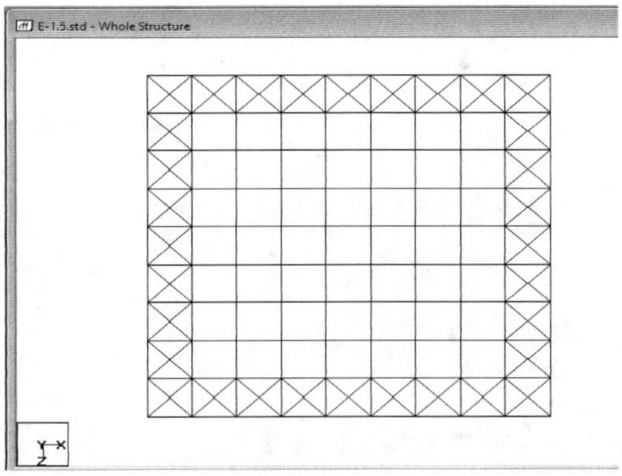

Figure 6. Four-way on Periphery

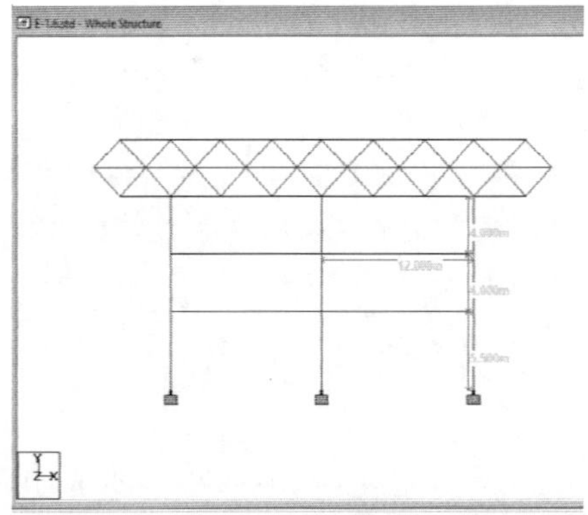

Figure 7. Two-way three-layered grids

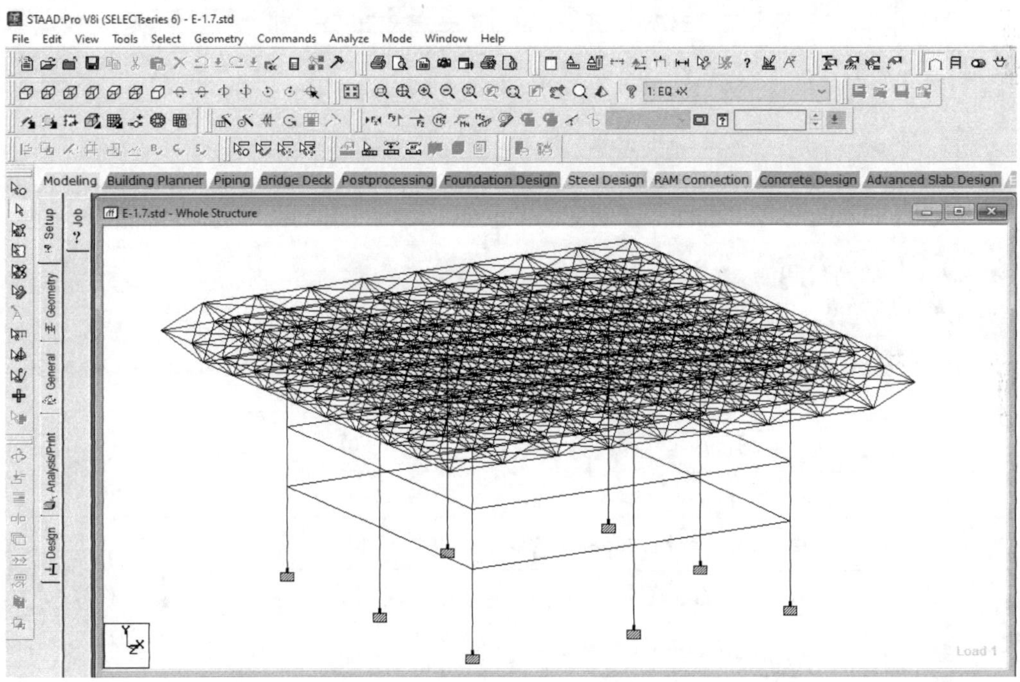

Figure 8. Two-way three-layered grids in 3D view

2.1.Time-History function

In the time history analysis, an earthquake motion is applied directly to the base of the structure. For the full duration of the earthquake, instantaneous stress throughout the structure are evaluated at small intervals. From the output records we can obtain maximum stress in any member. In this study, Bhuj earthquake data is selected for analysis. The graph below shows acceleration vs time graph of Bhuj earthquake data.

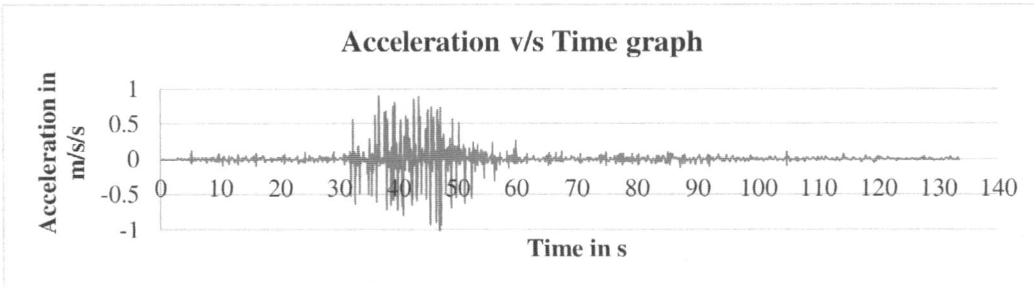

Figure 9. Acceleration v/s Time graph of Bhuj earthquake data

Table 4. Time History Definitions: Type I

Type	Acceleration
Data source	File Bhuj.txt

Table 5. Time History Parameters

Damping	0.05
Time step	-
Arrival Time (sec)	1,2,3

3. Results and discussion

3.1 Results of dynamic analysis of grid space structure

Table 6. Maximum horizontal displacements (mm) for 2 m, 3 m and 4 m grid size

Sl.no	Configuration	2.0 m	3.0 m	4.0 m
1	Two-way double layered grids	52.08	52.65	50.56
2	Three-way double layered grids	62.85	46.36	46.55
3	Four-way double layered grids	63.44	62.64	47.49
4	Three-way on Periphery	55.42	49.41	48.49
5	Four-way on Periphery	60.09	46.29	49.36
6	Two-way three-layered grids	63.23	55.92	52.15
7	Three-way three-layered grids	64.37	64.42	61.99
8	Four-way three-layered grids	82.82	63.81	63.77

Table 7. Maximum vertical deflection (mm) for 2 m, 3 m and 4 m grid size

Sl.no.	Configuration	2.0 m	3.0 m	4.0 m
1	Two-way double layered grids	11.99	10.57	9.33

2	Three-way double layered grids	11.45	8.22	7.79
3	Four-way double layered grids	13.68	8.89	8.97
4	Three-way on Periphery	11.84	9.56	8.35
5	Four-way on Periphery	11.59	8.68	7.45
6	Two-way three-layered grids	8.82	5.67	5.84
7	Three-way three-layered grids	9.23	5.48	5.79
8	Four-way three-layered grids	10.85	6.45	7.15

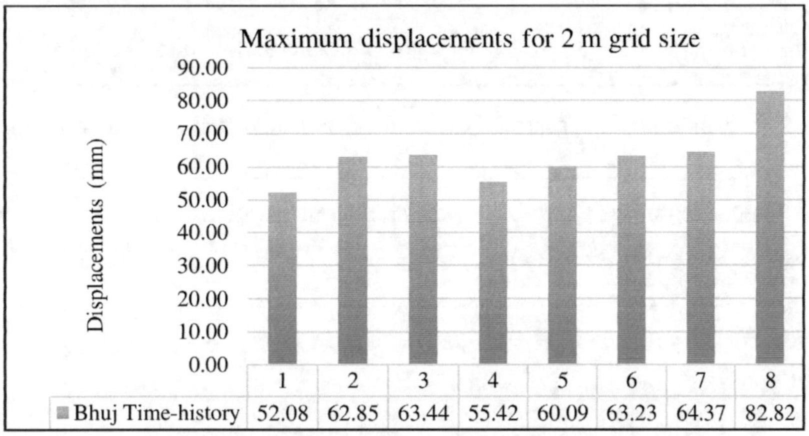

Figure 10. Variation of maximum displacement of 2 m grid size space structure

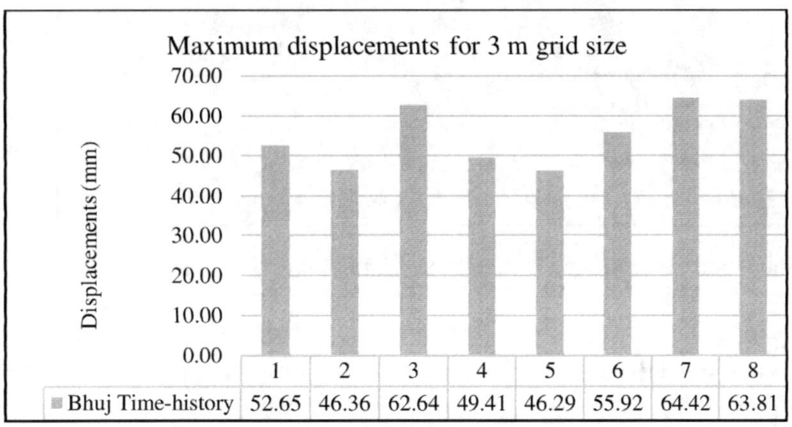

Figure 11. Variation of maximum displacement of 3 m grid size space structure

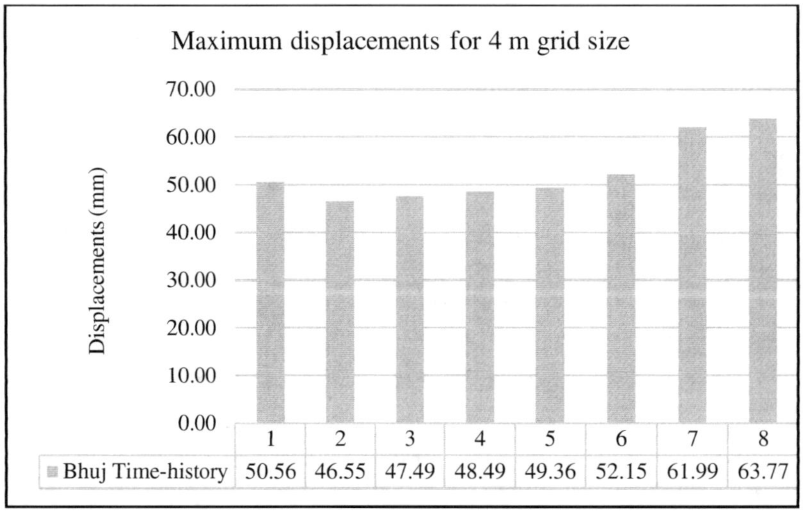

Figure 12. Variation of maximum displacement of 4 m grid size space structure

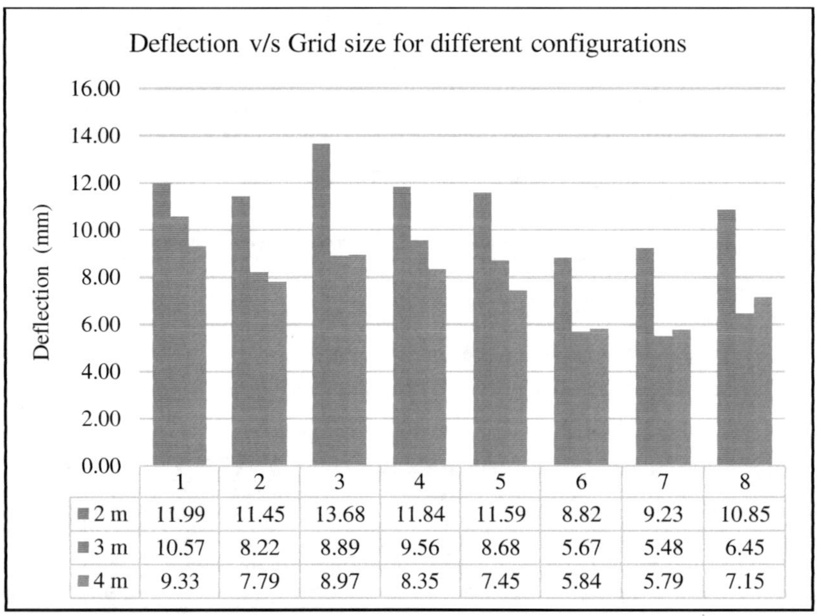

Figure 13. Variation of maximum deflection of 2 m, 3m and 4m grid size space structure

3.2 Modal Analysis Results

The modal analysis is carried out for different configurations of grid space structures for grid spacing of 2 m, 3 m and 4 m. The results of analysis are tabulated in the table 8. A maximum of 6 number of modes is considered. A maximum time period of 2.317 seconds is obtained for Four way three layered grids for 2 m grid size. A maximum frequency of 6.705 Hz is obtained for Four way double layered grids for 4 m grid spacing.

Table 8. Frequency and Time period for 2 m, 3 m and 4 m grid size space structures

1. Two-way double layered grids

	Frequency (Hz)		
Mode No.	**4 m**	**2 m**	**3 m**
1	0.706	0.633	0.713
2	0.706	0.633	0.713
3	0.795	0.751	0.872
4	5.373	5.215	5.484
5	5.39	5.339	5.511
6	5.39	5.339	5.511

Time Period (Seconds)		
4 m	**2 m**	**3 m**
1.417	1.579	1.403
1.417	1.579	1.403
1.259	1.331	1.147
0.186	0.192	0.182
0.186	0.187	0.181
0.186	0.187	0.181

2. Three-way double layered grids

	Frequency (Hz)		
Mode No.	**4 m**	**2 m**	**3 m**
1	0.657	0.579	0.658
2	0.665	0.583	0.666
3	0.748	0.693	0.814
4	5.409	5.224	5.531
5	5.435	5.341	5.551
6	5.478	5.351	5.604

Time Period (Seconds)		
4 m	**2 m**	**3 m**
1.522	1.728	1.52
1.503	1.715	1.501
1.337	1.444	1.229
0.185	0.191	0.181
0.184	0.187	0.18
0.183	0.187	0.178

3. Four-way double layered grids

	Frequency (Hz)		
Mode No.	**4 m**	**2 m**	**3 m**
1	0.654	0.516	0.583
2	0.654	0.516	0.583
3	0.742	0.618	0.72
4	6.674	5.23	5.613
5	6.705	5.349	5.613
6	6.705	5.349	5.615

Time Period (Seconds)		
4 m	**2 m**	**3 m**
1.528	1.938	1.715
1.528	1.938	1.715
1.348	1.619	1.389
0.15	0.191	0.178
0.149	0.187	0.178
0.149	0.187	0.178

4. Three-way on Periphery

	Frequency (Hz)		
Mode No.	**4 m**	**2 m**	**3 m**
1	0.685	0.618	0.693
2	0.69	0.618	0.699

Time Period (Seconds)		
4 m	**2 m**	**3 m**
1.461	1.617	1.443
1.449	1.617	1.431

3	0.764	0.723	0.839	1.309	1.383	1.192
4	5.401	5.216	5.519	0.185	0.192	0.181
5	5.405	5.337	5.524	0.185	0.187	0.181
6	5.439	5.337	5.566	0.184	0.187	0.18

5. Four-way on Periphery

	Frequency (Hz)			Time Period (Seconds)		
Mode No.	4 m	2 m	3 m	4 m	2 m	3 m
1	0.646	0.596	0.662	1.549	1.679	1.51
2	0.646	0.596	0.662	1.549	1.679	1.51
3	0.699	0.682	0.779	1.430	1.467	1.284
4	5.435	5.216	5.553	0.184	0.192	0.18
5	5.441	5.333	5.573	0.184	0.188	0.179
6	5.441	5.333	5.573	0.184	0.188	0.179

6. Two-way three-layered grids

	Frequency (Hz)			Time Period (Seconds)		
Mode No.	4 m	2 m	3 m	4 m	2 m	3 m
1	0.634	0.519	0.616	1.578	1.925	1.622
2	0.634	0.519	0.616	1.578	1.925	1.622
3	0.741	0.625	0.774	1.350	1.599	1.291
4	5.382	5.217	5.493	0.186	0.192	0.182
5	5.382	5.322	5.494	0.186	0.188	0.182
6	5.383	5.322	5.494	0.186	0.188	0.182

7. Three-way three-layered grids

	Frequency (Hz)			Time Period (Seconds)		
Mode No.	4 m	2 m	3 m	4 m	2 m	3 m
1	0.583	0.475	0.565	1.715	2.105	1.77
2	0.59	0.479	0.572	1.694	2.089	1.748
3	0.688	0.577	0.717	1.453	1.734	1.394
4	5.399	5.225	5.514	0.185	0.191	0.181
5	5.443	5.326	5.558	0.184	0.188	0.18
6	5.467	5.337	5.586	0.183	0.187	0.179

8. Four-way three-layered grids

	Frequency (Hz)		
Mode No.	4 m	2 m	3 m
1	0.523	0.432	0.511
2	0.523	0.432	0.511
3	0.619	0.525	0.649
4	5.477	5.231	5.600
5	5.477	5.338	5.600
6	5.501	5.338	5.621

Time Period (Seconds)		
4 m	2 m	3 m
1.911	2.317	1.958
1.911	2.317	1.958
1.616	1.906	1.541
0.183	0.191	0.179
0.183	0.187	0.179
0.182	0.187	0.178

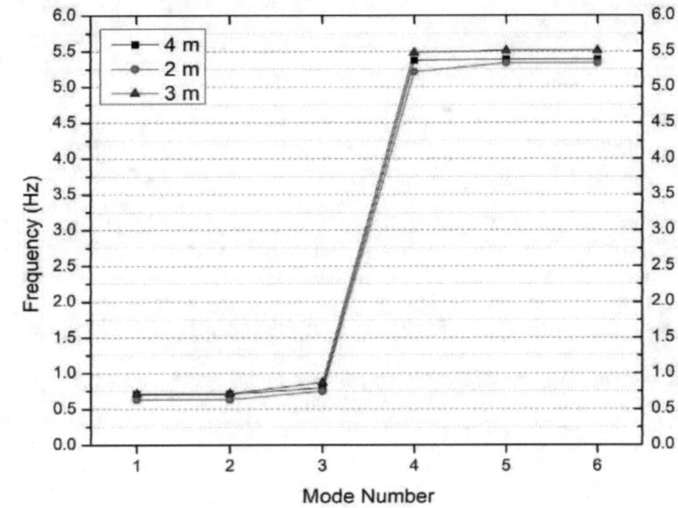

Figure 14. Frequency in Two Way Double layer grids for different grid spacing

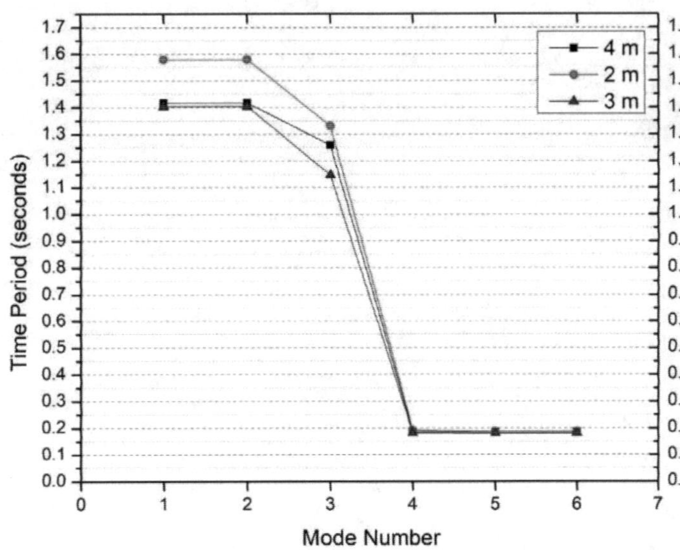

Figure 15. Period in Two Way Double layer grids for different grid spacing

4. Conclusions

The analytical study on grid space structure is carried in this study. From the results and comparative study, following conclusions are drawn:

- From the modal analysis results, a maximum time period of 2.317 seconds is obtained for Four way three layered grids for 2 m grid size. A maximum frequency of 6.705 Hz is obtained for Four way double layered grids for 4 m grid spacing.
- The deflection is reduced by 46.3% for three-layered grids compared to double layered grids. Hence three-layered grids are suitable for long span space structures.
- From the dynamic analysis results of 2 m, 3 m and 4m grid spacing, the displacement values are less for 3 m compared to 2 m and 4 m. Hence 3 m grid spacing is optimum.
- From the dynamic analysis of grid space structures, the value of maximum horizontal displacement for zone 2 and zone 5 for 3.0 m grid size is 21.40 % greater for Two-way three-layered grids compared to Two-way double layered grids. Hence double layered grids are more suitable for earthquake regions.

References

[1] Fu F and Parke G A R 2018 Assessment of the progressive collapse resistance of double-layer grid space structures using implicit and explicit methods *International Journal of Steel Structures* **18(3)** pp.831-842
[2] Moghaddam H A 2000 Seismic behaviour of space structures *International Journal of Space Structures* **15(2)** pp.119-135
[3] Yeda L, Bing Z, Qiang M and Renqiang W 2018 Seismic response analysis of single-layer cylindrical reticulated shell with different length *AIP Conference Proceedings* **1973(1)** pp.020006
[4] Zhang M, Parke G and Chang Z 2018 The dynamic response and seismic damage of single-layer reticulated shells subjected to near-fault ground motions *Earthquake and Structures* **14(5)** pp.399-409
[5] Yu Z W, Zhi X D, Fan F and Lu C 2011 Effect of substructures upon failure behavior of steel reticulated domes subjected to the severe earthquake *Thin-walled structures* **49(9)** pp.1160-1170
[6] Space Structures: Principles and Practice 2006 Volume **1** Subramanian Narayanan. Multi-Science Publishing
[7] John Chilton 2000 Space Grid Structures 1st Edition Architectural Press.
[8] IS: 800 - 2007 Indian Standard general construction in steel - code of practice.
[9] IS 1161 1998 Steel Tubes for Structural Purposes – specification.

Influence of Waste Brick Powder on Properties of Masonry Mortar and its Impact on the Masonry Strength

P Murthi[1], K Poongodi[2] and N Kottiswaran[3]

[1.] Centre for Construction Methods and Materials, S R University, Warangal, India
[2.] Department of Civil Engineering, S R Engineering College, Warangal, India
[3] Department of Civil Engineering, Info Institute of Engineering, Coimbatore, India

Email: dr.murthi@srecwarangal.ac.in

Abstract. The rapid urban development leads the exponential growth of construction and demolition waste in India. The handling and disposal strategies of building debris are not properly developed and create more threats to environment. Brick and concrete materials are the main wastes generated during demolition. As an alternate measures of recycling the brick waste, it is proposed reusing the waste brick powder particles in to the development of masonry mortar. The waste brick powder is used as cement replacement material and four different level of replacement was considered in this investigation as 5%, 10%, 15% and 20%. The strength of masonry mortar with three ratios (1:4, 1:6 and 1:8)at 7 days, 28 days and 90 days were determined. Further, the impact of innovative masonry mortar on masonry prism strength and bond strength were determined. The results obtained from this investigation had concluded that the cement can be replaced 12% to 15% by brick powder without affecting the compressive strength and shear bond strength performance of brick masonry.

1. Introduction

The solid waste management is one of the important and inevitable social task in order to avoid the environmental imbalance situation [1].In India, a huge amount of construction activities are in progress to meet the housing needs and infrastructural demand according to the population growth and urbanization. The infrastructural development has been accounted as an important contribution in the Indian economy and amounted to more than 9% of the GDP. Owing to the rapid growth of construction sector, the demand of natural resources for construction is exponentially increased. On the other hand, construction and demolition (C&D) waste materials are generated when demolishing the existing building for re-construction or renovation works. The regulation mechanism of handling the C&D waste is not yet properly adopted and hence it develops the environmental threat [2]. The majority of the C&D waste materials are dumped in the low level areas, road side and illegally dump in rivers and water bodies. This kind of practice is naturally affecting the sustainable development of environment and causing an extra burden to the local administration on solid waste management [3]. In India, it was estimated as per the statistics presented by the Union Ministry of Urban Development that the annual construction waste generated is around 175 million tons. It is estimated that the C&D waste is accounted as 14% of total municipal solid waste generation. The C&D waste is a composite nature which comprises of concrete, bricks, steel, timber, bitumen and a mixture of site clearance materials.

The brick and masonry waste alone is estimated as 25% - 31% of total wastes generated in C&D and found as the highest waste from C&D debris [3]. In some places, the brick residues reaches more than 50% [4].As a part of recycling this waste brick in the form of fine powder by replacing cement without compromising the properties of concrete and mortar is an alternate way to make the sustainable concrete and mortar. The studies performed by various authors with recycling of waste bricks as lightweight aggregate [5, 6, 7] and supplementary cementitious materials in concrete [8, 9, 10]. The substitution of more than 10% waste brick powder (WBP) causes reduction of fresh concrete properties significantly. As the replacement level increases, the compressive strength of WBP blended mortar in the early period was decreased [10]. However the compressive strength of WBP blended mortar improved gradually during the later curing period and the mortar contain 20% of WBP had achieved the higher compressive strength at 90 days curing [8-10].The pozzolanic activity of WBP and properties of WBP blended cement was investigated by Zhao et al. and concluded that while increasing the grinding time, the WBP particles refined in to spherical shape and increase the specific surface area and its pozzolanic activity [11]. Strength and durability properties of WBP blended mortars were evaluated by Schackow et al [12] and their conclusions show that the improvement in durability properties of mortar up to 40% replacement level of WBP was found as a result of combined physical and pozzolanic pore filling effect of WBP. The WBP could also effectively minimize the mortar expansion due to alkali silica reaction [13].

The partial replacement of cement by various mineral admixtures were experimented to minimize the environmental unkind effect and also reduce the utilization of natural resources. The mineral admixtures are suitably blended for making dense concrete and mortar due to the pozzolanic reaction [14, 15, 16]. Enormous research findings were published for finding the suitability of WBP as partial replacement of cement in concrete and mortar similar to other mineral admixtures. The application of WBP blended mortar in the development of masonry mortar is currently limited. However, the consumption of cement is dominant in masonry construction from rural sector to skyscrapers [17]. With this background, this investigation is intended to evaluate the impact of WBP blended cement on compressive strength and shear bond strength of brick masonry and predict the optimum replacement level of WBP without affecting the mechanical properties.

2. Experimental program

2.1 Materials

This experimental investigation utilized Portland Pozzolana cement (PPC) specified in IS:1489-1991 (Part-1) [18]. The discarded waste clay brick was obtained from the local brick chamber. The collected waste bricks were crushed and ground in to fine particles using a laboratory model ball mill. The crushed WBP was sieved using 75 micron sieve. The SEM image and EDS spectrum of cement and WBP are shown in Figure 1 to Figure 4. The physical properties and chemical compositions of cement and WBP are shown in Table 1. The grading limit of cement and WBP is shown in figure 5. The local river was employed as fine aggregate whose specific gravity of 2.67 and the fineness modulus of 2.79. Three different category of clay bricks were used and the selection of bricks was made based on the compressive strength. The strength of brick units were determined in accordance with procedure laid down in IS 3495 (Part I):1992 [19] and the mean value of 12 specimens were reported with the standard deviation and co-effiecnt of variations as presented in Table 2. The water absorption and efflorescece of selected bricks were determined and the results are mentioned in Table 2 [19].

Figure 1. SEM image of cement

Figure 2. SEM image of WBP

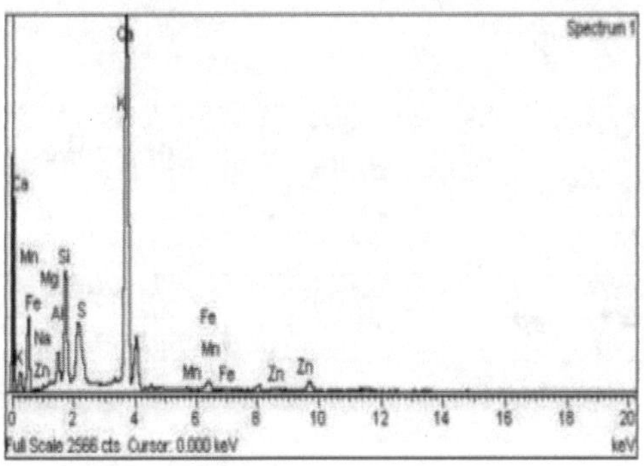

Figure 3. EDS spectrum of cement

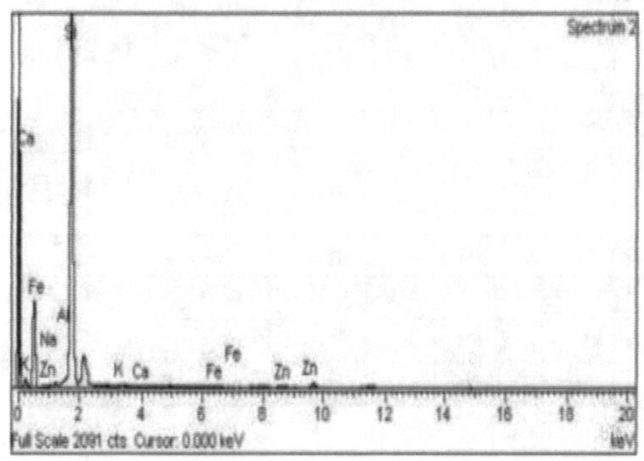

Figure 4. EDS spectrum of WBP

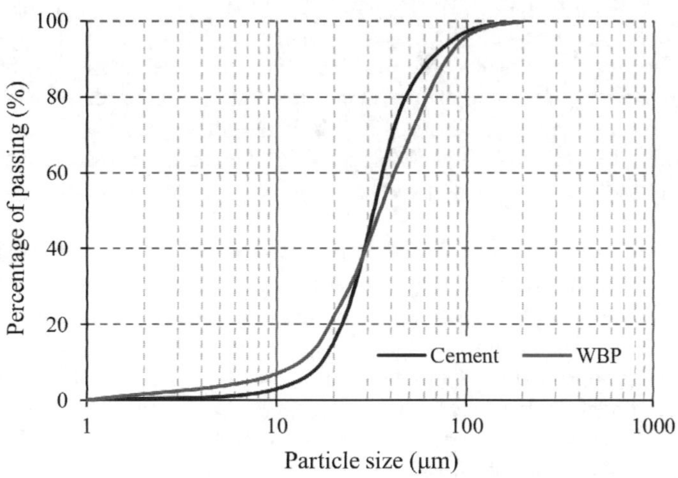

Figure 5. Grading limit of cement and WBP

Table 1. Chemical compositions and physical properties of PPC and WBP

Characteristics	OPC	WBP
Chemical Compositions (%)		
SiO_2	23.42	24.57
Al_2O_3	4.87	2.43
Fe_2O_3	3.78	0.76
CaO	63.16	29.16
MgO	1.85	1.53
LoI	1.16	29.28
Physical properties		
Specific gravity	3.15	2.75
Fineness (cm^2/g)	2950	3170

Table 2. Properties of clay bricks

Brick	Size (mm)	Compressive strength (MPa)			Water absorption (%)	Efflorescence (%)
		Mean	SD	COV(%)		
B1	190 x 90 x 80	7.042	0.333	4.73	10.33	Less than 10
B2	190 x 90 x 80	5.708	0.293	5.10	11.87	Less than 20
B3	190 x 90 x 80	4.414	0.221	5.35	14.05	Less than 25

2.2 Experimental design
The masonry mortar was deveoped by replacing PPC with 5%, 10%, 15% and 20% of WBP by weight of cement. Three mortar mixes (1:4, 1:6 and 1:8) were considered as a variable parameter in this investigation. The strength of control mortar specimens with 0% replacement in all the three mixes were determined to compare the impact of WBP blended masonry mortar. The mortar specimens were prepared in four different occurrence when masonry prisims were prepared and three specimes were casted in each occurrences. The mean value of 12 mortar cube specimens and its standard deviations were reported. The masonry prism and triplet specimens were prepared for finding the compressive strength and shear bond strength of masonry [20]. The height to thickness (h/t) ratio and the mortar thickness were maintained as 1.5 and 12 mm respectively. The tests for masonry specimens were tested after 7 days, 28 days and 90 days of curing period.

2.3 Testing methods
The compressive strength of masonry mortar was tested using mortar cubes of 70.7 mm size and it was determined after the curing period of 7, 28 and 90 days. The samples of the mortar cubes used in this study is shown in figure6. The strength of masonry prism was determined using compression testing machine. Thetesting of masonry prism is shown in Figure 7. The mean value of compressive strength of three prism were determined and the normalized compressive strength of prism (f_p) were calculated considering a shape modification factor based on the slenderness ratio as per IS: 1905-1987 [21].

Normalized compressive strength of prism = shape modification factor x mean strength of prism. The shear bond strength of brick triplet was tested to evaluate the suitability of the masonry mortar devloped in this investigation [22]. An average of three triplet results were reported and testing of triplet specimen is shown in Figure8.

Figure 6. Masonry mortar samples

12 mm thikness mortar

(a) Masonry prism

(b) Testing under progress

Figure 7. Compressive strength test on masonry prism

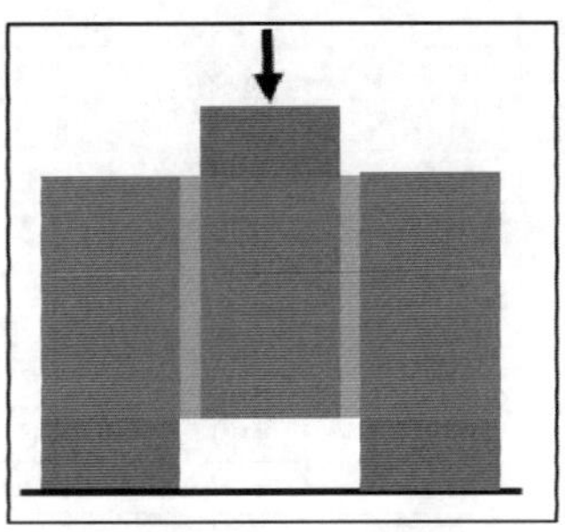

12 mm thikness mortar

(a)Triplet specimen

(b) Shear bond strength test in progress

Figure 8. Shear bond strength test on masonry triplet

3.Results and discussion

3.1 Effect of WBP on compressive strength of masonry mortar

The compressive strength of masonry mortar mixes selected in this investigation with and without WBP after the curing period of 7 days, 28 days and 90 days are presented in Table 3. The results were obtained as the mean value of 12 specimens and hence the standard deviation (SD) and coefficient of variation (COV) were also determined in order to check the consistency in results from the mortar cube specimens casting in different occasions. Though the mortar cube specimens were prepared whenever the masonry prism casted, the uniformity of strength development was observed from SD and COV since SD varies mostly less than 0.126 MPa and COV varies less than 1.2%. It can be seen from the results that the mortar strength at 7 days curing was significantly lower than the brick unit strength due to the slow strength development of PPC in the early age. Further it can also be noted that the compressive strength of mortar at 28 days curing and brick strength were mostly identical and the stronger mortar than the brick was obtained after 90 days curing period. This strength variations of the control cement mortar can be attributed due to the use of blended cement (PPC) and it can be concluded that the prolonged curing is essential when PPC is used for masonry mortar preparation.

The impact of WBP substitution as partial replacement of PPC on the compressive strength of masonry mortar after the curing period of 28 days and 90 days are shown in Figure 9 and Figure 10 respectively. The results evidenced that the sustained compressive strength of mortar was discovered up to 15% replacement of PPC by WBP in all the proposed mortar mixes in both 28 days and 90 days cured specimens. This may be attributed to the pore filling nature of WBP and promote the hydration of available cement particle by maintaining the water requirements for chemical reaction. Sajanthan et al (2019) found that the presence of fine particle in cement-soil mortar could leads to reduce the porous nature and make the mass as dense which promotes the strength of mortar [23]. A reduction of compressive strength of mortar with more than 15% WBP was noticed and it may due to lower the reaction products in the presence of higher amount of inert particles in the mortar.

Table 3. Compressive strength of masonry mortar

Mortar Mix	WBP Level (%)	7 Days			28 Days			90 Days		
		f_j (MPa)	SD (MPa)	CoV (%)	f_j (MPa)	SD (MPa)	CoV (%)	f_j (MPa)	SD (MPa)	CoV (%)
1:4	0	4.37	0.072	0.932	7.37	0.061	0.823	8.85	0.072	1.132
	5	4.31	0.083	0.905	7.47	0.078	1.051	9.03	0.081	1.092
	10	4.26	0.079	0.917	7.62	0.073	0.975	9.17	0.077	1.107
	15	4.22	0.093	0.922	7.61	0.085	1.093	9.23	0.082	1.135
	20	4.05	0.091	0.949	7.33	0.081	0.991	9.07	0.073	1.033
1:6	0	3.77	0.085	0.974	6.24	0.077	0.912	7.73	0.081	0.945
	5	3.59	0.089	0.937	6.29	0.083	0.838	7.97	0.079	0.897
	10	3.47	0.095	0.921	6.33	0.088	0.911	8.03	0.073	1.102
	15	3.35	0.102	0.978	6.37	0.092	0.934	8.11	0.083	1.095
	20	3.19	0.113	0.915	5.79	0.096	1.057	7.89	0.087	1.102
1:8	0	3.53	0.095	0.845	5.03	0.084	1.022	6.65	0.079	1.117
	5	3.37	0.099	0.822	5.15	0.093	1.017	6.73	0.071	1.109
	10	3.15	0.112	0.807	5.19	0.105	0.921	6.83	0.113	1.005
	15	3.09	0.119	0.811	5.21	0.112	0.933	6.71	0.126	1.121
	20	2.87	0.122	0.769	4.97	0.108	0.998	6.23	0.167	1.108

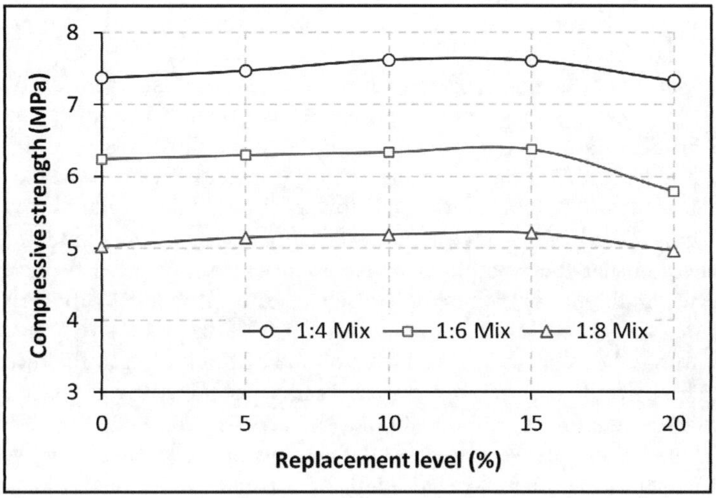

Figure 9. Strength development of 28 days cured masonry mortar

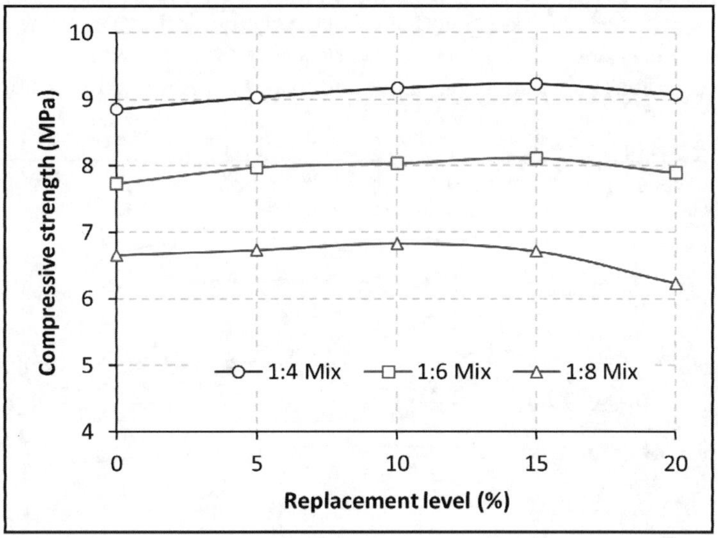

Figure 10. Strength development of 90 days cured masonry mortar

3.2 Effect of WBP blended masonry mortar on compressive strength of masonry prism

The effect of brick strength and WBP replacements in mortar on the masonry prism strength are shown in Figure 11 to Figure 13. From the results, it can be noticed that the curing period of the prism specimen had contributed the strength development of masonry in addition to the brick strength. The usage of blended cement can give the improvement in strength beyond 28 days in the presence of water due to the pozzolanic reaction of the mineral admixture presence in the cement. Accordingly the compressive strength of prism after 90 days curing period had shown higher values in all the mixes considered in this investigation. The results obtained from a single mix showed that the higher strength brick demonstrated the higher prism strength. The compressive strength of the prism when B1 bricks were used in all the mixes had shown the higher strength between the WBP replacement levels of 10% to 15%. Though the same mortar mix mortar with the optimum replacement of WBP between 10% to 15% were considered, the lower strength brick causes the significant reduction of masonry strength. The combination of the brick B1 and mix 1: 4 had shown the strength of prism at 10% WBP replacement level as 5.28 MPa at the age of 90 days and interestingly the results of the same 1:4 mortar mix and WBP replacement showed 3.79 MPa which is 28% lower. However, the reduction of prism strength due to poor mortar mix considered in this investigation (1:8) had shown less than 9% than the rich mortar mix. It can be concluded that the strength of masonry prism is governed by brick unit strength and mortar strength. As a result, the brick strength offers greater strength to the prism than the mortar. Thaickavil and Thomas [24]was found that an average of 70% in prism strength was attributed by the brick strength [12]. It may be endorsed to the bulk volume of brick offered the direct load transmission of brick masonry. The effectiveness of load transfer can be obtained by the brick-mortar bond in the masonry, however the roughness of the brick surface is credited the brick-mortar interface mechanism[25,26]. The results are validated by developing the relationship between the compressive strength of control mortar and compressive strength of prism with various brick strength as shown in Figure 14.The effect of 10% to 15% replacement of cement by WBP on the relationship between masonry strength and mortar are shown in Figure 15. The higher correlation coefficient in this relationship was noticed which showed the correctness of the relationship. It was found that the sustained compressive strength of prism and there is no negative influence on the strength of masonry prism due to the optimum usage of WBP in mortar. However, from the Figure 14 and Figure 15, it can be concluded that the compressive strength of brick unit was found to be more impact than the mortar strength in masonry.

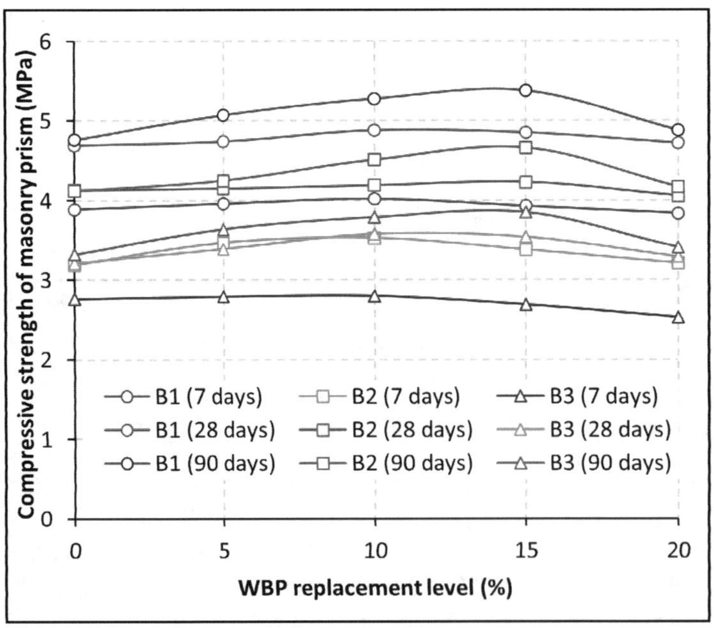

Figure 11. Strength development of masonry prism using 1: 4 mix

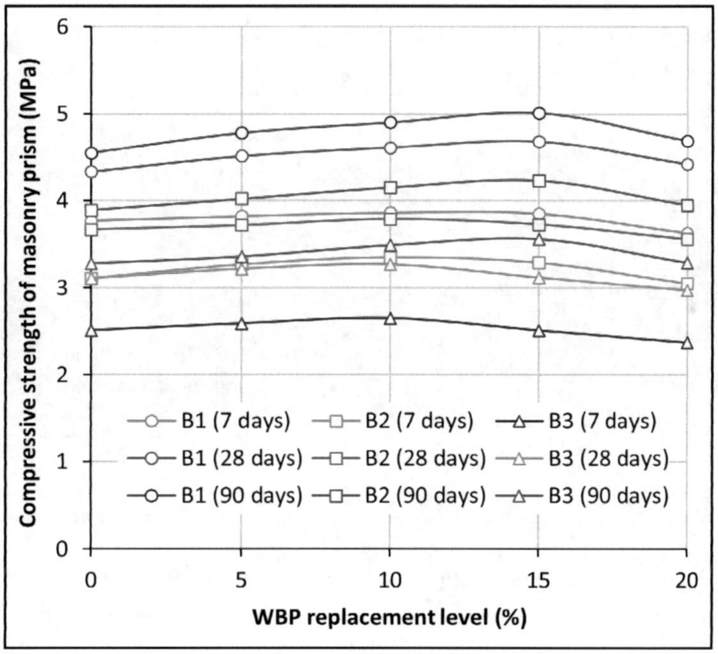

Figure 12. Strength development of masonry prism using 1: 6 mix

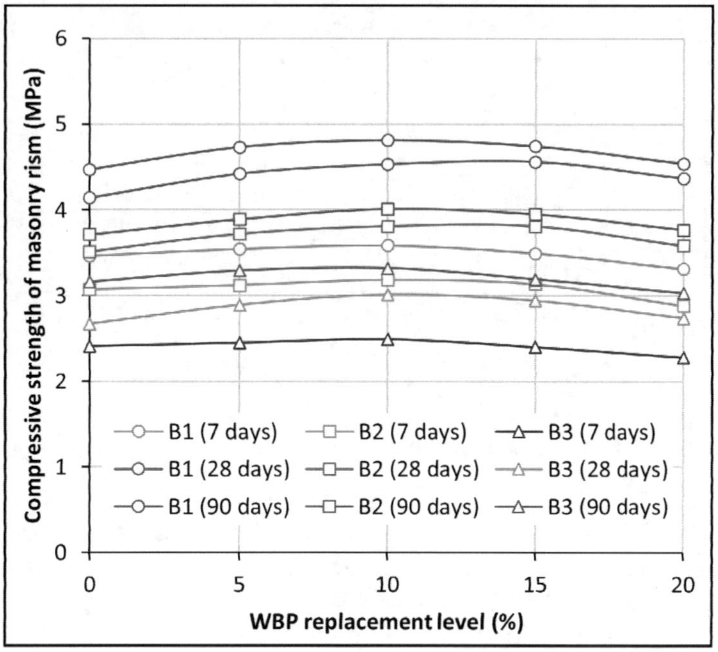

Figure 13. Strength development of masonry prism using 1: 8 mix

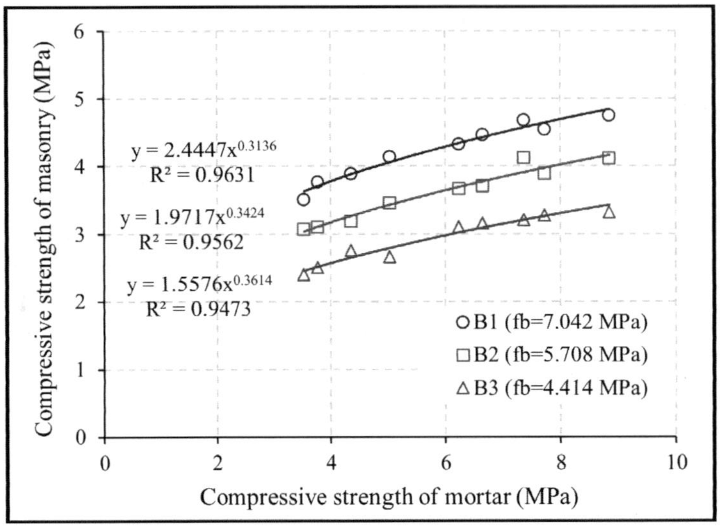

Figure 14. Relationship between compressive strength of control mortar and masonry prism

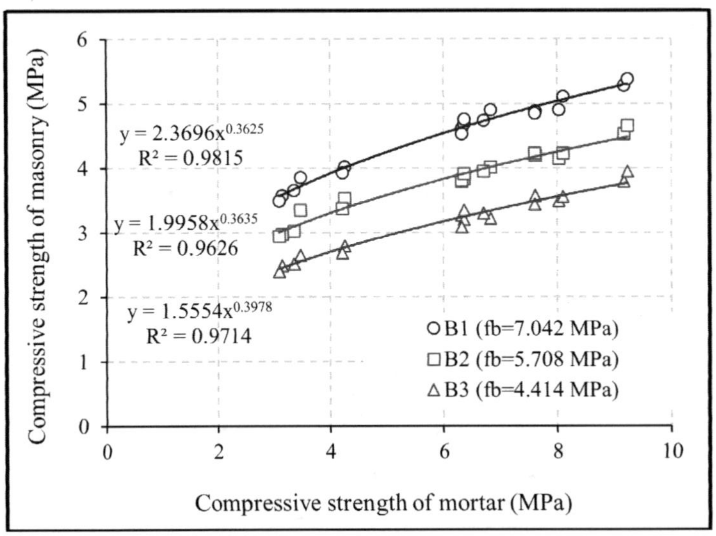

Figure 15. Relationship between compressive strength of WBP blended mortar and masonry prism

3.3 Influence of WBP blended masonry mortar on shear bond strength of masonry

Masonry triplet specimens were performed to ascertain the effects of WBP in shear bond strength of masonry. The 12 mm thickness of masonry mortar was considered. The shear bond strength variation were determined and compared with the results obtained from control mortar specimens. The results obtained from 28 days cured were reported as shown in Figure16. [23,28]. The shear bond strength of masonry using the brick B1 with 1:4 control mix was determined as 0.105 MPa. With the reduction of compressive strength of brick such as B2 and B3 bricks, the reduction of shear bond strength in the same mix was noticed. In the same way, the reduction of shear bond strength of masonry was noticed when decreasing the compressive strength mortar using 1:6 and 1:8 mixes.From the results it was found that shear bond strength values were governed by both brick strength and mortar strength in the control specimens [20].

Further, the investigation was carried out with the masonry mortar at 10% and 15% WBP replacement level of cement. From the Figure16, it can be observed that the bond strength of brick masonry were increased significantly with the substitution of 10% WBP in the masonry mortar in all the mixes and bricks considered in this investigation. In contrast, the bond strength of 15% WBP mortar decreases than the results of 10% WBP. However, the results obtained from 15% WBP had shown almost similar bond strength compared with control mortar specimen. The effect of bond strength was also measured with the lower strength brick specimen (B2 and B3) and observed that the reduction of bond strength was found in the same mix. The results show that 10% WBP blended masonry mortar were higher in all the category of bricks and mixes. This might be attributed to the pore filling effect of WBP in the early period and contribute the strength development in the later stage with the presence of water. The higher the WBP level leads to delay the cement hydration process causes the lower bond strength of brick masonry. The low strength bricks are relatively higher water absorption in nature. It causes the reduction of water content in mortar and routes to reduce the hydration process similar to higher WBP. This physical characteristics was ascribed to reduce the bond strength of masonry.

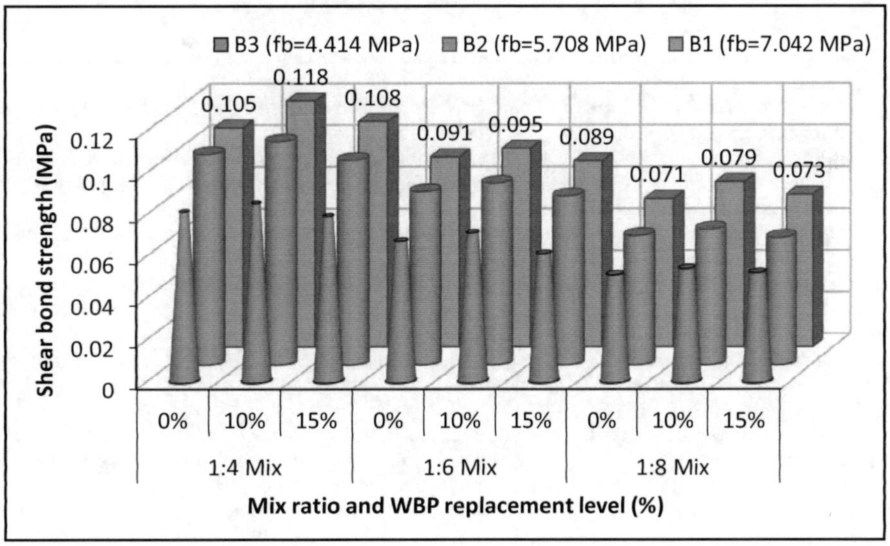

Figure 16. Shear bond strength variation of masonry triplet specimens.

4. Conclusions

The following concluded points were obtained from this investigation:

- There was no significant variations in compressive strength of WBP blended mortar detected after 28 days curing of mortar specimens up to 15% replacement level.
- The compressive strength of brick masonry prism was influenced by both brick strength and mortar strength. However the brick strength was dominate than the mortar strength and hence the

replacement of cement up to 15% is not affecting in the compressive strength of prism though the mortar was developed by PPC.

- The relationship between the compressive strength and mortar strength was established with higher correlation coefficient in all the mixes considered in this investigation.
- The results of shear bond strength of WBP blended had shown the maximum results up to 10% replacement of PPC. However, the results of 15% WBP replacement has shown an equivalent results to the control mortar.
- It can be concluded that the PPC was replaced up to 15% by WBP and found as optimum replacement level.

References

[1] Viviana Letelier, José Marcos Ortega, ID PedroMuñoz , Ester Tarela and GiacomoMoriconi 2018 Influence of Waste Brick Powder in the mechanical properties of recycled aggregate concrete *Sustainability* Vol. **10** doi:10.3390/su10041037.
[2] Sakshi Guptha, Malik R K 2018 The impact of C&D waste on Indian Environment: A critical review *Civil Engineering Research Journal* Vol. **5** No. 2 pp. 57-63 doi: 10.19080/ CERJ.2018.05.555658.
[3] AsokanPappu, Mohini Saxena, Shyam R Asolekar 2007 Solid waste generation in India and their recycling potential in building materials *Building and Environment* Vol. **42**, pp. 2311-2320, doi:10.1016/j.buildenv.2006.04.015
[4] Liu O, Tong T, Liu S, Yang D, Yu O 2014 Investigation of using hybrid recycled powder from demolished concrete solids and clay bricks as a pozzolanic supplement for cement *Construction and Building Materials* **73** 754–763
[5] Tiwari Darshita, Patel Anoop 2014 Study of strength and workability of different grades of concrete by partial replacement of fine aggregate by crushed brick and recycled glass powder *International Journal of science and research* Vol. **3** No. 6 pp. 141-145.
[6] Paulo B Cachim 2009 Mechanical properties of brick aggregate concrete *Construction and Building Materials* Vol. **23** pp. 1292-1297, doi:10.1016/j.conbuildmat.2008.07.023
[7] Debieb F, Kenai S 2008 The use of coarse and fine crushed bricks as aggregate in concrete *Construction and Building Materials* Vol. **22** No.5 pp.886-893 doi:10.1016/j.conbuildmat.2006.12.013
[8] Jiahu Shao, JianmingGao, Yasong Zhao, Xuemei Chen 2019 Study on the pozzolanic reaction of clay brick powder in blended cement pastes *Construction and Building Materials* Vol. **213** pp. 209-215 doi.org/10.1016/j.conbuildmat.2019.03.307
[9] Mehmet S Kırgız Fresh and hardened properties of green binder concrete containing marble powder and brick powder *European Journal of Environmental and Civil Engineering* doi.org/10.1080/19648189.2016.1246692
[10] ZhiGe, Yuanyuan Wang, Renjuan Sun, Xinsheng Wu, Yanhua Guan 2015 Influence of ground waste clay brick on properties of fresh and hardened concrete *Construction and Building Materials* Vol. **98** pp.128-136 doi.org/10.1016/j.conbuildmat.2015.08.100
[11] Yasong Zhao, JianmingGao, Chuanbei Liu, Xuemei Chen, ZhenhaiXu 2020 The particle-size effect of waste clay brick powder on its pozzolanic activity and properties of blended cement *Journal of cleaner production* Vol. **242** doi.org/10.1016/j.jclepro.2019.118521
[12] Schackow A, Stringari D, Senff L, Correia S L, Segadães A M 2015 Influence of fired clay brick waste additions on the durability of mortars, Cement & Concrete Composites Vol. **62** pp. 82-89 doi.org/10.1016/j.cemconcomp.2015.04.019
[13] Turanli L, Bektas F, Monteiro P J M 2003 Use of ground clay brick as a pozzolanic material to reduce the alkali-silca reaction *Cement concrete research* Vol. **33** pp. 1539-1542.
[14] Kavitha O R, Shyamala G, Iyappan G, Rajesh Kumar K 2020 Influence of fly ash and metakaolin on high performance concrete *International Journal of Scientific and Technology Research* Vol. **9** No 2 pp.5582-5586.
[15] Thirumalairaja K, Murthi P 2015 Bagasse ash and rice husk ash as cement replacement in self-compacting concrete *International Journal of Gradevinar (Civil Engineer)* Vol. **67** No: 1 PP.23-30.
[16] Murthi P, Poongodi K, Awoyera P O, Gobinath R, Saravanan R 2019 Enhancing the strength properties of High-performance concrete using Ternary blended cement: OPC, Nano-silica, Bagasse ash, Silicon doi: org/10.1007/s12633-019-00324-0.
[17] Vimala S, Kumarasamy K 2014 Studies on the strength of stabilized mud block masonry using different mortar proportions *International Journal of Emerging Technology and Advanced Engineering* Vol. **4** pp. 720–724.
[18] IS 1489 (part I): 1919 Specification of Portland Pozzolana cement (Fly ash based) *Bureau of Indian Standards*, 2000.
[19] IS 3495 (Part I-III):1992 Methods of tests of burnt clay building bricks *Bureau of Indian Standards* 1992.
[20] Murthi P, Md.Akib, Md. Imran, Shareeq Ahmed, Prasanna V 2020 Studies on the strength variation of brick masonry using novel blended masonry mortar mixes and mortar thickness *Materials Today: Proceedings* https://doi.org/10.1016/j.matpr.2020.06.395

[21] IS:1905-1987 Indian standard code of practice for structural use of unreinforced masonry *Bureau of Indian Standards* (New Delhi, India) 1987.

[22] Deepa A Joshi, R K Jain 2015 Evaluation of compressive strength and basic compressive stress of clay brick unreinforced masonry by prism test *International Journal of Science and Research* Vol. **4** No. 5 pp. 913-916.

[23] Kunasegaram Sajanthan, Balasingam Balagasan, and Navaratnarajah Sathiparan 2019 Prediction of compressive strength of stabilizes earth block Masonry *Advances in Civil Engineering* https://doi.org/10.1155/2019/2072430

[24] Nassifnazeer Thaickavi, Job Thomas 2018 Behaviour and strength assessment of masonry prisms, *Case studies in Construction Materials* Vol. **8** pp. 23-28, http://doi.org/10.1016/ j.cscm.2017.12.007

[25] Ravula M B, Subramaniam K V L 2017 Experimental investigation of compressive failure in masonry brick assemblages made with soft brick *Materials and Structures* Vol. **50** No.19 doi.org/10.1617/s11527-016-0926-1.

[26] Nagarajan S, Viswanathan S, Ravi V 2014 Experimental approach to investigate the behaviour of brick masonry for different mortar ratios *Proceedings of the International Conference on Advances in Engineering and Technology* (Singapore) pp. 586–592.

[27] Samarasinghe W, Lawrence S J 1992 Effect of high suction rate in low strength bricks on brick mortar bond *Proceedings of 4th International Seminar on Structural Masonry for Developing Countries* (Madras, India) pp. 43-50.

[28] Murthi P, Bhavani M, Md.Saqlain Musthaq, Md Osman Jauhar, Rama Devi V 2020 Development of relationship between compressive strength of brick masonry and brick strength Materials *Today: Proceedings* https://doi.org/10.1016/j.matpr.2020.07.040

IOP Conf. Series: Materials Science and Engineering 1006 (2020) 012026 doi:10.1088/1757-899X/1006/1/012026

Study on Environmental Contamination and Assessment of the Impact of Municipal Solid Waste on Soil at the Vellalore Dumping Yard, Coimbatore City.

S Pratheba[1], V Johnpaul[2]and N Balasundaram[3]

[1,2]Assistant Professor, Department of Civil Engineering, Karpagam Academy of Higher Education.
[3]Professor and Head, Department of Civil Engineering, Karpagam Academy of Higher Education, Coimbatore,Tamilnadu, India.

E-mail: pratheba.s@kahedu.edu.in

Abstract. The main objective of the work is to research the soil condition and quality of municipal solid waste from (2012-2017) Coimbatore region. Coimbatore is a popular town in the southern Indian state of Tamil Nadu with an area of 246.8 km / sq. of 1.60 million, according to the 2001 sense. It is the second largest city after Chennai and one of India's fastest growing cities with rapid developments in all fields. Due to the rapid increase in urbanization, industrialization and population, which leads to an increase in the generation of solid waste and encourages dumping on landfill sites, environmental issues are a major concern. The improper disposal of solid waste pollutes soil, water and air. The soil characteristics study was undertaken to assess the soil quality due to the solid waste dumped at the Vellalore dumping yard. This paper sets out the methodology used to find all the chemicals in the soil and, by means of a prospective test, we found the limits for their strength to be present in the soil. In order to assess the soil conditions in the dumping yard for the months of May and December 2016, soil samples from the Vellalore area were collected at different locations. In the current study, samples from five locations, namely S1 , S2 , S3, S4 and S5, are to be collected in all four directions and from the dumping site. This paper deals with the study of soil polluted. Physical – Chemical characteristics of soil samples were analyzed on the basis of a standard procedure. Characteristics of soil such as pH, EC, MC, OM, Ca2 +, Mg2 +, Na+, K+ and related parameters such as SAR, CEC and ESP have also been calculated. The characteristics study of soil samples from the above locations showed that the solid waste dump at these locations had changed the characteristics of the soil from the test data.

1. Introduction

Coimbatore city occupies an area of nearly 105.6 km^2 and it is the second city of being vast in Tamil Nadu. It is further expanded to about 257 km^2 because of the people's migration and thus the waste generation was high in this area. By revealing the past data, the waste generation varies per capita per day and the rate was increasing day by day. The annual solid waste was nearing above 40 million tons in our country and the wet content in the waste was also to be noted. It will reach 260 million tons by the next 20 or 25 years as soon, that is the huge problem going to face by our future generation because of not managing their wastes. The aim is to investigate and assess the properties of the soil due to the dumping of solid waste in Dumping site.

2. Experimental methods & materials

Coimbatore Corporation has four zones and each consists of 18 wards. From these, Vellalore dumping yard is the one, where the solid wastes are brought all together by Lorries or tractors. The yard consists of segregation area, Closure, Composting area, Lagoons and Landfill. The Vellalore dumping yard is about 643 acres totally and it has four lagoons and four ponds facilitated with drain pipe, in order to collect the drain water from all other locations near the yard. The quantity dumped every day in the site is about 8 lakh tons/m^3 due to the high generation of wastes and the height is about 9m it seems.

3. Study on environmental contamination

Table 1: Study on Environmental Contamination in Coimbatore City

AUTHOR	TITLE	FACTORS CONSIDERED
Times of India	Solid waste management project launched	Solid waste
Deccan Chronicle Correspondent	Irregularities alleged in waste Management	Solid waste
Deccan Chronicle Vidyashree Dhanmaraj	Poll fever mounts so does garbage	Solid waste
Times of India Subburaj	Poor waste management puts Coimbatore to shame	Solid waste
The Hindu Pankaja Srinivasan	Lesson in soil waste management	Solid waste
The Hindu Pankaja Srinivasan	Solid waste management wards that shows the way	Solid waste
The Hindu R. Sairam	Generating energy from waste	Solid waste
The Hindu Karthik Madhavan	Coimbatore initiative for effective waste management	Solid waste
Deccan Chronicle Laasyashekhar	NGT issue warrant to kovai corporation head	Solid waste
Times of India Komal gautham	Solid waste management Coimbatore city corporation to present power point before green tribunal	Solid waste
V.C. Sengodan, S. Manivannan	Biomedical waste minimization in the orthopedics ward in a tertiary care hospital	Solid waste
R. Prem Sudha, Dr.R.N. Uma, Meiaraj	Assessment of soil characteristic around municipal solid waste disposal site in sulur block	Solid waste
A. Benuel Sathish Raj, C. Deepathi	Recycling of municipal solid waste for electricity generation and green Earth	Solid waste
S. Aravind, R. Mousia, B. Ganesh, S. Vignesh	Optimization of bird hazard Reduction	Solid waste
K. Senthamil Selvan, M. Palanivel	Quantification and characteristics of the municipal solid waste Dharapura municipality	Solid waste
PA. Ganeshwaran, S. DeepaShri	A solid waste management in Coimbatore	Solid waste
T.Murugan, R. Chandrasekaran	A study and development of comprehensive solid waste management for panchayat	Solid waste

4. Sampling and methodology

In order to determine the soil conditions in the dumping yard, soil samples were obtained from the Vellalore region in different locations for the months of May and December 2016. In this analysis, samples from five sites – S1 , S2 , S3, S4, and S5 – are expected to be obtained in all four directions and from the dumping site. Samples from surface and sub-surface shall also be collected in May and December as 10 samples and are closed, labeled and transmitted to the laboratory for inspection. Soil dry sample of 25 g is taken and the result is analyzed by means of a sieve of 2.36 mm. It should be dissolved in 50mL of water after sieving and well stirred for an hour. The use of Wattman's filter paper No.42 and the sample are to be filtered and the filtrate is taken for analysis. The samples have been analyzed and the characteristics and nutrients present will be revealed and the data collected below.

Table 2: Sampling Location

S.No.	Direction	Depth	Distance
S1	North	0-15 cm	5m
S2		16–30 cm	5m
S3	East	0-15 cm	5m
S4		16–30 cm	5m
S5	West	0-15 cm	5m
S6		16–30 cm	5m
S7	South	0-15 cm	5m
S8		16–30 cm	5m
S9	Dumpsite	0-15 cm	Initial Point
S10		16–30 cm	Initial Point

Table 3: Methods Adopted for Physic Chemical Analysis

S.No.	Parameters	Symbols	Testing Methods
1	pH		pH analyzer
2	Moisture Content (%)	Mc	Oven dry method
3	Electrical Conductivity (µ ohms/cm)	Ec	EC analyzer
4	Organic Matter (%)	OM	Titration method
5	Chloride (mg/l)	Cl–	Chromatography
6	Calcium (mg/l)	Ca2+	(0.05 N) titration
7	Sodium (mg/l)	Na+	Flame photometer
8	Magnesium (mg/l)	Mg2+	(0.05N)titration
9	Potassium (mg/l)	K+	Flame photometer
10	Total Alkalinity (mg/kg)	TA	IS:3025 (Part 23)
11	Total hardness (mg/l)	CaCO3	Sum of Calcium and Magnesium
12	Iron (mg/l)	Fe	Spectrophotometer
13	Cadmium (mg/l)	Cd	Trace Metal Analyzer
14	Copper (mg/l)	Cu	Trace Metal Analyzer
15	Manganese (mg/l)	Mn	DTPA
16	Lead (mg/l)	Pb	Trace Metal Analyzer
17	Chromium (mg/l)	Cr	XRF Analyzer
18	Zinc (mg/l)	Zn	Trace Metal Analyzer
Soil hazards parameters			
19	Sodium Adsorption Ratio (SAR) (%)		Na+/ (Ca2+ +Mg2+/2)1/2
20	Cation Exchange Capacity (CEC)	Meq/l	Ca2++Mg2++ Na++ K+
21	Exchangeable Sodium Percentage (ESP) (%)		Na+/ (Ca 2++Mg2++ Na++ K+) *100

Table 4: Max. Permissible limits of BIS (IS: 10500:1991)

S.No.	Parameters	Allowable Value (mg/l)	Permissible Value mg/l)
1	pH	6.5 to 8.5	No relaxation
2	Electrical Conductivity (mS/cm)	1	No relaxation
3	Moisture Content (%)	80	100
4	Organic Matter (%)	30	50
5	Calcium (mg/l)	110	120
6	Magnesium (mg/l)	70	100
7	Sodium (mg/l)	-	-
8	Potassium (mg/l)	-	0.75
9	Chloride (mg/l)	250	1000
10	Total Alkalinity(mg/kg)	200	600
11	Total hardness (mg/l)	300	600
12	Iron (mg/l)	0.3	1.0
13	Cadmium (mg/l)	0.01	No relaxation
14	Copper (mg/l)	0.05	.5
15	Manganese (mg/l)	0.1	0.3
16	Lead (mg/l)	0.1	0.3
17	Chromium (mg/l)	0.05	No relaxation
18	Zinc (mg/l)	5	15

5. Result and discussion

Table 5: Soil samples collected at depth of 15 cm and 30 cm (May)

Soil Sample	Location	pH	EC	MC	OM	Ca	Mg	Na	K	Cl
S1 0-15 cm	North	10.04	0.932	5.83	4.537	103.94	4.3	21	130	124
S2 16–30 cm		9.28	1.031	6.97	3.27	106	10.67	43	76	177
S3 0-15 cm	East	9.82	1.004	4.95	4.002	82.48	21.63	44	10	230
S416–30 cm		9.12	1.038	6.592	3.004	93.28	10.44	64	15	266
S5 0-15 cm	West	9.57	0.544	6.17	3.15	63.82	13.11	53	23	283
S6 16–30 cm		8.68	1.032	7.141	2.58	82.64	13	15	10	453
S7 0-15 cm	South	9.98	0.144	5.05	3.596	88.59	38.18	56	114	653
S8 16–30 cm		9.41	1.037	7.0	2.8	97.52	14.67	79	91	763
S9 0-15 cm	Dumpsite	8.93	1.382	9.973	2.067	83.79	9.98	91	33	983
S10 16–30 cm		9.22	1.258	9.104	2.005	88.34	8.94	100	45	1199

Table 6: Soil samples collected at depth of 15 cm and 30 cm (May)

Soil Sample	Location	TA	$CaCO_3$	Fe	Cd	Cu	Mn	Pb	Cr	Zn
S1 0-15 cm	North	1344	1190	3.10	0.73	8.24	17.68	1.23	2.3	0.5
S2 16–30 cm		1344	783	5.10	1.02	8.4	18.62	3.2	3.2	1.2
S3 0-15 cm	East	1264	1134	8.42	0.01	2.07	11.22	1.2	1.31	4
S416–30 cm		1236	1342	11.54	0.01	3.8	13.36	1.5	1.3	6
S5 0-15 cm	West	673	1782	23.28	0.01	6.24	10.98	4.3	1.01	1.3
S6 16–30 cm		782	1785	24.64	0.02	7.79	12.36	4.7	1.1	2.3
S7 0-15 cm	South	1285	1562	20.3	1.72	9.54	15.76	4.8	0.8	3.04
S8 16–30 cm		1263	1563	20.14	1.76	9.54	15.78	5.3	0.8	6.80
S9 0-15 cm	Dumpsite	1435	1850	25.26	2.12	10.48	18.74	22.1	3.14	7.66
S10 16–30 cm		1654	1723	26.04	2.13	11.54	19.81	23.7	4.12	9.23

Table 7: Soil samples collected at depth of 15 cm and 30 cm (December)

Soil Sample	Location	pH	EC	MC	OM	Ca	Mg	Na	K	Cl
S1 0-15 cm	North	10.41	0.836	4.65	5.387	119.48	12.02	50	164	170
S2 16–30 cm		9.32	1.043	7.0	3.7	110.24	13.66	62	96	192
S3 0-15 cm	East	10.21	0.909	5.461	4.353	91.04	38.87	58	62	248
S416–30 cm		9.23	1.039	6.843	3.184	97.52	17.77	64	40	301
S5 0-15 cm	West	9.5	0.866	5.533	2.294	71.78	34.81	12	30	345
S6 16–30 cm		8.92	1.033	7.238	2.65	89.04	16	20	15	567
S7 0-15 cm	South	9.6	1.055	6.142	4.178	98.06	10.8	102	206	672
S8 16–30 cm		9.6	1.044	7.128	3.094	99.64	15.72	115	111	873
S9 0-15 cm	Dumpsite	8.93	1.765	9.145	3.639	108.49	20.09	150	32	986
S10 16–30 cm		9.43	1.634	9.344	2.324	99.30	20.34	124	58	1298

Table 8: Soil samples collected at depth of 15 cm and 30 cm (December)

Soil Sample	Location	TA	$CaCO_3$	Fe	Cd	Cu	Mn	Pb	Cr	Zn
S1 0-15 cm	North	1335	1209	5.10	1.76	8.18	18.64	1.6	2.1	0.6
S2 16–30 cm		1398	983	9.64	1.076	8.18	18.68	2.1	2.1	0.7
S3 0-15 cm	East	1243	1289	6.24	1.89	0.81	14.72	1.2	1.4	3.1
S416–30 cm		1234	1293	7.79	1.98	1.67	14.82	1.23	2.1	3.4
S5 0-15 cm	West	678	1830	21.86	0.04	3.98	10.96	3.89	1.1	2.3
S6 16–30 cm		900	1852	21.96	0.05	5.10	11.22	4.5	1.3	2.45
S7 0-15 cm	South	1293	1672	11.54	1.25	8.42	15.96	5.9	0.56	6
S8 16–30 cm		1299	1562	17.54	1.28	8.44	18.70	6.4	0.78	6.2
S9 0-15 cm	Dumpsite	1532	1983	24.88	2.87	9.82	24.36	23.7	4.2	10.54
S10 16–30 cm		1529	1900	25.40	2.92	9.64	32.40	34.9	5.23	11.50

Table 9: Soil hazards parameters collected at depth of 15 cm and 30 cm (May)

Soil Sample	Location	SAR	CEC	ESP
S1 0-15 cm	North	2.03	258.24	8.10
S2 16–30 cm		4.07	235.67	18.25
S3 0-15 cm	East	4.55	157.11	27.83
S416–30 cm		6.44	182.72	35.03
S5 0-15 cm	West	6.31	151.93	34.66
S6 16–30 cm		1.56	120.64	12.43
S7 0-15 cm	South	5.39	295.77	18.87
S8 16–30 cm		7.70	282.19	28.00
S9 0-15 cm	Dumpsite	9.65	216.77	41.79
S10 16–30 cm		10.37	242.28	41.27

Table 10: Soil hazards parameters collected at depth of 15 cm and 30 cm (December)

Soil Sample	Location	SAR	CEC	ESP
S1 0-15 cm	North	4.45	346.5	15.47
S2 16–30 cm		5.63	280.9	21.89
S3 0-15 cm	East	5.42	248.91	23.21
S416–30 cm		6.10	219.29	29.19
S5 0-15 cm	West	1.27	148.59	8.08
S6 16–30 cm		2.03	140.04	14.28
S7 0-15 cm	South	10.03	416.86	24.47
S8 16–30 cm		11.09	341.36	33.69
S9 0-15 cm	Dumpsite	13.78	310.58	48.30
S10 16–30 cm		11.85	301.64	41.11

5.1 Ph

The pH values at Vellalore dumping yard varies between 10.04 to 8.68 on May 2016 and 10.41 to 8.92 on December 2016 with a mean value 9.35 and 9.51 where in 2012 a mean value obtain 8.63 and during 2013 the value was 8.53 and it was observed that pH values around dumpsite become alkali soil when compare with 2012 & 2013respectively.

5.2 Moisture content

Vellalore disposal site moisture content ranges from 9.973% to 4.95% on May 2016 and 9.344% to 4.65% on December 2016 with the mean value 6.87% and 6.84% where in 2012 a mean value obtains 6.55% and during 2013 a mean value obtains 6.5%. It is marked that moisture content around dumpsite increases when compare with 2012 & 2013 respectively.

5.3 Organic Matter

The organic matter collected from bio degradable materials is the important parameter of the soil. The organic matter of soil sample is varied from 4.537 to2.005 in the month of May 2016 and 5.387 to 2.294 in the month of December 2016 with the mean value 3.101 and 3.480. Where in 2012 the mean value obtained is 2.63% and during 2013 the mean value obtained is 2.52%. It is observed that the organic matter of the soil sample is increased as compared with 2012 and 2013 respectively.

5.4 Calcium(Ca)

Calcium is a chemical element and it is an alkaline earth metal with pale yellow color. The calcium value of soil sample is varied from 106 to 82.48 on May 2016 and 119.48 to 71.78 on December 2016 the mean value obtained is 89.04 and 98.46. Where in 2012 the mean value obtained is 79.79mg/l and 76.28mg/l in 2013respectively.

5.5 Magnesium(Mg)

Magnesium is a chemical element which has same electron configuration and it is a shiny gray solid. The magnesium value of soil sample is varied from 38.18 to 4.3 on May 2016 and 38.87 to 10.8 on December 2016 with the mean value of 14.49 and 20.00 where in 2012 the mean value obtained is 9.27mg/l and 8.08mg/l for 2013respectively.

5.6 Sodium(Na)

It is soft, silvery white, highly relative metal. It is alkali metal because it has a single electron. Sodium changes the properties of soil and it affects the availability of crop water. The sodium value of soil sample on May 2016 is varied from 100 to 15 and 150 to 12 for December 2016. The mean value obtained is 56.6 and 75.7 where in 2012 the mean value obtained is 22.00mg/l and 29.5mg/l in 2013. This shows the random increase in value.

5.7 Potassium(K)
Potassium is an alkali metal and it is considered as the macro element for soil and crop productivity. The potassium value of soil sample is varied from 130 to 10 on May 2016 and 206 to 15 on December 2016. The mean value obtained is 54.7 and 565.2 where in 2012 the mean value is 33mg/l and 29.5mg/l in 2013. This shows the value is randomly increased.

5.8 Chlorine(Cl)
Chlorine is an extremely relative element yellow green gas at room temperature and a strong oxidizing agent. The chlorine value of soil sample is varied from 1199 to124 on May 2016 and 1298 to170 on December 2016. The mean value obtained is 1313.1 and 565.2.

5.9 Total Alkalinity(Ta)
Alkalinity refers to the water to neutralize acid and it is has buffering capacity. The total alkalinity of soil sample is varied from 1654 to 673 at the month of May 2016 and 1532 to 678 at the month of December 2016. The mean value obtained is 1228 and 1244.1, this is due to over liming acid in soil.

5.10 Total Hardness
Total hardness is determined by multivalent cation's concentrations. The total hardness of soil sample is varied from 1850 to 783 at the month of May 2016 and 1983 to 983 at the month of December 2016. The mean value obtained is 1471.4 and 1557.3 respectively.

5.11 Iron(Fe)
The iron content of soil sample is varied from 26.04 to 3.10 at the month of May 2016 and 25.40 to 5.10 at the month of December 2016. And the mean value obtained is 16.782 and 15.195 respectively. The high iron content in soil sample may cause high level of toxicity.

5.12 Cadmium(Cd)
The cadmium content in soil sample is varied from 2.13 to 0.01 at the month of May 2016 and 2.92 to 0.04 at the month of December 2016 the mean value obtained is 0.953 and 1.5116 respectively. This shows high level of toxicity in soil.

5.13 Copper(Cu)
The copper content in soil sample is varied from 11.54 to 2.07 in the month may 2016 and 9.82 to 0.81 in the month of December 2016.The mean value obtained is 7.764 and 6.424 respectively. The copper content in soil sample is because soil has texture between loan and clay.

5.14 Manganese (Mn)
The manganese content in soil sample is varied from 19.81 to 10.98 in the month of May 2016 and 32.40 to 10.96 in the month of December 2016. The mean value obtained is 15.431 and 18.046 respectively.

5.15 Chromium(Cr)
The chromium content in soil sample is varied from4.12 to 0.8 at the month of May 2016 and 5.23 to 0.56 at the month of December 2016. The mean value obtained is 1.908 and 2.087 respectively.

5.16 Zinc(Zn)

The zinc content in soil sample is varied from9.23 to 0.05 in the May 2016 and 11.50 to 0.6 in the month of December 2016. The mean value obtained is 4.203 and 4.679 respectively. High content of zinc in soil may produce leached acids which is very harmful to plants.

5.17 Lead(Pb)

The lead content in soil sample is varied from23.7 to 1.2 in the month of May 2016 and 34.9 to 1.2 in the month of December 2016. The mean value obtained is 72.03 and 85.42 respectively. High lead content in soil sample leads to high toxicity which may cause damages to plants which grows over there.

5.18 Electrical Conductivity(Ec)

Around Vellalore dump yard it varies between 1.01 mS/cm - 1.0 mS/cm with average value of 1.01 mS/cm in 2012 and the average value of 1.0 mS/cm in 2013 respectively

6. Conclusion

- Soil contamination in the yard impacts groundwater and induces pollution near the disposal site, causing groundwater contamination by excessive percolation.
- As a result of this contamination of ground water, the effects of drinking water in nearby areas are more humane.
- There is no proper disposal system for the above-mentioned problem, the soil is highly polluted and destroys the growth of plants and nutrients in the soil.
- Made to be highly toxic components and the results tested reveal the truth behind this, that it concludes that the problem will be more to nature, animals and people's health.

Reference

[1] Nithya R, Velumani A 2012 Optimal location and proximity distance of municipal Volume **8** Issue 4 ISSN:2224-3496.

[2] Johnpaul V, Dr N Balasundaram 2019 Environmental impact of dumping GGBFS on lands in Coimbatore Volume **8** Issue 2S2 Pages 332-334 .

[3] Velumani 2013 GIS based optimal collection routing model for municipal solid waste-case study in Singanallur ISSN:2277-3754 volume **3**.

[4] Prem Sudha, R N Uma and C Meiaraj 2016 Impact assessment of soil characteristics in and around Vellalore dumpyard in Coimbatore City, Tamilnadu Volume **20** Issue 2 ISSN 0971–765.

[5] R Prem Sudha, Dr R N Uma and Dr C Meiaraj 2016 Assessment of soil quality at municipal solid waste dump site in Coimbatore-Tamilnadu, India Volume.**7** Issue. 2 E-ISSN 0976-3945

[6] R Prem Sudha, R N Uma 2015 Adverse impact on soil and ground water due to solid waste open dump in Coimbatore town ISSN0972-768.

[7] SenthamilSelvan K, Palanivel M 2016 A Case study approach on muncipal solid waste generation and its impact on the soil environment in Dharapuram muncipality, Tamil Nadu, India. Volume.**9** No.02 Pg 196-204, ISSN: 0974-4290.

[8] Ganeswaran, DeepashrS s 2015 A Solid Waste Management in Coimbatore City Volume **10** No. 14 ISSN 1819-6608.

[9] Mohan and Lekeshmanaswamy M 2014 Environmentally and economically viable manager sustainable solid waste management in Coimbatore, Tamil Nadu, India Volume.**1** Issue 7 ISSN 2348 –5787.

[10] Wajahat Nazif, Sajida Perveen and Iftikhar Saleem 2006 Status of micro nutrients in soils of district Bhimber (Azad Jammu And Kashmir) Volume.1 No.2 ISSN 1990-6145.

[11] G Sivapraveena, P V V Prasada Rao, 2016 Impact of leachate on soil properties in the dump site Volume. **4** Issue.1 ISSN 2091-2730235.

[12] Nisha Gupta, Manisha Agrawal A 2014 Characterization and utilization of municipal solid waste soil of open dump site of Potiya Drug District in (C.G) India. Volume.**3** Issue.11 ISSN 2278 –0211.

Effects of Corn Cob Ash as Mineral Admixture on Mechanical and Durability Properties of Concrete – A Review

P Murthi[1], K Poongodi[2] and R Gobinath[2]

[1] Centre for Construction Methods and materials, S R University, Warangal, India
[2] Department of Civil Engineering, S R Engineering College, Warangal, India

E-mail: dr.murthi@srecwarangal.ac.in

Abstract. Concrete is the second most consumed product next to water on earth. Cement is the key ingredient for making concrete. India is the second largest cement producer in the world. The production of cement is emitting an equal amount of CO_2 in to the atmosphere and is accounted as about 8% of the world's total emissions. The addition of mineral additives as supplementary cementitious materials in concrete is one of the solution to avoid such environmental pollution. The natural mineral admixture obtained from the novel agricultural by-product, namely corn cob ash (CCA), is being utilized in cement manufacturing and concrete construction. The similarities in both physical and chemical properties of CCA were established when compared to the decade old agro-based mineral admixtures like rice husk ash (RHA) and sugar cane bagasse ash (BA) through research findings. The results are mentioned that the CCA can be used as chemically active mineral admixture and micro-filler in concrete. With this background, the paper presents a review on the effect of substituting CCA in fresh, mechanical and durability properties of concrete.

1. Introduction

Cement and its products are the highly necessitated man-made products in the world. Indian cement production is positioned in the second place next to China. India has a huge potential of cement demand during the forthcoming decades due to recent major initiatives of Government such as development of 98 smart cities and concrete pavement projects. The cement production in India had been raised at a level of 28.3 million tons per annum as on June 2019 statistics. The production is expected to raise up to further 20 million tons per annum in the year 30-31. The projected cement production strategy is expected as more than 50 million tons per annum during 2050 and expected to reach the top most cement producer in the world [1]. The production of cement is emitting an equal amount of CO_2 in to the atmosphere and is accounted as about 8% of the world's total emissions.

There have been three main ways to reduce CO_2 emissions from cement industry:

- Increasing energy efficiency by optimizing processes and modernizing factories by introducing new cement plants using the best available technology or upgrading old plants.

IOP Conf. Series: Materials Science and Engineering 1006 (2020) 012027 doi:10.1088/1757-899X/1006/1/012027

- Substituting fossil fuels with other energy sources by using alternative fuels such as biomass and industrial waste and to replace fossil fuels in cement kilns.
- Using additives in cement clinkers to produce a large range of cement according to their application by utilizing the natural products such as pozzolanic materials or industrial by-products such as fly ash, slag and ashes from agricultural waste.

The pozzolanic material mentioned in the third option is defined as a siliceous and aluminous material which possesses little or no cementitious value however it chemically reacts with the calcium hydroxide at ordinary temperature in the presence of water in finely divided form to develop the compound possessing cementitious properties [2]. Due to the depletion of raw materials and focusing towards the reduction of CO_2 emission, the usage of Portland Pozzolana Cement (PPC) is emerging all over the world. In the last few decades, several technological and scientific developments were achieved after using pozzolanic material as one of the raw materials for the production of factory-blended composite cement and as partial replacement of cement in cement concrete and mortar [3, 4, 5].

The by-products, obtained from mechanical and agricultural industry having pozzolanic characteristics which can impart the advantages of making blended cement. The research findings were proved that the usage of blended cement is improved durability of concrete in addition to the strength and hence the pozzolanic materials are called as Supplementary Cementitious Materials (SCM). Their properties are varied considerably depending upon their origin because of the variable physical and mineralogical characteristics. It can exist in various forms ranging from amorphous reactive materials to crystalline products. The presence of amorphous silica can reactwith calcium hydroxide and alkaline more rapidly than the silica in crystalline form [6]. The suitability and benefits of natural SCM such as volcanic tuff [7], pumice [8], diatomite [9] and metakaolin[10] were investigated in all over the world more than a century. Later industrial by-products like fly-ash (FA), ground granulated blast furnace slag (GGBS) and silica fume (SF) [10, 11, 12] and the by-product from agricultural industry such as rice husk ash [12, 13], palm oil clinker powder [14] and sugar cane bagasse ash [12, 13] were considered as artificial SCM.The suitability of these SCMs are examined in mortar and concrete and found suitable for enhancing the performance in mechanical and durable properties.

2. Corn cob ash

Corn cob is the hard core portion of the corn and base of the kernel as shown in figure 1. Maize (corn crop) is the third most important cereal crop in the world. In India, Telangana state is the second largest leading producer of maize which is about 16% next to the leading producer Karnataka about 18% as per the production statistics of the year 2017 as shown in Figures 1,2 [15]. More than 30 Million MT maize being produced in India every year and simultaneously an equal amount of corn cob waste was generated. The corn cob is generally used as fire wood in various industries. The ash generated from this corn cob is then dumped in low-level arrears or land filling which causes the environmental problems particularly the habitat in nearby areas. Corn Cob Ash (CCA) is one among the agricultural by-products. This can be considered for recycling or reusing in to various ways. One of the opportunity is reusing as pozzolanic material in cement or concrete.It is rich in silica similar to BA and RHA and can be used as SCM in cement and concrete. The CCA was considered as pozzolanic material during manufacturing of blended cement by adding with cement clinkers during grinding process [16] and also added as partial replacement of OPC while making concrete or mortar [17]. Researchers are doing investigation from the last few decades for suitability of CCA as pozzolanic materials and finding the behaviour of CCA blended cement concrete [17, 18]. In this background this paper is intended to highlight the results published by various authors across the globe.

3. Production of CCA

The composition of corn cob is expected to differ from place to place depends up on the soil composition where the maize cultivated [1]. Naturally the corn cobs are wet condition and dried in sun light for a few days to remove the moisture content and avoid the growth of fungus. The dried corn cob is then allowed calcination process which is transform the crystalline stage of ash in to amorphous stage. Olafusi et al prepared the CCA from the sun-dried corn cob in open air to reduce moisture and milled in to pieces. The scraped corn cob were placed in a porcelain crucible inside the muffle carbonite furnace at a temperature of 650^{0}C for 5 hours to completely remove the carbonaceous material until a white substance was formed from the original corncob sample. The sequence of this process is shown in Figure 3(i) to 3(iv) [19,20].

Figure 1. Pieces of corn cob after removing kernel

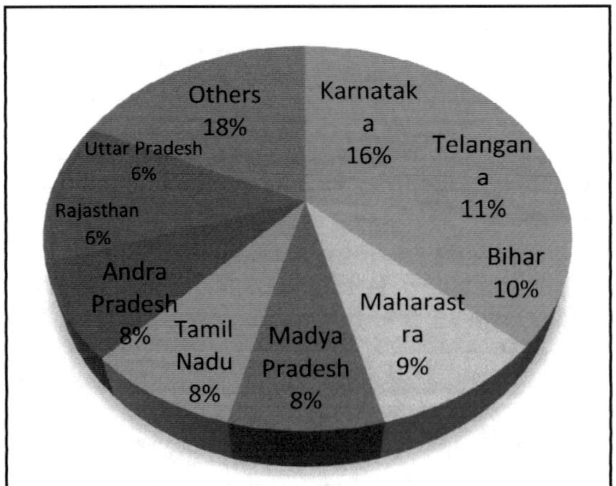

Figure 2. State wise Maize production share in India based on 2010-2016 data [15].

(i) Corn cob (ii) Crushed corn cob pieces

(iii) Heat treatment in muffle furnace (iv) Carbonized corn cob ash

Figure 3. Incineration of corn cob [19]

Kamau et al [17] had conducted incineration process of dried corncobs under uncontrolled conditions in open air at about 650^0C to 800^0C for over 8 hours until they turned to ashes. In order to facilitate calcination, the samples were grinded for 30 min using a porcelain ball mill and transferred into muffle furnace at selected temperatures ranged from 550^0C-800^0C with 50^0C intervals and burning times ranging from 1-4 h with 1 h intervals [21]. Shazim Ali Memon conducted investigation to find the optimum incineration temperature by varying the temperature from 400^0C to 800^0C in muffle furnace at the rate of 10^0C for 2 hours and also determined optimum incineration time by conducting the investigation with 15, 30, 60 and 120 minutes intervals [22]. A measured quantity of corn cob was chopped to maximum size of 12mm put in an electric muffle furnace and the temperature of the furnace was to 600^0C, which was attained in a period of 15 minutes and maintained for a continuous period of 10hours. After the cob has visibly turned to ash, the residue was recovered and weighed. This was repeated 12 times for getting the correct CCA [16].

4. Physical characteristics of CCA

The physical characteristics of mineral admixture, such as particle size, specific surface area and specific gravity are plays essential role in the properties of blended cement, concrete and mortar. Adesanya and Raheem was found the specific surface area of CCA as 272 m^2/kg [23] and noticed that it is lower than the other agro-based mineral admixtures such as Bagasse Ash (BA) and Rice Husk Ash (RHA) and comparable with Fly-ash (FA). Suwanmaneechot et al.[21] examined the specific surface area using the Brunauer, Emmett and Teller (BET) method and found that the specific surface area of CCA was higher (9050 m^2/kg) than that of OPC (3200 m^2/kg). After calcining at 600^{0}C for 4 h, specific surface area of CCA reduced insignificantly (7441 m^2/kg). The agglomeration of CCA particles initiates due to calcining at 200^{0}Cand further extending the calcining temperature increases the particle size due to the melting of smaller particles and amalgamate with larger particles. At 600^{0}C calcining temperature, the complete agglomerate effect was developed and increase the mean particle size of CCA [19]. An average particle size 10-30 µm was developed after 650^{0}C calcinating temperature [19-20]. Suwanmaneechot et al.[21] was found the particle size of CCA at 600^{0}C ranging from 23.56 to 28.45 µm. The specific gravity of CCA was noticed from the literature ranging from 1.95 to 2.55 [22, 24,28,37] and the values are almost equivalent to BA and RHA. The bulk density of CCA was found as 800 kg/m^3 by Ettu et al.[37].Interestingly it is slightly lower than the specific gravity of FA. The results of SEM image, as shown in figure 4, had shown that the raw CCA are polygonal shape with various particle size distribution [21].

(a) CCA (b) CCA calcined at 200^{0}C for 4 hours,
(c) CCA calcined at 400^{0}C for 4 hours and d) CCA calcined at 600^{0}C for 4 hours [21]
Figure 4. SEM Image of CCA and calcined CCA

5. Chemical compositions

The chemical compositions of CCA from various references are shown in Table 1. The EDS spectrum evaluation of calcined CCA at 650^{0}C for 2 hours is shown in Figure 5, which shows the range of major chemical compounds in CCA.The silica (SiO_2) content were noticed more than 50% and it varies up to a maximum of 79% [21]. However, ASTM C618 specified the requirement of sum of SiO_2 + Al_2O_3 + Fe_2O_3 should be more than or equal to 70% and all the references selected for review are satisfied this requirement [25,26]. Most of the CCA contains more than 10% CaO content which is essential for pozzolanic and cementitious properties [17]. These results had shown that CCA can be used as supplementary cementitious materials in concrete. Even though the chemical compositions are satisfied the ASTM C618 criteria, variations in chemical compositions are reported which could be attributed to the source of CCA and incineration process and later the pulverized methods employed by the various researchers. The research finding on the development of CCA blended factory made cement was reported after grinding along with OPC clinkers in a milling machine. The silica content of the CCA blended cements increased from 21.48% to 23.69% and the CaO content decreased from 63.65% to 61.46% when 25% CCA replacement respectively as shown in Figure 6 [14, 21]. This reduction of CaO content enhance the soundness of cement due to the reduction of free lime in the blended cement [23,24]. The variations in major oxide compounds of CCA blended cement agrees with the literature evidences for the requirements of Portland pozzolana cement [27].

Table 1 Chemical compositions of CCA from various sources

References	**Sio_2**	**Al_2O_3**	**Fe_2O_3**	**CaO**	**MgO**	**Na_2O**	**K_2O**	**LOI**
ASTM C618 [26]	$Sio_2 + Al_2O_3 + Fe_2O_3 > 70\%$			-	$\leq 4\%$	≥ 0.70	-	$\leq 10\%$
Adesanya and Raheem [16]	66.38	7.48	4.44	11.57	2.06	0.41	4.92	-
Udoeyo and Abubakar [35]	79.29	0.98	0.64	8.97	1.46	0.0003	3.83	10.8
Oluborode and Olofintuyi[28]	56.39	17.57	9.07	11.47	0.98	1.91	1.98	-
Oyebisi et al. [29]	64.50	6.48	4.03	16.23	1.99	0.90	1.05	5.95
Komalpreet Singh [30]	64.56	9.42	5.12	12.0	3.01	-	-	-
Akila et al. [31]	67.33	7.34	3.74	10.29	1.82	0.39	4.20	-
Desai[32]	62.30	6.25	4.40	10.57	1.86	0.36	3.89	-
P.Suwanmaneechot [21]	63.91	4.01	3.95	4.13	2.91	-	12.12	
Nnochiri[33]	64.90	10.79	4.75	10.24	2.08	0.43	4.23	

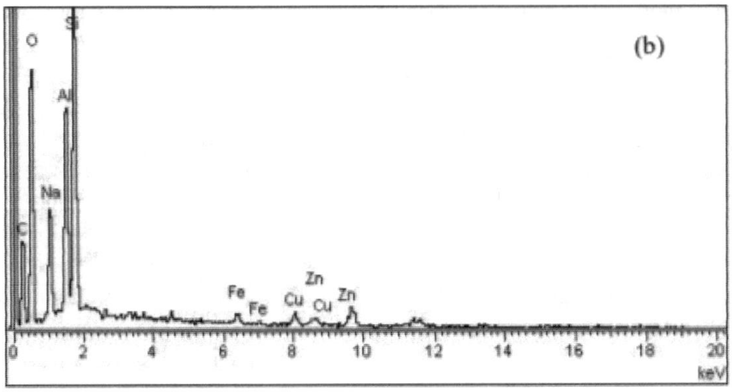

Figure 5 EDS spectrum of calcined CCA at 650^{0}C for 2 hours [21]

6. Effect of CCA on fresh concrete properties

Almost majority of the research findings clearly stated that the slump cone test was employed for finding the workability of concrete however few authors were conducted compacting factor tests for finding the effect of CCA in workability of fresh concrete. The results showed that the slump value and compaction factor decreased when CCA content increased which indicated that concrete becomes less workable as the CCA percentage increases [24, 34]. Udoeyo and Abubakar [35] conducted investigation of 1:2:4: mix concrete with water cement ratio of 0.6 using the CCA replacement level of 0% to 30% and found that the average slump value was decreased gradually from 25 mm at 0% to 2 mm at 30% [23]. Adesanya [34] and Olafusi et al. [19] reported their results that the decreasing the slump value was attributed due to the increased size of irregular shape particle after heat treating more than 600^{0}C and higher amount of silica presence in CCA. The trend line effect of OPC replacement by CCA on slump value and compaction factor is shown in Figure 7 and Figure 8 respectively [16, 35]. In contrast, Kamau et al [17] found that the addition of CCA with SiO_2 content of 38.8% resulting an increase of slump value. Further the author explained that the use of inferior chemical composition of the selected CCA with uncontrolled incineration process caused the increase in slump value.

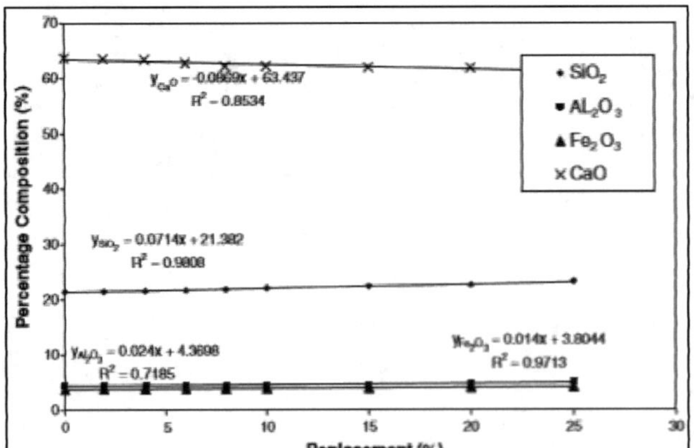

Figure 6 Effect of CCA percentage replacement on the chemical composition of blended cement [21]

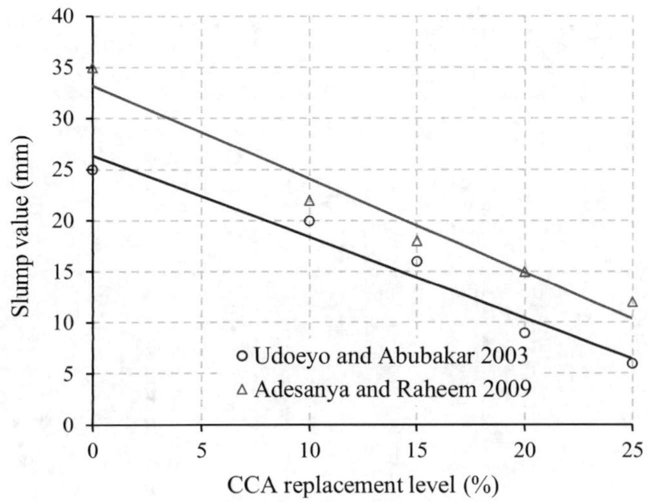

Figure 7 Effect of OPC replacement by CCA on Slump value [16,35]

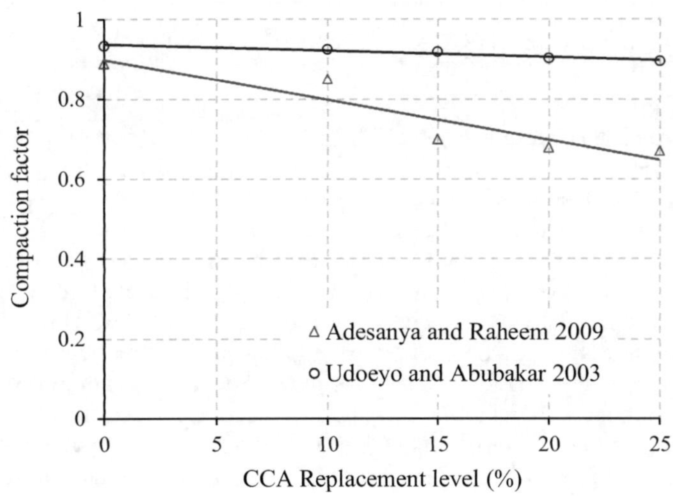

Figure 8 Effect of OPC replacement by CCA on compaction factor [16, 35]

7. Effect of CCA on hardened concrete properties

7.1. Density

Density of 10% CCA blended 28 days cured concrete was observed as 2373 kg/m3 [36] which is slightly lesser than the control concrete. Similar to other agro based SCMs, the density of CCA added concrete was decreasing due to the lower specific gravity and density of CCA particles. Further decrease in density was also observed with increasing curing periods due to the consumption of free lime obtained from hydration of cement during early periods and also the SiO2 portion of CCA. The density variation observed in 28 and 90 days by Kamau et al [17] are shown in Figure 9.

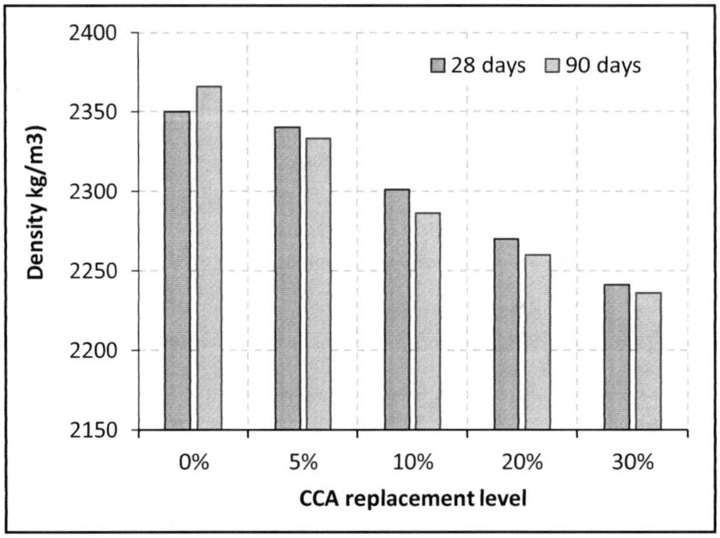

Figure 9 Effect of CCA replacement on Density concrete [17]

7.2 Compressive strength

Adesanya and Raheem (2009) examined the CCA blended cement concrete using 1:1 ½ : 3, 1:2:4 and 1:3:6 mix with w/c ratio of 0.5, 0.6 and 0.7 respectively and found that the compressive strength was more inferior to the control concrete at early curing periods [34]. After the prolonged curing, the compressive strength was improved significantly and had shown higher than the control concrete. The authors found that the substitution of 8% CCA found to be the optimum replacement level in all the mixes considered for investigation. The same trend was found in Ettu et al. investigation [37]. Based on the 90 days strength of OPC with CCA blended concrete, 10% CCA+90%OPC blended concrete had shown higher compressive strength than other proportions including the control concrete. The rate strength gain up to 21 days was found as very slow due to the slower rate of pozzolanic reaction. The strength development of concrete during the early period was exclusively due to the hydration of cement and supplementary cementitious materials like CCA are acting only filler materials and not contributing the development of strength. Further the creation of $Ca(OH)_2$ in concrete during hydration react with the SiO_2 presence in CCA and create secondary hydration products to contribute the strength development of later period [38, 39]. Price et al [24] found that CCA blended concrete was suitable based on the compressive strength up to 10% replacement level of CCA. However, all most all the literature evidences had clearly mentioned that the optimum replacement of CCA was restricted to 8 – 10% of OPC replacement with prolonged curing [23, 24, 34, 35]. The comparison of few the findings at 28 days curing are shown in Figure 10.

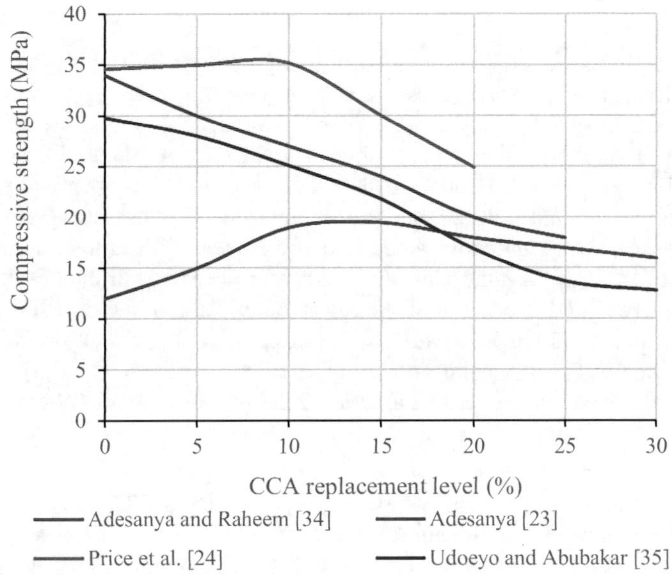

Figure 10 Effect of CCA replacement on compressive strength of concrete

7.3 Tensile strength

The tensile strength of CCA blended concrete decreased with increasing OPC replacement [17, 20, 35]. Desai [32] was examined the splitting tensile strength of CCA blended concrete and concluded that the tensile strength was gradually reduced similar to the compressive strength at the age of 28 days when compared to the control concrete however an increasing trend was observed while increasing the curing period. The tensile strength variations observed from various literature are shown in Figure 11.

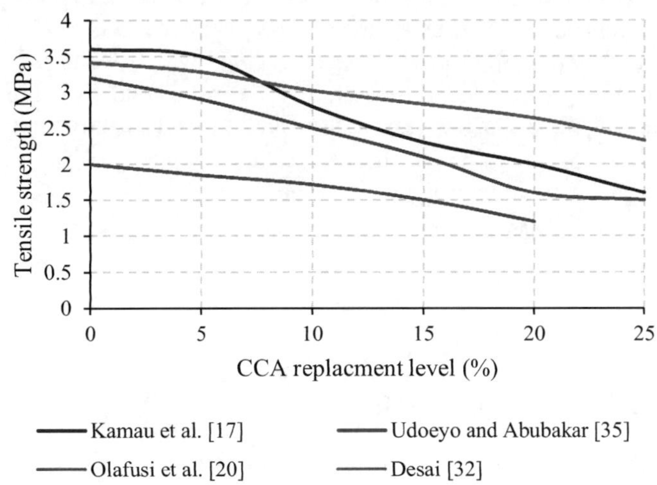

Figure 11 Effect of CCA replacement on tensile strength of concrete

7.4 Durability properties

Adesanya and Raheem [40] conducted water absorption and permeability investigation of CCA blended concrete in three different mix proportions as shown in Figure 12 and found decrease in water absorption up to 10% CCA substitution. The reduction of water absorption may be attributed to the pore filling effect of CCA before pozzolanic reaction starts. The CCA blended cement is contributed to improve the permeability properties of concrete [17]. The incorporation of CCA improves the resistance against chemical attack significantly in blended cement mortar. The loss of weight due to acid attach of CCA blended concrete was investigated by Adesanya and Raheem [38] and concluded that the presence of CCA reduces the loss of weight due to acid attack which indicates the improvement of impermeability of concrete and reduction in strength deterioration factor (SDF) when immersed in sodium sulphate solution. Kamau et al (2016) reported that the reduction of elongation of CCA specimen when immersed in Na_2SO_4 solution [17]. The filler effect of CCA during initial stage and the pozzolanic reaction in the later stage could be attributed the improvement in chemical attack.

8. Conclusions

After assessing the various research findings relating to substituting CCA as part of cement by various authors, the following conclusions are developed:

- In order to develop CCA as mineral admixture, the scraped corn cob should be calcinated at a temperature of 600^0 - 650^0C for 5 hours to completely remove the carbonaceous material until a white substance was formed.
- An average particle size of 10 - 30 μm was able to develop after 650^0C calcinating temperature with the specific gravity of 1.95 to 2.55.
- The chemical compositions of fully incinerated CCA was satisfied the limitations specified in ASTM C618 and possible to utilize as pozzolanic material similar to other agro based mineral admixtures.
- The results of all the literature evidenced that the workability was decreased due to the addition of CCA as cementitious material in concrete.
- There was reduction of mechanical properties such as density, compressive strength and tensile strength of concrete due to the substitution of more than 10% CSA.
- The improvement in durability properties of concrete with 10% CCA was observed from the literature and it can be concluded that the CCA can be used as supplementary cementitious material for making factory made cement or added as partial replacement material at work site.
- The proper awareness has to be encouraged for better utilization of CCA and also possible to convert as wealthy material in order to avoid the environmental threats.

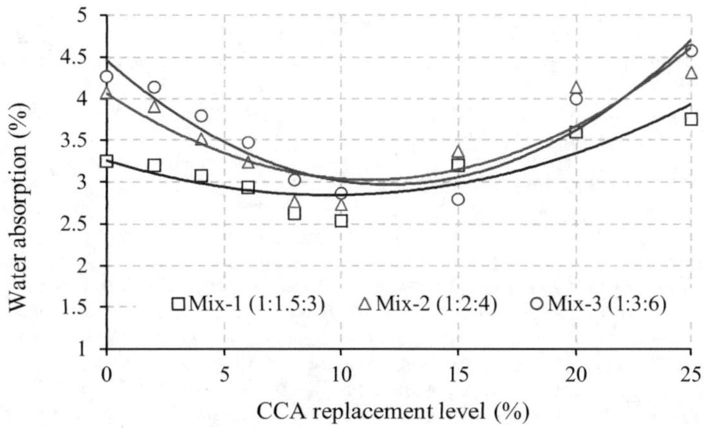

Figure 12 Effect of CCA replacement on water absorption of concrete [40

References

[1] Md S Imbabi, C Carrigan, S McKenna 2012 Trends and developments in green cement and concrete technology *International journal of sustainable built environment* Vol.**1** pp. 194-216, doi.org/10.1016/j.ijsbe.2013.05.001

[2] Murthi P, Poongodi K, Gobinath R, Saravanan R 2020 Evaluation of material performance of coir fibre reinforced quaternary blended concrete *IOP Conf. Series: Materials Science and Engineering* 872 doi:10.1088/1757-899X/872/1/012133.

[3] John Kamau and Ash Ahmed 2017 A review of the use of corncob ash as a supplementary cementitious Material *European Journal of Engineering Research and Science* Vol. **2** No. 8 pp. 1-6 DOI: http://dx.doi.org/10.24018/ejers.2017.2.8.415.

[4] Murthi P, Poongodi K, Awoyera P O, Gobinath R, Saravanan R, 2019, Enhancing the strength properties of High-performance concrete using ternary blended cement: OPC, Nano-silica, Bagasse ash, Silicon doi: org/10.1007/s12633-019-00324-0.

[5] Kavitha O R,Shyamala G, Iyappan G, Rajesh Kumar K 2020 Influence of fly ash and metakaolin on high performance concrete *International Journal of Scientific and Technology Research* Vol **9** No 2 pp.5582-5586.

[6] Sadaqat Ullah Khan, Muhammad Fadhil Nuruddin, Tehmina Ayub, and Nasir Shafiq 2014 Effects of Different mineral admixtures on the properties of fresh concrete *Hindawi Scientific World Journal* pp 1-11, http://dx.doi.org/10.1155/2014/986567.

[7] Abali Y, Bayca S U, Targan S 2006 Evaluation of blends tincal waste, volcanic tuff, bentonite and fly ash for use as a cement admixture *Journal of hazardous materials* Vol.**131** pp 126-130 doi.org/10.1016/j.jhazmat.2005.09.031.

[8] Serdar Aydin, Bulentbaradan 2007 Effect of pumice and fly-ash incorporation on high temperature resistance of cement mortars *Cement and concrete research* Vol. **37** No 6 pp.988-995 doi.org/10.1016/j.cemconres.2007.02.005

[9] Ali Ergun 2011 Effects of the usage of diatomite and waste marble powder as partial replacement of cement on the mechanical properties of concrete *Construction and building materials* Vol. **25** No. 2 pp. 806-812 doi.org/10.1016/j.conbuildmat.2010.07.002

[10] Sujjavanich S, Suwanvitaya P, Chaysuwan D, Heness G 2017 Synergistic effect of metakaolin and fly ash on properties of concrete *Construction and building materials* Vol. **155** pp. 830-837 doi.org/10.1016/j.conbuildmat.2017.08.072

[11] Oner A, Akyuz S 2007 An experimental study on optimum usage of GGBS for the compressive strength of concrete, *Cement and concrete composites* Vol. **29** No.6 pp. 505-514 doi.org/10.1016/j.cemconcomp.2007.01.001

[12] Murthi P, Poongodi, K, Awoyera P, Gobinath R, Sivaraja M, Thirumalai raja K, Viloria A 2020 Permeability properties of lightweight self-consolidating concrete made with coconut shell aggregate *Journal of Materials Research and Technology* https://doi.org/10.1016/j.jmrt.2020.01.092.

[13] Thirumalai Raja K, Murthi P 2015 Bagasse ash and rice husk ash as cement replacement in self-compacting concrete *GRADEVINAR* Vol. **67** No.1 pp. 23-30 doi:https://doi.org/10.14256/JCE.1114.2014

[14] Mohammad Razaul Karim, Huzaifa Hashim, Hashim Abdul Razak 2016 Assessment of pozzolanic activity of palm oil clinker powder *Construction and building materials* Vol. **127** pp. 335-343 doi.org/10.1016/j.conbuildmat.2016.10.002

[15] National commodity market monitor (NCoMM) 2017 *Special report* www.nceml.com.

[16] Adesanya D A, Raheem A A 2009 Development of corn cob ash blended cement *Construction and Building materials* Vol. **23** pp.347-352 doi:10.1016/j.conbuildmat.2007.11.013

[17] John Kamau, Ash Ahmed, Paul Hirst, Joseph Kangwa 2016 Suitability of corncob ash as a supplementary cementitious material *International Journal of Materials Science and Engineering* pp. 215-228 doi:10.17706/ijmse.2016.4.4.215-228
[18] Orlando R Bagcal, Melitob A Baccay 2019 Effect of maize cob ash as supplementary cementitious material on the mechanical properties of concrete *International journal of materials Engineering Innovation* Vol. **10** No. 2 pp.83-97.
[19] Olafusi O S, Adewuyi A P, Sadiq O M, Adisa A F, Abiola O S 2017 Rheological and Mechanical Characteristics of Self-compacting Concrete Containing Corncob Ash *Journal of Engineering research* Vol. **22** No.1 pp. 72-85.
[20] Olafusi O, Olutoge F 2012 Strength properties of corn cob ash concrete *Journal of Emerging Trends in Engineering and Applied Sciences* vol. **3** pp. 297-301.
[21] Suwanmaneechot P, Nochaiya T, Julphunthong P 2015 Improvement, characterization and use of waste corn cob ash in cement-based materials *IOP Conf. Series: Materials Science and Engineering* Vol. **103** doi:10.1088/1757-899X/103/1/012023.
[22] Shazim Ali Memon, Muhammad Khizar Khan 2017 Ash blended cement composites: eco-friendly and sustainable option for utilization of corn cob ash *Cleaner Production* doi:10.1016/j.jclepro.2017.12.050
[23] Adesanya D A 1996 Evaluation of blended cement mortar, concrete and stabilized earth made from ordinary Portland cement and corn cob ash *Construction and Building Materials* Vol. **10** pp. 451-456.
[24] PriceA, Yeargin R, Fini E, Abu-Lebdeh T 2014 Investigating effects of introduction of corncob ash into Portland cements concrete: Mechanical and thermal properties *American Journal of Engineering and Applied Science* Vol. **7** pp. 133-144.
[25] American Society for Testing and Materials 1994 Specification for Portland cement, ASTM C 150.
[26] ASTM C618 Standard specifications for coal fly ash and raw of calcined natural pozzolan for use in concrete.
[27] British Standard Institution 1991 Specification for Portland Cements BS 12, London, British Standard Institution.
[28] Kayode Oluborode, Ilesanmi Olofintuyi Self-compacting concrete: Strength evaluation of corn cob ash in a blended Portland cement,
[29] Oyebisi S O, Olutoge F A, Ofuyatan O O, Abioye A A 2017 Effect of corncob ash blended cement on the propeeties of laterite interlocking blocks *Proceedings of the 2nd International Conference of Engineering for a Sustainable World ICESW* (Ota, Nigeria) IJESD-193349.
[30] Komalpreet Singh, Jaspal Singh and Sarvesh Kumar 2017 A sustainable environmental study on corn cob ash subjected to elevated temperature *Current World Environment* Vol. **13** No. (1) pp. 144-150 doi.org/10.12944/CWE.13.1.13
[31] Akila S, Manila A, Meera D, Nathigamani G, Ramya S 2018 A partial replacement of cement with corn cob ash in concrete production *International Journal of Advanced Research Trends in Engineering and Technology* pp.226-233.
[32] Pranav H Desai 2018 Experimental study on corn cob ash powder as partial replacement of cement in concrete, *International research Journal of Engineering and Technology* Vol. **5** No.6 pp. 724-728.
[33] Nnochiri E S 2018 Effect of corn cob ash on lime stabilized lateritic soil *Journal of Civil Engineering* (Special Issue) pp. 73-85 DOI: 10.1515/sspjce-2018-0007.
[34] Adesanya D A and Raheem A A 2009 A study of the workability and compressive strength characteristics of corn cob ash blended cement concrete *Construction and Building Materials* vol. **23** pp. 311-317.
[35] Udoeyo F F, Abubakar S A 2003 Maize-cob ash as filler in concrete *Journal of materials in civil engineering* vol. **15** pp. 205- 208.
[36] Toryila T M 2016 Partial Replacement of cement with corn cob ash *International Journal for Innovative Research in Multidisciplinary Field* Vol. **2** No.7 pp.159-169.
[37] Ettu L O, Anya U C, Arimanwa J I, Anyaogu L, Nwachukwu K C 2013 Strength of binary blended cement composites containing corn cob ash *International Journal of Engineering and Development* Vol. **6** No.10 pp.77-82.
[38] Shetty M S Concrete Technology Seventh revised edition 2017 S Chand and Company Limited.
[39] Santhakumar A R Concrete Technology Ninth edition 2012 Oxford university press.
[40] Adesanya D A, Raheem A A 2010 A study of permeability and acid attack of corn cob ash blended cements *Construction and Building materials* Vol. **24** pp.403-409.

Forecasting Future Traffic Trend by Short-Term Continuous Observation

Vikas Mendi[1] and I Srinivasula Reddy[2]

[1]Assistant Professor,Department of Civil Engineering, RV College of Engineering,India
[2]Research Scholar,Department of Water Resources and Ocean Engineering, NITK Surathkal,India
E-mail: sr0863699@gmail.com

Abstract. Annual Average Daily Traffic (AADT) is a key parameter to understand the traffic flow rates, traffic density and to design any highway. Generally, short period observed traffic data mainly depends on that season in which the traffic surveys were conducted, which may be high or low compared to the other seasons. So, the behavior of seasonal variation of traffic must be considered for the AADT analysis. These seasonal variations can be found out using the past recorded data of that selected location. But in the case of a location where the past annual traffic data is not available, an alternative method is required to calculate the seasonal variation of the traffic data. The present study deals with the analysis of seasonal variation factors to estimate the AADT from the fuel sale data collected from the nearby petrol stations at the traffic survey point. This work explains how Annual Average Daily Traffic (AADT) can be estimated from a week's limited traffic data when there is a scarcity of automatic traffic data collecting systems.

1. Introduction

Traffic volume often measured concerning Annual Average Daily Traffic (AADT), is an important output of traffic monitoring systems [1]. Annual Average Daily Traffic (AADT) can be defined as the average amount of a year's traffic that passes through a certain location of the corridor. Information on (AADT) is essential for assessment of parameters majorly to be considered in highway planning, fuel-tax revenue projections, and pavement design [2].

David Albright [3], studied the history of estimating and evaluating annual traffic volume statistics in a report published by Transportation Research Record 1305, stated that in the 1930s the only mode of traffic survey available was the manual counting. In the early 1940s, the changeover to mechanical methods of traffic measurements simplified the process of traffic measurement. In the 1950s and 1960s, many authors and researchers proposed theoretical methods to perform traffic studies and traffic summary statistics. The procedures that were developed at that period were debatable at times, also cemented themselves strongly in the collection of traffic data for at least three decades.

Conventionally, AADTs can be estimated for most of the corridors by conducting short-term surveys (STTCs or coverage counts) and by applying some expansion factors which may be derived from Permanent Traffic Counts (PTCs) [4].

The Federal Highway Administration Traffic Monitoring Guide [5] recommends to various bodies involving and managing transportation systems about the design of effective monitoring systems, depending on the amalgamation of transportable counters used for a few days (Short Period Traffic Counts, or SPTCs) or at least one week (Seasonal Traffic Counts, or STCs) per year, and Automatic Traffic Recorders (ATRs) which give Permanent Traffic Counts (PTCs).

Generally, the collection of traffic data can be done by manual counting or automatic data recording computerized systems located at the required places of the corridor. The data collection through manual counting is hectic work to recruit people and monitor the counting team throughout the year. The data collection through automatic counting systems involves money and collecting data without flaws is a challenging job. So the observation of annual traffic data may not prove economical for the developing countries like India. Estimation methods for calculation of the AADT from the known seasonal traffic data of a particular location of the corridor are required and this method of approach is acceptable all over the world [6]. The present study focuses on the estimation of AADT from the traffic data of consecutive seven days of a year.

The traffic growth depends on several factors such as the economy of the country as well as the state economy, population growth, local industrial growth, economic status of the population, etc. Generally, the growth rate of public transport depends on population growth and commercial traffic growth depends on the growth of industries connecting the particular corridor.

2. Study Area and data collection

Traffic surveys were conducted on NH 8A (presently NH 47) locating a station having a distance of 28 kilometers from Ahmedabad city, near a village named Rajoda, Ahmedabad district, Gujarat state, India. The considered traffic survey point is the main station on the corridor which is a major transport national highway between north Gujarat & adjacent states and south Gujarat.

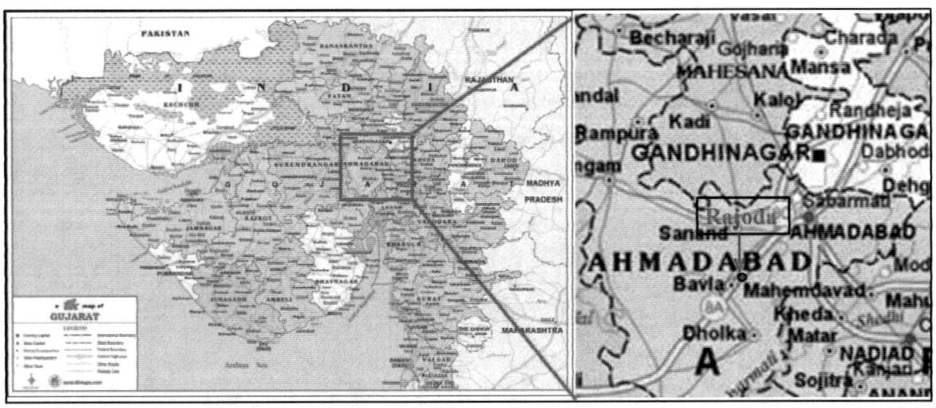

Fig. 1: Location of the traffic count survey point on the corridor.

The traffic data was collected by manual counting method. The data was collected by counting the vehicles in prescribed formats containing both passenger traffic and goods traffic vehicle segregation. The passenger traffic format contains cars (white board and yellow board separately), buses (minibus, RTC buses, private buses and school buses separately) and tata magic. The goods traffic format contains light goods vehicles, two-axle, three-axle and multi-axle vehicles. The data was collected continuously throughout the day and night for seven consecutive days of September month, year 2012.

3. Methods and analysis

The estimation of AADT of any corridor involves the following stages: First, selection of location to estimate the AADT which depends on application of estimated AADT, identification of major junction of the corridor is required in order to analyze the inflow and out flow traffic of the corridor and to fix the maximum traffic location if the objective is to predict the toll revenue of the corridor through AADT analysis. Second, preparation of traffic counting formats containing different types of vehicles which are flowing through the corridor. Third, counting the traffic round the clock of seven days and enquiring the reasons for abnormal traffic during to survey period if the traffic data is not reliable. Fourth, Collection of petrol and diesel sales data for Seasonal Variation Factor analysis, investigate the reason for major variations of fuel sales data to minimize the errors in Seasonal Variation Factors. Fifth, estimation of ADT and AADT from the collected data.

Massimiliano Gastaldi et. al [2] demonstrated method based on fuzzy theory that represents measures of uncertainty and road groups by fuzzy boundaries. Predefined road groups were assigned by road segment using Neural network methods. Nale Zhao et al., [7] proposed a working method for determining characteristics of traffic flow by the using toll data.

Study by Jung-Ah Ha and Ju-Sam Oh [8] estimated AADT from traffic volume data measured in short-term survey in a year using different methods and investigated more precise methods than previous studies which requires adjustment factors. The methods used can reduce error due to daily traffic variation while estimating AADT.

Joe Flaherty [9], Lee et al., [10] and Ha et al., [11] estimated AADT by clustering PTC points to apply adjustment factors. Ha et al., [12] proposed a method to assign them to the classified groups. The proposed method cannot reflect daily characteristics by group.

Lim et al., [13] analyzed different methods based on the same section, grouping of adjustment factors to estimate AADT of STC points and compared. The methods based on same section showed small error. Daily traffic variations cannot be reflected using this method even though monthly factors and weekly factors applied. In addition to the method applying adjustment factors AADT can be estimated using regression models, artificial neural networks and spatial statistics models.

Xia et al., [14] and Zhao et al., [15] performed multi-variate regression analysis to estimate AADT using spatial data. socio-economic variables and road characteristics were applied to the multi-variate regression analysis and found that the applied variables affect AADT.

A spatial regression analysis was used by Eom et al., [16] and Heo et al., [17] to estimate AADT. Eom et al. [16] estimated the parameters of variogram using Euclidean distance in the application of spatial regression analysis (generalized kriging method). Instead of Euclidean distance Heo et al. [17] applied the shortest path to estimate the parameters of variogram which resulted improvement in accuracy of estimated AADT.

Spatial statistics models (kriging models) were used by Kim [18], Selby et al., [19] and Ha et al., [12] estimate AADT. Traffic volume of nearby roads and traffic volume of the previous year were used as secondary variables in Kim's and Ha et al. studies respectively.

These methods depends on the previous year traffic data, the present year data is not useful for estimation of AADT.

In the present study, traffic surveys were conducted for a week continuously which includes the daily traffic variations of weekdays and weekends days. The seasonal traffic variations were taken care by traffic seasonal variation analysis.

3.1.Seasonal Variation Factor (SVF)

The traffic on any corridor varies day to day and season to season in a particular year due many reasons like climatic conditions, yield of industries, festival season etc. The Seasonal Variation Factor (SVF) defines the variation of traffic on monthly basis or seasonal basis. The SVF can be found out from the available monthly traffic data of previous year; but alternative methods are required to calculate SVF when the data is not available. In the present study, traffic surveys were done for a duration of one week, so the monthly variation factors were estimated from the fuel sale data collected from nearby petrol stations. Both the petrol and diesel sales data of previous one year was collected for the analysis. Fuel sales in the month (September) when the traffic surveys were done was taken as base month; to calculate the SVF of the rest of the months of that particular year and the detailed summary of the SVF are presented in Table 1 below.

Table 1: Seasonal variation factor of fuel sales data.

Month	Petrol (kL)	Seasonal Variation Factor	Diesel (kL)	Seasonal Variation Factor
Aug, 2011	19.0	1.06	88.0	1.57
Sep, 2011	**18.0**	**1.00**	**56.0**	**1.00**
Oct, 2011	20.0	1.11	85.0	1.52
Nov, 2011	13.0	0.72	41	0.73
Dec, 2011	14.0	0.78	51	0.91
Jan, 2012	16.0	0.89	57	1.02
Feb, 2012	16.0	0.89	47	0.84
Mar, 2012	17.0	0.94	45	0.80
Apr, 2012	19.0	1.06	50	0.89
May, 2012	20.0	1.11	52	0.93
Jun, 2012	18.0	1.00	59	1.05
Jul, 2012	16	0.89	58	1.04

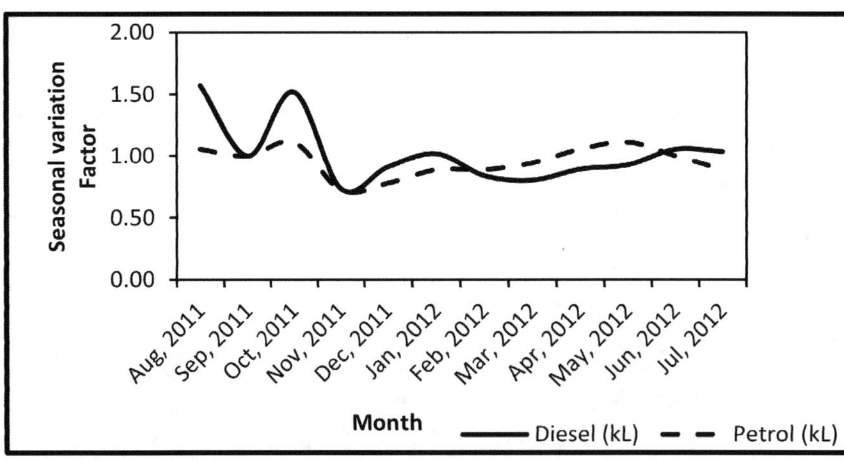

Fig. 2: Seasonal variation factor of fuel sales data.

The selection of SVF depends on the type of the vehicle. The SVF obtained from the petrol sales data was used for the estimation of two wheel vehicles as they consume only petrol. All the goods vehicles and buses consume diesel so the SVF obtained from the diesel sales data was used for the estimation of the same. But the cars run with petrol and diesel. So the average SVF obtained from the petrol and diesel sales data was used for the estimation of cars.

3.2. Passenger Car Units (PCU)

Traffic contains different types of vehicles flowing with different speeds and different space consumptions. The estimation of this traffic requires a common unit to represent all the vehicle types flowing in the corridor. Passenger Car Unit (PCU) is a metric unit to measure the total traffic flowing through the corridor in terms of one single passenger car unit. The conversion of different vehicle modes to PCU requires a conversion factor which depends on rate of flow of that particular vehicle type.

Gecchele et al.,[20] proposed fuzzy set theory and tried to represent the fuzzy boundaries of road groups and uncertainty measures to address the difficulty of classifying a particular type of vehicle into groups in the given road section. However, the conversion factors used for the present study are tabulated below (IRC 64: 1990 [21]).

Table 2: Passenger car unit conversion factors of different vehicle types. [21]

Passenger Car Unit factors			
Passenger vehicles		**Goods vehicles**	
Vehicle type	**PCU factor**	**Vehicle type**	**PCU factor**
Bike	0.5	LCV	1.5
Auto	1	Mini LCV	1
Car/Jeep/Tatamagic	1	2 Axile / 3Axle	3
Bus	3	Multiaxile	4.5
Minibus	1.5	HEM	4.5
Cycle	0.5	Tractor	1.5
Cycle rickshaw	2	Tractor with trailer	4.5

3.3. Calculation of ADT and AADT

The traffic analysis was carried out using one-week data collected at selected location in the corridor. The data was collected in various classes of goods and passenger vehicles as detailed in Figure 3. The data was collected on hourly basis; continuously, for one week. The average hourly vehicular traffic of one-week is depicted in Figure 3. From the average hourly vehicle data, it was observed that the total traffic at night was less compared to day time and goods traffic was more than the passenger traffic at night.

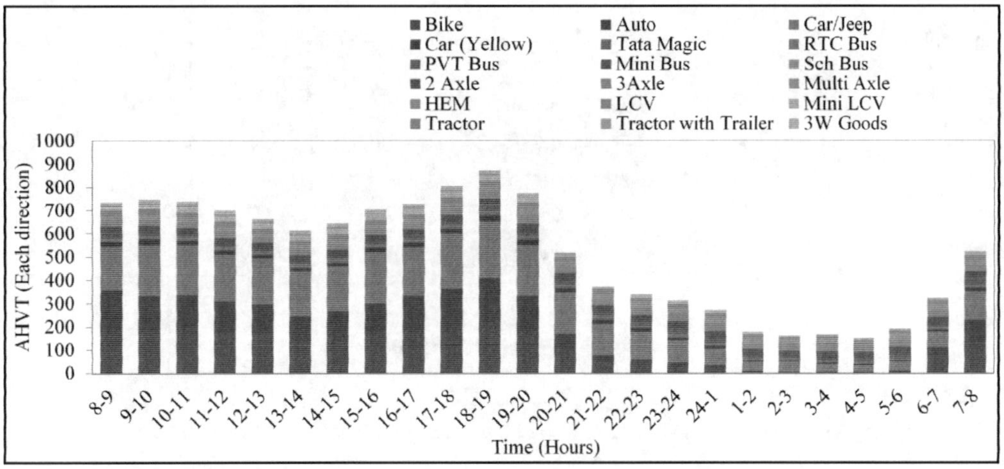

Fig. 3: Average hourly vehicle traffic calculated from one-week data.

The analysis was done using MATLAB to calculate each component of the analysis. ADT and AADT values of the selected location of year 2013 were calculated as illustrated below: [22]

$$\mathrm{ADT}_i = \frac{1}{7}\left\{\sum_{k=1}^{7}\left[\sum_{j=1}^{24} \mathrm{C}_{ijk}\right]\right\} \dots\dots\dots\dots\dots\dots (1)$$

The Average Daily Traffic (ADT) was calculated using the equation (1),
Here,
C_{ijk} = Number of vehicles of ith vehicle type, j^{th} hour and k^{th} day
ADT_i = Average Daily Traffic if i^{th} vehicle type

$$\mathrm{AADT}_{Total} = \frac{1}{12}\left\{\sum_{j=1}^{12}\left(\sum_{i}^{n} \mathrm{W}_{\mathrm{PCU}_i}\mathrm{W}_{\mathrm{SVF}_{ij}}\mathrm{ADT}_i\right)\right\} \dots (2)$$

The Annual Average Daily Traffic (AADT) was calculated using the equation (2),
Here,
W_{PCU_i}= Weight factor of ith vehicle type to convert into passenger car units
$W_{SVF_{ij}}$ = Weight factor of ith vehicle type and j^{th} month to consider the seasonal variation
n = Number of vehicle type data collected
ADT_i = Average Daily Traffic if i^{th} vehicle type
$AADT_{Total}$ = Total AADT in passenger car units

4. Results

The comparison of ADT and AADT is shown in Figure 4. The calculated ADT is less compared to the estimated AADT, this is due to the variation of traffic in different months. The ADT calculations does not include the seasonal variation effect, ADT purely depends on the observed data in that particular week, the analysis may go wrong if the observed data is abnormal due to disturbances in local activities. So to estimate the AADT analysis of seasonal traffic variations are very important.

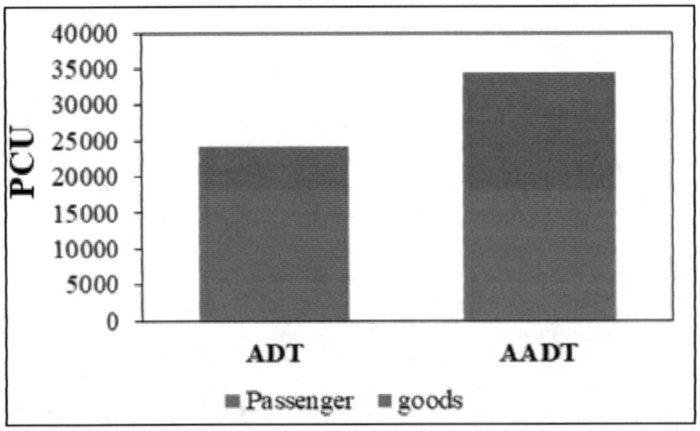

Fig. 4: Comparison of ADT and AADT at the selected location of the corridor.

Gujarat Infrastructure Development Board along with LEA Associates South Asia Pvt. Ltd., New Delhi had conducted traffic surveys in the year 2007 at a location 15 km away from the present point of study of the same corridor (NH 8A) [23]. They determined the AADT of the year 2007 by conducting surveys of one week duration and estimated the future AADT till the year 2040 (Figure 5). From the projected AADT of 2007 traffic has reached 33605 PCUs in 2012 year which is very closer to the estimated AADT obtained (34632 PCUs) from the present study.

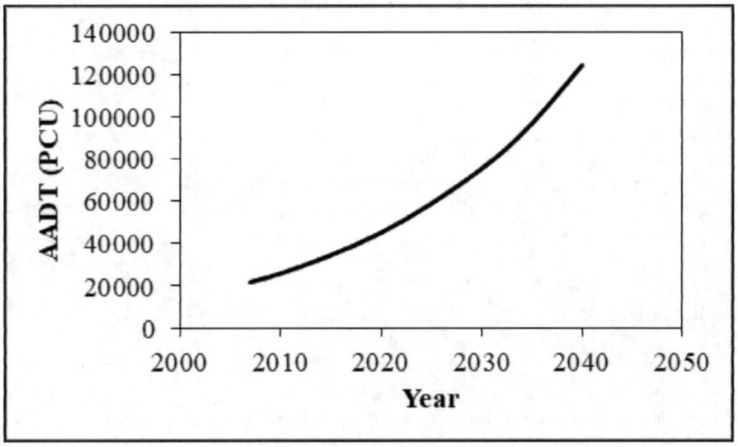

Fig. 5: Estimated AADT from the year 2007 traffic data (Gujarat Infrastructure Development Board) [23].

5. Conclusions

By comparing the AADT estimated from 2007 traffic data by Gujarat Infrastructure Development Board (35424 PCU in 2013), the obtained AADT from the present analysis (34632 PCU) is having good aggregation.

The day time total traffic is more compared to night time traffic but the goods traffic is more at night time compared to passenger traffic in the present study area.

From the estimated ADT and AADT, it is concluded that the seasonal effect is more on the traffic in the present study area.

Data can be used to predict the future AADT which is useful to calculate the revenue can be generated from that particular corridor.

Notations/abbreviations

AADT = Annual Average Daily Traffic
ADT = Average Daily Traffic
AHVT = Average Hourly Vehicle Traffic
LCV = Light Cargo Vehicle
PCU = Passenger Car Unit
SVF = Seasonal Variation Factor

References

[1] Giuseppe Grande, Steven Wood, Auja Ominski and Jonathan D Regehr 2017 *Transportation Research Record: Journal of the Transportation Research Board* **2644** 30–38.

[2] Massimiliano Gastaldi, Riccardo Rossi, Gregorio Gecchele and Luca Della Lucia 2013 *Elsevier Procedia - Social and Behavioral Sciences* **87** 279 – 291.

[3] David Albright History of estimating and evaluating annual traffic volume statistics technical report *transportation research record* 1305.

[4] Ming Zhong, Ehsan Bagheri and Jim Christie 2012 *Elsevier Procedia - Social and Behavioral Sciences* **43** 607 – 617.

[5] Federal Highway Administration (FHWA) *Traffic Monitoring Guide* 2013

[6] Traffic Data Collection and Analysis 2004 *Roads Department, Ministry of Works and Transport* Roads Department, Private Bag 0026, Gaborone, Botswana.

[7] Nale Zhao, Tongyan Qi, Lei Yu, Juwen Zhang and Pengpeng Jiang 2014 *Elsevier Procedia - Social and Behavioral Sciences* **138** 632 – 640.

[8] Jung-Ah Ha and Ju-Sam Oh 2014 *Journal of Emerging Trends in Computing and Information Sciences* **5(7)**
[9] Joe Flaherty 1993 *Transp. Res. Rec.* **1410** 93-99.
[10] Lee S J, N C Baek and H J Kwon 2002 *Journal of Korean Society of Transportation* **20(6)** 59-68.
[11] Ha J A and S C Oh 2012 *The Journal of the Korea Institute of Intelligent Transportation Systems* **11(2)** 10-20.
[12] Ha J A, T Y Heo, S C Oh and S H Lim 2013 *The Journal of the Korea Institute of Intelligent Transportation Systems* **12(1)** 1-14.
[13] Lim S H and J S Oh 2004 *Journal of Korea Society of Civil Engineers* **24** 19-29.
[14] Xia Qing , Fang Zhao, Zhenmin Chen, L David Shen and Diana Ospina 1999 *Transp. Res. Rec.* **1660** 32-40.
[15] Zhao F and S Chung 2001 *Transp. Res. Rec.* **1769** 113-122.
[16] Eom J K, M S Park, T Y Heo and L F untsinger 2006 *Transp. Res. Rec.* **1968** 20-29.
[17] Heo T Y, M S Park, J K Eom and J S Oh 2007 *The Korean Journal of Applied Statistics* **20(3)** 459-473
[18] Kim H Y 2010 *Journal of the Korean Association of Geographic Information Studies* **13(4)** 138-147.
[19] Selby Brent and Kara M Kockelman 2011 Spatial Prediction of AADT in unmeasured locations by universal kriging *the 90th Annual Meeting of the Transportation Research Board*
[20] Gecchele G, Rossi R, Gastaldi M and Kikuchi S 2012 *Transp. Res. Rec.***2308** 148-156.
[21] IRC 64: 1990: *guidelines for capacity of roads in rural areas*
[22] Giuseppe Grande, Steven Wood, Auja Ominski, and Jonathan D Regehr 2017 *Transportation Research Record: Journal of the Transportation Research Board* **2644** 30–38.
[23] Manual of specifications and standards for six laning of national highways through public private partnership, *Department of Road Transport and Highways*, Government of India 2008

Future Traffic Prediction from Short Period Traffic Data

I Srinivasula Reddy [1]and Vikas Mendi [2]

[1]Research Scholar, Department of Water Resources and Ocean Engineering, NITK Surathkal ,India
[2]Assistant Professor, Department of Civil Engineering, RV College of Engineering, India
E-mail: sr0863699@gmail.com

Abstract. Prediction of future traffic is a complicated process which requires logical prediction methods and experience. The present study focused on the prediction of future traffic from the AADT data which is estimated from the one week continuous traffic data collected from the selected study area. State wise traffic contributions are estimated from the one-day number plate survey data and major states effecting the traffic in the study area are observed. The population data, vehicle registration data and economic data of the major states which are observed through number plate survey are used for the prediction of future traffic data. The vehicle growth rates were calculated using average vehicle registration growth rate method and nonlinear regression (power regression). Finally, the growth rates were decided according to the traffic design norms provided by NHAI. The year wise future traffic was predicted and presented in this paper from the year 2013 to 2040. The changes in Level of Service (LOS) of the NH 8A for four lane and six lane configuration is detailed for the future scenario.

1. Introduction

Roads are the major transport mode of any country's transport system for both passenger and freight transport. The development of any country depends on the good connectivity of each and every corner of the country. India is a country where the 85% of total passenger traffic and more than 60% of freight traffic use road transport system [1]. As the country grows faster, a good strategic plan of developments in road systems are required. National and state high ways are the major corridors in Indian road transportation system. The growing countries like India, funding is a major problem in any development projects like high way construction. So, good statistical methods are required to predict the future traffic and to make development plans. Prediction of any natural phenomenon is not simple and straight forward. Even though prediction of future traffic though statistical methods is not an exact scientific process, many organizations are using the same methods with some adjustments in the growth rates of future traffic which requires proper experience and knowledge on the traffic flow of that particular corridor.

Generally, the future traffic can be estimated either from the past traffic record or by predicting the future traffic growth rates from the dependent parameter such as population, Per capita Income (PCI), Gross State Domestic Product (GSDP), Net State Domestic Product (NSDP) and growth of vehicle registration. Counting the continuous traffic data of number every year requires automatic counting systems, which is not economical and viable method for a wide country like India. So, prediction of future traffic from the limited traffic data is a better choice and economical.

Present study concentrated on prediction of future traffic from the AADT estimated from the one week traffic data. The future traffic growth rates were estimated using growth rates of vehicle registration data and nonlinear regression process.

2. Study area and data collection

Revenue analysis is the ultimate study of any national development projects. The proposed development project should be in a position to generate the amount spent for the project. In the development of national high ways selection of toll plaza location is a key decision. The selected toll plaza location should have maximum toll-able traffic. In this present study, location nearer to the village named Rajoda, Ahmedabad district, Gujarat state, India which is 28 kilometers from Ahmedabad city on NH 8A (present NH 47), four lane National Highway was considered for the traffic analysis, presented in Fig. 1. Traffic volume count survey was conducted for a week and number plate survey was conducted for one working day of September month, year 2012. The collected traffic volume data was used to find out the AADT of the year 2012 and number plated survey data was used to find out the state wise traffic contribution. The estimated AADT data used as base input data for the present study.

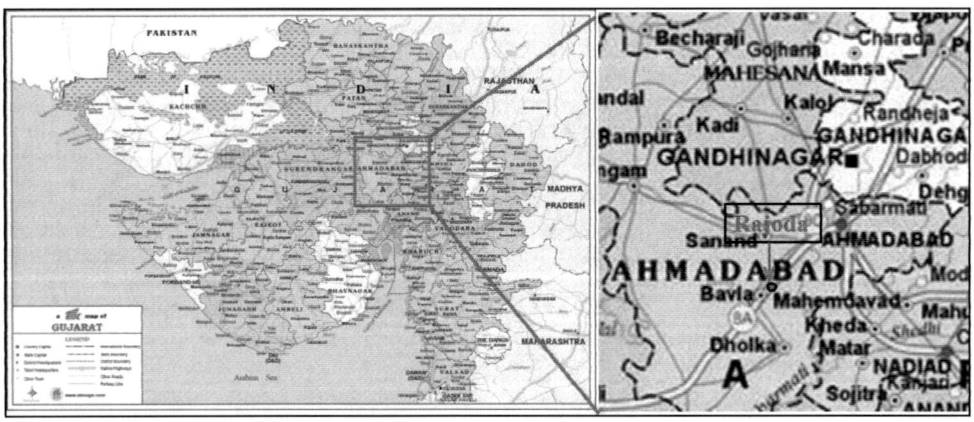

Fig. 1: Location of the traffic volume survey point.

3. Methods and analysis

The traffic growth of any nation depends on the economic and cultural development of the nation. The economic development can be assessed from SGDP, NSDP and PCI of the respective state or nation. The cultural development and population growth effect mainly the passenger traffic modes. The present analysis is concentrated on economic growth and population growth as main key parameter.

Massimiliano Gastaldi et. Al, [2] demonstrated method based on fuzzy theory that represents measures of uncertainty and road groups by fuzzy boundaries. Predefined road groups were assigned by road segment using Neural network methods. Study by Jung-Ah Ha and Ju-Sam Oh [3] estimated AADT from traffic volume data measured in short-term survey in a year using different methods and investigated more precise methods than previous studies which requires adjustment factors. The methods used can reduce error due to daily traffic variation while estimating AADT.

Joe Flaherty [4] and Ha et al., [5] estimated AADT by clustering PTC points to apply adjustment factors. Ha et al., [5] proposed a method to assign them to the classified groups. The proposed method cannot reflect daily characteristics by group.

Lim et al., [6] analyzed different methods based on the same section, grouping of adjustment factors to estimate AADT of STC points and compared. The methods based on same section showed small error. Daily traffic variations cannot be reflected using this method even though monthly factors and weekly factors applied. In addition to the method applying adjustment factors AADT can be estimated using regression models, artificial neural networks and spatial statistics models.

Xia et al., [7] and Zhao et al., [8] performed multi-variate regression analysis to estimate AADT using spatial data. socio-economic variables and road characteristics were applied to the multi-variate regression analysis and found that the applied variables affect AADT.

A spatial regression analysis was used by Eom et al., [9] and Heo et al., [10] to estimate AADT. Eom et al. [9] estimated the parameters of variogram using Euclidean distance in the application of spatial regression analysis (generalized kriging method). Instead of Euclidean distance Heo et al. [10] applied the shortest path to estimate the parameters of variogram which resulted improvement in accuracy of estimated AADT.

Spatial statistics models (kriging models) were used by Kim [11], Selby et al., [12] and Ha et al., [13] estimate AADT. Traffic volume of nearby roads and traffic volume of the previous year were used as secondary variables in Kim's and Ha et al. studies respectively.

These methods depends on the previous year traffic data, the present year data is not useful for estimation of AADT.

The estimated AADT data from the one week traffic data was considered as main input traffic data to the present analysis. The population growth data, vehicle registration data, and economic growth data of the states which were contributing the majorly are collected from the Open.

Government Data (OGD) Platform India. The collected data was used to calculate the future traffic growth factors which are key factors in prediction of future traffic. The growth rates were analyzed using vehicle registration growth rate method and regression analysis and the final decision of growth rates were considered as per the norms given in manual of specifications and standards for six laning of national highways, government of India, 2008. The detailed step wise procedure of prediction of future traffic is presented in the flow chart given in Fig. 2:

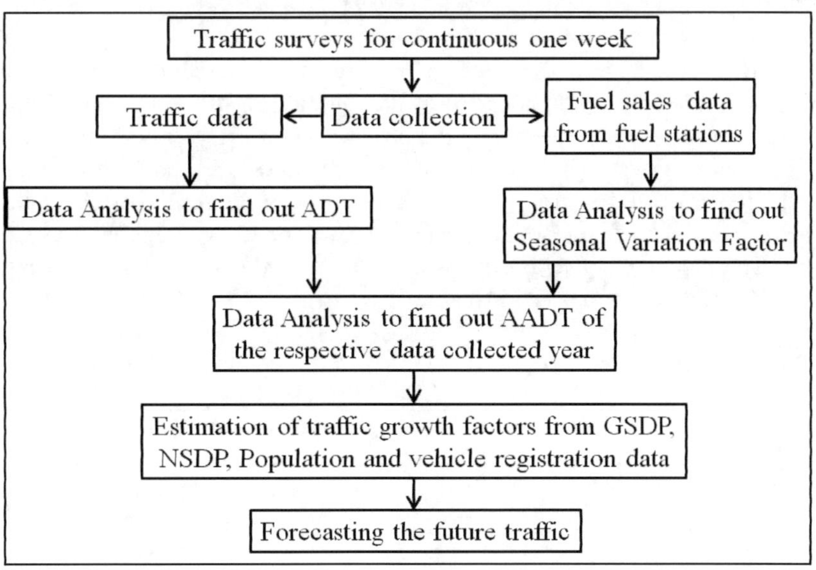

Fig. 2: Flow chart of the methodology for the present study

3.1. Estimation of AADT

Estimation of AADT form the one week traffic volume requires seasonal traffic volume variation data. The seasonal variations traffic volume can be found out by the past traffic volume data or from the fuel sales data of the petrol stations nearer to the survey point. For the present study, the seasonal variations were incorporated by finding the seasonal variation factors calculated from the petrol sales data. The month in which traffic volume surveys were conducted was considered as base month for calculating the monthly seasonal variation of traffic volume. The obtained AADT of individual vehicle was converted to passenger car units (PCU) and used to predict the future traffic volumes.

3.2. State contribution

Analysis of state wise contribution of traffic flowing through the study area is important to analyses the impact of state growth on the traffic growth. Gujarat, Rajasthan, Maharashtra and Haryana states are identified as the major traffic contributing states on the present study area from number plate survey. The percentage of state wise traffic contributions are calculated from the data collected by conducting one day number plate survey on working day of a week and the data is presented in Table 1. The major passenger and goods traffic were contributed by Gujarat state, Rajasthan and Maharashtra states goods traffic contribution was more compared to Haryana state contribution.

Table 1: State wise contributions of passenger and goods traffic

State	State - wise traffic contribution (%)								
	Motor bike	Auto	Car	Bus	2 Axle	3 Axle	M Axle	LCV	Mini LCV
Gujarat	100	100	96	95	86	75	59	94	99
Rajasthan	0	0	2	2	3	8	23	2	0
Maharashtra	0	0	2	2	8	12	11	3	1
Haryana	0	0	0	1	3	5	7	1	0

3.3. Growth rate analysis through vehicle registration data

Identifying the traffic growth rates from the vehicle registrations data is a direct method. Even though the analysis of traffic growth rates from the vehicle registrations data is straight forward method, it involves lots of uncertainties. From the observation of 11 years (from year 2001 to 2011) vehicle registration data, it was observed that the growth rates are not following the continuous increase in trend, the growth rates may follow negative trends for some period of years. Average growth rates were used for the present analysis to overcome the uncertainties in the vehicle registration data. The calculated average growth rates from the vehicle registration data of the major traffic contributing states are given in Table 2.

Table 2: Average growth rates obtained from vehicle registration data.

State	Average Growth Rate (%)				
	Motor bike	Auto	Car	Bus	Trucks
Gujarat	9.15	8.13	9.94	3.59	6.74
Rajasthan	11.30	9.00	11.6	4.70	8.98
Maharashtra	10.00	3.95	10.33	1.88	7.13
Haryana	11.50	14.25	15.77	26.50	8.06

3.4. Regression model

The uncertainties in the average growth methods, some suitable methods are required to overcome the drawbacks of average growth methods and to predict the growth rates in a scientific way. Regression models are one of the most popular methods of analysis of complicated physical problems which involves lots of uncertainty. The regression models basically correlate the factors effecting the physical problem and estimates the effect of that particular factor or factors on the physical problems. Generally researchers tried linear regression, multiple linear regression and different nonlinear regression models to predict the growth rates of vehicles. The nonlinear regression models are better models when compared to linear regression models to analyze the natural physical phenomenon. The present study is followed the nonlinear regression (power regression) model to analyze the vehicle growth rates. The regression analysis basically needs proper understanding of the physical problem and the effecting parameters. In the present analysis population growth, PCI, NSDP and GSDP are considered as the basic parameters which are effecting the growth of vehicles, but the growth of truck vehicles does not depends on the population growth. So, population growth was not considered to calculate the truck vehicle growth through regression analysis. The following power regression model (equation 1) is considered to estimate the vehicle growth rates.

$$V = a(x)^b \dots\dots\dots\dots (1)$$

Here,
V = Vehicle type (dependent variable)
x = Population/ PCI/GSDP/NSDP (independent variable)
a = Constant
b = Elastic constant
The nonlinear models can be transformed to linear models by using simple transformation technics. The above equation 1 has transformed to linear model by introducing Ln on both sides of the equation as shown in equation 2.

$$\text{Ln}(V) = \text{Ln}(a) + b\,\text{Ln}(x) \dots\dots\dots (2)$$

The elastic constants of the each vehicle type can be find out using the least square error method with respect to all the depending variables. The elastic constants of cars with respect to population, PCI and NSDP are plotted in Fig. 3.

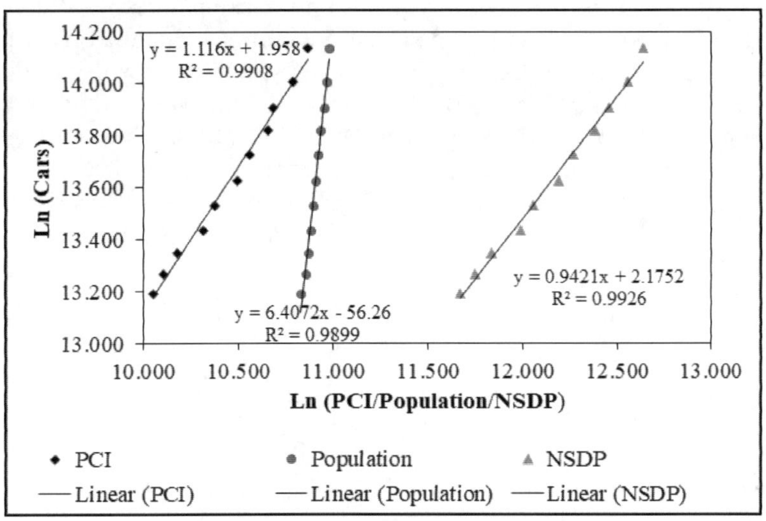

Fig. 3: Elastic constants form regression analysis of cars in Gujarat state.

The population, PCI, NSDP and GSDP data was collected for 11 years (from year 2001 to 2011) to calculate the average growth rates of respective data. The obtained average growth rates cannot be used directly to predict the vehicle growth rates due to the uncertainties in the population, PCI, NSDP and GSDP growth data. So the obtained growth rates should be corrected by multiplying the corresponding elastic constant obtained from the regression analysis. The obtained growth rated and the corresponding correlation coefficients of Gujarat state are given in Table 3. Similarly, the growth rates of different vehicle types were calculated for the each state which identified earlier state wise contribution analysis. The obtained growth factors of each state are used to calculate the effective growth factors of individual vehicle types.

Table 3: Growth rates form the regression analysis of Gujarat state

Vehicle Type	Growth Rate			R^2 Value		
	Population	PCI	NSDP	Population	PCI	NSDP
Motorbike	9.29	9.10	9.15	0.99	1.00	1.00
Auto	8.45	8.29	8.34	0.98	0.99	0.99
Car	9.80	9.58	9.63	0.99	0.99	0.99
Bus	4.76	4.86	4.86	0.73	0.80	0.98
Trucks	---	6.20	6.25	---	0.97	0.98

4. Results

Selection of growth rates is a tricky process which involves the knowledge on vehicle type and economical and social parameters. In the present analysis vehicle growth rates were obtained from the population, PCI, NSDP and GSDP growth data. While selecting the obtained growth rates, the growth rate obtained from the PCI data was selected for both motorbike and car vehicle types because the personal transport modes always depends on per capita income. The growth of public transport depends mainly on population growth, the growth rates obtained from population data were used to find out growth rates of auto and bus vehicle types. The growth of any goods vehicles depends on the industrialization and economic development of any nation which impacts the NSDP and GSDP. The growth rates obtained from the NSDP and GSDP data were used to calculate the growth rates of truck vehicle types. Finally the average of growth rates obtained from the vehicle registration data and regression analysis are considered along with state wise traffic contribution factors. The final growth factors were used from Table 4 which were obtained from equation 3.

$$FGF_i = \sum_1^n S_i AGF_i \text{ .... } (3)$$

Here,
FGF_i = final growth factor of i^{th}vehicle type
S_i = State wise traffic contribution factor (From Table 1) of i^{th}vehicle type
AGF_i = Average growth rare obtained from vehicle registration data and regression analysis of i^{th}vehicle type
n = Number of vehicle types

Table 4: The final growth rates Obtained from the analysis using the data base from year 2001 to year 2011.

Data Year	Motor bike	Auto	Car	Bus	2 Axle	3 Axle	M Axle	LCV	Mini LCV
2001-2011	9.2	8.3	9.8	4.6	6.6	6.8	7.3	6.5	6.4

4.1. Growth rates adopted according to NHAI norms

Even though the growth rates were calculated from the analytical methods, the actual future growth rates never follow the analytical growth rates due to several reasons like transition of vehicle types from one mode to another mode, change of mode of transport from road to rail mode due to improvements in railway systems, new government policies, changes in fuel rates, changes in vehicle rates, developments in port systems etc. So, the prediction of future traffic requires some adjustments in future growth rates to overcome the uncertainties in future. The National Highway Authority of India has given some norms to design the traffic in manual of specifications and standards for six laning of national highways, government of India, 2008. According to the NHAI manual the growth rates obtained from the analysis can be used to the next five tears turn from the year of traffic data considered for the analysis. The future growth rates should reduce by 2% for all the vehicle types in every five years turn and the growth rates of any vehicle type should not be less than 5% in any year. By incorporating the NHAI traffic design norms, the future vehicle growth rates are adopted as given in Table 5.

Table 5: The modified future vehicle growth rates (%) as per NHAI norms.

Year	Motor bike	Auto	Car	Bus	2 Axle	3 Axle	M Axle	LCV	Mini LCV
2012-2015	9.2	8.3	9.8	5.2	6.6	6.8	7.3	6.5	6.4
2015-2020	7.2	6.3	7.7	5.2	5.1	5.3	5.6	5.1	5.0
2020-2025	5.2	5.0	5.7	5.1	5.0	5.0	5.0	5.0	5.0
2025-2030	5.0	5.0	5.0	5.1	5.0	5.0	5.0	5.0	5.0
2030-2035	5.0	5.0	5.0	5.1	5.0	5.0	5.0	5.0	5.0
2035-2040	5.0	5.0	5.0	5.1	5.0	5.0	5.0	5.0	5.0

4.2. The predicted future traffic

Prediction of future traffic is required to study the technical and financial viability of the national highway. Design of highways requires stage wise construction to save the economy. The predicted traffic will be used to develop the future construction plans and to estimate the revenue can generate through the project. In the present study, the estimated AADT of the year 2013 was used as base traffic data to predict the future traffic. The future traffic for each vehicle type in terms of passenger car units was calculated using the equation 4 and presented in Table 6.

$$AADT_{ij} = GR_{ij} * AADT_{ij-1} \text{..........(4)}$$

Here,

$AADT_{ij}$ = AADT of ith vehicle type and j^{th} year ($2014 \leq j \leq 2040$)

GR_{ij} = Growth rates of ith vehicle type and j^{th} year from Table 5

$AADT_{ij}$ = AADT of ith vehicle type and $(j-1)^{th}$ year

Table 6: The predicted future traffic for each vehicle type in PCUs.

Year		Motor bike	Auto	Car	Bus	2 Axle	3 Axle	M Axle	LCV	Mini LCV	Total Traffic (PCU)
2012	**2013**	**2600**	**3936**	**7744**	**3790**	**4039**	**3813**	**5646**	**1224**	**788**	**33580**
2013	**2014**	2838	4263	8502	3986	4306	4073	6055	1303	839	36166
2014	**2015**	3099	4616	9333	4192	4591	4352	6495	1388	892	38959
2015	**2016**	3383	5000	10246	4409	4895	4649	6966	1478	949	41975
2016	**2017**	3626	5315	11040	4636	5147	4894	7356	1553	997	44564
2017	**2018**	3888	5649	11895	4875	5411	5153	7767	1632	1047	47317
2018	**2019**	4167	6005	12817	5126	5689	5425	8202	1714	1099	50245
2019	**2020**	4467	6384	13809	5390	5981	5711	8662	1801	1154	53360
2020	**2021**	4789	6786	14879	5667	6288	6013	9147	1892	1212	**56674**
2021	**2022**	5038	7125	15734	5958	6603	6314	9604	1987	1273	59635
2022	**2023**	5300	7481	16638	6264	6933	6629	10084	2086	1336	62752
2023	**2024**	5576	7856	17594	6585	7280	6961	10588	2191	1403	66033
2024	**2025**	5866	8248	18605	6923	7644	7309	11118	2300	1473	69486
2025	**2026**	6171	8661	19674	7278	8026	7674	11674	2415	1547	73119
2026	**2027**	6479	9094	20660	7650	8427	8058	12257	2536	1624	76785
2027	**2028**	6803	9548	21695	8041	8848	8461	12870	2663	1706	**80635**
2028	**2029**	7143	10026	22782	8452	9291	8884	13514	2796	1791	84678
2029	**2030**	7501	10527	23923	8884	9755	9328	14189	2936	1880	**88923**
2030	**2031**	7876	11053	25122	9338	10243	9794	14899	3083	1974	93382
2031	**2032**	8269	11606	26378	9813	10755	10284	15644	3237	2073	98059
2032	**2033**	8683	12186	27697	10312	11293	10798	16426	3399	2177	102971
2033	**2034**	9117	12796	29081	10837	11858	11338	17247	3569	2286	108129
2034	**2035**	9573	13436	30535	11389	12451	11905	18109	3747	2400	113545
2035	**2036**	10051	14107	32062	11969	13073	12500	19015	3934	2520	**119232**
2036	**2037**	10554	14813	33665	12575	13727	13125	19966	4131	2646	125202
2037	**2038**	11082	15553	35349	13213	14413	13782	20964	4338	2778	131471
2038	**2039**	11636	16331	37116	13883	15134	14471	22012	4554	2917	138054
2039	**2040**	12218	17148	38972	14587	15891	15194	23113	4782	3063	144966

4.3. The predicted traffic data by Gujarat infrastructure development board (GIDB)

Gujarat Infrastructure Development Board (GIDB) had conducted traffic surveys in the year 2003 to predict the future traffic in the National Highway (NH 8A). The GIDB conducted many locations on the NH 8A, the data provided for the location 15 km away (to words Rajkot, Gujarat) from the present study area is used to validate the present study. The vehicle growth rates were calculated using the Econometric Method, Trend Growth of past traffic data and Confusion Assessment Method (CAM). The results obtained by the present study are having good aggregation with the results obtained using Econometric Method provided by Gujarat Infrastructure Development Board. The comparison of the present study results and GIDB results are plotted in Fig. 4.

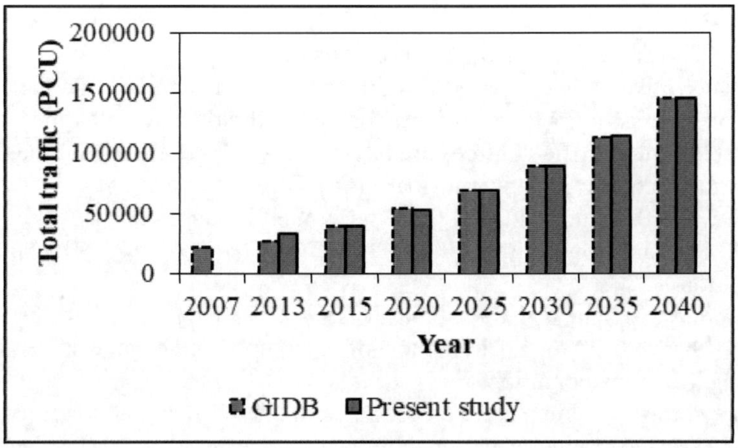

Fig. 4: Comparison of predicted traffic from the present study and GIDB [14].

4.4. Volume study

The traffic volume studies are important to improve the stage wise capacity of the any highway. Stage wise development plan can be prepared because construction of any highways requires lot of money. Generally, the capacity of any highway can be improved by providing, earthen shoulders in first stage, paved shoulders in second stage and extra lane in final stage of construction. The stage wise development plans requires the future traffic data which is predicted from the present study. From the standards provided in Highway Capacity Manual, Transportation Research Board, Washington, D.C. (2016) [15], the existed four lane (NH 8A) with paved shoulders is having the maximum capacity of 79600 PCU/day. According to the NHAI standards any national highway should not cross the level of service 'C' (LOS-C) to maintain the free flow (100 km/h) traffic service. From the predicted traffic volume data, the changes in level of services and the requirement of six lane configuration in future scenario of the NH 8A is detailed in Table 8.

Table 7:Daily service volume and capacity table for two way high ways [15]

Area type	Terrain	AADT (PCUs/day/Lane)		
		LOS (A-C)	LOS (D)	Capacity / LOS - E
Urban	**Level**	14,400	17,500	19900
Urban	**Rolling**	13,700	16,700	19000
Rural	**Level**	12,100	14,800	16800
Rural	**Rolling**	11,000	13,400	15200

Table 8: The predicted level of services of NH 8A (NH 47) at free flow condition.

Level of service	Year of Reaching Level of service	
	4 - Lane highway	6 - Lane highway
LOS (C)	2021	2030
LOS (D)	2015	2034
Capacity / LOS - E	2028	2036

5. Conclusions

The following conclusions are drawn from the present study.

- The average growth rates estimated from the population data, PCI data and NSDP data are having good aggregation with each other for each individual vehicle type.
- The predicted future traffic volumes are having good aggregation with the data provided by Gujarat Infrastructure Development Board (GIDB).
- The existed NH 8A attains LOS - C in the year 2021 and reaches its full capacity in the year 2028. So transition of four lane to six lane is required in 2021 to maintain the free flow condition.
- The developed six lane NH 8A attains LOS - C in the year 2030 and reaches its full capacity in the year 2036. So transition of six lane to eight lane is required in 2030 to maintain the free flow condition.
- The predicted traffic volume can be used to estimate the financial viability development of four lane to six lane of NH 8A national highway.

Notations/abbreviations

AADT= Annual Average Daily Traffic
GIDB = Gujarat Infrastructure Development Board
GSDP = Gross State Domestic Product
LOS = Level of Service
NHAI = National Highway Authority of India
NSDP = Net State Domestic Product
Ln = Natural logarithm ($\log_e$)
PCI = Per capita Income
PCU = Passenger Car Unit
R^2 = Correlation coefficient

References

[1] www.worldbank.org/en/news/feature/2011/09/23/india-transportation accessed on 20-08-2020
[2] Massimiliano Gastaldi, Riccardo Rossi, Gregorio Gecchele and Luca Della Lucia 2014 *Elsevier Procedia - Social and Behavioral Sciences* **87** 279 – 291
[3] Jung-Ah Ha and Ju-Sam Oh 2014 *Journal of Emerging Trends in Computing and Information Sciences* **5(7)**
[4] Joe Flaherty 1993 *Transp. Res. Rec.* **1410** 93-99.
[5] Ha J A and S C Oh 2012 *The Journal of the Korea Institute of Intelligent Transportation Systems* **11(2)** 10-20.
[6] Lim S H and J S Oh 2004 *Journal of Korea Society of Civil Engineers* **24** 19-29.
[7] Xia Qing , Fang Zhao, Zhenmin Chen, L.David Shen and Diana Ospina 1999*Transp. Res. Rec.* **1660** 32-40
[8] Zhao F and S Chung 2001*Transp. Res. Rec.* **1769** 113-122.
[9] Eom J K, M S Park, T Y Heo and L F untsinger 2006 *Transp. Res. Rec.* **1968** 20-29.
[10]Heo T Y, M S Park, J K Eom and J S Oh 2007 *The Korean Journal of Applied Statistics* **20(3)** 459-473
[11] Kim H Y 2010 *Journal of the Korean Association of Geographic Information Studies* **13(4)** 138- 147.
[12] Selby Brent and Kara M Kockelman 2011 Spatial Prediction of AADT in unmeasured locations by universal kriging *the 90th Annual Meeting of the Transportation Research Board*
[13] Ha J A, T Y Heo, S C Oh and S H Lim 2013 *The Journal of the Korea Institute of Intelligent Transportation Systems* **12(1)** 1- 14.
[14] Consultancy services for the preparation of feasibility report for developing an access controlled corridor between Ahmedabad and Rajkot, Gujarat Infrastructure Development Board (GIDB), LEA Associates South Asia Pvt. Ltd., New Delhi, 2008.
[15] Planning and Preliminary Engineering Applications Guide to the Highway Capacity Manual, *Transportation Research Board* (Washington, D.C) 2016 (www.TRB.org)

Compressive Strength and Durability Properties of M35 Grade Concrete by Replacing Sand Partially with Vermiculite and Granite Powder and Coarse Aggregate by Re-Cycled Aggregate

N Kiran Kumar[1], B Damodhara Reddy[2] and C Sasidhar[3]

[1]Research Scholor (PhD),Jawaharlal Technological University, Anantapur, Andhra Pradesh, India
[2]Professor, SVCET (Autonomous), Chittoor, Andhra Pradesh, India
[3]Professor, Jawaharlal Technological University, Anantapur, Andhra Pradesh, India

E-mail: kirsir.kumar@gmail.com

Abstract. The Construction Industry is one of the industries contributing highest GDP in Indian economy. The material that is most chosen in construction Industry is concrete. Concrete is a material made with Cement, Fine aggregate in the form of sand, Coarse aggregate in the form of gravel and water. With increasing scarcity of sand, construction works are coming to jolt and thereby increasing the need for choosing an alternative material. Vermiculite is a material which after exfoliation can be used as a filler material replacing sand partially without affecting strength much. One more material Granite (by product which is a waste is causing lot of environmental Issues), produced from granite industry, is available in India in several million tons. Coarse aggregate generally used is a crushed aggregate obtained from rocks like granite, basalt and soon. Recycled aggregate is an aggregate which is obtained after demolishing of an existing building, which is a waste causing economical in balance and which if put into use can not only decrease the construction cost but also make this waste into a better use. In the present study, for a M35 grade of concrete, Vermiculite (0, 5%, 10% and 15% of weight of sand) and granite powder (fixed at 10% of weight of sand used) is used to replace sand partially and in the place of normal coarse aggregate, recycled aggregate is used which is 20 mm passing and 12 mm retained After preparation of Mix-Design(1:1.83:2.69 with w/c ratio of 0.38) the concrete cubes are casted to test for compressive strength after curing for 28 days, 56 days and 90 days. Forsec brand super plasticizer is used to take care of workability requirements. Durability test in the form of resistance to attack of sulphuric acid was conducted along with compressive strength. The test results were promising at 10% replacement levels of Vermiculite.

1. Introduction

The construction industry is consisting of unorganized companies working as subcontractors making significant contribution to national economy both by increased GDP as well as in providing employment. In construction sector there are three segments like real estate combined with residential and commercial constructions. In the construction industry, in the presence of concrete elements

depending on work inherent nature of the infrastructure or capital formation industry, capital goods are used directly and indirectly with all other types of resources.
Concrete is combination of material made from sand, gravel and cement. The cement is a mix of various minerals which when mixed with water, hydrate and roughly become strong binding the sand and gravel into a solid mass. Cement, normally in powder form, acts as a binding agent when mix with water and aggregates. This combination, or concrete mix, will be dipped and harden into the durable material.
Workability is the most important one of these properties. The workability is necessary in a concrete mix and it is totally depends upon the purpose for which is used and the equipment and methods used in placing and handling it in work. The concrete strength is the next important attribute to be considered. Unit volume of concrete with a fixed amount of cement, the hardened and most impermeable concrete is one that has the greatest density, i.e., the given unit volume is that which has the largest percentage of solid materials.

2. Review of literature

According to Lakshmi Kumar Minapu et.al (2014), an attempt has been made to study the Mechanical Properties of a structural grade light weight concrete M30 using the light weight aggregate. Pumice stone is used as a partial replacement to coarse aggregate and mineral admixture materials like Fly Ash and Silica Fume are adopted. By using 20% of light weight aggregate as a partial replacement to natural coarse aggregates the compressive strength results are promising. Syed Abdul Rahman et.al (2016) in their study on structural light weight aggregate concrete revealed that the 10% replacement of vermiculite of fine aggregate when compared to control mix yielded better results.

According to Dr D.V. Prasad Rao(2017), this study combined influence of Ston dust and Metakaolin on compressive strength, split tensile strength, flexural strength and modulus of elasticity of M30 grade of concrete is investigated. Finally they found that the various strength characteristics of concrete can be improved by the addition of 15% of Metakaolin and 60% of quarry dust content.

According to study conducted by Dr.K.Chandrasekhar Reddy and G.Ashok (2017) on " A study of durability properties of concrete using Geo cement and Vermiculite", he revealed that for M35 grade concrete after experimentation that at 10% replacement of Vermiculite the water absorption is very less when compared with other mixes and conventional samples after soaking for 28 days.

According to Gumma Soumya(2018),this work intended to study the impact of concrete by partially replacement of cement with silica fume and fine aggregate with quartz sand. The optimum percentages of substitution of silica fume and quartz sand in concrete was found to be 10% and 60%.

According to study of C.Sangeetha and S.Manikandan (2018), on 'Performance of Concrete using Dolomite and Vermiculite as Partial Replacement of Cement and Fine Aggregate', for M30 grade concrete with 20% replacement of Vermiculite has given slightly better compressive strength when compared with Conventional concrete.

2.1 From the literature review, the objectives drafted are:

To find whether any improvement of compressive strength is there with partial replacement of Natural sand with fixed percent of Granite and variable percentages of Vermiculite when recycled coarse aggregate is used.

Further this study also aims to compare weights of various mixes of vermiculite with Conventional Concrete when recycled coarse aggregate is used.

This study also aims in studying the ability of Vermiculite based mixes in resisting Acid attack when prepared with recycled coarse aggregate.

IOP Conf. Series: Materials Science and Engineering 1006 (2020) 012030 doi:10.1088/1757-899X/1006/1/012030

3. Materials and methods

In this study Ordinary Portland cement (with fineness as retained on 90 micron sieve) of 53 grade was used as per IS code. In the production of concrete, River sand of size of that is 4.75mm and less is used as Fine Aggregate. The Re-Cycled coarse aggregate an important material of concrete with size of 20mm is used in this concrete. Water used in this study is free from acid, dust, suspended materials and having PH value ranging between 6 to 8.

Vermiculite material is a low density material of phyllo-silicate mineral group and good in insulating and acoustic properties. Even though Vermiculite is a costlier material, it has superior properties of thermal insulation and good acoustic behavior and can be adopted.

Granite powder is Quartz sand that is recently weathered from granite or gneiss quartz crystals will be angular. It is called Grus in geology or *sharp sand* in the building trade where it is preferred for concrete, and in gardening where it is used as a soil amendment to loosen clay soils. The properties of Cement, Fine aggregate (Natural Sand, Vermiculite and Granite) and Coarse aggregate are shown in Table 1to Table 3.

Table 1: Properties of cement

S.NO	Properties	Results	Permissible Limits as per IS 12269-1987
1	Specific gravity	3.15	3.10-3.25
2.	Fineness of modulus	2.1%	<10%
3.	Normal consistency	27%	-
4.	Initial setting time	32 minutes	30 minutes (minimum)
5.	Final setting time	271 minutes	600 minutes (maximum)

Table 2: Properties of fine aggregate

Properties	Results			Permissible Limits as per IS 383-1970
	Natural Sand	Vermiculite	Granite	
Specific gravity	2.48	2.55	2.6	2.5-3.0
Bulking	18 at 6%	20 at 6.5%	21 at 6.5%	-
Fineness of modulus	3.48	3.5	2.35	2-3.5
Grade	Zone-I			
Size	< 4.75mm			

Table 3: Properties of Recycled Coarse aggregate (RCA)

S.NO	Properties	Results	Permissible Limits as per IS 383-1970
1	Specific gravity (20mm)	2.9	2.5-3.0
2	Flakiness index	21.96%	35% maximum
3	Elongation index	19.19%	35% maximum
4	Aggregate impact value	14.98%	45% maximum
5	Fineness modulus	2.65	
6	Shape	Angular or flaky	

The granite fines at 10% of weight of fine aggregate is fixed and vermiculite at 5%, 10%, and 15% of weight of Natural sand is fabricated and Tested as depicted in Table 4 as Mix Proportions .

Table 4: Mix Proportions

Mix	Cement	Fine Aggregate	Coarse Aggregate
Conventional Concrete	100% Cement	100% Fine Aggregate	100% Normal Aggregate
M1	100% Cement	90% FA+10% Granite+ 0%	100% Recycled Aggregate
M2	100% Cement	85% FA+10% Granite+ 5%	100% Recycled Aggregate
M3	100% Cement	80% FA+10% Granite+ 10%	100% Recycled Aggregate
M4	100% Cement	75% FA+10% Granite+ 15%	100% Recycled Aggregate

After preparation of Mix-Design(1:1.83:2.69 with w/c ratio of 0.38) the concrete cubes are casted and cured for 28 days, 56 days and 90 days. Forsec brand super plasticiser is used to take care of workability requirements. The tests conducted on concrete cubes include Compressive strength, water absorption and durability test (Resistance to Acid attack).

4. Results and discussions

The Vermiculite is a special material which not only decreases structural weight but also exhibits inertness and is good thermal insulator. When compared with Weight of Conventional Mix and all other Mixes M1, M2, M3, the M4 exhibited fewer weights, as shown in Table 5. This study is in accordance to that study results of Dr. K. Chandra Sekhar Reddy and G. Ashok at 10% of Vermiculite replacing Fine Aggregate. They achieved the results using Geo cement but instead in this study along with Vermiculite, Granite powder is used to replace partially natural sand and Recycled coarse aggregate is used in the Place of Gravel. Results shown that at 10% of Vermiculate (M3) is effective in improving compressive strength of Concrete cubes prepared.

Table 5: Weight of concrete mixes

S.No	Mix	Weight of Concrete at 28 Days (KG)
1	CC	8.71
2	M1	8.20
3	M2	7.94
4	M3	7.68
5	M4	7.17

From Table 5, it is clear that Weight of Concrete when compared with conventional sample is less. The compressive strength of cubes of 15 cm x 15 cm x 15 cm made with M35 grade both with conventional concrete and with Mixes of M1, M2, M3 and M4 are conducted using compression testing machine.

Table 6: Compressive strength of Concrete Cubes

Concrete Mix	End of 7 Days	End of 14 Days	End of 28 Days	End of 56 Days	End of 90 Days
Conventional Concrete	29.86	32.71	38.73	40.91	41.87
M1	30.84	33.92	39.77	41.38	42.89
M2	31.70	34.55	40.14	41.92	43.02
M3	32.83	35.61	40.53	42.41	43.96
M4	31.45	34.39	39.04	40.92	42.85

Here M3 is giving better results as depicted in Table 6 when compared with other mixes as well as conventional mixes. The reason may be attributed to reduction in voids at 10% of vermiculite replacing the Fine aggregate at fixed 10% of Granite Fines. The test results are supported by the results of study of Sangeetha et.al and Abdul Rahman et.al. However unlike them, in this present study Recycled aggregate along with fixed percentage of Granite Fines is used.

Table 7: Durability of Concrete Cubes (Resistance to Acid Penetration)

Mix	Weight of Cubes after 28 days before immersing in Acid (KG)	Weight of Cubes after 28 days After immersing in Acid (KG)	% Loss	Compressive Strength of Cubes after 28 days before immersing in Acid	Compressive Strength of Cubes after 28 days before immersing in Acid	% Loss
Conventional Concrete	8.71	8.12	6.78	38.73	37.62	2.87
M1	8.20	7.73	5.73	39.77	38.87	2.26
M2	7.94	7.57	4.65	40.14	39.35	1.94
M3	7.68	7.37	4.03	40.53	39.76	1.89
M4	7.17	6.62	7.67	39.04	38.21	2.04

From the Table 7, it is clear that M3 mix is absorbing less water content and also showing less reduction in compressive strength when compared with Other Mixes and conventional samples. The reason may be attributed to improved resistance to Acid attack due to usage of Recycled aggregate at an appropriate usage of Vermiculite content in filling the air voids and improving the Compressive strength.

5. Conclusion

From the Study it can be inferred that there is an improvement in Compressive strength at 10% replacement of Natural sand with Vermiculite and at Fixed 10% replacement of Natural sand with Granite fines. By increasing the vermiculite content in Natural sand at a fixed 10 percentage of Granite, the weight of concrete cubes decreased and recorded lowest at 10% replacement levels. Vermiculite based mixes exhibited better durability when compared with conventional mixes, even though Recycled aggregate is used in the place of normal aggregate. The reason may be attributed to improved texture at later dates thus improving compressive strength of specimens.

References

[1] Lakshmi K M, Ratnam M K M V, Rangaraju U 2014 Experimental study on light weight aggregate concrete with pumice stone, silica fume and fly ash as a partial replacement of coarse aggregate *International Journal of Innovative Research in Science Engineering and Technology* Vol **3** pp 18130-18138

[2] Syed A R S, Gijo K B 2016 An experimental investigation on light weight cement concrete using vermiculite minerals *International Journal of Innovative Research in Science, Engineering and Technology* Vol **5** pp 2389-2392

[3] Prasada Rao D V, Yamuna B 2017 A study on strength of concrete by partial replacement of cement with metakaolin and fine aggregate with stone dust *International Journal of Innovative Research in Science, Engineering and Technology* Vol **6** pp 899-906

[4] Chandrasekhar R K, Ashok G 2017 A study of durability properties of concrete using Geo cement and Vermiculite *International Journal of Application or Innovation in Engineering & Management* Vol **6** pp 60-66
[5] Soumya G, Karthiga S 2018 Study on mechanical properties of concrete using silica fume and quartz sand as replacements *International Journal of Pure and Applied Mathematics* Vol **119** pp 151-157
[6] Sangeetha C, Manikandan S 2017 Performance of concrete using dolomite and vermiculite as partial replacement of cement and fine aggregate *International Journal of Engineering Research & Technology* Vol **5** pp 1-3

Studies of Concrete Mechanical Properties with Basalt Fibers

Sangmesh V Biradar[1], M Sai Dileep[2] and Dr T Vijaya Gowri[3]

[1] Assistant Professor, Department of Civil Engineering, BVRIT, Narsapur, Hyderabad, India
[2] Student, Department of Civil Engineering, BVRIT, Narsapur, Hyderabad, India
[3] Associate Professor, Department of Civil Engineering, BVRIT, Narsapur , Hyderabad, India

E-mail: sangmesh.b@bvrit.ac.in

Abstract. Current situation of construction industry demands competency in inventing new materials to enhance the mechanical properties of concrete in its strength point of view as well as durability. Basalt fibers are formed due to melting crushed volcanic basalt rock in a single stage phase. These are non-toxic and environmentally safe. Basalt fibers (BF) are highly resistant to heat, and possess insulating properties apart from having an elastic structure. In this context, the present study has been taken up to study the performance of concrete blended with basalt fibers and their influence on mechanical properties of concrete. The Basalt fibers of different proportions 0%, 0.1%, 0.3% and 0.5% in volume of concrete added to M40-grade concrete (BFRC)and tested for compressive strength, tensile strength and bending strengths. The results showed that concrete with 0.3% fibers exhibited maximum strength values.

1. Introduction

Concrete, a construction material is most widely used to build in structural elements with or without reinforcement and exhibits low tensile strength and tensile strain. Normally, addition of fibers is one of the methods to improve strength properties of the concrete such as flexural strength, tensile strength, impact strength, abrasion resistance, load bearing capacity, toughness, and deformation capability. Microstructure of concrete with BF showed good bond at early days than later days [1-6]. More recently, many types of fibers such as cellulose, asbestos, steel, cellulose, glass and polypropylenes are being used as reinforcing cement products. Addition of BF to the concrete could provide remarkable improvement in flexural strength, fracture energy and abrasion resistance of plain concrete [7-8]. Basalt is being used as composite material in plain concrete, polypropylene composites, geopolymer concrete, epoxy composites, inorganic binder like phosphate and polysiloxane based matrices by exploiting its bonding capability [9-14].Basalt fiber is the 21st century green industrial material and has advantage of safety of ecological system, sustainability of nature and several other properties. It can be applied innovatively due to its better chemical and mechanical performances than traditional carbon and glass fiber in many economic and industrial areas, from automotive to aeronautic and from building constructions to energy efficiency. Increase of percentage of weight of BF and Nano clay particles (up to 5%) increase the yield strength and the Young's modulus of clay reinforced polypropylene due to its reinforcing effect.[15-16].

IOP Conf. Series: Materials Science and Engineering 1006 (2020) 012031 doi:10.1088/1757-899X/1006/1/012031

Properies of PCC or RCC for Pavement concrete slabs are enhaced by addition of basalt fibres due to its physical features and chemical features like resistance to exteme temperature variations and resistance to alkali reactions [17]. Fracture energy is improved by adding basalt fibres and glass fibres at the dosage of 1%. Specifically, composite materials like Cement based or polymeric with basalt fibers of different geometrical features and their configurations can be used in the construction of structures as well as for retrofitting and strengthening of damaged structures [19].

Fig.1.1. Fiber Basalt.

2. Methodology

2.1. Objectives

Objectives of the experimental investigation are as follows:

To calculate the Workability, compressive strength, split tensile strength and Flexural strength of Concrete beams with Basalt fiber used to be 0%, 0.05%, 0.1%, 0.3% and 0.5% of the whole volume of the mix compared with the plain concrete of M40 Grade.

2.2. Materials

53 grade Ordinary Portland Cement was used as binding material. The characteristics of cement are shown in Table 1. River sand of size 3mm dia. was used as fine aggregate in BFRC. The specific gravity and fineness modulus of sand is 2.47 and 2.62 respectively. 20 mm size crushed aggregate was used as coarse aggregate and its specific gravity is 2.86. Table 2 describes the details of the mix of proportions.

Fibers: Fiber of 12mm length and diameter of 20 μm were used. The chopped BF was supplied by Nickunj Eximp Enterprise Private Limited. The mixing of the BF used to be 0%, 0.05%, 0.1%, 0.3% and 0.5% of the whole volume of the mix added to M40-grade concrete (BFRC). Table 3 shows physical and mechanical characteristics of basalt fibers.

Table 1. Cement characteristics.

Chemical characteristics	%	Physical Characteristics		
CaO	63.40	Specific gravity		3.10
SiO_2	19.96	Initial setting time (min)		30
Al_2O_3	5.17	Final setting time (min)		400
Fe_2O_3	3.41	**Mechanical properties**		
MgO	1.61		3 Days	30
SO_3	2.13	Compressive strength(MPa)	7 Days	41
K_2O	0.80		28 Days	55
Na_2O	0.13			
Loss on ignition	1.12			

2.3 Mixing details and curing process

In mixing of BFRC, dry fine and coarse aggregates, cement, basalt fibers and water had been added to the mixer successively and mixed well for one minute for each addition of ingredient shown Fig.2.1. The duration for concrete mixing was taken as three minutes to ensure uniform dispersion of fiber in the concrete shown Fig.2.2. The concrete was cast in 100 mm x 100 mm x 500 mm prismatic molds for testing bending strength, in 150 mm cube molds for compressive strength testing and separately in 100 mm diameter and 300 mm height of cylinder molds for tensile strength testing. Sample size was taken as 3. The samples were placed for curing after 24 hours for 3 days, 7 days and 28 days.

Table 2. Mix proportions of BFRC for One Cubic Meter

Grade of Concrete	Cement (kg)	Water (Litres)	Fine Aggregate (kg)	Coarse Aggregate (kg)
M40	438.13	197.16	598.60	1180.17

Table 3. Properties of Basalt fibers.

Fibre	Length (mm)	Density (g/cc)	Modulus of Elasticity (GPa)	Tensile Strength (MPa)	Break Elongation (%)	Absorption of water
BF	12	2.63	75-90	3200-3850	3.1	< 0.5

Fig. 2.1. Dry mix of concrete with basalt fiber

Fig.2.2. Concrete mixing

Fig. 2.3. Compressive strength test

Fig. 2.4. Split tensile strength test

2.4 Testing methods
The cubes of BFRC were tested for compressive strength and the split tensile strength according to the Indian standard IS 516–1959 [20] shown Fig.2.3. and Fig.2.4. respectively. The bending strength at 3, 7 and 28 days was tested using the third-point loading. The 400 mm bending span is used. The load was regulated by displacement up to the samples failed at a rate of 0.2 mm/min.

3. Results and discussion

3.1. Workability of BFRC
The BFRC's slump as shown in Table 4. The table indicates that the slump decreases due to the addition of fiber to some extent. The slump for the normal concrete is 165 mm. If BF adds 0.05%, 0.1%, 0.3% and 0.5% to volume mixtures, the workability of BFRC decreases to 147 mm, 125 mm, 84 mm and 59 mm respectively. It shown that adding fibers to the concrete leads to a reduction in the concrete's workability. It may be to the large surface areas of basalt which accumulate more cement paste to cover over and this lead to raise the viscosity of mix. Hence it contributes to decreases in slump.

3.2. Compressive strength of BFRC
The value of the compressive strength and strength ratio of BFRC samples at 3, 7 and 28 days is shown in Table 4. In this experiment, the compressive strength of basalt FRC is not clearly improved.
Sivakumar and Santhanam [20] concluded that the polyester, nylon, and GFRC compressive increased by 1.8%, 6.3%, and 3.0% respectively, at usage of fiber of 0.6 kg / m3 for 28 days. Similar to normal concrete, the compressive strength of concrete reinforced with BF with volume mixtures of 0.05%, 0.1%, 0.3% and 0.5% at 3 days increase by 9.26%, 19.86%, 27.37% and 18.28% respectively. At a later age, FRC's compressive strength improves relative to the previous

age. The compressive strength of BFRC is reduced at 7 and 28 days. At 7 and 28 days, the BF reinforced concrete compressive strength ranges from 4.21% to 11.41% and 3.36% to 9.82% respectively. The reduction in the fiber concrete's compressive strength is at 28 days. Due to the aging of the interface between the concrete mix and fibers, the binding potential is decreased. However, BFRC 3 (0.3%) showing higher compressive strength than the remaining mixes including PC.

Table 4. Compressive Strength of different BFRC mixes.

Mixture no.	V_m (%)	Slump (mm)	Compressive strength (MPa) 3 Day Measured (MPa)	3 Day Strength ratio (%)	7 Day Measured (MPa)	7 Day Strength ratio (%)	28 Day Measured (MPa)	28 Day Strength ratio (%)
PC	-	165	17.72		31.37		44.28	
BFRC1	0.05	147	19.36	9.26	32.69	4.21	45.77	3.36
BFRC2	0.1	125	21.24	19.86	34.14	8.83	47.47	7.21
BFRC3	0.3	84	22.57	27.37	34.95	11.41	48.63	9.82
BFRC4	0.5	59	20.96	18.28	32.94	5	46.12	4.16

Note: Strength ratio (%) = [(Fibre concrete strength - normal concrete strength) / normal concrete strength] x 100%.

3.3. Tensile strength of BFRC

Table 5 list the strength ratio of the fiber concrete tensile strength at 28 days.

The tensile strength of the BFRC tends to increase relative to the normal concrete. The tensile strength of BFRC increases by 8.38–36.70%.

Previous studies indicated that tensile strength is increased by adding fiber to concrete.Song et al. [21] concluded that the polypropylene and nylon FRC tensile strengths were 9.7% and 17.1% higher than that of normal concrete at quantity of fiber of 0.6 kg / m3, respectively. Yurtseven et al. [22] concluded that adding polypropylene fiber and carbon fiber at a volume of mixtures of 0.5% can increase concrete tensile strength by 19.5% and 31.6% at 28 days. It is due to the joining operation of the three directed dispersed fibers around cracks, which in the early stage effectively prevents the spread of micro cracks. The strain is move to at joining fibers after the flexural crack in the concrete and thus the macro crack growth is regulated and the tensile strength is increased.

Table 5. Tensile strength of BFRC.

Mixture no.	V_m (%)	Tensile strength (MPa) 28 Day Measured (MPa)	Strength ratio (%)
PC	-	3.46	
BFRC1	0.05	3.75	8.38
BFRC2	0.1	3.92	13.29
BFRC3	0.3	4.73	36.70
BFRC4	0.5	4.26	23.12

3.4. Bending strength

The bending strength and strength ratio FRC was reported in Table 6 at 3, 7 and 28 days. Previous studies have demonstrated that adding fiber to concrete could enhance concrete's flexural behavior. Song and Hwang [21] concluded that as steel fibers were added to the mix, the bending strength of concrete increased. Sivakumar and Santhanam [20] concluded that, at a fiber of 0.6 kg / m3, reinforced concrete bending of polypropylene, polyester and glass fiber was 7.6%, 21.3% and 14.4% respectively. FRC samples show an improvement in bending strength relative to normal concrete.

Table 6. Bending strength of Different BFRC mixes.

Mixture no.	V_m (%)	Bending strength (MPa)					
		3 Day		7 Day		28 Day	
		Measured (MPa)	Strength ratio (%)	Measured (MPa)	Strength ratio (%)	Measured (MPa)	Strength ratio (%)
PC	-	4.78		5.24		6.16	
BFRC1	0.05	5.21	8.99	5.73	9.35	6.58	6.82
BFRC2	0.1	5.56	16.32	6.03	15.07	7.05	14.45
BFRC3	0.3	5.97	24.89	6.67	27.29	7.32	18.83
BFRC4	0.5	5.71	19.47	6.17	17.75	7.13	15.75

For BF reinforced concrete with the volume mixtures of 0.05%, 0.1%, 3% and 0.5%, the increased bending strength ranges from 8.99% to 24.89%, 9.35% to 27.29% and 6.82 to 18.83 at 3, 7 and 28 days respectively. The BFRC in the tension zone will bear the tensile stress continuously after flexural cracking until the basalt fiber bridging fracture crosses the cracks. Hence, BFRC shows higher bending strength. From the above results, it was concluded that the increase in bending strength due to the increase in BF's volume mixtures. However, when the fiber volume mixtures increase to 0.5 percent in BFRC4, the bending strength decreases slightly compared to BFRC3 with 0.3 percent basalt fiber volume mixtures. It is due to difficulties in fiber dispersion with mixtures of high volume of 0.5%. It is also indicate that normal concrete showed brittle behavior than BFRC.

4. Conclusions

This study demonstrates measurements of mechanical characteristics on BFRC. The results concluded that addition of Basalt Fiber to the concrete implies reduction of concrete workability with increase of fiber quantity. Compared to PC, the BF-reinforced concrete shows better strengths in bending and tension whereas lesser improvement in compression even reduces at a later age. The Strengths in compression, tension and bending of BFRC increase by 3.36-9.82%, 8.38-36.7% and 6.82-18.83% for 28 days respectively compared with the strengths of PC. Mechanical characteristics of BF concrete are increasing with the rise of BF.

References

[1] Tassew ST, Lubell AS 2014 Mechanical properties of glass fiber reinforced ceramic concrete *Construct Build Mater* **51** 215–24.
[2] Uygunog ˇlu T 2008 Investigation of microstructure and flexural behavior of steel fiber reinforced concrete *Mater Struct* **41(8)** 1441–9.
[3] Faiz UAS 2013 Review of mechanical properties of short fibre reinforced geo-polymer composites *Construct Build Mater* **43**: 37–49.
[4] YazıcıS, InanG, TabakV 2007 Effect of aspect ratio and volume fraction of steel fiber on the mechanical properties of SFRC *Construct Build Mater* **21(6)**1250–3.
[5] Mohammadi Y, Singh SP, Kaushik SK 2008 Properties of steel fibrous concrete containing mixed fibres in fresh and

hardened state *Construct Build Mater* **22(5)** 956–65.
[6] Chaohua Jiang, Ke Fan, Fei Wu, Da Chen 2014 Experimental study on the mechanical properties and microstructure of chopped basalt fibre reinforced concrete *Materials and Design* **58** 187–193.
[7] Hannant DJ 2003 Fibre reinforced concrete. In: Newman J, Choo BS, editors. *Advanced concrete technology-processes. Oxford: An Imprint of Elsevier* p. 146–163.
[8] Nihat Kabay 2014 Abrasion resistance and fracture energy of concretes with basalt fiber *Construction and Building Materials* **50** 95–101.
[9] Vladmir B Brik 2003 Advanced concept concrete using Basalt/BF composite rebar reinforcement *Final Report for Highway IDEA Project* **86**.
[10] Mingchao Wang, Zuoguang Zhang. Li yubin, Li Min and Sun Zhijie 2008 Chemical durability and mechanical properties of alkali –proof basalt fiber and its Reinforced epoxy composites *Reinf Plast Compos* **27(4)** 393–407.
[11] Glogar Petr, Cerny Martin, Tolde Zdenek 2007 Fracture behaviour of the basalt fiber reinforced composites with polysiloxane derived matrix *Acta Geodyn Geomater* **42(146)** 27–37.
[12] Ding Zhu, Zhanxi Lu, Li Yuan 2011 Feasibility of basalt fiber reinforced inorganic adhesive for concrete strengthening *Adv Mater Res* 287–290 :1197–200.
[13] Li Weimin, Jinyu Xu 2009 Mechanical properties of basalt fiber reinforced geopolymeric concrete under impact loading *Mater Sci Eng A* **505**:178–86.
[14] Smriti Raja, V Ramesh Kumar, B H Bharath Kumar, Smitha Gopinath, Nagesh R Iyer 2015 Flexural studies on Basalt Fiber Reinforced Composite sandwich panel with profile sheet as core *Construction and Building Materials* **82** 391–400.
[15] Parnas R, Shaw M, Liu Q 2007 Basalt fiber reinforced polymer composites *The New England transportation consortium* Project No. 03–7.
[16] Reza Eslami - Farsani, S Mohammad Reza Khalili, Ziba Hedayatnasab, Neda Soleimani 2014 Influence of thermal conditions on the tensile properties of basalt fiber reinforced polypropylene–clay nano composites *Materials and Design* **53** 540–549.
[17] Maria Wlodarczyk, Igor Jedrzejewski 2016 Concrete slabs strengthened with basalt fibres – experimental tests results *Science Direct, Procedia Engineering* **153** 866 – 873.
[18] Almet B Kizilkanat, Nihat Kabay, Veysel Akyuncu, Swaptick Chowdary, Abdullah H Akra 2015 Mechanical Properties and Fracture behavior of Basalt and Glass fibre reinforced Concrete: An experimental Studies *Construction and Building Materials* **100** 218-224.
[19] Elisabetta Monaldo, Francesca Nerilli, Giuseppe Vairo 2019 Basalt-based fiber reinforced materials and structural applications in civil engineering *Composite Structures* **214** 246-263.
[20] Sivakumar A, Santhanam Manu 2007 Mechanical properties of high strength concrete reinforced with metallic and non-metallic fibers *Cem Concr Compos* **29(8)** 603–8.
[21] Song PS, Hwang S, Sheub BC 2005 Strength properties of nylon-and polypropylene fiber-reinforced concretes *Cem Concr Res* **35** 1546–50.
[22] Yurtseven AE, Yaman IO, Tokyay M 2006 Mechanical properties hybrid fibre reinforced concrete *Monit Model Concr Prop* 207–214.

Studies on Flexural Behaviour of Concrete Beams Reinforced with Basalt Rebars

Sangmesh V Biradar[1] and Sidram S Injaganeri[2]

[1] Assistant Professor, Department of Civil Engineering, BVRIT, Narsapur, Hyderabad, India
[2] Professor, Department of Civil Engineering, BEC Bagalkot, Karnataka, India

E-mail: sangmesh.b@bvrit.ac.in

Abstract. The use of reinforcement of Fibre- Reinforced Polymer (FRP) in concrete structures has increased rapidly in the last 25 years due to its excellent corrosion resistance, fire resistance, high alkali resistance, high tensile strength to weight ratio, durability, flexural strength, insulation, good fatigue properties and magnetic transparency properties. Basalt concrete reinforcement bars called Basalt Fiber Reinforced Polymer (BFRP) are a new material, so it is important to define the mechanical properties and drawback of their use in the concrete structures in comparison to conventional reinforced concrete members.

In this study Flexural behaviour of concrete beams reinforced with basalt rebars will be tested and compared with the Flexural strength obtained from experimental result with RC Beams and ACI-440 codal provision. It has been shown that much lower cross-sectional stiffness of the BFRP basalt bars produces higher deflections and crack widths compared to the steel bars beams of the same cross-section. Comparisons between the flexural strength obtained from the ACI-440 codal provision and that experimentally values in the current investigation elsewhere indicates disagreement.

1. Introduction

The use of fibre reinforced polymer (FRP) reinforcement in concrete structures has rapidly increased because of excellent corrosion resistance, high strength to weight ratio and good non-magnetizing properties. As a result it can have many applications in structures used in marine environments, in chemical plants, in temporary structures. Bridge decks, floor slabs and wall type structures are examples of such structural components. However, the reinforced concrete members of the FRP behave differently from those reinforced with steel due to the linear elastic stress-strength relationship of the FRP bars to rupture. In addition, the low elasticity modulus of the FRP induces a significant decrease in the flexural stiffness of the reinforced concrete members of the FRP after cracking. Due to their mechanical characteristics, deflections and cracking in FRP RC beams are greater than those seen in conventional RC members. As a result, the construction of reinforced concrete FRP members is often controlled by the serviceability limit rule. Therefore, the use of reinforcement of the FRP requires a better understanding of the behaviour of reinforced concrete FRP members. [1].

Basalt bars are the latest type of FRP reinforcement used in civil engineering. The mechanical characteristics of basalt bars are identical to those of glass, so it can be assumed that BFRP and GFRP RC members can be designed in accordance with the same design rules. However, BFRP bar is a relatively new material, the most attractive feature of BFRP is its integrated properties that are superior to traditional FRP composite, including its strength being 20% higher and modulus, comparable cost, and higher chemical stability relative to GFRP, a wider range of working temperatures and a significantly lower cost than CFRP. While BFRP has competitive characteristics compared to conventional materials. Therefore, the actions of BFRP RC members should still be thoroughly analysed [2].

The objective of this study is to test the mechanical characteristics of BFRP bars as well as the flexural behaviour of concrete beams reinforced with basalt rebars and comparing with the Flexural strength obtained from experimental result with the ACI-440codal provisions.

2. Experimental program

2.1. Objectives

Objectives of the experimental investigation are as follows:

- To study the Mechanical characteristics of BFRP bars and compare with Steel bars.
- To calculate the Flexural strength of Concrete beams reinforced with Basalt rebars for different grades of concrete viz. M20, M40, M60 and M80 with a constant reinforcement of 0.243% and compare with beams reinforced with Steel by experimentally
- To compare the Flexural strength obtained experimentally with ACI-440 codal provisions.

2.2. Materials used

The materials used in this experiment were cement, fly ash, fine aggregate, coarse aggregate, water, BFRP bar and Steel bar. Cement of OPC of 43 and 53 grade will be used. The chemical composition of fly ash is as Sio_2-51.12%, Al_2O_3-29.53%, Fe_2O_3-5.57%, CaO-2.99%, K_2O-2.38%, SO_3-1.34%, MgO-1.03%, Na_2O-0.5%, BaO-0.06%, Others-2.42% and Loss of ignition-3.06%. Fineness modulus of sand 2.74 was used as fine aggregate and coarse aggregate was 20 mm size crushed stone and fineness modulus of 6.55.

2.3. BFRP and steel bars characteristics

The reinforcement materials used in the beam was BFRP and steel bars. The diameter of the BFRP bars was 8 and 10 mm shown in Fig. 2.1. It was supplied by Nickunj Eximp Enterprise Privet Limited. According to the manufacturer the properties BFRP rebar are listed in Table 1. The *characteristics* of BFRP and steel bars of all beams are tested in Universal Testing Machine but it was failed in slip of BFRP bar shown in Fig. 2.2.

2.4. Mix proportion of concrete

All the beams were made of M20, M40, M60 and M80 concrete grade. The mix proportion presented in Table 2.

2.5. Testing program

In the experimental work the Flexural behaviour of M20, M40, M60 and M80 mix design will be analyzed. The mix is designed by using IS 10262-2009. The cubes of size 150 mm × 150 mm × 150 mm for this mix design will be cast to analyze the compressive strength for 7 and 28 days respectively.

Eight slender beams size of 230mm X 300mm X 1200mm with shear span to depth ratio of 1.85 (a / d = 500/ 270) and effective length to effective depth ratio 3.70 (l / d = 1000/270) are will be cast by providing constant Flexural reinforcement ratio (0.243%) with different grade like M20,M40,M60 and M80. The Flexural behaviour is tested in 500kN capacity displacement controlled actuator under three point loading as shown in Fig.3.5. A linear deflection transducers (LVDT) is used at centre-span section to measure the displacement during the test. The load is applied through a 500 kN servo controlled actuator. A data acquisition system is employed for measuring the displacements. Table 3 Shown Designation of four RC beams and Table 4 Shown Designation of four BFRP beams along with characteristics of main bar and concrete. Fig. 2.3 shows reinforcement details of SB2, BB2, SB6, BB6, SB8 and BB8. Fig. 2.4 shows reinforcement details of SB4 and BB4.

Table 1. Mechanical characteristics of BFRP and Steel bar.

Sl. No	Bar of Diameter	Weight/m (gm)	Young's modulus (GPa)	Yield stress (Mpa)	Ultimate stress (Mpa)	Breaking stress (Mpa)
1	8 mm (BFRP)	104	55	-	1200	-
2	10 mm (BFRP)	162	52	-	1200	-
3	6 mm (Steel)	228	200	250	325.26	260
4	8 mm (Steel)	420	200	504.12	547	507.39

Table 2. Mix Proportion of Concrete.

Concrete Grade	M 20	M 40	M 60	M 80
Mix Proportion	1: 1.83:2.8	1:1.55:2.4	1:1.35:2.19	1:1.06:1.89
w/c ratio	0.5	0.43	0.29	0.27
Cement (kg/m^3)	387	438	450	450
Fly ash (kg/m^3)	-	-	54.21(12%)	140.436(31%)
Sand (kg/m^3)	708.21	678	683.24	626.2176
Coarse aggregate (kg/m^3)	1083.6	1051	1108.13	1117.136
Super plasticizers	-	1% of cementitious material	0.8% of cementitious material	1% of cementitious material
Compressive strength at 28 days in MPa	26	45	66.5	84.5

Fig. 2.1. BFRP bars.

Fig.2. 2. Universal Testing Machine.

3. Results and discussion

3.1. Mechanical characteristics of BFRP and Steel bars

According to mechanical *characteristics* of BFRP and Steel bar are listed in Table 1 shows that the Modulus of elasticity (55 GPa) and Density (2680 Kg/m^3) of BFRP bar is 3.64 and 2.93 times less than that of Steel bar (200 GPa,7850 Kg/m^3) respectively. The tensile strength (1200 MPa) of BFRP bar is 2.4 times more than that of Steel bar (500 MPa).

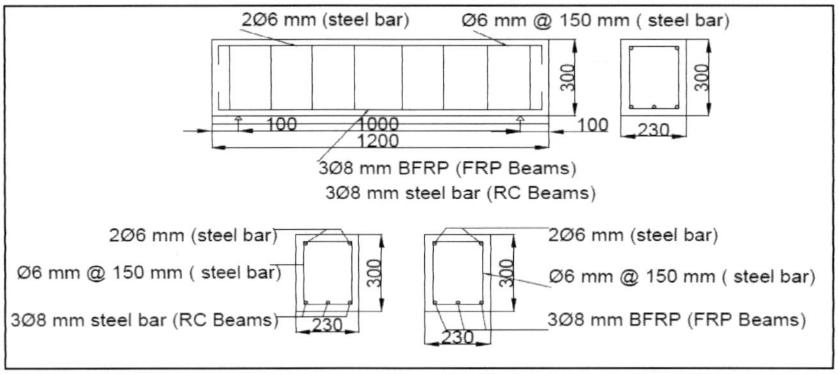

Fig. 2.3. Reinforcement details of SB2, BB2, SB6, BB6, SB8 and BB8 Beams.

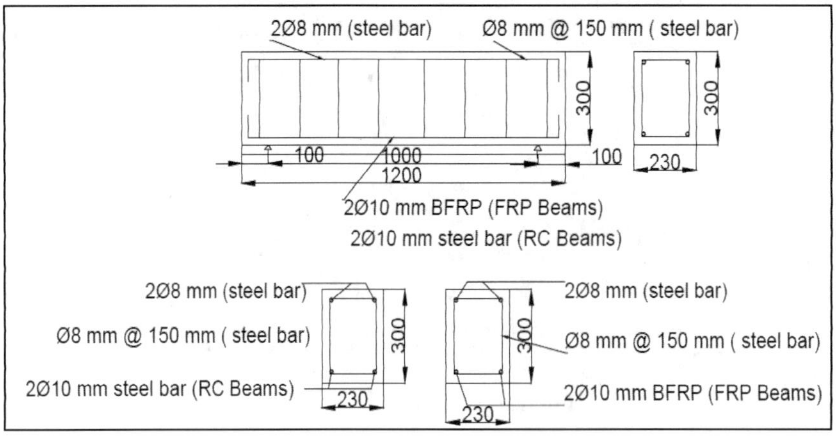

Fig. 2.4. Reinforcement details of SB4 and BB4 Beams.

Table 3. Designation of RC beams, features of main bar and concrete.

Beam notation	Main bar (mm)	Reinforcement Ratio, ρ_f(%)	Balanced section Ratio, ρ_{fb} (%) (ACI 440)	Balanced section Ratio, ρ_{fb} (%) (IS 456)	Concrete compressive Strength, f_{cK} (MPa)
SB2	3#8	0.243	1.610	0.962	26
SB4	2#10	0.243	2.810	1.929	45
SB6	3#8	0.243	3.612	2.895	66.5
SB8	3#8	0.243	4.821	3.859	84.5

Table 4. Designation of BFRP beams, features of main bar and concrete.

Beam notation	Main bar (mm)	Reinforcement Ratio, ρ_f(%)	Balanced section Ratio, ρ_{fb}(%) (ACI 440)	Balanced section Ratio, ρ_{fb}(%) (IS 456)	Concrete compresive Strength, f_{cu} (MPa)
BB2	3#8	0.243	0.212	0.137	26
BB4	2#10	0.243	0.382	0.274	45
BB6	3#8	0.243	0.482	0.411	66.5
BB8	3#8	0.243	0.641	0.548	84.5

3.2. Load- displacement response at midspan

Fig. 3.1 shows load vs displacement graph for M20 grade RC and BFRP beam. The RC beam failed at an ultimate load of 130 kN and BFRP beam of 104.985 kN. The RC beam has 1.238 times the ultimate load of BFRP beam. The displacement BFRP beam at an ultimate load of 104.985 kN is 8.745 mm i,e 3.102 times the displacement of RC beam (2.819 mm) at the same load because of lower elasticity modulus BFRP bar.

Fig. 3.2 shows load vs displacement graph for M40 grade RC and BFRP beam. The RC beam failed at an ultimate load of 120kN and BFRP beam of 100 kN. The RC beam has 1.2 times the ultimate load of BFRP beam. The displacement BFRP beam at an ultimate load of 100Kn is 12.952 mmi,e1.785 times the displacement of RC beam (7.162 mm) at the same load because of lower elasticity modulus BFRP bar.

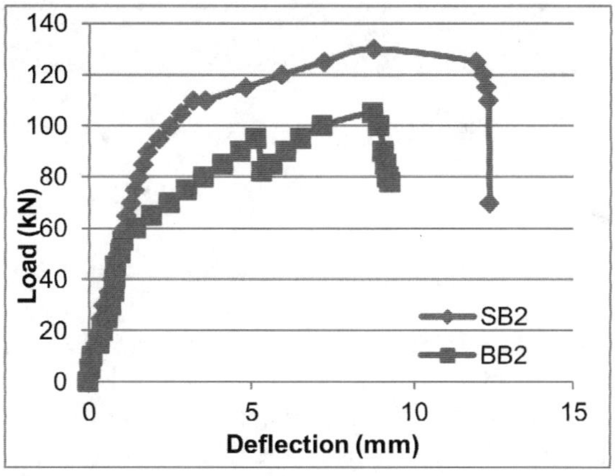

Fig. 3.1. Load v/s deformation for M20 grade RC and BFRP beam

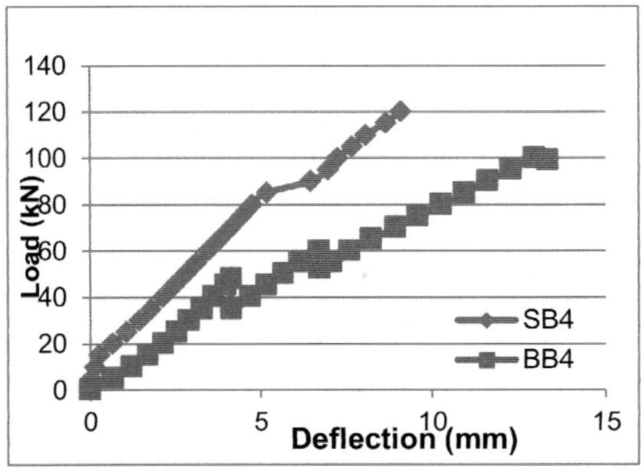

Fig. 3.2. Load v/s deformation for M40 grade RC and BFRP beam.

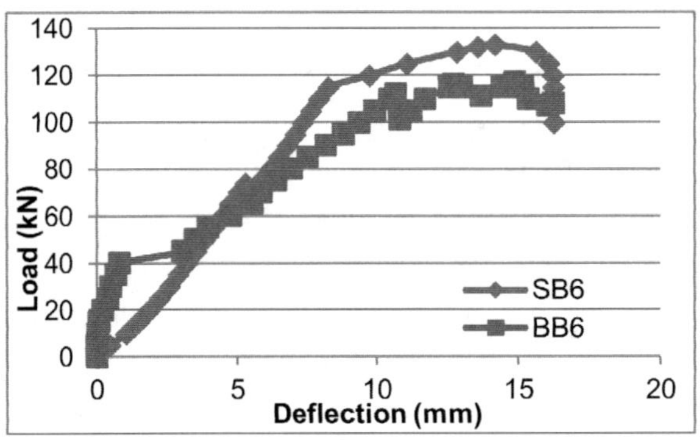

Fig. 3.3. Load v/s deformation for M60 grade RC and BFRP beam

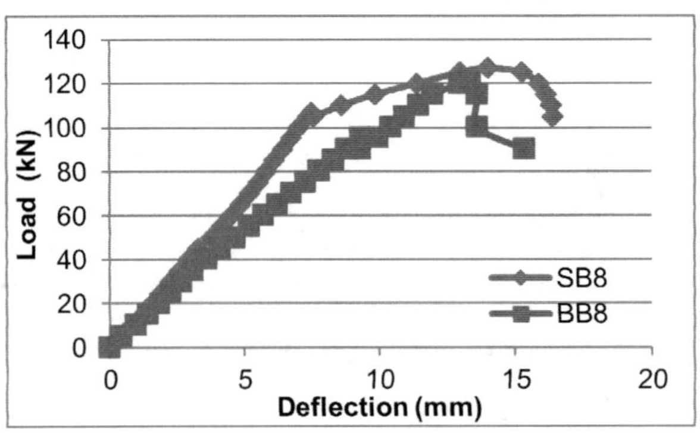

Fig. 3.4. Load v/s deformation for M80 grade RC and BFRP beam.

Fig. 3.3 shows load vs displacement graph for M60 grade RC and BFRP beam. The RC beam failed at an ultimate load of 133 kN and BFRP beam of 117 kN. The RC beam has 1.137times the ultimate load of BFRP beam. The displacement BFRP beam at an ultimate load of 117 kN is 14.919 mmi,e1.742 times the displacement of RC beam (8.567 mm) at the same load because of lower elasticity modulus BFRP bar.

Fig. 3.4 shows load vs displacement graph for M80 grade RC and BFRP beam. The RC beam failed at an ultimate load of 127 kN and BFRP beam of 121 kN. The RC beam has 1.050 times the ultimate load of BFRP beam. The displacement BFRP beam at an ultimate load of 121 kN is 13.357 mm i,e1.159 times the displacement of RC beam (11.525 mm) at the same load because of lower elasticity modulus BFRP bar.

3.3. Comparisons between theoretical and experimental flexural strength

BFRP reinforcement was chosen to analyze two types of flexural failure, namely BFRP reinforcement break and concrete crushing. The first mode is accomplished by using a reinforcement ratio ρ_f less than the balanced section ratio ρ_{fb} according to the guidelines ACI 440.1R-06 [7], whereas the second mode by using a reinforcement ratio greater than ρ_{fb}. The reinforcement ratio, ρ_f, and balanced section ratio, ρ_{fb} respectively, calculated from Eqs. (1) and (2), as defined in the guidelines ACI 440.1R-06:

$$\rho_f = \frac{A_f}{bd} \qquad (1)$$

$$\rho_{fb} = 0.85\beta_1 \frac{f_{cu}}{f_{fu}}(E_f\varepsilon_{cu}/\{E_f\varepsilon_{cu}+f_{cu}\}) \qquad (2)$$

where A_f is the area of BFRP reinforcing bars, f_{cu} is the compressive strength of concrete (MPa), b is the width of the beam (mm), d is the effective depth of the beam (mm), f_{fu} is the ultimate tensile strength of BFRP bars (MPa), ε_{cu} is the ultimate strain in concrete, E_f is the elasticity modulus of BFRP bars (MPa), and β_1 is the strength reduction factor that can be calculated on the basis of the ACI 440.1R-06 in SI units:

$$\beta_1 = 0.85 - 0.05(\{f_{cu}-27.6\}/6.7) \qquad (3)$$

The experimental and theoretical moment capacities for the beams tested are shown in Table 5. Experimental moment failure at centre-span is calculated. The theoretical moment capacities, M_{the}, of mid-span are calculated based on ACI 440.1R-06 Eqs. (4) and (5) when $\rho_f \geq \rho_{fb}$ and Eqs. (6) and (7) when $\rho_f < \rho_{fb}$:

$$M_{the} = \rho_f f_f(1-0.59\frac{\rho_f}{f_{cu}})bd^2 \qquad (4)$$

$$f_f = \text{Squareroot}[(\{E_f\varepsilon_{cu}\}^2/4)+(\{0.85\beta_1 f_{cu}E_F\}/\rho_f)0.5E_f\varepsilon_{cu} \leq f_f \qquad (5)$$

$$M_{the} = A_f f_{fu}(d-\{\beta_1 C_b\}/2) \qquad (6)$$

$$C_b = (\varepsilon_{cu}/\{\varepsilon_{cu}+\varepsilon_{fu}\}) \qquad (7)$$

where ρ_f, ρ_{fb} and β_1 are defined in Eqs. (1) – (3), f_f is the BFRP stress at which concrete crushing failure mode occurs, and C_b is the depth of neutral axis for balanced failure as defined by Eq. (7). Comparisons between the theoretical and the flexural strength of the beams tested in this study are shown in Table 5. The present study indicates disagreement between the predicted value according to ACI 440.1R-06 and experimental results.

3.4. The failure load and crack width of the beams

The nature of cracking and crack patterns in BFRP beams under flexural loading was similar to the RC beams. Crack width of BFRP beams was 1.2 to 1.55 times higher then the RC beams, because of the low modulus of BFRP bars relative to steel bars. Table 6 shows failure loads and crack width of the beams.

IOP Conf. Series: Materials Science and Engineering 1006 (2020) 012032 doi:10.1088/1757-899X/1006/1/012032

Table 5. Comparisons of theoretical and experimental flexural strength.

No	Beams name	M_{exp} (kNm)	M_{the} (kNm) ACI 440	$\frac{M_{exp}}{M_{the}}$	$\frac{M_{exp}}{f_{ck}bd^2}$	$\frac{M_{the}}{f_{ck}bd^2}$	failure Mode
Beams failed in bending							
1	BB2	26.246	46.234	0.568	0.078	0.138	FRP rupture
2	BB4	25	46.630	0.536	0.038	0.069	FRP rupture
3	BB6	29.250	47.035	0.622	0.029	0.047	FRP rupture
4	BB8	30.250	47.324	0.639	0.023	0.035	FRP rupture

Table 6. Failure loads and crack width of the beams.

Beams	Failure Load (kN)	Crack Width at failure (mm)
SB2	70	2
BB2	78	3
SB4	120	2
BB4	99	3
SB6	100	3.5
BB6	108	4.5
SB8	105	4
BB8	90	5

3.5. Crack profile and Failure mechanisms

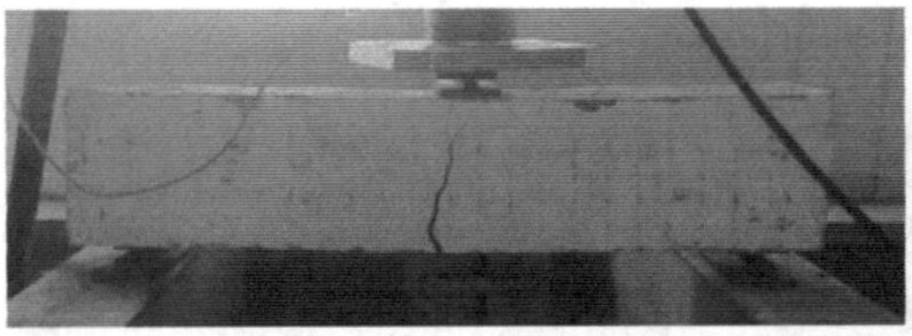

Fig 3.5 Crack Profile & Failure of Beam SB2

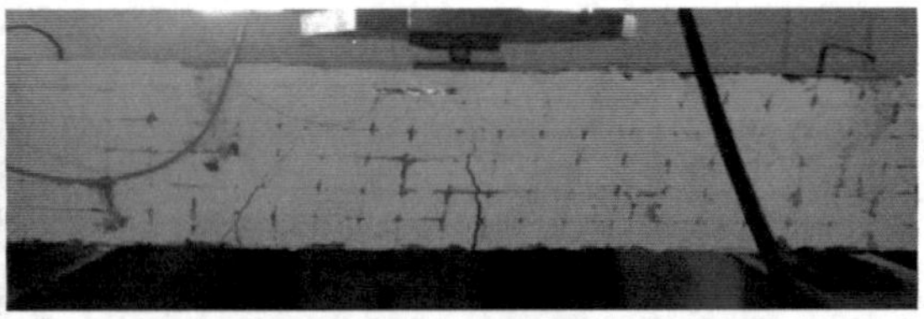

Fig. 3.6 Crack Profile & Failure of Beam BB2

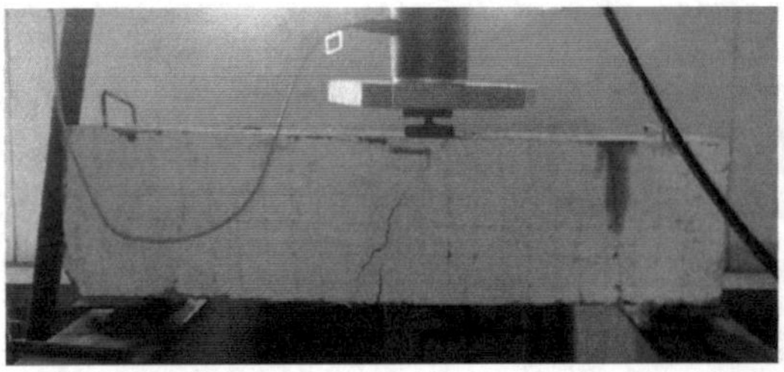

Fig 3.7 Crack Profile & Failure of Beam SB4

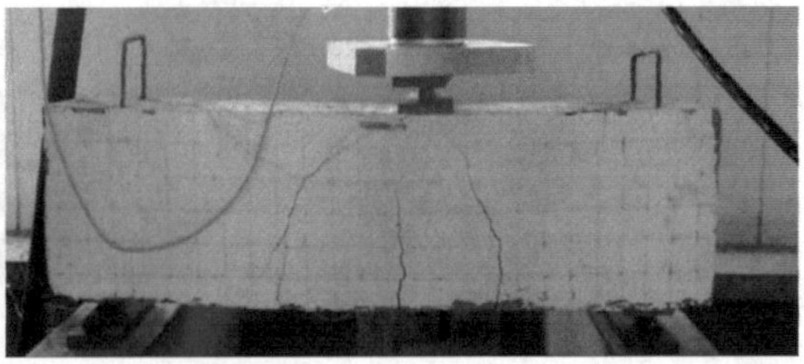

Fig. 3.8 Crack Profile & Failure of Beam BB4

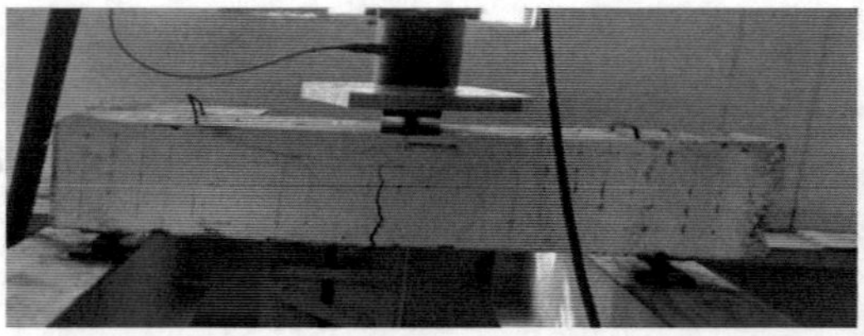

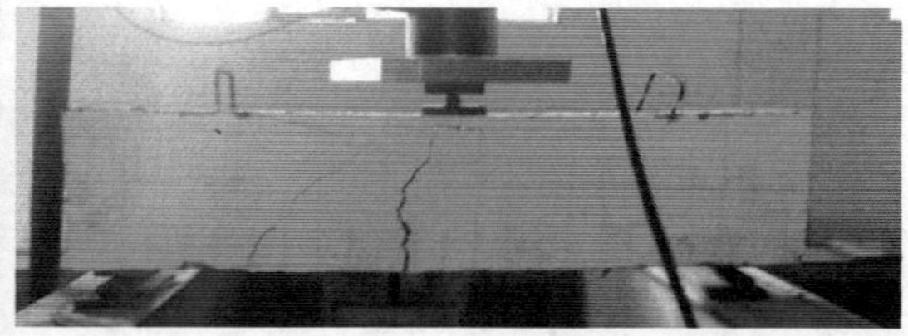

Fig 3.9 Crack Profile & Failure of Beam SB6 **Fig. 3.10**. Crack Profile & Failure of Beam BB6

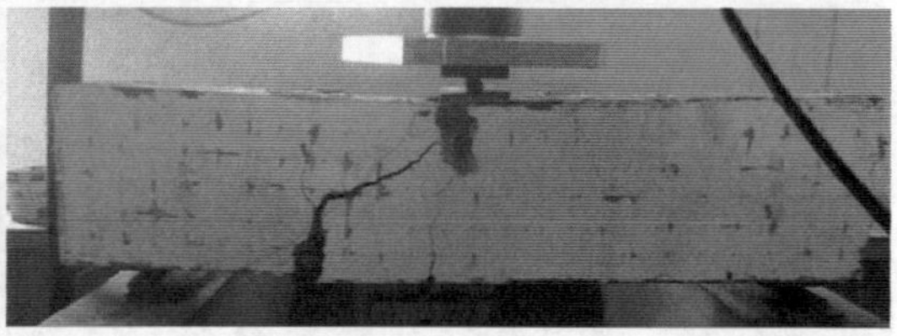

Fig 3.11 Crack Profile & Failure of Beam SB8 **Fig. 3.12**. Crack Profile & Failure of Beam BB8

4. Conclusions

The following conclusions have been drawn on the basis of the objectives set in the present study and the experimental work conducted in the laboratory.

- The Modulus elasticity (55 GPa) and Density (2680 Kg/m^3) of BFRP baris 3.64 and 2.93 times less than that of Steel bar (200 GPa,7850 Kg/m^3) respectively. The Tensile strength (1200 MPa) of BFRP bar is 2.4 times more than that of Steel bar (500 MPa).
- The Flexural strength of BFRP beams was 0.87 times the RC beams.
- Deflection of the BFRP beams was significantly higher (average of 1.95 to 3.10) than that of the RC beam because of the low modulus of the BFRP bars relative to the steel bars.
- The nature of cracking and crack patterns in BFRP beams under flexural loading was similar to the RC beams. Crack width of BFRP beams was 1.2 to 1.55 times higher then the RC beams, because of the low modulus of BFRP bars relative to steel bars.
- The present study indicates disagreement between the predicted value according to ACI 440.1R-06 and experimental results.

References

[1] Mahroug M E M, A F Ashour, D Lam 2014 Experimental response and code modelling of continuous concrete slabs reinforced with BFRP bar *Composite Structures* **107** pp 664–674.
[2] Dawid Pawlowskia, Maciej Szumigalaa 2015 Flexural behaviour of full-scale basalt FRP RC beams – experimental and numerical studies *Procedia Engineering* **108** pp-518 – 525.
[3] Douglas Tomlinson and Amir Fam 2015 Performance of Concrete Beams Reinforced with Basalt FRP for Flexure and Shear *Journal of Composites for Construction* Vol. **19** No. 2 .
[4] A Lapko, M Urbanski 2015 Experimental and theoretical analysis of deflections of concrete beams reinforced with Basalt rebar *Archives of Civil and Mechanical Engineering* **15** pp – 223 - 230.
[5] Marek Urbanski, Andrzej Lapko, Andrzej Garbacz 2013 Investigation on Concrete Beams Reinforced with Basalt Rebars as an Effective Alternative of Conventional R/C Structures *Procedia Engineering* **57** pp 1183 – 1191.
[6] A F Ashour 2006 Flexural and Shear capacities of concrete beams reinforced with GFRP bars *Construction and Building Materials* **20** pp 1005–1015.
[7] American Concrete Institute (ACI) 2006 Guide for the design and construction of structural concrete reinforced with FRP bars ACI 440.1R-06, ACI Committee 440, Detroit.
[8] IS 456:2000 *Indian Standard Code of practice for Plain and Reinforced Concrete* Fourth Revision, Bureau of Indian Standards, New Delhi.
[9] ISIS Canada: 2007 Design Manual Fibre reinforced polymer, Canada.
[10] ACI 318 M-05 Building Code requirements for Structural Concrete American Concrete institute, Michigan.
[11] Toutanji HA, Saafi M 2000 Flexural behavior of concrete beams reinforced with glass fiber-reinforced polymer (GFRP) bars *ACI Structural Journal* Vol. **97** No.5 , pp: 712–9.
[12] Pecce M, Manfredi G, Cosenza E 2000 Experimental response and code models of GFRP RC beams in bending. *Journal of Composites for Construction* Vol. **4** No.4 pp: 182–90.
[13] H A Abdalla 2002 Evaluation of deflection in concrete members reinforced with fibre reinforced polymer bars *Composite Structures* Vol. **56** pp: 63-71.

An Experimental Investigation on Concrete by Partial Replacement of Cement by Fly Ash and Fine Aggregate by Quarry Dust

B Damodhara Reddy[1], S Aruna Jyothy[2], N Kiran Kumar[3] and K Hemanth kumar[4]

[1]Professor &Head, SVCET (Autonomous), R.V.S. Nagar, Chittoor, Andhra Pradesh, India.
[2]Assistant Professor, S.V.University College of Engineering, S.V.University, Tirupati.
[3]Assistant Professor, SVCET (Autonomous), R.V.S. Nagar, Chittoor, Andhra Pradesh, India.
[4]PG Student, SVCET (Autonomous), R.V.S. Nagar, Chittoor, Andhra Pradesh, India.

Email: damu2007@gmail.com

Abstract The concrete is a composite construction material. The necessary ingredients like cement, Natural Sand, Gravel and potable Water are important components of concrete. The issue of environmental degradation and expensive nature of the river sand makes us to switch on to the alternative sources. One such material having strength and durability is quarry dust which is a by-product from quarries which is causing lot of environmental issues. In this present study the replacement of the fine aggregate is done with quarry dust and cement with fly ash. Replacements of cement are partially done with Fly Ash at 0%, 5%, 10%, 15%, 20% and 25%. The experimental work primarily concentrates with compressive strength, split tensile strength and sturdiness properties of M40 grade of concrete at completely different ages i.e., 7 days, 28 days, 56 days and 90 days. The Split tensile strength and sturdiness checks like Acid attack and Chloride penetration test for 28 day were done. The Test results were promising when cement was partially replaced by 10% of Fly ash for all the days.

1. Introduction

Concrete is popular material prepared combining Cement, Natural Sand, Gravel and potable water. The cement normally hydraulic cement and different building material like oxide fume function as binder for the mixture. Various chemical admixtures also are additional to realize varied properties. The additives usually a chemical when mixed with the concrete at the blending stage to realize changes within the properties of the concrete square measure referred to as admixtures. Use of admixture enhances the character of concrete in getting good mix design, good workmanship and good durability without altering the physical characteristics of concrete. The additives chosen for this study are Fly ash and Quarry dust.

2. Review of literature

Based on reports and articles in this area of study by earlier researchers involving use of fly ash and quarry dust added to concrete and we may assume quarry dust may be suitable. As per work done by Hmaid Mir (2015) the optimum compressive strength is achieved when natural sand is replaced by quarry dust by40%. Later Felekoglu et al. (2006) observed that by using Quarry dust one can reduce content of usage of Super Plasticizer in SCC mixtures.. The same is supported by Ho et al. (2002) who used the granite fines in the SCC preparation to solve durability problems, such as silica- alkali reactions.

According to Muhit et al.(2014) determined that Cement was partially replaced with fly ash by 3%, 5%, and 7 percentages. Likewise stone dust partially replaces sand by 15% to 50% at an increment of 5 percent. At levels of 35% of sand replacement and 3% of cement replacement Compressive Strength and tensile strength values are promising. Ukpata and Ephraim (2012) identified that the lateritic sand and quarry dust may be chosen for construction with condition that mixture content of quarry dust and laterite should be below 50%. The outcome of the study is stone dust can replace Natural sand up to 50% without compromising mechanical andeconomy.

From the literature Review the objective chosen is to estimate the strength and sturdiness properties of concrete incorporating silicon dioxide fume and Quarry Dust.

3. Materials & methods

The materials utilized in this study are Cement, Fly Ash, Quarry dust, Natural Sand, Gravel, and potable water whose characteristics are shown in Table 1 to Table 2. Potable water is having pH around 6 to 8 and free from organic compounds. The minimum water required in concrete is 38 percentage of weight of cement. The poly carboxylate based super-plasticizer; ConplastSP430 was used whose properties are shown in table 3. The Concrete mix selected for present study is M40. All the samples were casted and cured by the above designations and each designation mix composes of 12 cubes tested on 7, 28, 56, and 90 days curing periods, whereas cylinders are tested on 28 days curing periods respectively.

Table 1: Characteristics of Cement and Fly Ash

Characteristics of Cement			**Physical Characteristics of Fly Ash**	
Property	**Test Value**	**Standard Value (IS:4031:1988)**	**Physical Properties**	**Results**
Specific gravity	3.10	3.10-3.15	Physical state	Micronized powder
Standard consistency	32%	30-35	Odour Moisture	Odourless 0.058%
Initial setting time(min)	45	>30	Appearance Specific Gravity	White colour powder, 2.63
Final setting time(min)	380	<600	Pack density	0.76 gm/cc
Fineness	6.5%	<10	Colour, $P^H(5\%)$	White, 6.90

Table 2: Characteristics of Quarry Dust, Fine and coarse Aggregate

S. No	Characteristics of Quarry Dust	Quarry Dust	Natural Sand	Gravel
1	Fineness modulus	3.1	3.2	7.31
2	Specific gravity	2.85	2.61	2.738
3	Loose density	1460 kg/m^3		
4	Compacted density	1610 kg/m^3	1516.6 kg/m^3	
5	Grading zone	IV		

Table 3: Characteristics of Super-plasticizer (Conplast SP430)

Characteristics	Results
Colour & Relative density	Light brown liquid & 1.09-0.01 at 25° C
p^{H} value	> 6
Chloride content	< 0.2%

Table 4: Mixes for concrete specimens

Mix Details	Combination Of Mix
M1	100% Cement (Conventional Mix)
M2	100% C+0% SF+30% ES+0.6% SP
M3	95% C+5% SF+30% ES+0.6 SP
M4	90% C+10% SF+30% ES+0.6 SP
M5	85% C + 15% SF+30% ES+0.6 SP
M6	80% C+20% SF+30% ES+0.6 SP
M7	75% C+25% SF+30% ES+0.6 SP

In table 4, 'C' represents cement, 'SF' represents Silica Fume, 'ES' stands for replaced Ennore sand by Granite Fines and 'SP' represents super plasticizer. The Tests conducted on cement include preliminary tests like measurement Fineness, specific gravity, and standard consistency of Cement. The tests on initial and final setting time of cement are also performed. The Tests conducted on Fine aggregate are Specific Gravity, Bulk Density and Fineness Modulus and tests on Coarse aggregate include Specific Gravity and Fineness Modulus. The Tests conducted on Hardened concrete include Compressive Strength Test, Split Tensile Strength, Acid attack test, Chloride Penetration Test and water Absorption Test. Information of the strength of concrete in structure cannot be obtained from tests on an individual basis created specimens. Design mix is made for M40 grade and the concrete specimens with replacement of cement by various percentages of Fly Ash and with constant 30% of Quarry Dust.

4. Results and analysis

The Results of the tested concrete by various tests like Compressive Strength, Split Tensile strength, Durability tests like Acid resistance, chloride resistance are as follows:

As per IS 516-1959 concrete cubes was casted as cube size 150 x 150 x150 mm and are tested at different ages with the help of compression Testing Machine. These samples are tested in a 200 T compression testing machine and maximum compressive strength has been recorded. The compressive strengths for the samples which are partially replaced by SF and conventional concrete are compared and tested after curing at the age of 7, 28, 56 and 90days. The compressive strength test results shown in below table, explains that all cubes casted using Fly Ash with different percentages (0%, 5%, 10%, 15% , 20%, and 25%) shown high strength at all ages compare to conventional concrete but it shown that with the increase in percentage of Fly Ash in concrete there is decrease in strength. Up to 10% replacement level the strength of cubes has increased.

Table 5: Compressive strength for 7 days, 28 days, 56 days and 90 days

Mix	% of Cement	% of Fly Ash	% of fine aggregate	% of Quarry Dust	% of Super plasticizer	Compressive strength (N/mm^2)			
						7 days	28 days	56 days	90 days
M-1	100	0	100	0	0	32.89	42.22	47.11	52.45
M-2	100	0	70	30	0.6	33.78	43.6	49.78	54.22
M-3	95	5	70	30	0.6	36.89	49.78	55.56	58.67
M-4	90	10	70	30	0.6	41.33	56	60.89	61.33
M-5	85	15	70	30	0.6	35.11	53.78	56	59.65
M-6	80	20	70	30	0.6	33.78	52.44	54.22	57.22
M-7	75	25	70	30	0.6	34.22	49.65	53.23	55.56

As per IS 5816-1976, concrete cubes were casted as cylinder of size 150 mm diameter and 300 height and they are tested on compressive testing machine and load of 200 tonnes is applied to get tensile strength. The split tensile strengths for the Samples partially replaced with Fly Ash and conventional concrete are tested after 28 days of curing and values increased up to 15% of Fly Ash and later decreased.

Table 6: Split tensile strength for 28 days

Mix	% of Fly Ash	% of sand	% of Quarry Dust	% of super plasticizer	Split Tensile Strength N/mm^2
M1	Conventional	100	0	0	3.395
M2	0	70	30	0.6	3.395
M3	5	70	30	0.6	3.536
M4	10	70	30	0.6	3.678
M5	15	70	30	0.6	3.961
M6	20	70	30	0.6	3.678
M7	25	70	30	0.6	3.536

The Acid attack test is conducted at 28 days curing cubes in normal water again the cubes curing in 5% of diluted sulphuric acid solution for 28 days. After the 28 days curing period, the cubes were removed and determine the % of weight loss and compressivestrength.

Table 7: Weight-loss of Concrete Cubes after Acid attack

Mix	% of Fly Ash	% of Quarry Dust	% of super plasticizer	% of Weight loss	Compressive strength N/mm^2
M1	Conventional	0	0	6.27	18.22
M2	0	30	0.6	6.89	21.33
M3	5	30	0.6	6.17	26.22
M4	10	30	0.6	5.12	30.22
M5	15	30	0.6	8.9	28.45
M6	20	30	0.6	9.383	27.56
M7	25	30	0.6	9.530	25.33

From the acid attack test results M4 shows less weight loss than all other concrete mixes after acid attack. Weight loss is less, and then compressive strength is more. Weight loss of M4 mix contains 10% Fly Ash concrete cube gives less weight loss i.e. 5.12%and M1 mix i.e. conventional concrete weight loss is 6.27%.

The Chloride penetration test was carried out on cubes after curing in normal water for 28 days and later kept for curing in 3.5% of diluted sodium chloride solution for 28 days and the cubes were removed and determine the chlorine depth penetration. By observing the below figure at 0% of SF concrete gives the highest penetration depth is 16mm and minimum penetration depth occurred at 10% replacement of Fly Ash is 14mm. The maximum penetration depth of chlorine in concrete cubes with Fly Ash is 16mm and of conventional concrete is 14mm. In the present investigation water absorption test were conducted for conventional concrete and concrete with constant 30% of Quarry Dust and Fly Ash at different proportions. The results are shown below. The water absorption test for the Samples is conducted after 28 days of curing of samples.

Table 8: Chlorine penetration depth of Concrete cubes

Mix	% of Fly Ash	% of Quarry Dust	% of super plasticizer	% of Weight loss	Average Chlorine Penetration Depth (mm)
M1	Conventional	0	0	0.107	14
M2	0	30	0.6	0.195	16
M3	5	30	0.6	0.082	15
M4	10	30	0.6	0.058	14
M5	15	30	0.6	0.094	15
M6	20	30	0.6	0.104	15
M7	25	30	0.6	0.082	16

The Table 9 explains the water absorption percentage decreases as the % of Fly Ash replacement increases in concrete. Due to its more compactness Fly Ash absorbs less water. The minimum percentage of water absorbed by concrete specimens is obtained with 10% Fly Ash replacement.

Table 9: Water Absorption test results for Different concrete mixes

Mix Design	Water Absorption(%)
M1	2.78
M2	2.93
M3	2.72
M4	2.58
M5	2.64
M6	2.68
M7	2.77

5. Conclusion

Depending upon the experimental investigation conducted following conclusions were made. The size of Quarry Dust varies between nm to μm and this physical property enables us to use the Quarry Dust to fill the voids and is suitable for partially replacing Natural sand. Quarry Dust replacement with fine aggregate with 30% gives optimum results but after that, the strength got slowly decreases. With addition of Quarry Dust up to 30% the degree of workability in concrete is improved for M40 grade concrete. The results of Compressive strength and Split tensile strength are promising for water cement ratio of 0.4 on addition of fly ash at 10% replacementlevels. Fly Ash is much cheaper than cement and because of its fineness reduces the pores in concrete. Hence it can be concluded that concrete with Quarry Dust and Fly Ash can be utilized for all regular construction activities.

References

[1] Hmaid M A 2015 Improved concrete properties using quarry dust as replacement for natural sand *International Journal of Engineering Research and Development* **11** 46–52.

[2] Felekoglu K T, Baradan B, Altun A, and Uyulgan B 2006 The effect of fly ash and limestone fillers on the viscosity and compressive strength of self-compacting repair mortars *Cement and Concrete Research* **36** 1719–1726.

[3] Ho D W S, Sheinn A M M, Ng C C and C T Tam 2002 The use of quarry dust for SCC applications *Cement and Concrete Research* **32** 505–511.

[4] Muhit I B, Raihan M T and Nuruzzaman M 2014 Determination of mortar strength using stone dust as a partially replaced material for cement and sand *Advances in Concrete Construction* **2** 249–259.

[5] Ukpata J O and Ephraim M E 2012 Flexural and tensile strength properties of concrete using lateristic sand and quarry dust *ARPN Journal of Engineering and Applied Sciences* **7** 324–331.

IOP Conf. Series: Materials Science and Engineering 1006 (2020) 012034 doi:10.1088/1757-899X/1006/1/012034

Laboratory Evaluation of Engineering and Strength Properties for Semi Rigid Mixtures

Sunil S[1] and M S Naga kumar[2]

[1]Assistant Professor, Department of Civil Engineering, RV College of Engineering, Bengaluru, India
[2]Professor, Department of Civil Engineering, RV College of Engineering, Bengaluru, India

E-mail:sunils@rvce.edu.in

Abstract. Semi rigid mixture is a unique type of mixture in recent years, which has two components:Open graded bituminous mix and cement mortar grout. In this study, engineering and strength properties of semi rigid mixtures with three selected gradations are investigated. The flowability, compressive strength and flexural strength are identified to measure the optimum cement mortar grout. Meanwhile for Open graded bituminous mix, volumetric analysis, drain down and cantabro abrasion loss tests are carried out for selected gradations by varying binder content to obtain optimum mixture. Further the compressive strength, flexural strength, stability, tensile strength and durability tests are evaluated for the selected gradations of optimised Semi Rigid mixtures. The results showed that cement mortar grout containing cement to M-Sand ratio of 1:1, water-cement ratio of 0.35 with super plasticizer (polycarboxylic ether polymer) dosage of 0.45% resulted in desired flow of 12 seconds, compressive strength and flexural strength of 27 MPa and 3.5 MPa respectively. The optimum binder contents of Open graded bituminous mix for the selected gradations are found to be in the range of 4.5-5.0%. Furthermore the obtained 4.5% binder content in Semi Rigid mixture have higher engineering and strength properties compared to 5.0% binder content .

1. Introduction

Over the past decades due to urbanization heavy trucks have increased enormously in the total traffic volume in developing countries like India where more than 90 percent of roads are bituminous pavement, demanding stronger and long lasting pavements. Bituminous pavements are considered for their satisfactory performance against traffic loads and also high resistance to skidding, good serviceability and riding quality, are used as highways [1]. On the other hand it suffered from major distresses like rutting, moisture damage and cracking due to high traffic volumes, overloading and variation in climatic condition. The disadvantages of bituminous pavement can be overcome by concrete pavements but initial cost is high, requires skills for construction and maintenance and also due to joints in concrete pavements riding quality is poor as well as the noise generated by the joints, the differential expansion/contraction needs to be accommodated [1]. Thus, a new technology of pavement need to be arise to overcome these distresses. One such type of pavement is Semi Rigid mixture, which is a reliable , joints free pavement, highly load carrying capacity, longer service life, easy to construct and repair and desirably having the nature of both rigidity and flexibility, moreover, offers riding quality and satisfactory serviceability. It has been adopted to furnish a durable and water proof surface for the heavily loaded areas such as gas stations, bus terminal, airport runways, traffic intersection, parking lots and the like. This new type of pavement has prospects to be designed as a high performance pavement because it integrates the advantages of both rigid and flexible pavement, which appears to be a promising approach but less researched.

2. Background

In the 1960s, France introduced a new technology for pavement surface similar to Semi Rigid mixes known as 'Resin Modified Pavement' and in USA it was applied in 1987 [2]. Later, Semi rigid mixes technology was implemented in various countries like France, USA, Germany, Japan, Spain, Portugal, Sweden, Norway, Finland, Saudi Arabia, Great Britain, China, Denmark , Malaysia, South Africa and Austria had been used in bus stations, parking areas, intersections, warehouses, roads, aprons, etc. However in UK officially there is no BSEN (British Standards, Europe) specification on Semi rigid mixes yet. But ADEPT (Association of Directors of Environment, Economy, Planning and Transport) UK has developed a code of practice for grouted Macadam which is approved by the Highway Authorities Product Approval Scheme (HAPAS) in 2006 [3].This technology was implemented by an organization called CIP Joint less Surfacing, primarily in Harbors, Airports, Warehouse Distributions, Bus-Depot Stations, Manufacturing-Production areas etc. In European countries like Denmark, Netherlands, Sweden etc. Semi Rigid mixes technology has been extensively used. The Copenhagen Airport, Denmark already laid approximately 3 lac m2. M/S BREMAT, Rosmalen, Netherlands carried out the execution and maintenance of this technology in Roundabouts, Intersections and Signals. Stockholm, Sweden implemented at different Bus stops and Bus- Depot stations, to bear the static loads and to diminish the chances of wearing and tearing. In USA, this technology incorporated in Army bases, Air force Stations and Airports. For durable surface, fuel damage resistant surface and heavily trafficked at two signalized intersections investigation was carried on McCord Air force Base, Washington and Logan International Airport, Malmstrom Air Force base, Montana and Mississippi Department of Transportation. In the Asian countries, this technology has been widely implemented in Malaysia, China, Japan, Saudi Arabia etc. Kualalampur City Council, Malaysia has implemented and maintained this technology since 2001 in heavily trafficked surface and Bus Lanes in the city area. In China, Ministry of Construction of Chinese Technologies had executed several projects in late 20th Century. In India, this technology has been still in research stage. The performance studies on this technology is carried out by CSIR-CRRI, New Delhi, National laboratory at Chennai and IIT Kharagpur, West Bengal. The Surat Municipal Corporation in collaboration with Sardar Vallabhbhai National Institute of Technology, Surat, Gujarat laid on two test sections in the city of Surat, of 100m length each for resurfacing the pavement [4].

3. Materials

The materials incorporated for the present study includes Conventional aggregates, Bitumen, Construction & Demolition waste aggregates, Fibers (Sugarcane fibers), Cement, M Sand and Super plasticizer.

3.1 Aggregate

The aggregates used in this investigation are procured from quarry near Bidadi, Bengaluru. Table 3.1 gives the Summary of aggregates physical properties.

IOP Conf. Series: Materials Science and Engineering 1006 (2020) 012034 doi:10.1088/1757-899X/1006/1/012034

Table 3.1: Summary of aggregates physical properties

Tests	Result Obtained	Standard Specification	As Per Code
Aggregate Impact (%)	19	Max24	IS: 2386-1963(Part 4)
Los Angeles Abrasion (%)	20.76	Max30	IS: 2386-1963(Part 5)
Flakiness and Elongation (%)	25.12	Max30	IS: 2386-1963(Part 1)
Water Absorption (%)	0.2	Max2.0	IS: 2386-1963(Part 3)

Aggregate Specific gravity

Sieve size, mm	Specific gravity	
19/12.5	2.88	IS: 2386-1963(Part 3)
12.5/9.5	2.87	
9.5/4.75	2.87	
4.75/2.36	2.88	
2.36/0.075	2.81	

3.2 Bitumen

VG 30 grade bitumen used in this study and the basic tests were carried out according to IS 73 2013.Table 3.2 shows summary of VG-30 bitumen physical properties, the mixing and compaction temperature for this bitumen is 152°C-156°C and 138°C-142°C respectively. The temperatures corresponding to viscosity ranges of 0.17Pa-sec and 0.27Pa-sec were plotted in the graph and the resulting temperatures are chosen for compacting and mixing.

3.3 Cement

In this investigation Ultratech 53 grade cement were used. Table 3.3 shows summary of 53 grade cement properties.

3.4 M- Sand

The M-Sand was procured from a plant near Bidadi, Bengaluru. The sand should pass 0.6mm size and retained 0.075mm so that the grout will enter into the air voids easily. Table 3.4 shown the summary of M sand properties of as per IS 1542.

Table 3.2: Summary of VG-30 bitumen physical properties

Tests	Result Obtained	Standard Specification	Standard test method
Penetration@ 25°C, 0.1 mm	58	45 minimum	IS:1203
Ductility @ 25°C, cm	91	40 minimum	IS:1208
Specific Gravity	0.98	0.97-1.02	IS:1202
Softening Point Test ,°C	50.75	47 minimum	IS:1205
Kinematic Viscosity, cSt	368	350 minimum	IS:1206 – Part 3

Table 3.3: Summary of 53 grade cement Properties

Properties evaluated	Obtained Value	Standard Specification
Fineness, m^2 /kg	260	225
Standard Consistency,%	33	--
Initial Setting Time(min)	80	30
Final Setting Time(min)	470	600

Table 3.4: Summary of M sand properties

Properties evaluated	Obtained Value	Standard test method
Specific Gravity	2.68	IS:1542-1992
Water absorption, (%)	18	IS:1542-1992
Bulk Density (Kg/m^3), loose	1264	IS: 2386-1963(PART 3)
Bulk Density (Kg/m^3), compacted	1500	IS: 2386-1963(PART 3)

3.5 Super-Plasticizer

Super plasticizer helps in reducing the water requirement without reduction in strength of the grout, thus improving the flowablity and helps in early gain of grout strength. In this investigation we used Sulphonated Naphthalene Formaldehyde and Poly Carboxylic Ether Polymer for the cement grout.

3.6 Fibres

In this study, natural sugar cane fibre is used to enhance crack resisting property to the mixtures, 2.36mm passing size fibres were selected for preparing of samples.

3.7 Selection of aggregate gradations

The three aggregate gradations were identified for preparation of open graded bituminous mixes which allow cement mortar grout flows freely into the voids of mixture with selected quantity of bitumen and compaction. The gradation for the aggregate were adopted according to master gradation provided by the ASTM D7064 (OGFC), New Jersey MOGAC (12.5-mm NMAS) and Transit New Zealand (PA 14 HV) as shown in graph Figure 3.1. Generally, open gradations are preferred for Semi Rigid mixture so that minimum air void content of 20-30% is available in the bituminous mix skeleton. The specimens of different binder content mixes prepared for midpoint aggregate gradations were compacted.

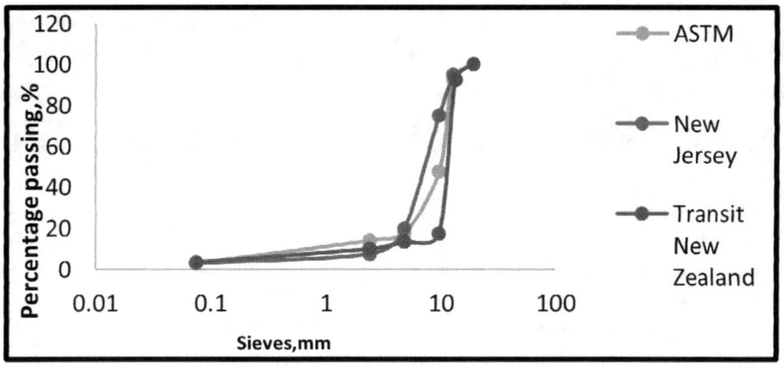

Figure 3.1: Aggregate gradation graph

4. Methodology

The first stage is to develop a cement mortar grout. The second stage is to open graded bituminous mix and finally the production of the Semi rigid mixtures. The following methods were carried on Semi rigid mixtures.

- Selection of Cement Mortar Grout
- Preparation of Open Graded Bituminous Mix Specimens
- Mix design for Semi rigid pavement
- Strength properties
- Durability test.

4.1 Selection of cement mortar grout

Around 22 cementious grouts were evaluated to identify the proportions of grouts for the suitability of semi rigid mixtures. Flowability and strength of the grouts were measured and proportions were changed progressively based on the obtained results. The ratio 1:1 proportion of cement and M sand was fixed and two types of super plasticizer with different dosage were used for water cement ratio of 0.30-0.40. Mixing of the grout was done in a mechanical mixer for three minutes. A summary of cement mortar grout mixtures are shown in Table 4.1.

Table 4.1: Summary of cement mortar grout mixtures

Cement Mortar (Cement: M Sand)	Type of Super Plasticizer (SP)	w/c ratio	% of SP dosage
1:1	Sulphonated Nepthalene Formaldehyde	0.30,0.35, 0.40	0.25,0.30,0.35, 0.40,0.45,0.50
	Poly Carboxylic Ether Polymer		

4.1.1 Flowablity of cement mortar grout

The flowable properties of cement mortar grouts for two different types of superplasticizer were examined. The flow times of grout are analyzed using effect of superplasticizer dosage and superplasticizer type of flow characteristics are shown in Figure 4.1. From the obtained results, dosage of superplasticizer significantly influence the flowablity of cementious grout. According to the trails, the optimal flow time shall be 12 seconds.

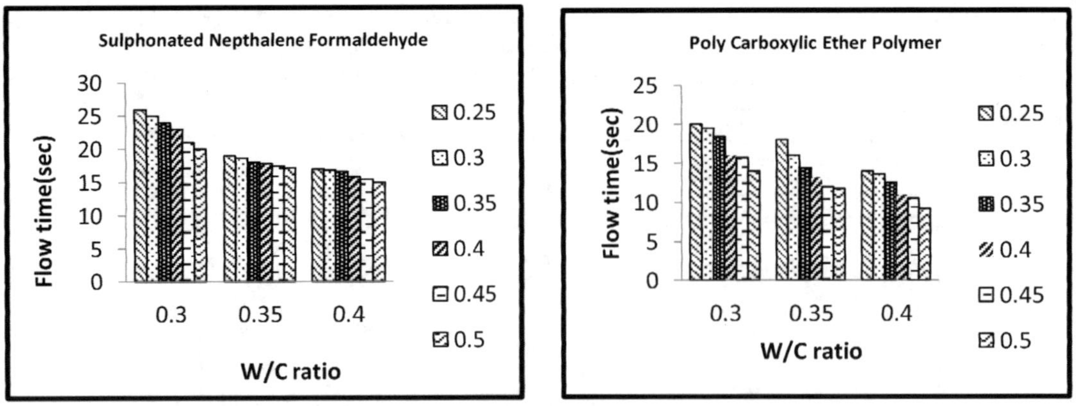

Figure 4.1: Flowablity of grout containing SP- Sulphonated Nepthalene Formaldehyde and Poly Carboxylic Ether Polymer

4.1.2 Strength properties of cement mortar grout

The strength results for selected grout mixtures containing superplasticizer dosage with flowablity 12 seconds are shown in Table 4.1.The compressive and flexural strength of mix for 28 days was carried out. From the results obtained, the grout mix containing 1:1 Cement: M-Sand ratio, 0.35 w/c ratio and 0.45% Poly Carboxylic Ether Polymer dosage produced the highest compressive strength and flexural strength is 27.5 MPa and 3.5 MPa respectively. Table 4.2 shows summary of strength properties of Cement Grouts.

Table 4.2: Summary of strength properties of Cement Grout

Cement grout	Super Plasticizer dosage	w/c ratio	Compressive Strength , MPa	Flexural Strength, MPa
Cement: M Sand	Sulphonated Nepthalene Formaldehyde			
1:1	0.50	0.40	21.3	2.1
	Poly Carboxylic Ether Polymer			
	0.50	0.30	21.4	1.9
	0.35	0.35	22.5	2.2
	0.40	0.35	24.2	2.4
	0.45	0.35	27.5	3.5
	0.5	0.35	25.6	3.0
	0.25	0.40	25.2	2.8
	0.30	0.40	24.6	2.5
	0.35	0.40	22.3	2.1
	0.40	0.40	21.2	1.7
	0.45	0.40	20.6	1.5
	0.50	0.40	19.4	1.2

4.2 Preparation of open graded bituminous mix specimens

Requisite binder content is determined by the method of Marshal Mix design according to codal provision conforming to ASTM-D6927-06. The mixtures were compacted by giving 50 no. of blows on only one face of the specimen using Marshall hammer, for 4 binder contents at an increment of 0.5%. Around 1000gm of aggregates is allowed to heat to a temperature of l60-l75°C. Asphalt is melted to a temperature of 150-165°C with the initial experimental percentage of binder (say 4.0 to 5.5 percent by weight of the aggregates). The heated aggregates and binder is carefully batched at 160-165°C temperature. The mixture is then poured inside the heated mould & compacted using a rammer with the 50 number of blow at temperature 130°C to 140°C. The weight of mixed aggregates used for the casting of the sample can be suitably changed to get the desired compacted thickness of 63.5mm. Increase the bitumen content in next trial by +0.5% and repeat the above technique.

4.2.1 Determination of optimal bitumen content (OBC)

The optimal bitumen content was determined for all three gradations by (i) Volumetric analysis of mix, (ii) Abrasion loss, and (iii) Drain down test for the prepared open graded bituminous mix samples. For all gradations, bituminous mixtures were prepared at different binder contents in increment of 0.5 %. The minimum criteria for Air void content is of 20% to meet the requirements of open graded bituminous mix samples mixes, The particle loss for the unconditioned samples should not exceed 20% and Drain down loss criteria of 0.3% (maximum) for high void bituminous mix is considered for selecting the optimum binder content.

4.2.1.1 Volumetric analysis of mix

Bulk densities, Gmb (ASTM D2726, 2010) and theoretical maximum specific gravities, Gmm

(ASTM D2041, 2011) of the mix were determined and obtained values were used for volumetric analysis of mix. Voids in Mineral Aggregate (VMA) and Voids in Mix (VIM) and were estimated using equations 4.1 and 4.2 respectively. Figure 4.2 and 4.3 shows VIM and VMA values for binder contents.

$$VMA= (1-\frac{Gmb}{Gsb}* Ps)*100 \qquad (4.1)$$

$$VIM= (1-\frac{Gmb}{Gmm})*100 \qquad (4.2)$$

Where,
Gmb= bulk specific gravity of the mix
Gsb= bulk specific gravity of aggregates
Gmm=theoretical specific gravity of the mix
Ps= fraction of aggregates present, by total mass of mix

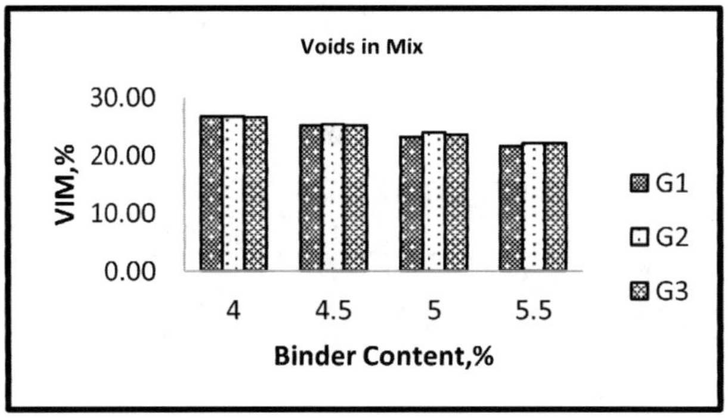

Figure 4.2: VIM values with binder content

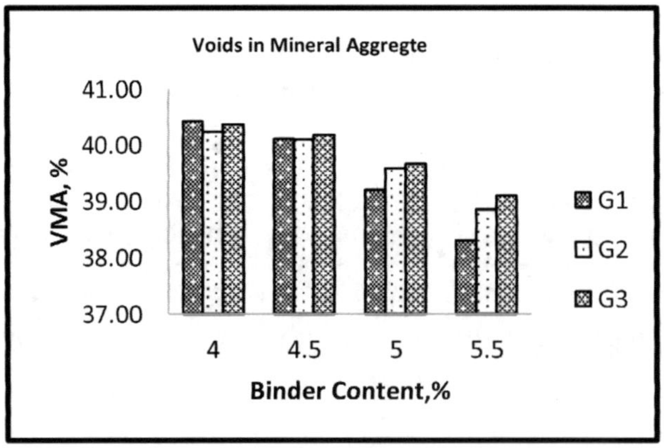

Figure 4.3: VMA values with binder content

4.2.1.2 Cantabro Abrasion loss

Cantabro abrasion loss test was used to determine the abrasion loss value of unaged open graded bituminous mixes using the Los Angeles abrasion machine. The test specimen was placed in a Los Angeles abrasion drum without any abrasive charges, and the machine was operated at a speed of 30–33 revolutions per minute for 300 revolutions. The percentage of weight loss in the specimen when compared to its initial weight was expressed as the unaged abrasion loss (UAL). The temperatures recorded during the test procedure were within the range of 25 ± 5°C as specified in the ASTM D 7064 -08. The result of the test is determined using the Equation 4.3 and Figure 4.4 shows Cantabro abrasion loss for different binder content.

$$P = \frac{(P1 - P2)}{P1} * 100 \quad (4.3)$$

Where,
P – Cantabro abrasion loss,
P1-Initial weight of the sample,
P2-Final weight of the sample.

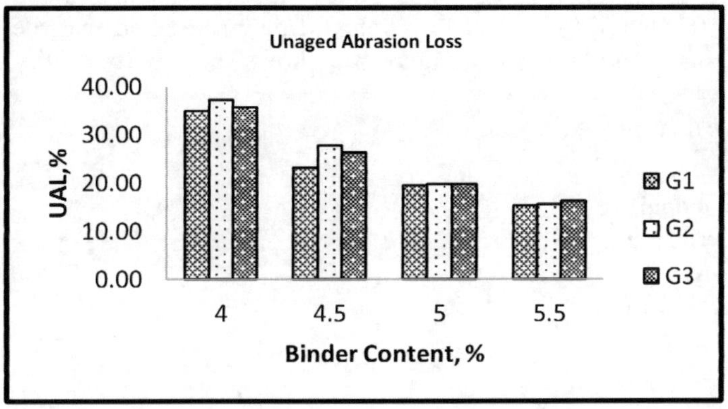

Figure 4.4: Cantabro abrasion loss for different binder content.

4.2.1.3 Drain down test

The drain-down test is used to determine the correct amount of fiber needed for the mix. The test measures the potential for asphalt binder to drain from the coarse aggregate structure while the mix is held at an elevated temperature. The test is performed in accordance with ASTM D6390. The mass of the drained material is determined to calculate the amount of drain down as a percentage of the mass of the total asphalt mix sample are determined using Equation 4.4 and Figure 4.5 shows drain-down test results for binder contents.

$$\text{Drain down, \%} = \frac{Wi - Wf}{Wi} * 100 \quad (4.4)$$

Where,
Wi - Initial weight of the sample,
Wf - Final weight of the sample.

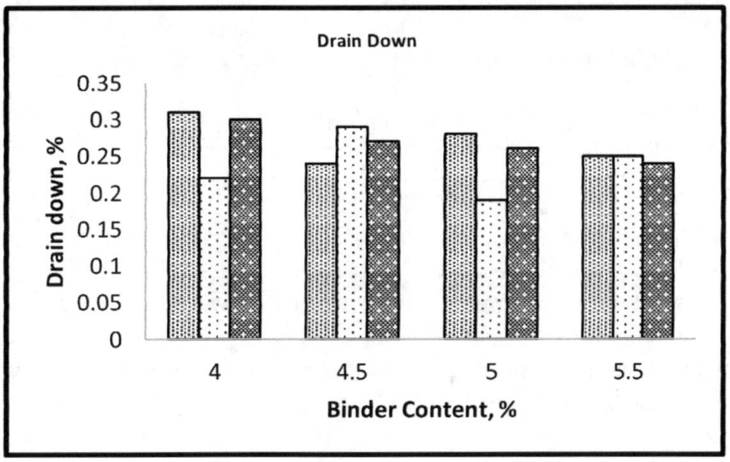

Figure 4.5: Drain-down test results for binder contents

4.2.3 Permeability test

Falling head permeameter was used for conducting the permeability test as per IS Code 2720 part 17. The specimen for determining permeability is1prepared by Marshal Method and the specimen is1sealed around its periphery on both the1sides to prevent the loss of water from sides. Water is allowed1to flow through the specimen for1a certain head, the time1required for the water to pass through the1specimen is recorded using stop watch. The resulting permeability cm/s is converted to m/day is determined1using the Equation 4.5. Following the typical guidelines for OGFC's, a minimum value of 100 m/day for permeability should be ensured, so that the cement grout can penetrate the voids in the bitumen mix. Figure 4.6 shows the f permeability results for binder contents.

Permeability=$\frac{aL}{At}\log(\frac{h1}{h2})$(4.5)

Where,

a = area of the top standpipe (cm2),
L = height of the sample (cm),
A = cross-sectional area of the sample (cm2);
t = time taken for the water to fall from h1 to h2 (s);
h1 = initial water head and
h2 = final water head.

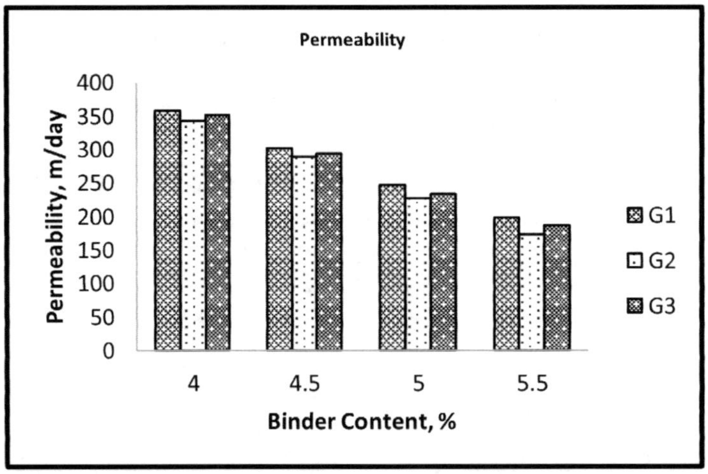

Figure 4.6: Permeability results for binder contents

4.3 Preparation of Semi Rigid Mixtures

As discussed in the above sections the optimum range of binder content (OBC 4.5-5.0%) for the selected open graded gradation (G1,G2 and G3) and optimized cement mortar grout (Cement: M Sand: w/c: SP-1:1:0.35:0.45 proportion) will be used to obtain the augmented Semi Rigid mixture samples.

- The open graded bituminous mix were prepared as discussed in the above section.
- The optimized cement mortar grout is identified as explained above section.
- The open graded bituminous mix specimens cooled down to ambient temperature, prepared cementious grout should be poured gradually on the surface of the specimen.
- The care should be taken to prevent spill of grout in mould while pouring.
- The electric vibrator is used for 30 seconds to provide mild vibration to ensure grout penetrated into the mix.

- The specimen can be demould after 24 hours and shall be cured for 28 days, direct water curing is not recommended. Wet jute bags can be used for curing.
- Photos 4.1 to 4.4 show different stages of preparation of the Semi Rigid mixtures

Photo 4.1:Sampleprepared mix before **Photo 4.2**: Vibration of the mould Grouting

Photo 4.3: Sample after Grouting **Photo 4.4**: Sample after demould

4.3.1 Properties of grouted mix

The saturation degree and remaining air void (after filling the air void of open graded bituminous mix with grout) in Semi Rigid mixtures are critical parameters concerning the performance and durability of Semi Rigid mixtures under traffic loading and climate conditions. Therefore remaining air void of 3 to 6% for Semi Rigid mixtures should be maintain to avoid bleeding of bituminous mixture. In order to calculate the parameters of grouting saturation degree and remaining air void are calculated using Equation 4.6. The density of grout 1.963 gm/cc is calculated as per BS EN 1015-6 (1999). The Table 4.3 shows summary of volumetric analysis of mixes before and after grouting.

Saturation degree $=\frac{W2-W1}{\rho * V * Va}$ (4.6)

Where,

W1 and W2 are specimen mass before and after grouting in gram, respectively,

$\rho$is the density of grout in gr/cm^3,

V is the specimen volume in cm^3,

Va is the air void of mixtures

Table 4.3: Summary of volumetric analysis of mixes before and after grouting

Aggregate Gradation	Binder Content (%)	Avg. weight in air before grouting (gms)	Air voids before grouting (%)	Average weight in air after grouting (gms)	Saturation degree	weight of grout (gms)	Air voids After grouting (%)	Average weight of grout penetrated (gms)
G1	4.5	1036	24.8	1231	0.79	248	5.25	234.97
	5	1042	23.1	1239	0.85	231	3.35	223.25
	5.5	1048	21.9	1241	0.88	219	2.56	213.40
G2	4.5	1039	24.5	1237	0.81	245	4.65	233.60
	5	1042	23.2	1241	0.86	232	3.25	224.45
	5.5	1050	22.3	1246	0.88	223	2.65	217.08
G3	4.5	1037	24.8	1234	0.80	248	5.05	235.46
	5	1041	23	1237	0.85	230	3.35	222.28
	5.5	1051	21.7	1247	0.91	217	2.05	212.54

5 Results and discussion

5.1 Compressive strength

Compressive strength of compacted asphalt mixtures is measured as per ASTM D1074 by Compressive Testing Machine (CTM). The same method was adopted to test Semi Rigid mixture samples, the sample preparation is explained in above section and dimensions of the cylindrical samples are 101.6mm diameter and 63.5mm thickness. Figure 5.1 shows compressive strength values of binder contents and Equation 5.1 shows compressive strength values in N/mm^2.

Compressive strength= $\frac{P}{A}$ N/mm^2 (5.1)

Where,

P= maximum applied load in N

A= area of the sample in mm^2

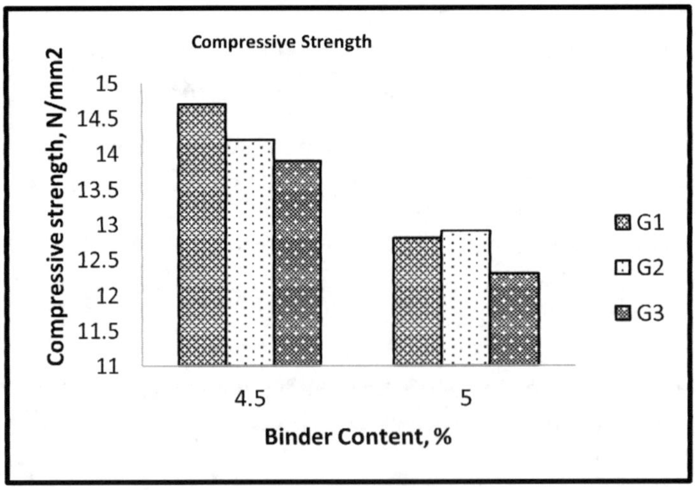

Figure 5.1: Compressive strength values of binder contents

5.2 Flexural Strength

ASTM C 78 test method provides a method for measuring flexural strength of the concrete. The same method was adopted to test Semi Rigid mixture samples, the dimensions of mould size are 500mm*100mm*100mm.The open graded bituminous mix for the optimum binder contents (4.5-5.5%) is poured on the beam mould, CTM is used to compact the mix to obtain target air voids 20-30% and optimized cement mortar grout (1:1:0.35:0.45) is poured on the mixture when it cools down to ambient temperature. After 24 hours the beam sample were demould and kept for 28 days water curing. Two point loading test carried on the sample, point of failure at maximum load recorded for calculating flexural strength test. Figure 5.2 shows the flexural strength values of binder contents and Equation 5.2 shows flexural strength values in N/mm^2.

Flexural strength = $\frac{Pl}{bd^2}$ (5.2)

Where,

P = Failure load in N
L = Effective span of the beam in mm
b = Breadth of the beam in mm
d = Depth of the beam in mm

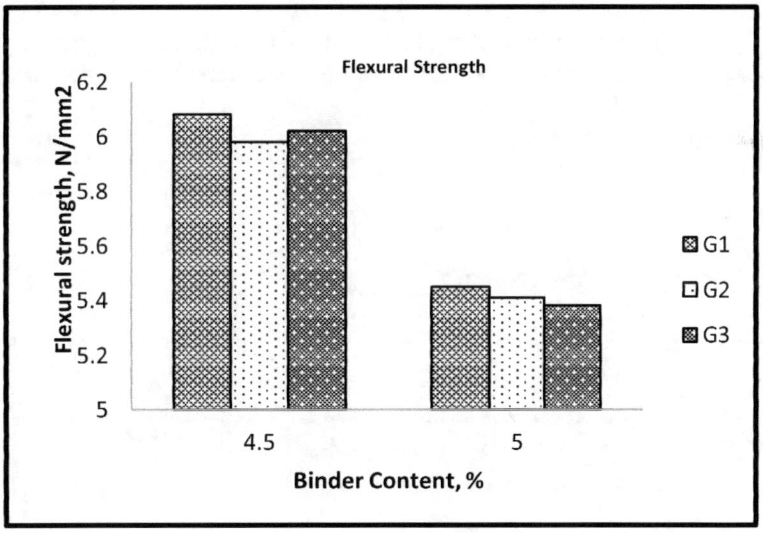

Figure 5.2: Flexural strength values of binder contents

5.3 Marshall Stability

Marshall Method is used almost everywhere in the world for its easy procedure and economy compared to other methods. Strength behavior of samples at optimum binder contents were studied using Marshall Stability. ASTM D6927 test method used for testing procedure of the samples. The cylindrical samples of 101.6mm diameter and 63.5mm thickness were cast to test the stability with 50blows of compaction on single side to obtain target air voids 20-30%. The point of failure is defined when the maximum load reading obtained and maximum load value is expressed in kN is recorded as the Marshall Stability value of the samples. Figure 5.3 shows the stability values of binder contents.

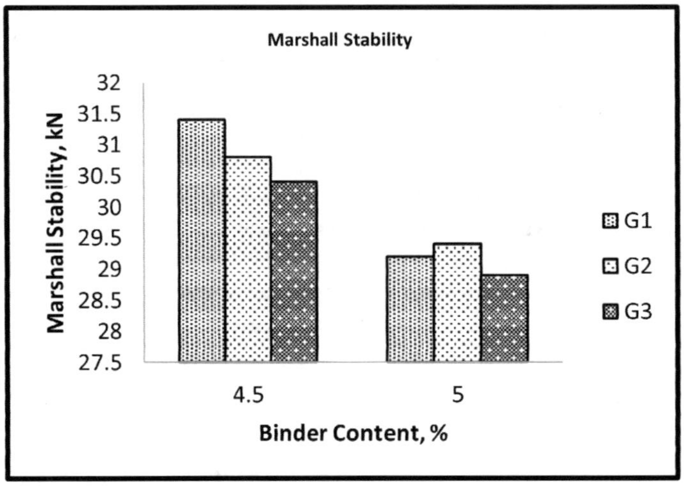

Figure 5.3: Stability values of binder contents

5.4 Indirect Tensile Strength

Indirect tensile strength test performed to study the tensile behavior of Semi rigid mixtures. The Indirect Tensile Strength (ITS) was determined as per ASTM D6931-17 on cylindrical samples by loading them along a diametrical axis at a deformation rate of 51 mm per minute. The same method was adopted to test Semi Rigid mixture samples. The maximum load P, taken by the specimen was used to calculate the indirect tensile strength. The ITS value can be determined by Equation 5.3 and Figure 5.4 shows the ITS value of conditioned and unconditioned for binder contents.

$$ITS=\frac{2000P}{\pi * t * d} \quad (5.3)$$

Where,

ITS= Indirect Tensile Strength in N/mm2
P = maximum applied load in N
t = mean thickness of sample in mm
d = diameter of sample in mm

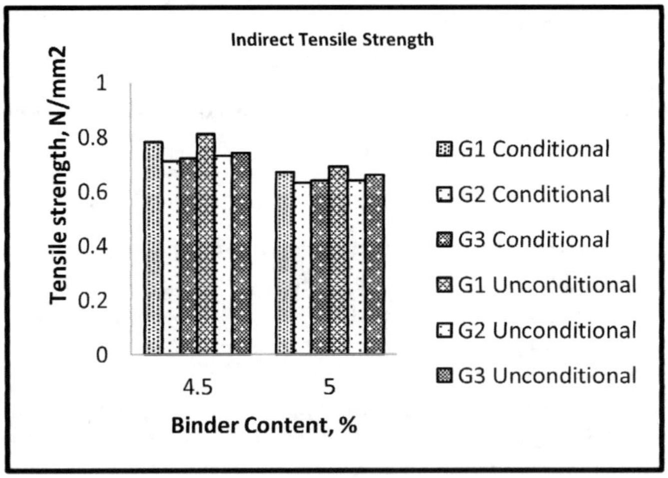

Figure 5.4: ITS value of conditioned and unconditioned for binder contents

5.5 Durability

Cantabro abrasion test conducted at temperature of 25ºC in Las Angeles drum, rotated for 300 revolutions. This test is generally conducted on both dry and wet conditioned specimen as specified in the ASTM D 7064 -08. To study worst condition, aged samples were used in the present study. The samples were conditioned for 168 hours at 40°C and samples are then cooled to 25°C and stored for 4 hours prior to abrasion. The result of the test is determined using the Equation 5.4. Figure 5.5 shows Cantabro abrasion loss for binder contents.

$$P = \frac{(P1 - P2)}{P1} * 100 \quad (5.4)$$

Where,

P – Cantabro abrasion loss,

P1 – Initial weight of the sample,

P2 – Final weight of the sample.

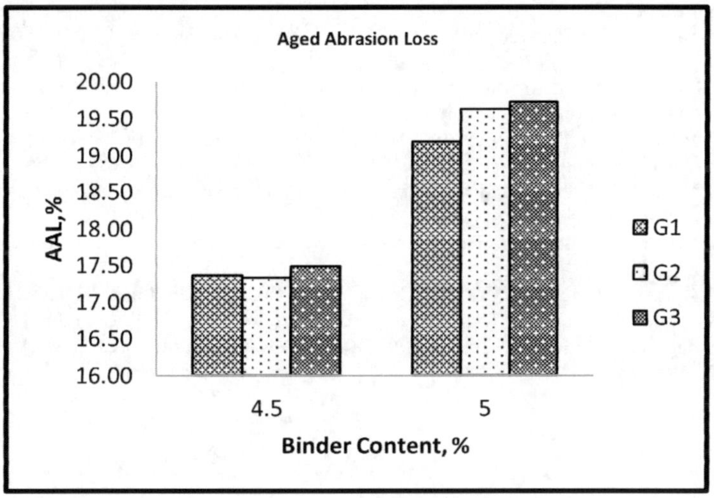

Figure 5.5: Cantabro abrasion loss for binder contents

6 Conclusions

This current study investigated the engineering and strength properties of Semi Rigid mixtures. In the first stage of this study, the effect of flowability and strength properties of cement mortar grout was evaluated. In the next stage, the open graded bituminous mix, which is the skeleton of the Semi Rigid mixtures, was designed to find optimum binder content by air voids drain down and abrasion loss was evaluated. Finally, Semi Rigid mixtures specimens were prepared by grouting cement mortar into an open graded bituminous mix. Mechanical properties including compressive strength, flexural strength, Marshall Stability, tensile strength and durability were evaluated for optimum binder content. From the test results it was clearly recommended that the air voids 20-30% and filling rate of Semi Rigid mixtures specimens were greatly influenced by the flowability of cement grout within 12 seconds. Through evaluating the strength characteristics compressive strength, flexural strength, Marshall Stability, tensile strength and durability it was found that the 4.5% binder content have higher values compare to 5.0% binder content due to increase in void content and cement grout.

References

[1] A Hassani, M Taghipoor and M M Karimi 2020 A state of the art of semi-flexible pavements : Introduction, design, and performance *Constr. Build. Mater* vol. **253** p. 119196.

[2] Battey R L and Whittington J S 2007 Construction Testing and Performance Report on the Resin Modified Pavement Demonstration Project *Mississippi Department of Transportation, USA*.
[3] Anderton G L 1996 User's Guide: Resin Modified Pavement. U.S. Army Engineer Waterways Experiment Station Vicksburg.
[4] G Bharath, Manoj Shukla, M N Nagabushana, Satish Chandra & Amit Shaw 2019 Laboratory and field evaluation of cement grouted bituminous mixes *Road Materials and Pavement Design* pp 1-19.
[5] Anderton G L 2000 Engineering Properties of Resin Modified Pavement (RMP) for Mechanistic Design. U.S Army Corps of Engineers, Vicksburg, Mississippi.
[6] Satyarno I, Permana A, Meyarto C and Hadiyatmoko D 2014 Practical method for mix design of cement-based grout *Procedia Eng.*, **95** (Sustainable Civil Engineering Structures and Construction Materials) 356–365.
[7] Suhana Koting, Mohamed Rehan Karim, Hilmi Bin Mahmud, N. A. A. H. 2014 Mechanical Properties of Cement-Bitumen Composites for Semi-Flexible Pavement Surfacing. *Balt. J. ROAD Bridg. Eng.* ISSN **9(3)** 191–199.
[8] Sun Y, Cheng Y, Ding M, Yuan X and Wang J 2018 Research on properties of high-performance cement mortar for semiflexible pavement *Adv. Mater. Sci. Eng.*
[9] MoRTH 2013 Specification for Road Works and Bridges *Ministry of Road Transport and Highways* 5th Edition New Delhi, India.
[10] IS 73 2013 Paving Bitumen The Bureau of Indian Standards, New Delhi, India.
[11] MS-2 (Manual Series No.2) 7th Edition Mix Design Methods for Asphalt Concrete and Other Hot Mix Types. Asphalt Institute, Kentucky, USA.
[12] IS 12269-1987 Specification for 53 grade ordinary portland cement. The Bureau of Indian Standards, New Delhi, India.
[13] ASTM D 7064-04 2004 Standard practice for open-graded friction course (OGFC) mix design *American Society for Testing and Materials*, West Conshohocken.
[14] ASTM D6390-11 2011 Standard Test Method for Determination of Drain down Characteristics in Un-Compacted Asphalt mixtures *American Society for Testing and Materials*, Pennsylvania, USA.
[15] ASTM D 6931 2012 *Standard* Test Method for Indirect Tensile (IDT) Strength of Bituminous Mixtures *American Society for Testing and Materials*, Pennsylvania, USA.
[16] IS 2720-17 1986 Methods of test for soils, Part 17Laboratory determination of permeability, The Bureau of Indian Standards, New Delhi, India.
[17] AASTHO T 283 2014 Standard Method of Test for Resistance of Compacted Asphalt Mixtures to Moisture-Induced Damage *American Association of State Highway and Transportation Officials* Washington DC, USA.
[18] ASTM D 1074 Standard Test Method for Compressive Strength of Asphalt Mixtures *American Society for Testing and Materials*, Pennsylvania, USA.

Prediction of Compressive Strength with the Variation of Steel Fiber in the Concrete by Using ANN

Sandela Haripriya[1], Md Ikramullah Khan[2], D Swetha[2], G Alok[3], Bandi Bhaskar[4], P Sudheer Kumar[5], T Pranay Kumar[5]

[1, 3]Centre for Design, SR University, Ananthasagar, Hasanparthy, Warangal, Telangana, India-506371.
[1, 2]Department of Civil Engineering, SR University, Ananthasagar, Hasanparthy, Warangal, Telangana, India-506371.
[4]Department of Computer Science and Engineering, SR University, Ananthasagar, Hasanparthy, Warangal, Telangana, India-506371.
[5]Department of Civil Engineering, Balaji Institute of Technology and Sciences, Laknepally, Narsampet, Warangal, Telangana, India – 506 331.

Email: priya.hari700@gmail.com

Abstract. The compressive strength of concrete up on replacement by silica fume and the adding of steel fiber has been predicted by using artificial neural networking in this investigation. Compressive strength was measured at 7, 14, and 28 days by using these data ANN model has been constructed. The input parameters which are considered as the particulars in the artificial neural networking model have been set out in a form are cement, silica fume, and steel fiber and the compressive strength of concrete is an output variable. The results exhibited that the envision of the compressive strength of concrete by ANN can be another approach.

1. Introduction

Concrete, which is an important material in the construction of civil engineering projects. Various studies have been performed on concrete to enhance its characteristic properties, by replacement or addition of fibers or various materials. In this study, the compressive strength of cement has been increased with the replacement of silica fume up to 10 %. Upon the addition of certain admixtures, many modern concrete mixes are modified. The tensile strength of concrete has been improved by the addition of steel fibers in various proportions. The compressive strength of concrete for 7, 14, and 28, days curing by the reaction of silica fume and steel fiber is the objective of the study. The applications of the artificial neural network have been implemented in this case to last the concrete characteristics.

Pawade has studied with the increase in SF (silica fume) the compressive strength also increased when compared with ordinary concrete[1,2]. Upon the addition of SFR (steel fibers), the weight density of concrete has been enhanced. Both Strength aspects of concrete, flexural strength; compressive strength raised upon the addition of SFR[3,4]. Khelan has investigated that the increased concrete's compressive strength performance with steel fibers containing silica fume at age of 28days[5]. Also, due to the lower aspect ratio, there is no problem with balling handling is easy. Among all replacing the silica fume with steel fiber 10 % silica fume with a 2% volume of fiber is the best combination. Cahit Bilim has forecast the

Compressive strength of ground granulated blast furnace slag concrete by using ANN [6]. Forty-five concrete samples were endured and avail in the ANN. the compressive strength of moist cured specimens was measured at 3,7,28,90,360 days. Fatin Ozan has studied an ANN and fuzzy logic study and occurred for the envision of compressive strength of silica fume concrete[7]. Faezehossadat has investigated the ANN in determining the 28 days compressive strength of concrete[7]. Hence, specific concrete characteristics[8,9]considering as inputs, the ANN has been developed and the compressive strength of concrete is determined[8,10–12]. Predicting the 28 days compressive strength of concrete by ANN indicates as a best replica.

2. Experimental methodology

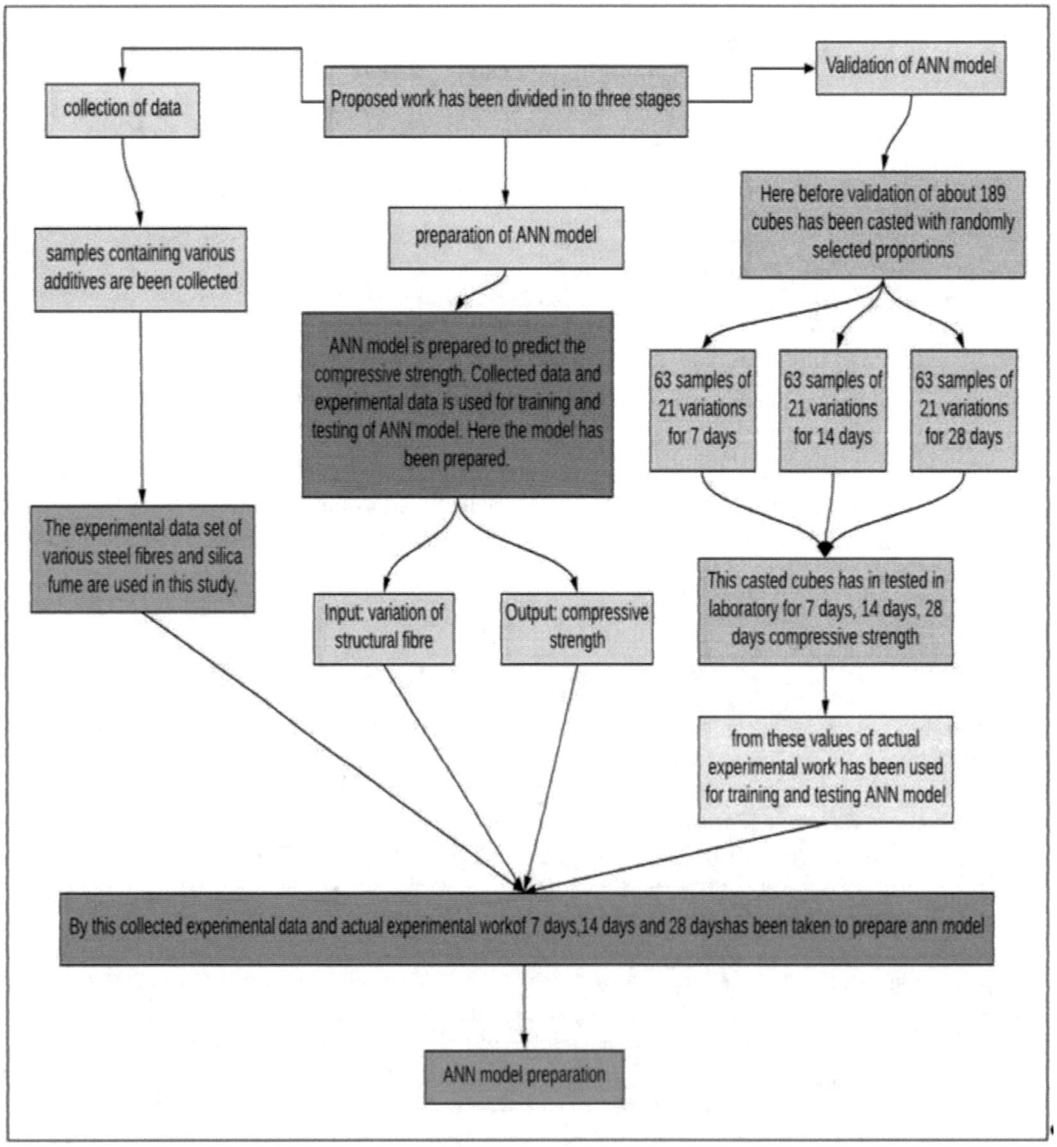

Figure 1: Representing the experimental methodology

3. Artificial neural network replica

It has the potential to analyze from studies to increase its overall performance. Similar to the human brain, an Artificial Neural Network receives facts through studying. ANN model is a gentle computing approach composed of simple factors that may be replaced with commonplace computations which cannot solve the troubles exactly. ANN contains an input layer, 1 or more hidden sheet(s), connected by neurons [13], and an output layer as shown in Figure1. A man-made brain-like network supported certain mathematical algorithms developed employing a numerical computing environment like MATLAB is named as an Artificial Neural Network (ANN).

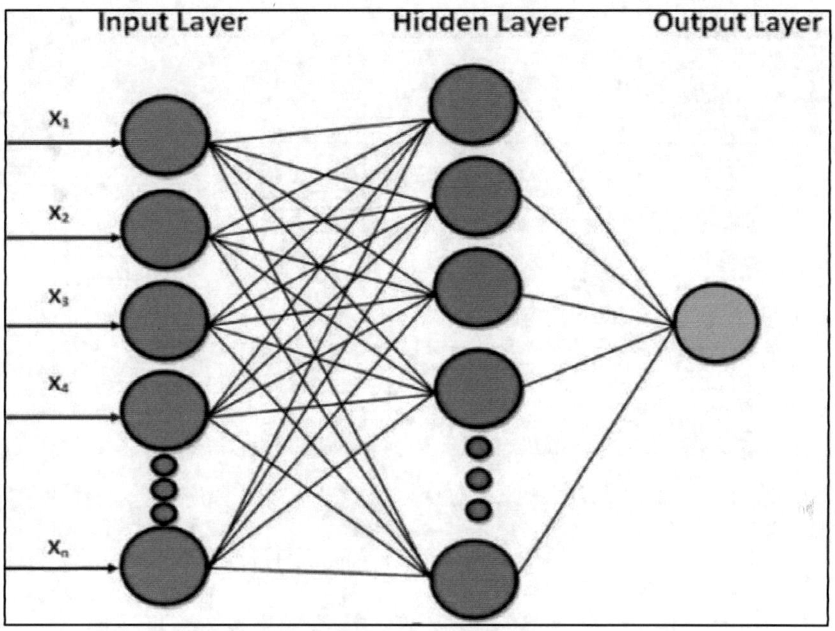

Figure 2: Basic Structure of the Artificial Neural Network model

Standard steps caused in designing a Neural Network in Mat lab are:

a. Data Collecting
b. Network Creating
c. Network Configuring
d. Initialize the weights and biases
e. Network Training
f. Network Validating
g. Network Use

In the first step, the data is collected through the literature and the experimental results conducted in the laboratory. This data is imported into the MatLab workspace in the matrix format with the input parameters as input data matrix and experimental &collected data values as the target matrix. Configuration of the network is done to create the network in the neural network fitting tool. The fitting device within the neural community is used to map between a dataset of numerical inputs and a fixed of numerical targets. The Neural Network Fitting Tool will assist to select statistics, create and teach a community, and examine its overall performance the use of implying mean square error and regression evaluation.

A total of 189 numbers of samples data is used to prepare the ANN model. Inputs 'input' is an a1×21 matrix, representing static data, I .e.21 samples of 1 element. Target's' target' is a a1×21 matrix, representing static data, i.e. 21 samples of 1 element. The 21 samples data are randomly divided into three kinds of samples.

- **Training:**Out of the 21 samples, 70% of samples i.e.15 samples are given for training of the network.
- **Validation:**For validation 15% of the samples i.e. 3 samples are used for the validation of the network.
- **Testing:**The network testing is done to 15% of the samples i.e.3 samples are used. The samples used for the training process should not repeat during the testing process as testing is done to evaluate the network.

4. Results and conclusion

4.1 7 -Day compressive strength

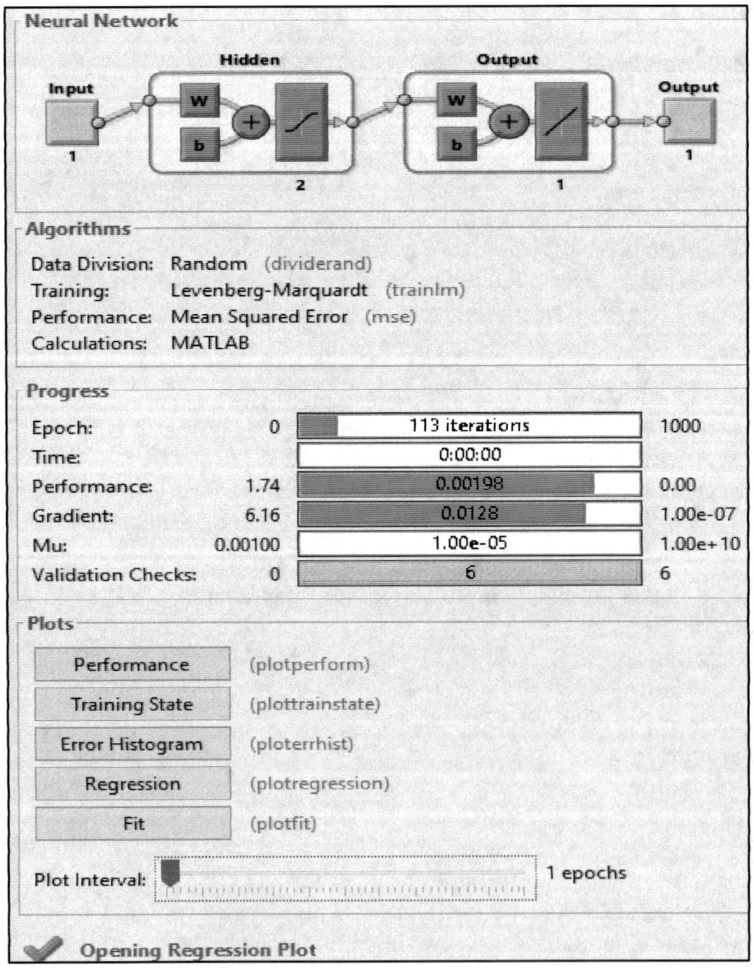

Figure 3: Neural network training

The network architecture is trained during the modeling process for the given hidden layer is shown in Figure 3 for 7 days of strength. It has been observed that training is done for 113 iterations since 113 times inputs have been trained and it has been compared with all the samples and an output is given.

IOP Conf. Series: Materials Science and Engineering 1006 (2020) 012035 doi:10.1088/1757-899X/1006/1/012035

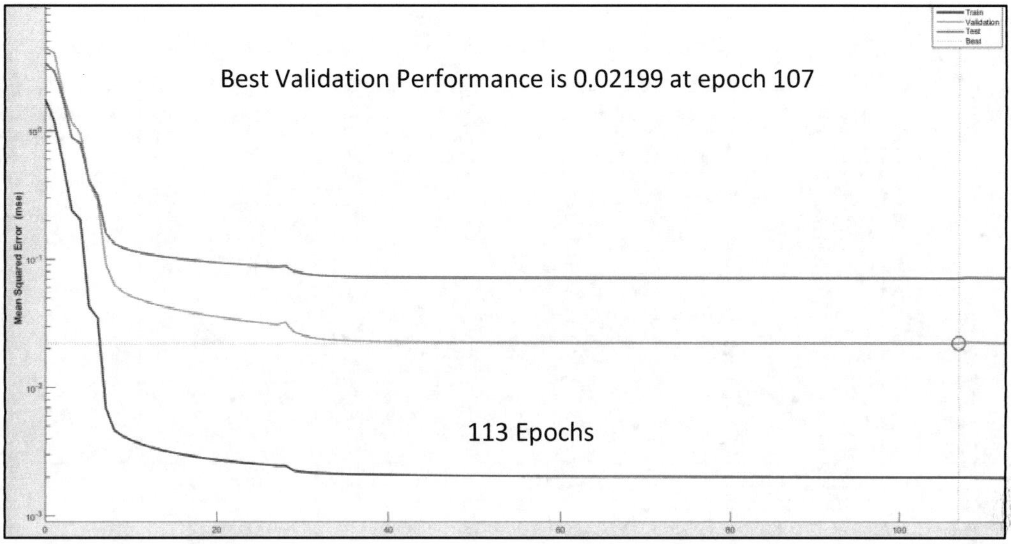

Figure 4 Best Validation Performance in Artificial Neural Network Model

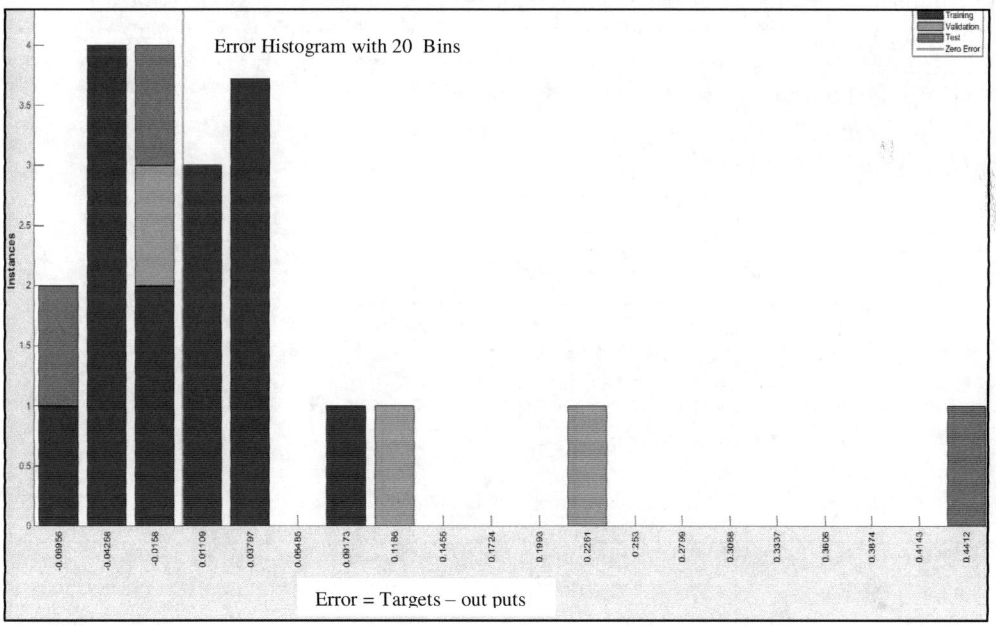

Figure 5 Error histogram for the Training, Validation, and Testing

In Figure 5 the zero error is illustrated with a yellow line in the middle with 6 instances in the training set. Zero means no error.

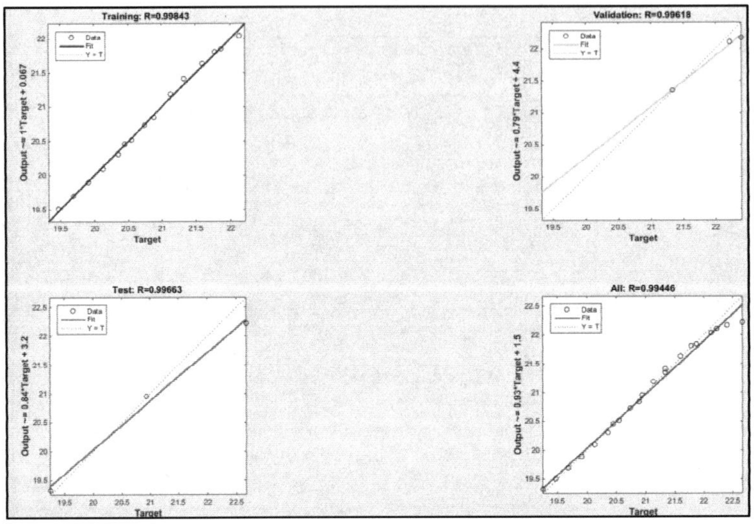

Figure 6 Training, validation, and testing set in the ANN model

Figures 6 illustrate the coincidence amidst the goal and exhibiting parameters for training, and validation steps, respectively. The objective values gives the calculated Compressive Strength and the yield values suggest the Prophesy Compressive Strength employing Matlab Software. The term R (Regression) is acquired utilizing Matlab Software which gives the version performance. Mean Squared Error (MSE) is the common squared distinction between yields and aims. An R-value of 1 means a close relationship, 0 a random relationship.

4.2 14 -Day compressive strength

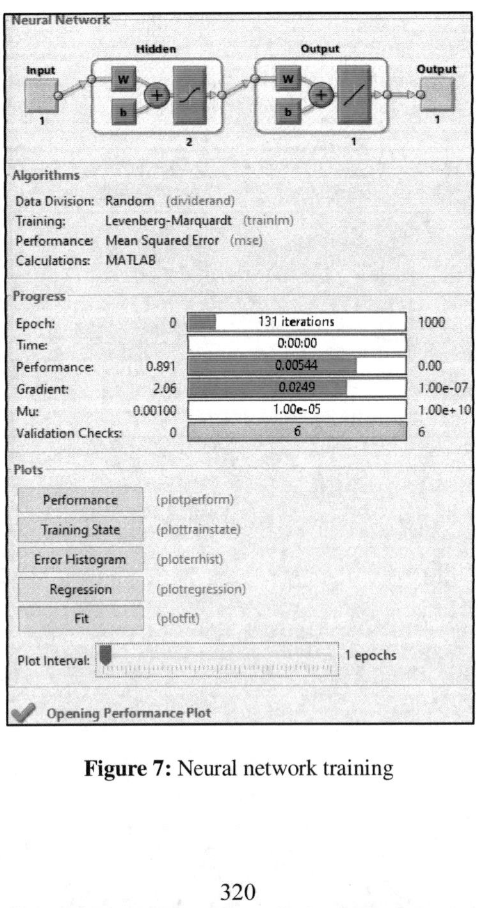

Figure 7: Neural network training

The network architecture is trained during the modeling process for the given hidden layer is shown in Figure 7 for 14 days of strength. It has been observed that training is done for 131 iterations since 131 times inputs have been trained and it has been compared with all the samples and an output is given.

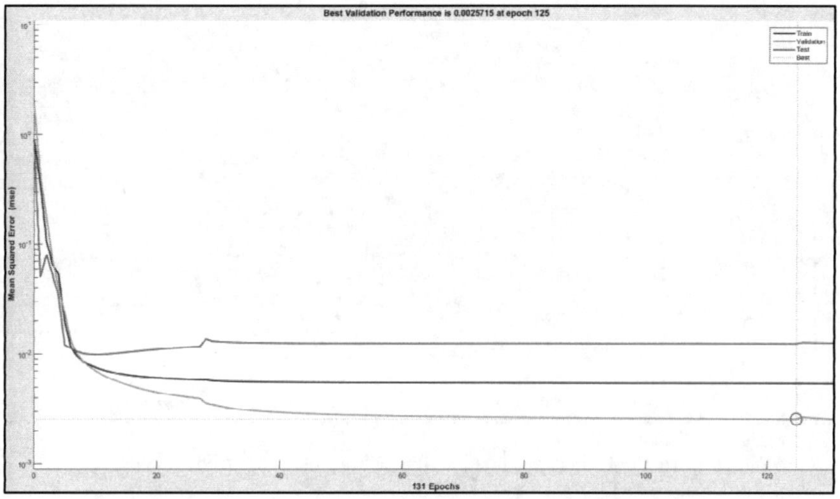

Figure 8: Best Validation Performance in Artificial Neural Network Model

Figure 8 indicating the validation role and mean squared error of the network starting at a higher value and lowering to a small value for 14 days compressive strength. As it is given in Figure 8, the best validation performance happens at epoch 131, and after 131 error repetitions, the process is stopped at epoch 131. The best validation obtained is 0.0025715 at epoch 125

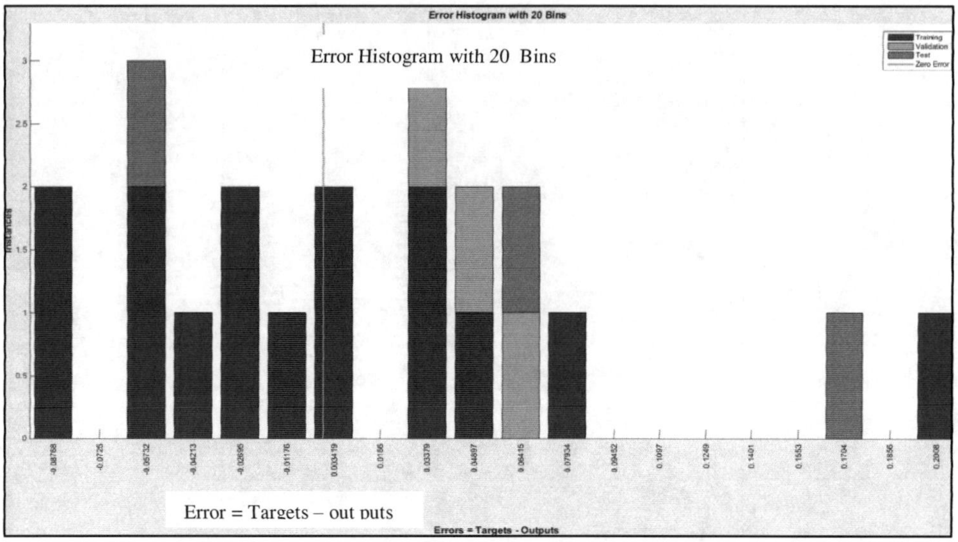

Figure 9 : Error histogram for the Training, Validation, and Testing

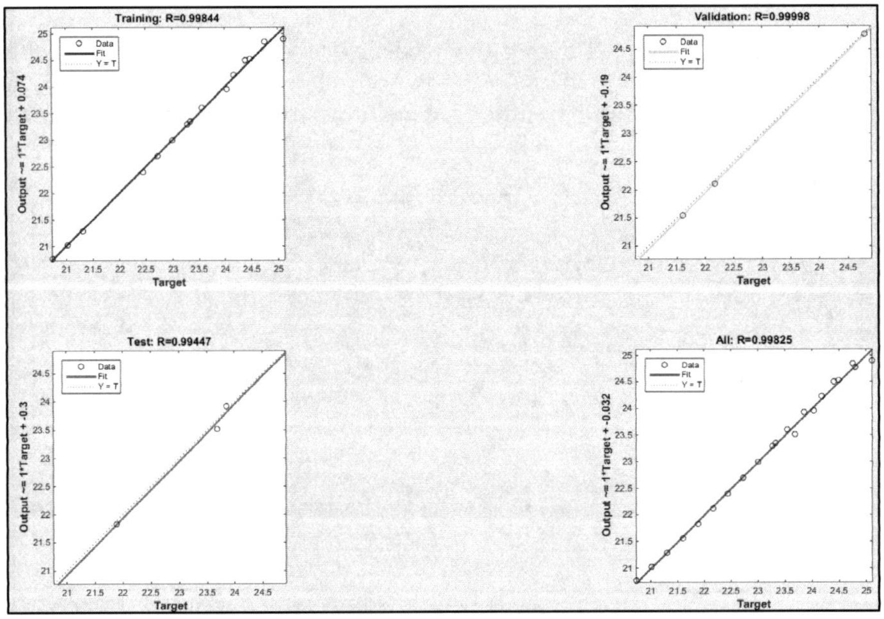

Figure 10: Training, validation, and testing set in the ANN model

4.3 28 -Day compressive strength

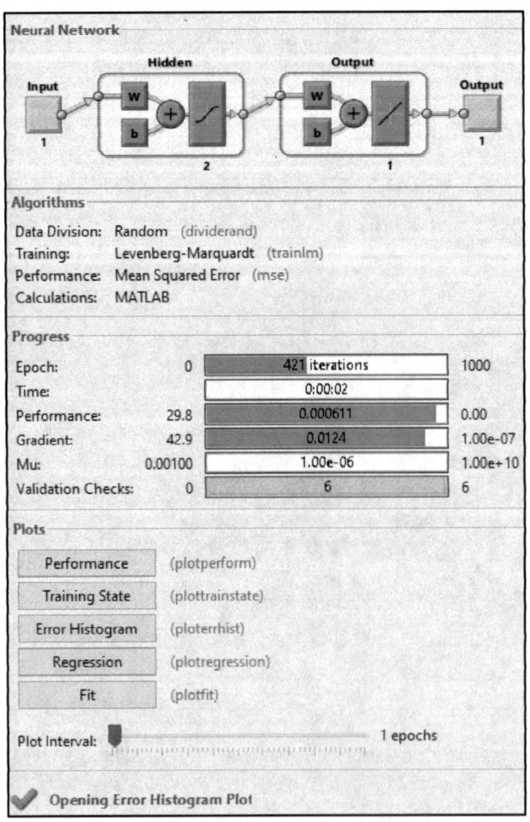

Figure 11: Neural network training

The network architecture is trained during the modeling process for the given hidden layer is shown in Figure 11 for 28 days of strength. It has been observed that training is done for 421 iterations since 421 times inputs have been trained and it has been compared with all the samples and an output is given

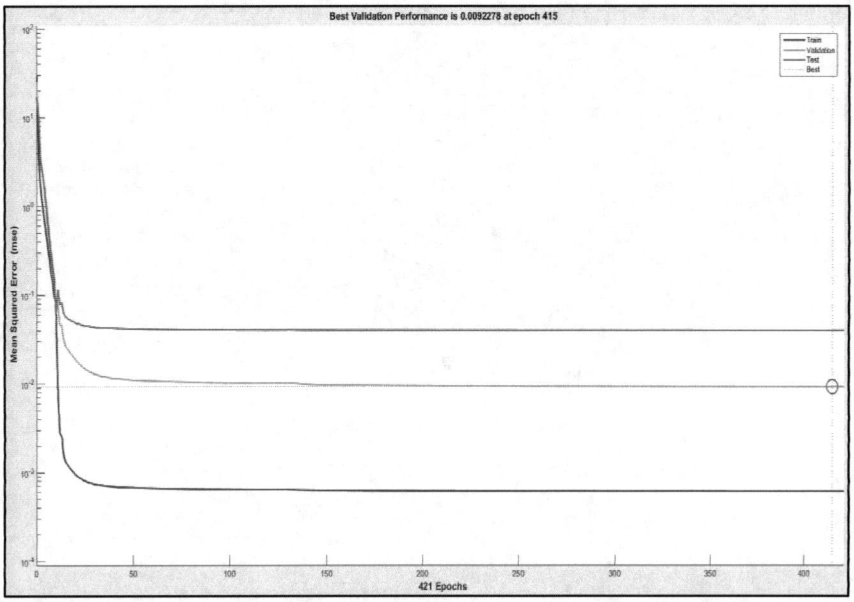

Figure 12 : Best Validation Performance in Artificial Neural Network Model

Figure 12 represents the validation performance and mean squared error of the network starting at a higher value and lowering to a small value for 28 days compressive strength. As it is given in Figure 8, the better validation production has happened at epoch 421, and after 421 error recurrence. The better validation got is 0.00922278 at epoch 415.

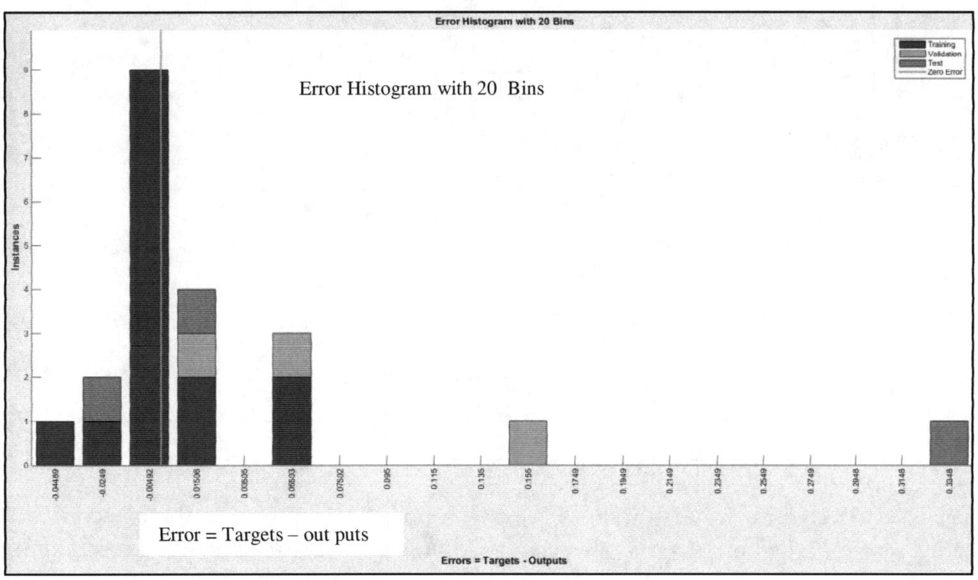

Figure 13: Error histogram for the Training, Validation, and Testing

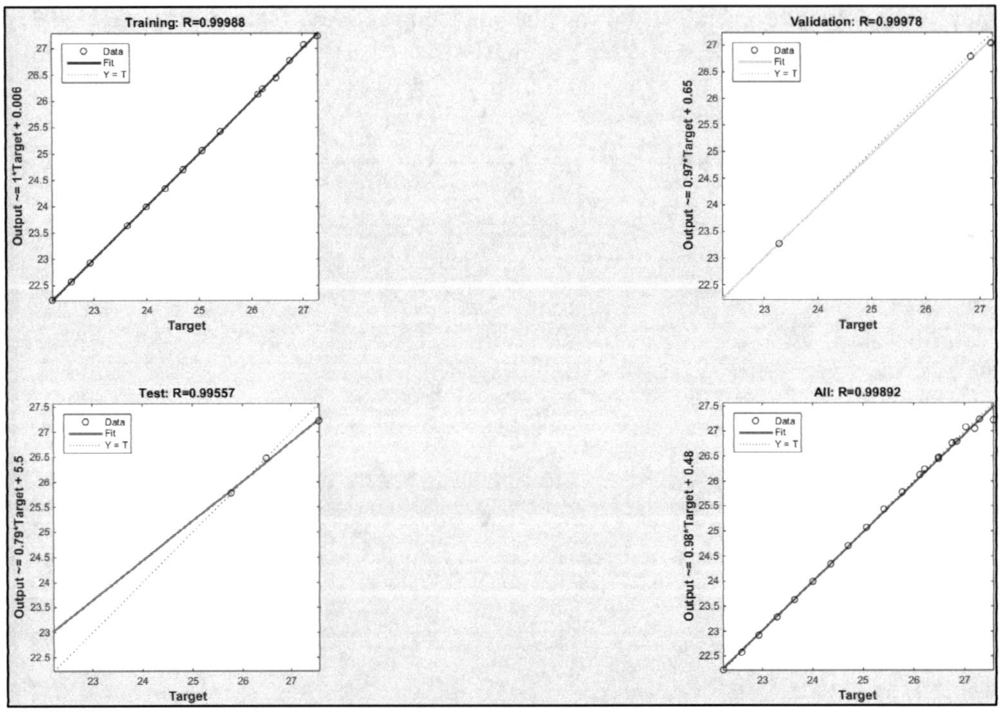

Figure 14: Training, validation, and testing set in the ANN model

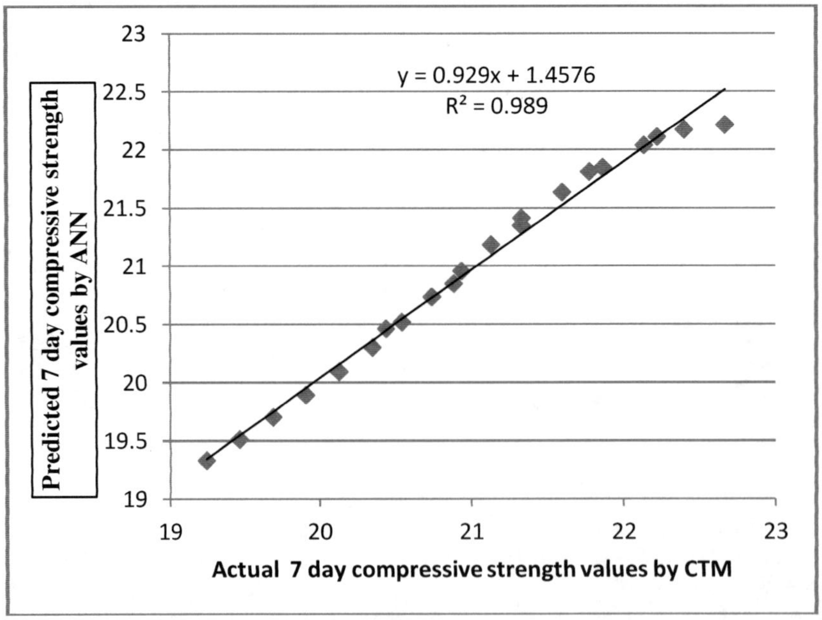

Figure 15: Comparison between the actual and predicted 7 days Compressive Strength by ANN model.

The term R^2=0.989 gives the effectiveness of the replica. The lined similarity is given by an equalization Y=0.929X +1.457 Here in this equalization X is the actual compressive strength and Y is the prophecy compressive strength. It was examine that the percentage error got the actual and envision values are low. The details of percentage error values are given in the annexure.

Therefore, the ANN is a reliable method for predicting the 7 days compressive strength of concrete and can be utilized.

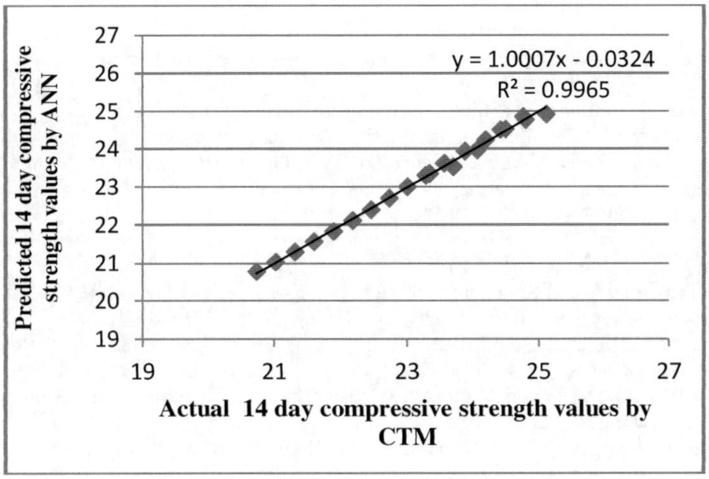

Figure 16: Comparison between the actual and predicted 14 days Compressive Strength by ANN model

The term R^2= 0.996 which gives the effectiveness of the replica. The lined similarity is given by an equalization Y= 1.000X− 0.032 Here in this equalization X is the actual compressive strength and Y is the envision compressive strength. It was examined that the percentage error got from the actual and exhibited values are low. The details of percentage error values are given in the annexure. Therefore, the ANN is a reliable method for predicting the 14 days compressive strength of concrete and can be used widely utilized.

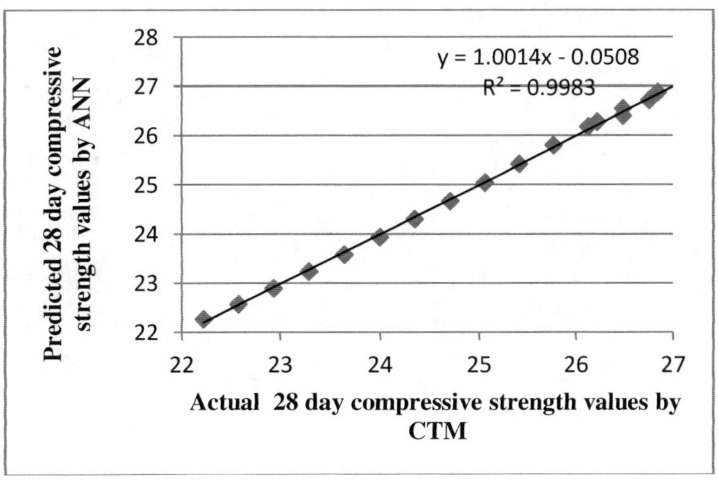

Figure 17: Comparison between the actual and predicted 28 days Compressive Strength by ANN model

The term R^2= 0.998 which gives the effectiveness of the replica. The lined similarity is given by an equalization Y= 1.001X− 0.050. Here in this equalization X is the actual compressive strength and Y is the predicted compressive strength. It was examined that the percentage error got through the actual and forecast values are low. The details of percentage error values are given in the annexure. Therefore, an ANN is a reliable method for predicting the 28 days compressive strength of concrete and can be utilized.

IOP Conf. Series: Materials Science and Engineering 1006 (2020) 012035 doi:10.1088/1757-899X/1006/1/012035

Therefore, from the top observations, it can be finally gather that the made replica is effective in the built up of the ANN replica for 7days, 14days, and 28 days compressive strength.

References

[1] Reddy D V and Pawade P Y 2012 Combine effect of silica fume and steel fiber on mechanical properties on standard grade of *Int. J. Adv. Eng. Technol.* **3** 361–6

[2] Pradhan D and Dutta D 2013 Influence of silica fume on normal concrete *Int. J. Eng. Res. Appl.* **3** 79–82

[3] Praveen G V and Shekhar L 2017 A study on characterization of cement concrete by the addition of steel fibre as an admixture *International Conference on Emerging Trends in Engineering, Science and Management* pp 240–5

[4] Singha Roy D K and Sil A 2012 Effect of partial replacement of cement by silica fume on hardened concrete *Int. J. Emerg. Technol. Adv. Eng.* **2** 472–5

[5] Parikh K and Dhyani D J 2013 Use of micro steel fibers in concrete with innovative supplementary material silica fume- A review *Paripex - Indian J. Res.* **2** 143–5

[6] Bilim C, Atiş C D, Tanyildizi H and Karahan O 2009 Predicting the compressive strength of ground granulated blast furnace slag concrete using artificial neural network *Adv. Eng. Softw.* **40** 334–40

[7] Faezehossadat K and Sayed Mohammadmehdi J 2016 Predicting the 28 days compressive strength of concrete using artificial neural network *i-manager's J. Civ. Eng.* **6** 1–6

[8] Poongodi K, Murthi P, Awoyera P O and Gobinath R 2019 Effect of mineral admixtures on early age properties of high performance concrete *IOP Conference Series: Materials Science and Engineering, ICMSMT* p 561 012067

[9] Prasanna E L, Tipraj B, Haripriya S, Khan M and Gobinath R 2019 Mechanical properties of fly ash based concrete aided with recycled aggregates and manufactured sand *Int. J. Recent Technol. Eng.* **8** 9848–51

[10] Awoyera P O, Adesina A and Gobinath R 2019 Role of recycling fine materials as filler for improving performance of concrete - a review *Aust. J. Civ. Eng.* **17** 85–95

[11] Khan I, Challa B, Haripriya Varma S and Sayyed M A A 2019 Sorptivity and durability assessment of dolomite impregnated ternary concrete *Int. J. Recent Technol. Eng.* **8** 5676–81

[12] Yadav G S, Sudarshan D S, Sahithi G and Prasanna E L 2020 Predicting the characteristics of pond ash concrete using artificial neural networks *IOP Conf. Ser. Mater. Sci. Eng.* **872** 012176

[13] Haripriya Sandela, Rajesh B and Khan M I 2020 Strength Characteristics of Recycled Aggregate Concrete by Ann *Int. J. Innov. Technol. Explor. Eng.* **9** 1210–4

IOP Conf. Series: Materials Science and Engineering 1006 (2020) 012036 doi:10.1088/1757-899X/1006/1/012036

Effect of Metallic and Non-Metallic Fibre and Recycled Aggregate on High Strength Concrete

S Sebastin[1], S M Murali Ram Kumar[2]and M Franchis David[3]

[1,2,3]Assistant Professor, National Engineering College, Kovilpatti

E-mail: ssebastin1981@gmail.com

Abstract. The most prominent building material utilized within the field of civil engineering was concrete. Among structural elements concrete have very poor scrap value, so that it cannot be used anywhere. This paper deals about the possibility of utilizing the structural waste as replacement for ingredients used in concrete. Recycled aggregate were replaced of about 20%, 30%, 40% in the coarse aggregate. In order to increase the strength of concrete, the fibres like steel fibre, polypropylene fibre and glass fibre were used. They are all added of 0.5% each from the total weight of cement. Additional admixtures like lime and silica fume were incorporated to expand the strength of the concrete. The specimens were casted to consider its mechanical properties. By comparing the consequences of various proportions the specimens having 20% of Recycled aggregate has accomplished the best compressive strength of 62.2 N/mm^2. The result is of 8.86% not exactly typical mix. This proportion also shows great outcome for split tensile strength and flexural strength (i.e., 5.10 N/mm^2 and 5.6N/mm^2). Thus by expanding the level of reused aggregate, the strength of concrete is tends to decrease.

1. Introduction

The concrete could be a compound material comprised of crushed stone aggregate reinforced along with cement paste which solidifies over the period. There are numerous sort of concrete accessible made by changing the proportion of important additives by means of using the substitution materials that having the same properties of materials that normally used in field. The wrapped up item was custom fitted to the area where the desired properties required. The strategies and innovation for creating this concrete of high strength were not considerably diverse from the ordinary conventional concrete.

W/C ratio proportion ought to be within the value of 0.3 – 0.35 or indeed it may lesser. A wide extend of aggregates can be utilized in spite of the fact that pulverized crushed aggregates (appropriately having crushing value of high) were ideal. And a few of 20%, 30%, and 40% of recycled aggregate are replaced within the coarse aggregate and in addition of admixtures like lime powder, silica fume, super plasticizer and fibre like Glass fibre, Steel fibre, Polypropylene fibre. High strength concrete was formed by reducing the w/c ratio to 0.35 and even lesser. Lower the water/binder ratio and by utilization of fine silica fume will reduced the concrete workability, it was one of the major problems while dealing with high strength concrete application where dense reinforcement cages were used. In order to compensate the decrease in workability, plasticizers were commonly included to high strength blends.

Brindhaet al. investigated the mechanical and durable property by adding the copper slag waste from copper refining industries as a partial replacement for fine aggregate. The different proportions of copper slag were being added and out of which upto 40% replacement, the concrete shows good strength. It shows good resistance to sulphate attack.

Fual Koksal et al. studied about the effect of adding the hybrid fibres on the mechanical properties of the concrete. In addition to that silica fume is also added as partial material replacement for cement. The addition of silica fume shows good results in terms of strength perspective, where the addition of fibre also good improvement in mechanical properties of concrete. However the addition of fibre affects the workability of concrete to some extent.

Manu Santhanam et al. perform the experimental investigation by trying the different combinations of metallic and non-metallic fibres and for different proportions. The addition of metallic fibres increase the impact strength of concrete and the addition of non-metallic fibres will delays the formation of cracks on structure. Out of different combinations, the steel and polypropylene fibre combination shows good results when compared to other combinations.

Vinay Kumar et al. studied about the effect of adding the recycled aggregate both in terms of fine and coarse aggregate in concrete. All the combination of four different mixes shows the satisfying results, but the addition of recycled aggregate will be somewhat reduced the workability of concrete. There will be slight decreased in compressive strength of concrete also. Wu Yao et al. studied about the effect of adding the hybrid fibres on strength of concrete, in which three different types of fibres were used. By comparing the mechanical properties of concrete having single fibre, the concrete contains two or more fibres shows good results in terms of strength.

2. Materials

Cement of OPC (53 grade) having specific gravity of 3.15 was used. The common sand with a specific gravity of 2.72 was used for fine aggregate. The coarse aggregate is of measure which retained in 12.5mm and having specific gravity of 2.78 was used. The broken concrete blocks from the demolished buildings were taken, in that coarse aggregate alone taken. The cement mortar on the aggregate surface was cleaned off. The specific gravity of recycled aggregate was found to be 2.76. The percentage of aggregate get replaced by recycled aggregate was mentioned in Table 1.

Table 1: % of Aggregate Replacement

Mix ID	Coarse Aggregate (%)	Recycled Aggregate (%)
S	100	-
S1	80	20
S2	70	30
S3	60	40

Silica fume was usually a finer round shape material having particle size < 1 micron as of diameter, normally the size may be of approximately 0.15 micron. Since of the fineness it acts as void filler and it makes a difference in maintaining a strategic distance from voids which by implication diminish the formation of cracks. The percentage by weight of cement get replaced by silica fume is mentioned in **Table 2.**

Table 2: % of cement replacement

Mix ID	Cement (%)	Silica Fume (%)
S	100	-
S1	90	10
S2	90	10
S3	90	10

The steel fibre used here having aspect ratios of about 30 - 250 and its diameters varies from 0.25 mm to 0.75 mm as shown in Figure 1. Generally compared to other fibres, glass fibres usually have high tensile strength as in the range of 1020 – 4080 N/mm^2 and length of 25mm were normally was utilized as shown in Figure 2.Polypropylene fibre was one of artificial fibre made from petrochemical and from textile industries as shown in Figure 3. Due to its high softening point and low modulus of elasticity it sort's applications in boards and shotcrete. The utilization of fine silica fume will increase the strength of concrete, but in turns decrease the workability of the concrete, so in order to maintain the workability of concrete, the super plasticizers were included to increase the strength of the mixtures.

Figure 1: Steel Fibre

Figure 2: Glass fibre

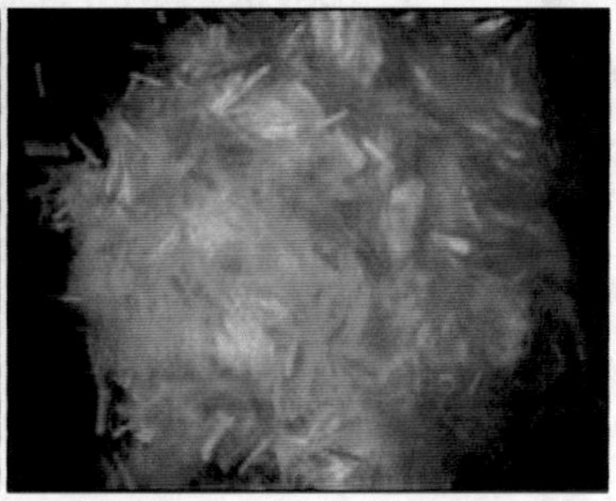

Figure 3: Polypropylene fibre

3. Mix proportions

The mix proportions used for the various concrete mix proportions were mentioned in Table 3

Table 3: Mix proportions for various combinations

ID	Cement %	Silica fume %	Sand %	Coarse Aggregate %	Recycled Aggregate %	Polypropylene Fibre %	Steel Fibre %	Glass Fibre %	Lime %
S	100	-	100	100	-	-	-	-	-
S1	90	10	100	80	20	0.5	0.5	0.5	1.5
S2	90	10	100	70	30	0.5	0.5	0.5	1.5
S3	90	10	100	60	40	0.5	0.5	0.5	1.5

4. Experimental procedure

The test samples were cast in cast-iron steel moulds of cube having size of 150 mm x 150 mm x 150 mm, cylinder having size of 150 mm diameter and 300mm height, prism having size of 100 mm width, 100 mm depth and of 500 mm length. The interior of the moulds were applied with oil to encourage the easy removal of specimens. The concrete was put within the moulds in three layers of equal thickness with appropriate tamping. After approximately for 24 hours of hardening, the samples were demoulded and were set promptly for water curing for period of 28 days as shown in Figure 4.

Figure 4: Concrete cubes after 28 days curing

The bearing surfaces of the testing machine will be cleaned off and any free sand or other material expelled from the surfaces of the specimens which are to be in contact with the compression platens. On account of cubes, the specimens will be set in the machine in such a way, that the load will be applied to inverse sides of the cubes as cast, that is and not to the top and base. The axis of specimens will be lined up with the focal point of push of the circularly situated square is brought to be in contact with the specimens, so that uniform seating might be gotten as shown in Figure 5. The load will be applied without stun and at a uniform pace until the resistance of specimens to the expanding load separates and no more loads can be supported. The ultimate load applied to the samples will at that point be recorded and used for further studies.

Figure 5: Compressive strength of concrete

5. Result and discussion

From the specimens that had been molded and cured for 28 days they were testing for the mechanical properties via compression, tension and flexure, the results were mentioned in the form of tables.

5.1 Compressive strength

The tested compressive strength results for all the test specimens were mentioned in the form of Table 4.

Table 4: Compressive strength of concrete

Mix ID	Compressive Strength (N/mm^2)
S	68.89
S1	62.20
S2	55.11
S3	49.33

From the testing results for compressive strength, it is apparently shows that there is decrease in strength of samples containing recycled aggregate in contrasted with the normal mix. This happens for the reason that the increase in addition of recycled aggregate results in decrease in strength, the reason behind was irregular shape of recycled aggregate affects the density of concrete and rough surface of aggregate absorbs large amount of water and thus it reduced its strength. The another one important factor is aggregate used here is already subjected to load, so that it can carry less load in contrast to that of normal aggregate.

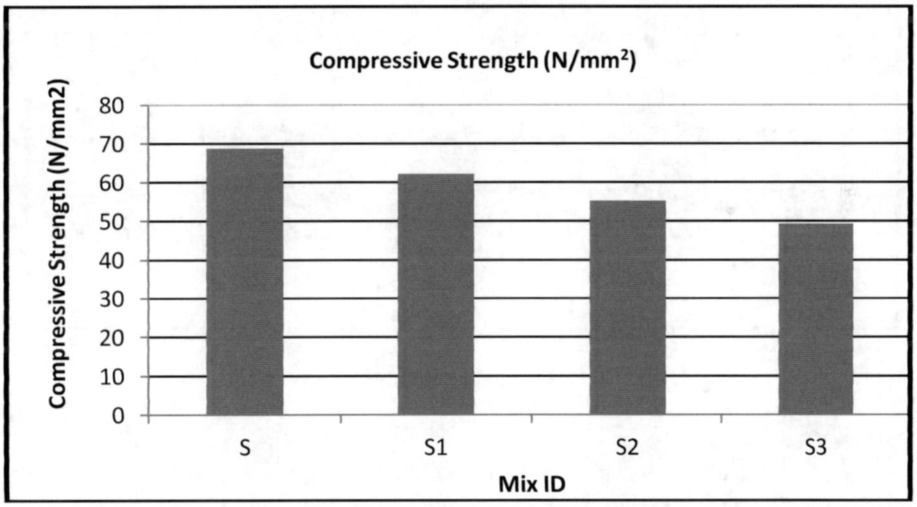

Graph 1: Compressive strength of concrete

The strength attained for 28 days were found to be about 8.86% lesser than of normal mix. The addition of silica fume 10% to the weight of cement appears to have valuable impact on improving the strength of mixes having recycled aggregate. The average 28 days strength for various mix proportions were presented in Graph 1.

5.2 Split tensile strength

The test consequences for the cylindrical samples were accounted for in Table 5 alongside their average split tensile strength.

Table 5: Split Tensile strength of concrete

Mix ID	Split Tensile Strength (N/mm^2)
S	5.51
S1	5.10
S2	4.55
S3	4.10

The split tensile strength of normal mix was viewed as higher diverged from reference to that of specimens having recycled aggregate. From Table 5, it will in general be seen that the concrete containing 20% substitution of recycled aggregate and 10% of silica fume shows strength almost like that of typical mix. Anyway for different proportions, the strength is lesser than that of typical mix. The average 28 days strength for various mix proportions were presented in Graph 2.

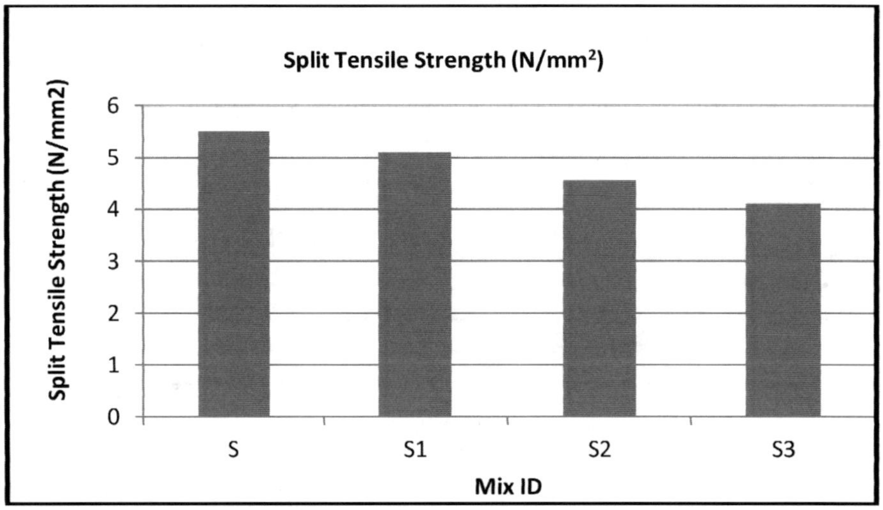

Graph 2: Split tensile strength of concrete

5.3 Flexural strength

The flexural strength results were presented in Table 6. In comparison to normal mix without recycled aggregate, all the mixes having recycled aggregate indicated a considerable decrement in flexural strength as appeared in Graph 3.

Table 6: Flexural strength of concrete

Mix ID	Flexural Strength (N/mm^2)
S	6.40
S1	5.60
S2	4.80
S3	4.40

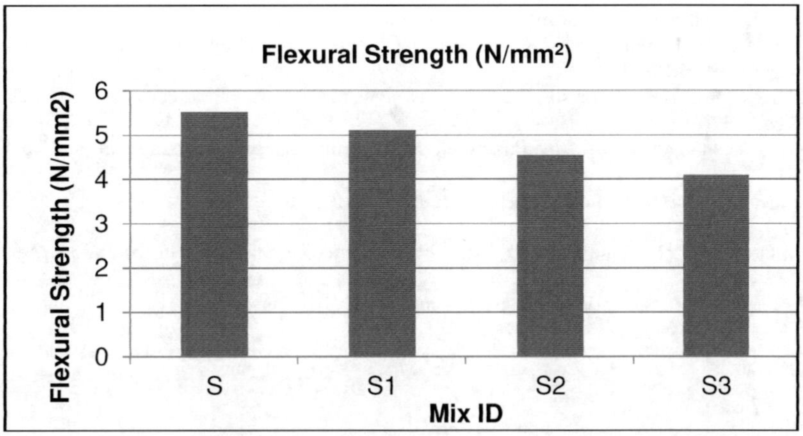

Graph 3: Flexural strength of concrete

From the over tables the proportion of S1 had accomplished the most extreme strength among the distinctive proportion specified. The greatest compressive strength of cube gotten 62.2 N/mm^2 and the most outrageous tensile strength of cylinder had gotten of 5.10 N/mm^2. By then the best flexural strength gotten is of 5.6 N/mm^2. Regardless these characteristics are all underneath the ordinary mix strength gotten.

6. Conclusion

It had been concluded that lime 1.5%, coarse aggregate 80% recycled total 20%, Silica fume 10% steel fibre 0.5%, glass fibre 0.5% ,polypropylene 0.5% accomplishes most extreme compressive strength at 28 days. It accomplishes 91.13% of the characteristic compressive strength. At that point lime 1.5%, coarse aggregate 80% recycled aggregate 20%, Silica fume 10% steel fibre 0.5%, glass fibre 0.5% ,polypropylene 0.5% achieves greatest tensile strength at 28 days. It accomplishes 92.5% that of control mix. Then lime 1.5%, coarse aggregate 80% recycled aggregate 20%, Silica fume 10% steel fibre 0.5%, glass fibre 0.5% ,polypropylene 0.5% achieves most extreme flexural strength at 28 days. Silica Fume was normally 100 times smaller in size that of ordinary binding material. Since of the fineness it acts as void filler and it makes a difference in maintaining a strategic distance from voids which by implication diminish the formation of cracks.The addition of steel fibres reduced the width of cracks and controls the further extension of crack firmly, in this way progressing toughness. Make strides that it can withstand the high impact and act as an abrasion resistance to concrete. Normally utilized for the high strength precast structures and highways, air terminal asphalts, headstrong and for canal linings etc. The glass fibres due to its high impact strength and having high expanded flexural strength increased its application over large areas. It accomplishes 100% of the flexural strength. The mechanical properties of concrete with encourage substitution recycled aggregate can be carried as a future study.

References

[1] Brindha D and Nagan S 2011 *Durability* studies on copper slag admixed concrete Vol.**12** No.5 563578
[2] Shetty M S 2010 Concrete technology – Theory and practice
[3] FualKoksalet al 2008 Combined effect of silica fume and steel fibres on the mechanical properties of high strength concrete **22** 1874–1880
[4] Manu Santhanam and Sivakumar A2007 Mechanical properties of high strength concrete reinforced with metallic and non-metallic fibre **2 9**603–608
[5] Vinay kumar BM and Ananthan H2017 Experimental studies on utilization of coarse and fine fraction of recycled concrete aggregate 88-90
[6] Wu Yaoand Jie Li 2003 Mechanical properties of hybrid fibre – reinforced concrete at low fibre volume fraction **33** 27–30
[7] Rao G A1998 Influence of silica fume replacement of cement on expansion and drying shrinkage Vol. **28** 1505–1509

[8] Rafat Siddique 2003 Performance characteristics of high-volume class F fly ash concrete 487–493
[9] Muhammad Aslam 2016 Benefits of using blended waste coarse lightweight aggregates in structural lightweight aggregate concrete **119** 108e117
[10] Elnaz Khankhaje 2016 Properties of sustainable lightweight pervious concrete containing oil palm kernel shell as coarse aggregate **126** 1054–1065
[11] Weiguoshen et al Characterization of manufactured sand: particle shape, surface texture and behavior in concrete **114** 595–601
[12] James Alexander S and Antony Godwin 2016 Study on partial replacement of fine aggregate with red soil in concrete vol**4** Issue V
[13] Paul Awoyera O and Joseph Akinmusuru O 2016 Green concrete production with ceramic wastes and laterite **117** 29–36
[14] Sudha C and Anand A 2016 Study on mechanical properties of High performance concrete using M-sand vol **9**(5)

Study of HSC by Using Fly Ash as Partial Replacement of Cement & Incorporating of Steel Fibers

Kotharu Srinivasa Rao[1] and Kuppala Srinivasa Rao[2]

[1]Associate Lecturer, Department of Civil Engineering, Sir C R Reddy Polytechnic, Eluru, Andhra Pradesh
[2]Assistant Professor, Department of Civil Engineering, Sir C R Reddy College of Engineering, Eluru, Andhra Pradesh

Email: ksrao.ce@gmail.com, seenukotharu@gmail.com

Abstract. Concrete is one in all the foremost common building materials used. The current study deals with high strength concrete by adding steel (metallic) fiber material in different volume fractions 0%, 0.3%, 0.6%, 0.9%, 1.2% and 1.5% to strengthen the engineering concrete properties and partial replacement of OPC with fly ash in percentages of 0-25% in various ratios. The specimen's exposure periods of 7, 28 and 56 days on numerous tests like compressive, flexural and split tensile strengths. The research has focused on mechanical and durability properties of high strength concrete. It was observed that 20% fly ash replacement produces the max compressive strength parameters with 1.2% crimped steel fibers, but, the split tensile and flexural strength of concrete was maximum at 25% fly ash along with 1.5% steel fibers. Furthermore, the durability of concrete has conducted such as acid attack due to HCl and H_2SO_4. The minimum loss of weight produces at the 25% fly ash is 2.98% and 2.81% by HCl and H_2SO_4.

1. Introduction

In our country, concrete business plays a serious role in the development of nation; government pay an enormous amount for building the nation like bridges, buildings and dams etc. The high strength concrete is incredibly expensive. However, it is utilized in the civil industry by substitution of different waste materials like fly ash, plastic, GGBS, E-Waste, fibers results to decrease the effective value increase the mechanical properties of concrete(Barbuta M et al. 2016). In this research work situation, the cement by fly ash and felt fibers in concrete with different percentages. We tend to success performed different experimental investigation tests for obtaining the mechanical features of high strength concrete. This type of research work protects the environment from the release of a high amount of CO2 due to making of cement production so presenting my research work carried on fly ash waste material with fibres. The manufacturing procedure for cement generates a massive quantity of CO2. Nearly 8% of these greenhouse gas emissions creates from cement production(Ahmaruzzaman M 2010). ACI 211 committee recommended that the 15-25% substitute of cement by fly ash(FA) in HSC (ACI Committee, 1993). The prosperous use of fly ash is not only going to lower the cement ingestion but in addition, eradicate the waste disposal expenses. Class F fly ash comprises a slight amount of lime. So, compressive strength falls with all the increment of fly ash material. But, as a result of pozzolanic action of fly ash, strength rise in after phases of curing (Sumer M 2012, Mohamed HA 2011). Concrete is a naturally brittle material i.e the concrete weak in tension but strong in compression, this drawback can overcome by the

IOP Conf. Series: Materials Science and Engineering 1006 (2020) 012037 doi:10.1088/1757-899X/1006/1/012037

inclusion of fibers like metallic, synthetic, glass fibers. This paper focused on fly ash along with steel fibers. The steel fibers play a vital role to improve the different properties of concrete like tensile strength and flexural strengths(Kaikea et al. 2014, Afroughsabet et al. 2015). The macro metallic steel fibers used as secondary reinforcement in concrete matrix in addition to the rebar in concrete structural and non-structural applications. They reported the enhanced the flexural strength, ductility and shear resistance in concrete (Lofgren 2005). As a result of the pozzolanic response, the strength of concrete increment lasts to get a lengthier duration of time evaluating into the concrete (Chindaprasirt et al. 2004). The reaction time of fly ash in concrete relatively requires a longer period of time to develop a greater strength. So very low compressive strength has been reported throughout the beginning phases of curing. Nevertheless, the mechanical parameters of concrete are depending upon the fly ash own physical and chemical qualities. According to (Kumar et al. 2007) found that the FA mixture of fine particle size distribution shows greater compressive strength comparing to the ordinary blend mix. The more proportion dose of flyash (FA) the greater the standards of concrete compressive ability until finally 30% of FA from treated in potable water up to 28days and also the smallest results observed to be specimens were cured in the air (Vinayagam 2012).Even the workability of concrete blends rises has been increased by the dose of fly ash 0-40%. The replacement-level of cement concrete can be raised to more than 10% since there has been an increase at compressive strength for 20-40% FA content material. The substitute material sample listed that the Maximum permeability, whereas the 40% FA sample attained exactly the less permeability (Bahedh et al. 2018). The research work done flexural Strength evaluation on standard size beams. A definite mixture M20 has designed depending on IS 10262:2009 guidelines. The water-cement ratio has maintained as 0.48. The Flexural strength evaluation was conducted at a regular interval. It had been concluded that the 28thday flexural Energy of concrete was raised up to 11.1% with 20% replacement-level of Portland cement by fly ash (Parmar et al).T.C Yet et al reported that the static and dynamic properties of concrete could be improved by the combined effect of steel fibers and fly ash (T.C yet et al. 2012).

2. Materials used

2.1 Cement

OPC 53Grade is available in India in 3-types-33, 43 and 53. The experimental investigation was done on ultra tech 53 Grade OPC confirming to IS:12269-2013. This brand and has maintained throughout project work. It is available in the locality of Eluru. The properties of cement were illustrated in table.1

Table 1: The physical parameter of cement

S.No	Physical Properties	Results of after performing the test
1	Fineness	Retained- 5.0% & Fineness-95.0%
2	Specific gravity	3.14
3	Setting times(Initial & Final)	43min and 535min
4	Soundness of cement(mm)	5
5	Consistency of cement	32.0%
6	Compressive Strength	54.84N/mm^2

2.2 Fine aggregate

In this study, locally available river sand can be used and confirming to IS:383-1970. The various fine aggregates tests were performed and shown in Table.2. The fine aggregates are free from silt, and clay was used for the investigation.

Table 2: Physical parameters of fine aggregate

S.No	Physical Properties	Results of FA after the test
1	Particle Size	FA <4.75mm & and Zone of sand is **II**
2	Fineness Modulus	2.61
3	Rodded Bulk density	1624Kg/m^3
4	Water absorption	1.05%
5	Specific gravity(G)	2.71

2.3 Coarse aggregate
The crushed stones of angular shaped aggregates were used and confirming to IS:383-1970. The aggregates have been from a locally available quarry and satisfy the standard Indian provisions. The coarse aggregates properties have found out by using the different tests and results illustrated in Table.3

Table 3: Physical properties of Coarse aggregate

S.No	Physical Properties	Results of CA after the test
1	Size of Aggregate	10mm and 20mm
2	Fineness Modulus(FM)	7.10
3	Rodded Bulk density	1599Kg/m^3
4	Water absorption	1.15%
5	Specific gravity(G)	2.67

2.4 Fly ash
According to the type of coal used for combustion, ASTM C 618 widely categorized fly ash into two different types. Inside my analysis class F fly ash can be used and having a specific gravity of 2.2, fineness modulus (FM) is 2.5 and colour is whitish-grey, in this current study, fly ash in NTTPS, Vijayawada can be used.

2.5 Crimped steel fiber
Steel fibers are defined as short, discrete length of steel having an aspect ratio (AR) from 20 to 100 with any cross-section. Aspect ratio means the ratio of length to diameter. The current study adopted an aspect-ratio 50 inside this undertaking. Within this endeavor, we utilize the fiber for increasing the strength of concrete. Crimped steel fibers are have been put into concrete to boost properties such as tensile and flexural strength. The iron content increases the flexural strength up to 25-100% depending upon the ratio of fiber included.

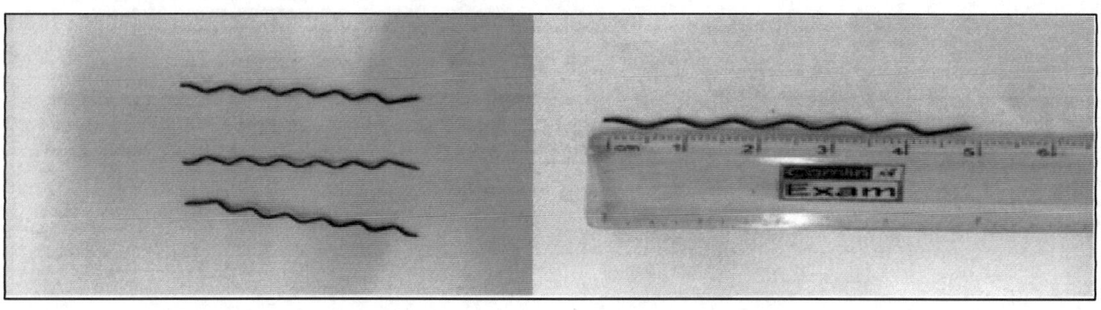

Fig.1. Crimped steel fibers with 1mm diameter & 50mm length.

Table 4: Mechanical properties of steel fibers

Type of fiber	**Shape of fiber**	**Length l (mm)**	**Diameter d (mm)**	**Aspect ratio (l/d)**	**Desity (Kg/m3)**	**Tensile strength (MPa)**
Steel	crimped	50	1	50	7.80	1098

2.6 Water

Portable drinking water was used in this work for all the purpose like mix design as well as curing.

3. Tests conducted

3.1 Compressive strength

The compressive strength of concrete is determined by using the CTM (compression testing machine) with a capacity of 3000KN depicted in fig. strength of the concrete found by using the specimens of size 150mm x 150mm x 150mm. further applied the uniform speed of 140kg/sq.cm/minute till the specimen failed. The machinery outcomes tabulated. Strength of the concrete can find out by using the following formula.

$$\text{Compressive Strength} = \frac{\text{load (P)}}{\text{cross} - \text{sectional area (A)}}$$

Fig.2. Compression Testing Machine of Capacity 3000KN.

3.2 Flexure strength of concrete

The flexural testing machine is shown in fig.3 used to determine the flexural strength of concrete. The failure of plain and blended mixes bending effect on beams of size 100mm × 100mm × 500mm. The specimens test at room temperature with a regular curing period of 7 and 28days. These prisms were tested four-point loading flexural testing machine after completion of specified curing time. The rate of load is applied uniformly with a face rate of 1.8kN/min till the Specimens break as per "IS 516-1959". The flexural strength evaluated by the following formula.

$$\text{Flexural Strength} = \frac{Pl}{bd^2} \left(\frac{N}{mm^2}\right)$$

Fig.3: Flexure Testing Machine in front and side view at the time of testing

3.3 Split tensile strength of concrete

It is an indirect test used to evaluate the tensile strength in concrete and uniformly apply the load axially. The concrete specimens in cylindrical shape of size 300mm(height)×150mm(diameter) was used to determine the splitting tensile strength. The samples are tested after a certain period of curing under the CTM at the age of 7, 28 & 56 days. The load is applied gradually at the loading rate of 1.2 MPa/min until the Specimens will be split/fail according to IS 516-1959.

$$\text{Split tensile Strength} = \frac{2P}{\pi LD} \left(\frac{N}{mm^2}\right)$$

4. Mix design

Mix design, depending on IS 10262-2009[18], the concrete mixture was prepared for high strength concrete with water-cement ratio of 0.32. The ratios from the weight of cement, fine aggregate and well-graded aggregate are accessed utilizing the Department of Environment (DOE) system is listed below. DOE method currently is your standard British procedure of concrete mixture design the task involved with this approach is clarified alternatively of outside dated Road Notice no 4 procedure [14]. These proportions are kept rigorously the same through the entire casting process to get a uniform standard and flowability concrete mixture. The compressive, split up tensile and flexure evaluations are investigated. Workability evaluations will also be done in this mixture. The mix proportions are shown in the table.5

Table 5: The mix proportions for concrete with and without replacement of fly ash.

S.No	% of fly ash	Cement	Fly ash	Fine aggregate	Coarse aggregate
1	0	1.00	0	1.04	1.57
2	5	0.95	0.05	1.13	1.69
3	10	0.90	0.10	1.23	1.82
4	15	0.85	0.15	1.35	1.95
5	20	0.80	0.20	1.46	2.06
6	25	0.75	0.25	1.58	2.19

5. Results and discussion

5.1 Workability test

The workability of concrete measure with the help of the slump cone apparatus. The slump range selected for this test is 90 to 110 mm. The workability of concrete can increase when using the fly ash as a mineral admixture with a water-cement ratio of 0.32.

5.2 Mechanical properties

Table 6: Compressive strength of cubes rate up to 56 days at different fly ash percentages 0-25%.

% of Replacement	Fiber dose %	7-d (MPa)	28-d (MPa)	56-d (MPa)
0	0	38.57	65.25	67.98
5	0.3	36.50	56.47	58.57
10	0.6.	34.10	60.45	62.65
15	0.9	32.33	64.32	66.02
20	1.2	31.58	68.12	70.32
25	1.5	29.51	66.22	68.72

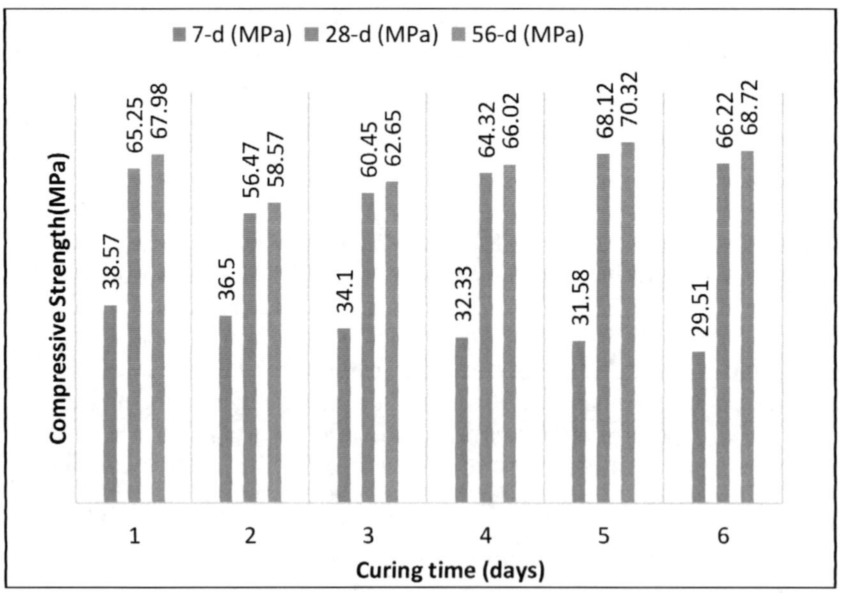

Fig.4 Compressive Strength of Concrete cubes using Fly ash and steel fibers

The compressive strength of concrete has been measured by the curing age of 7-d, 28-d and 56-d as shown in fig.4 At first, the conventional mix shows the compressive strength of 38.57MPa, 65.25MPa and 67.9Mpa at 7days,28days and 56days respectively. The graph indicates that the 7 days strength adopted the decreased trend. But, the 28 days of curing shows superior performance at optimum dosage of fly ash at 20% replacement along with crimped steel fibers of 1.2% volume are 68.12MPa which is better compared to the control mix. Even the excess dosage of fly ash beyond 20% also yielded satisfactory results compared to conventional mix properties at 28 days ageing. In the same way, the compressive strength of concrete blocks cured for 56 days had shown the maximum strength at 20% of fly ash and 1.2% of steel fiber, and it also satisfies the conventional mix properties.

Table7 : Split tensile strength of cylinders rate up to 56 days at different fly ash percentages 0-25%

S.No	% of Replacement	Fiber dose %	7-d (MPa)	28-d (MPa)	56-d (MPa)
1	0	0.0	3.30	5.98	6.87
2	5	0.3	2.61	5.15	5.81
3	10	0.6.	3.01	5.48	6.24
4	15	0.9	3.32	5.81	6.32
5	20	1.2	3.71	6.28	6.88
6	25	1.5	4.10	6.61	7.32

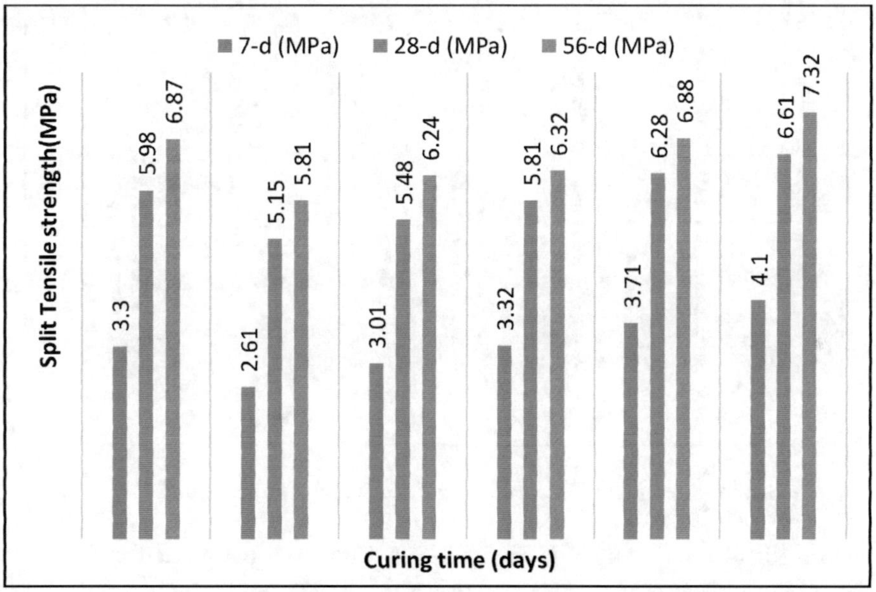

Fig.5 Split tensile Strength of Concrete using Fly ash and Steel fibers

Split tensile strength of concrete examines by compression testing machine by compressive load until it breaks. As per the experimental investigation of the split tensile strength of concrete, the trend of the results has significantly increased from 0.3, 0.6, 0.9, 1.2 & 1.5% volume fraction of rounded crimped steel fibers with fly ash of 5%, 10%, 15%, 20% and 25% replacements respectively for the curing period of 7, 28 and 56 days. The slight longitudinal hairline cracks (1mm) observed because of the crimped steel fibers exhibited the bridging action between the concrete matrix. Fig.5 depicted the split tensile strength adopted a gradually increasing trend for all the replacing percentages of fly ash along with incorporated steel fibers. The maximum split tensile strength was 4.10MPa at the age of curing for 7 days. It shows the upward trend, which means it gets increased with the addition of fly ash and round crimped steel fibers and it occurs at 25% of fly ash and 1.5% of steel fibers. The same pattern is followed by the remaining two curing ages that are 28 days and 56 days.

Table 8: Flexural strength of beams rate up to 56 days at different percentages of fly ash 0-25%.

% of Replacement	Fiber dose %	7-d (MPa)	28-d (MPa)	56-d (MPa)
0	0	4.45	6.96	7.21
5	0.3	3.74	6.46	6.88
10	0.6.	4.02	6.78	7.24
15	0.9	4.51	7.17	7.56
20	1.2	4.96	7.52	7.94
25	1.5	5.10	7.87	8.21

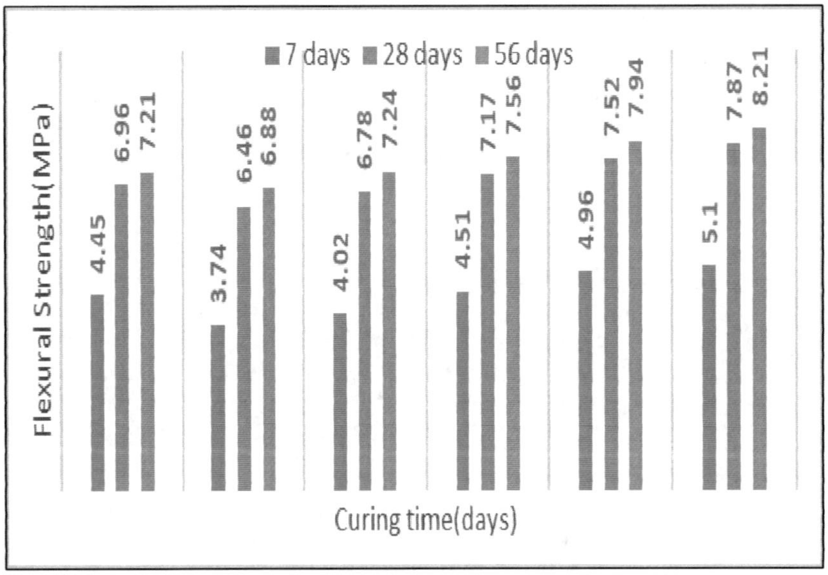

Fig.6 Flexure Strength of Concrete using Fly ash and crimped rounded steel fibers

The fibre-reinforced flexural strength of concrete was found out and plotted in fig. about 7, 28 and 56 days. The results indicated that the static flexural strength could significantly improve by the incorporation of round crimped steel fibers with a volume fraction of 0-1.5% along with fly ash. The flexural strength of concrete depicted in fig .6,which indicatesan increasing trend with the respective curing age of 7, 28 and 56 days. The highest flexural strength in curing age of 7 days was 5.1MPa at 25% of fly ash along with 1.5% of steel fiber volume in the same way the maximum flexural strength for curing ages of 28 days and 56 days were 7.87 MPa and 8.21 MPa at 25% of fly ash using 1.5% of steel fibers.

5.3 Water absorption test on concrete

Water absorption is one of the main factors effecting the durability of concrete. The water absorption test has performed on the cubical specimens (150×150×150) mm size. The water absorption percentage has a measure of porous material which can be allowed water from one layer of concrete to another layer, it corroding the embedded materials of fibers/steel/reinforcement in concrete due to water bringing the dangerous chemicals into the concrete. So this is one of the most severe problems in concrete. Initially, the samples collected after completion of curing period of 28days. Determine the standard cubical weight as W1; further, the specimens are dried in an oven at a temperature of nearly 100±5°C. this drying process has been continued until the alteration between the two consecutive weights becomes constant. The final weight of oven-dried specimens weighed after it reaches room temperature (W2). The formula used to found the water absorption of concrete is given below.

$$\text{Water absorption}(\%) = \frac{\text{W1} - \text{W2}}{\text{W2}} \times 100$$

The fly ash based concrete satisfy the conventional mix properties at proportional at 20% and 25% of fly ash. So the water absorption test is considered for only two dosages.

Table 9: The water absorption was conducted on plain concrete and fly ash along with crimped steel fibers

Mix id.	Weight before oven (Kg)	Weight after oven (kg)	% water absorption
0%Fly ash+0% SF	8.401	8.045	4.23
20%Fly ash+1.2%SF	8.656	8.410	2.84
25%Fly ash+1.5%SF	8.678	8.435	2.80

Table.9 reported the lower water value as 2.84, 2.80 at 20% and 25% fly ash respectively along with rounded crimped steel fibers. As per the results increase of fly ash from 20-25% more resistance to the water absorption of concrete.

5.4 Acid attack test

The concrete cubical blocks starting from the traditional concrete mix to the concrete mix combined with crimped steel fibers and fly ash in increased proportions like 0.3-1.5% and 0-25% respectively. The concrete may be deteriorated in the HCl and H_2SO_4 with 5% concentration. The results illustrated in table.9 were gradually decreased, the final result resolved after 28days. The necessary precaution followed was to dry the samples to avoid water content. After 28 days of 5% (HCl &H_2SO_4) the results as 2.98%, and 2.81% regarding loss in weight. Similarly, 15.13% and 15.10% considering the loss in compressive strength (25% fly ash +1.5% crimped steel fibers).

Table 10: Results of "Acid Attack (HCL)" at conventional and 0-25% fly ash along with steel fibers

Mix id.	Initial weight before oven (Kg)	weight after oven HCL attack(Kg)	% Loss in weight	% loss in Compressive strength
0%Fly ash+0% SF	8.32	8.01	3.87	28.31
5%Fly ash+0.3%SF	8.38	8.07	3.80	25.12
10%Fly ash+0.6%SF	8.42	8.13	3.56	22.13
15%Fly ash+0.9%SF	8.55	8.27	3.38	19.45
20%Fly ash+1.2%SF	8.63	8.37	3.10	16.07
25%Fly ash+1.5% SF	8.71	8.46	2.98	15.13

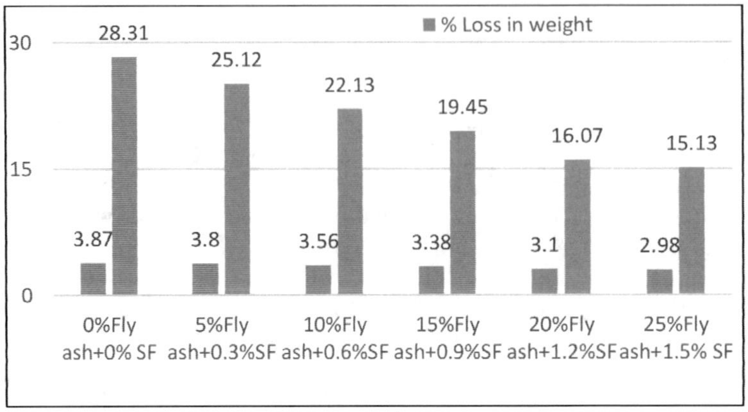

Fig.7: Acid attack (HCl) test for 0-25% fly ash along with steel fibers

Let the results to be discussed here basically on adding HCL to the concrete mix there is some amount reduction in weight and compressive strength of the samples. It is necessary to know the loss in weight and compressive strength, therefore the addition of 5% of HCL to the samples containing fly ash and steel fibers in quantities like 0%-25% and 0%-1.5% respectively. 3.87% and 28.31% loss by weight and compressive strength for traditional mix sample after that there is a significant decrease identified in both weight and compressive strength. They have approximately common decrement from 5% fly ash + 0.3% fibers to 20% fly ash + 1.5% fibers. But there is the nominal change in the loss of weight, between 20% fly ash and 25% fly ash were 3.10 and 2.98%

Table 11: Results of "Acid Attack(H_2SO_4)" at conventional and 0-25% Fly ash along steel fibers.

Mix id.	Initial weight before oven (Kg)	weight after oven HCL attack (Kg)	% Loss in weight	% loss in Compressive strength
0%Fly ash+0% SF	8.38	8.10	3.45	26.77
5%Fly ash+0.3%SF	8.49	8.22	3.28	24.31
10%Fly ash+0.6%SF	8.52	8.26	3.14	22.09
15%Fly ash+0.9%SF	8.60	8.35	2.99	18.89
20%Fly ash+1.2%SF	8.72	8.48	2.83	15.99
25%Fly ash+1.5% SF	8.76	8.52	2.81	15.10

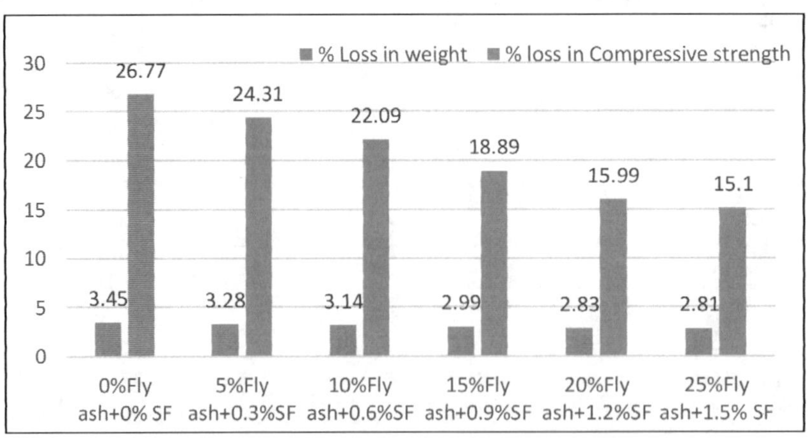

Fig.8 Acid attack (H_2SO_4)test for 0-25% fly ash along with steel fibers

In the same way, the acid attack caused by H_2SO_4 was also same as HCL, it follows the decreased trend, and the loss in weight was 3.45%, and loss in compressive strength is 26.77% for the conventional mix, and it shows a nominal change between 20% fly ash + 1.2% steel fibers and 25% fly ash +1.5% steel fibers. The comparison between the results obtained by the addition of HCL and H_2SO_4, there is a nominal change in weight and compressive strength occurred.

6. Conclusions

- The workability of concrete has increased by the addition of fly ash content from 0-25% with a water cement ratio of 0.32. when the fiber is used at 1.2% the target strength of concrete is reached at 20% fly ash compared to control mix, in addition, to get a good improvement in workability.
- The superior performance observed at 20% optimum of fly ash along round crimped steel fibers 1.2% dosage. The results indicated that the compressive strength of concrete is increasing gradually from 5-20% then starts to decrease.
- The early age strength of concrete decreased from 5-25% dose, but, the strength has significantly increased from 5-20% then start to decrease at 25% fly ash. The maximum compressive strength obtained at 20% optimum fly ash as 68.12MPa at 28days than conventional mix.
- The flexural and split tensile strength of the concrete increased by using rounded crimped steel fiber content of 0-1.5%. Engineering concrete properties like split and flexural strengths observed to rise for 7 and 28 days gradually, but particularly less improvement after a curing period of 56days. It is noticed that hairline cracks are formed in split tensile test longitudinally. Maximum all the mechanical properties of concrete is continuously increased for a curing period of 56 days.
- The more water resistance observed at 20 and 25% shown in table.9 The lower water resistance value as 2.92, 2.88 at 20% and 25% fly ash respectively along with fibers.
- The fly ash based concrete shows more resistance to the effects of acid attack plotted in fig.7 & 8.

References

[1] Ahmaruzzaman M 2010 A review on the utilization of fly ash *Progress in Energy and Combustion Science* **36** 327–63. https://doi.org/10.1016/j.pecs.2009.11.003.

[2] ACI Committee Alonzo: Guide for selecting proportions for high-strength - Google Scholar. https://scholar.google.com/scholar_lookup?title=Guide%20for%20selecting%20proportions%20for%20high-strength%20concrete%20with%20Portland%20cement%20and%20fly%20ash&author=ACI%20Committee%20211&publication_year=1993&pages=272-283 (accessed April 1, 2020).

[3] Barbuta M, Rujanu M, Nicuta A 2016 Characterization of polymer concrete with different wastes additions *Procedia Technology* **22** 407–12. https://doi.org/10.1016/j.protcy.2016.01.069.

[4] Sumer M 2012 Compressive strength and sulfate resistance properties of concretes containing Class F and Class C fly ashes *Construction and Building Materials* **34** 531–6. https://doi.org/10.1016/j.conbuildmat.2012.02.023.

[5] Mohamed HA 2011 Effect of fly ash and silica fume on compressive strength of self-compacting concrete under different curing conditions *Ain Shams Engineering Journal* **2** 79–86. https://doi.org/10.1016/j.asej.2011.06.001.

[6] Chindaprasirt P, Jaturapitakkul C, Sinsiri T 2005 Effect of fly ash fineness on compressive strength and pore size of blended cement paste *Cement and Concrete Composites* **27** 425–8. https://doi.org/10.1016/j.cemconcomp.2004.07.003.

[7] Kumar A, Kisku N 2007 Effect of silica fume and fly ash as partial replacement of cement on strength of concrete 5:7.

[8] P Vinayagam 2012 Experimental investigation on high performance concrete using silica fume and super plasticizer *International journal of computer and communication engineering* vol. **1** No.2 - Google Search n.d.

[9] Pitroda J 2013 Flexural strength of beams by partial replacement of cement with fly ash and hypo sludge in concrete 2:7.

[10] Parmar NR, Shah MS High Strength Concrete With Pond-Ash as Partially Replaced by Fine Aggregate and Fine Fly-Ash ,Alccofine as Cement n.d.:7.

[11] Siddique R 2004 Performance characteristics of high-volume Class F fly ash concrete *Cement and Concrete Research* **34** 487–93. https://doi.org/10.1016/j.cemconres.2003.09.002.

[12] Saha AK, Sarker PK 2017 Sustainable use of ferronickel slag fine aggregate and fly ash in structural concrete: Mechanical properties and leaching study *Journal of Cleaner Production* **162** 438–48. https://doi.org/10.1016/j.jclepro.2017.06.035.
[13] Bahedh MA, Jaafar MS 2018 Ultra high-performance concrete utilizing fly ash as cement replacement under autoclaving technique *Case Studies in Construction Materials* **9** e00202. https://doi.org/10.1016/j.cscm.2018.e00202.
[14] Löfgren I 2005 Fibre-reinforced concrete for industrial construction-A fracture mechanics approach to material testing and structural analysis *Chalmers University of Technology*.
[15] Kaïkea A, Achoura D, Duplan F & Rizzuti L 2014 Effect of mineral admixtures and steel fiber volume contents on the behavior of high performance fiber reinforced concrete *Materials & Design* **63** 493-499.
[16] Afroughsabet V & Ozbakkaloglu T 2015 Mechanical and durability properties of high strength concrete containing steel and polypropylene fibers *Construction and building materials* **94** 73-82.
[17] Chien Yet, Tan, R. Hamid, and Mudiono Kasmuri 2012 Dynamic Stress-Strain Behaviour of Steel Fiber Reinforced High-Performance Concrete with Fly Ash *Advances in Civil Engineering* 1–6.
[18] M. S. Shetty & A K Jain, Concrete Technology: Theory and Practice, S. Chand Publishing, 9789352533800, 2018.
[19] IS: 456-2000 Indian standard Specification for plain and reinforced concrete – code of practice. (Fourth revision), BIS, New Delhi.
[19]ASTM C989-99 Standard Specification for Ground Granulated Blast-Furnace Slag for Use in Concrete and Mortars.
[20] IS: 9103-1999 Specification for Concrete Admixtures" Bureau of Indian Standards, New Delhi, India.
[21] IS: 10262-2009 Guidelines for Concrete Mix Design", Bureau of Indian Standards, New Delhi, India.
[22] IS 12269-1987 Specification for 53 Grade OPC.
[23] IS: 516-1959 Indian Standard Methods of Test for Strength of concrete. Bureau of Indian Standards, New Delhi.
[24] IS: 2386-1963 Methods of Test for aggregates for concrete - Part 3: Specific gravity, Density, Voids, Absorption and Bulking, Bureau of Indian Standard, New Delhi.
[25] IS: 5816: 1999 Methods of tests for splitting tensile strength of concrete.
[26] IS: 383 – 1970 Indian standard specification for coarse & fine aggregates from natural sources for concrete, BIS, New Delhi.
[27] Indian standard IS:5816 Spliting Tensile Strength of concrete-Method of test,2004, New Delhi
[28] Indian standard IS:9013 Admixture- specification,2004, New Dehli
[29] Indian standard code on methods of test for aggregates for concrete IS 2368 Part 1-1963, Bureau of Indian standards, New Delhi.

IOP Conf. Series: Materials Science and Engineering 1006 (2020) 012038 doi:10.1088/1757-899X/1006/1/012038

Effects of Admixtures on the Self Compacting Concrete-State of the Art Report

S Christopher Gnanaraj[1], Ramesh Babu Chokkalingam[2], G LiziaThankam[3]

[1]Assistant Professor, Civil Engineering, KARE, virudhunagar, India
[2]Associate Professor and Head, KARE, virudhunagar, India
[3]Research Scholar, KARE, virudhunagar, India

E-mail: s.christophergnanaraj@klu.ac.in

Abstract. In concrete, many properties influence the strength parameters, among that most important parameter is proper compaction. Complete compaction is not possible using conventional concrete with hand compaction or machine vibration in structural element like long column and beam column joints. In these cases, self-compacting concrete (SCC) plays a vital role. It gives full compaction without any external effort. This type of concrete compacts by its own weight, another important advantage with SCC is the absence of segregation. Hence SCC usage has increased enormously in the last few decades. Having lot of merits SCC also has few demerits. One important demerit is that, to prepare SCC lot of cement content is needed compare to conventional concrete; this may increase the CO2 emission also it leads to higher heat of hydration this may cause shrinkage cracks if adequate curing is not provided. This study is an attempt to review the influence of different types of mineral admixtures and chemical admixtures used in self compacting concrete along with a brief knowledge about the change in mechanical behaviour with respect to the influence of these admixtures.

1. Introduction

Concrete, being one of most extensively used construction material because of its durability and cost effectiveness, however, concrete structure has some demerits also like weak tensile force, emitting more amount of CO_2, optimum strength will not obtain if it is not fully compacted and properly cured. To overcome this issues lot of research taking place to enhance the specialty of concrete. One of such research ended up with a new innovation called SCC in which full compaction will be achieved without any external effect. SCC has high workability which makes it pump able and SCC have high resistance against segregation. Chemical admixtures such as super plasticizer and viscosity modifying agent enhances the workability of concrete by minimizing the segregation and bleeding in fresh concrete. The chemical admixtures also improves the pumpability of the freshly prepared concrete and thereby mitigating the water to cement ratio, which results in highly impermeable concrete paving way for the overall resistance[1]. Mineral admixtures like fly ash and silica fume improves the compressive strength and flexural strength [2]. The yielding stress of a concrete is elevated with silica fume up to 16%, where as the plastic viscosity decreases at the start (0% - 4%) and then increases with respect to increase in silica fume content up to 16%[3]. Mechanical properties of a concrete can also be increased with the series effects of silica fume and recycled steel fiber [4]. Addition ofsilica fume develops the strength, porosity, bound water of the concrete specimen [5]. In SCC combined effect of fly ash and silica fume has high reduction in water absorption comparatively to that of SCC prepared with fly ash alone [6]. Rice husk ash as an alternative to cement also increase the mechanical behaviour of hardened concrete up to 15% replacement [7]. High degree of fine marble powder produces very good cohesiveness of concrete and Limestone powder enriches the mechanical as well as the durability parameters of concrete by exhibiting a good compact network [8]. Concrete mixture with more silica fume has smaller diameter of slump flow [9]. SCC mixes developed with bottom ash exhibits a minimal resistance to chloride permeability, abrasion resistance, water absorption and sorptivity at earlier age of concrete, however these properties decreases with increase in age and 30% rise in fly ash improved its flow properties [10]. The objective of this paper is to revels the effect of various chemical and mineral admixtures in the performance of the fresh and hardens concrete. In the production process of SCC Chemical admixtures has a vital role. The commonly used chemical admixtures are Polycarboxyle ether, synthetic copolymer, poly alcohol which technically know as super plasticizer, viscosity modifying agent, and anti-forming admixture respectively. These Chemical admixtures increase the workability of concrete and it also affect the strength parameter significantly. Classification of these admixtures, their physical and chemical parameters and the major effects with cement paste were been analysed in several studies [11,12]. Influence of these chemical admixtures increases the pump-ability, workability and resistance against segregation and bleeding and optimum compaction level. It also reduces the required water content, and air voids. This makes the concrete more impermeable, denser and durable [11].In the process of preparation of SCC we need to use more cement content which may cause high heat of hydration this may result in shrinkage cracks and high CO_2 emission. But by means of using mineral admixtures such as fly ash, GGBS, silica fume, Metakaolin may reduce these effects and also achieves more durable and denser concrete. Supposedly, the incorporation of metakaolin improves the penetration of chloride ions in the SCC specimens. When metakaolin is included as an alternative to cement content in different percentages, the mechanical properties such as the compressive strength and the tensile strength of the concrete specimens were noted to improve well at lower W/B ratio. Metakaolin tends to have better performance by means of splitting tensile strength. The absorption capacity of the control specimens were visualized to decrease with the addition of metakaolin. SCC is also developed with metakaolin by not including any viscosity modifying agent also [12,13]

2. Literature analysis

Erhan et al, 2015[14] in their experimental study detailed the fresh as well as the rheological behavior of the SCC when it is merged with nano silica and fly ash. Four different mixes with a w/c ratio of 0.3 and nano silica replaced with Portland cement at 0%, 2%, 4% and 6% to the weight of binder were formulated. Fly ash is incorporated for the 2%, 4% and 6% of nano silica containing self compacting concrete at 25%, 50% and 75% respectively to the weight of the total binder ratio. The incorporation of the nano silica in SCC is noted to enhance the slump flow and the V funnel flow at 25%, 50% and 75% also. Whereas for the L-Box test, the L-Box height ratio tend to increase with the fly ash content. The rheological tests for the same exposed the shear thickening of the self-compacting concrete by theaddition of the nano silica whereas the addition of the fly ash decreased the same. The fly ash content in SCC is also noted to decrease the mechanical properties of the SCC developed in this investigation

Benaicha et al,2015[15] experimentally studied the effects of silica fume and viscous modifying agents on the rheological and the mechanicalbehaviour of the SCC. In order to ascertain the rheological parameters slump test, L-box test, segregation test, V-Funnel test were conducted. In case of the mechanical properties, the compressive strength, tensile strength and modulus of elasticity were analysed. This study revealed the improvement in the plastic viscosity and the yield stress under a regulated condition of constant water/cement ratio along with the super plasticizer. In case of the mechanical properties, the SCC obtained with silica fume is noted to show better results than the SCC with VMA. Henceforth the paper concludes that based on the easy availability of a material, the VMA or the silica fume can be chosen accordingly.

Divya et al, 2015[7] experimentally investigated the fresh concrete studies, mechanical properties and durabiity properties of SCC developed with replacement of the binder with 0%, 5%,10%,15% and 20 % of rice husk ash (RHA). A rise of 33% is noted in the compressive strength whereas the split tensile strength is noted to elevate up to 15% of replacement with RHA. Notable reduction is spotted in chloride ion permeability. Adequate non permeability is visualized for the SCC developed with 15% of RHA replacement. Porosity of the specimens considerably reduced with longer curing time owing to the higher percentage of rate of hydration with the longer concrete age. Henceforth a much denser concrete is obtained by 15% replacement of RHA which in turn developed the maximum compressive strength than all the other specimens casted.

Rahmat et al,2015[16] detailed the influence of nano silica and carbon nano fibers in the self compacting concrete. A combined study with fresh and hardened characterization along with the hardened concrete study is conducted in this study. The SCC thus developed in this study is continuously monitored for the first 24 hours to clearly observe the accurate characterization of the SCC. The carbon nano fiber in the cement paste elevated the flexural strength by 30% which also caused surface cracking. However maximum shrinkage is observed for an ultrasonic pulse velocity value between 1500-2000m/s. The nano silica in SCC notably improved the compressive strength also improved the early age cracking and thereby affecting the durability of the SCC developed too.

Mehmet et al, 2012[8] included the industrial waste materials such as marble powder, lime-stone powder and fly ash. In this experimental analysis, industrial waste materials were included in SCC replacing the total binding material at 5%, i10% and 20% by weight. Though the use of marble powder and lime stone powder tend to increase the water content, the addition of fly ash neutralized the same in turn achieving a target slump flow. Thus, the fillers elevated the percentage of super plasticizer in concrete and also increased the initial and final setting time of SCC. The 28 days compressive strength of the ternary mixes were comparatively lower than that of the binary mixtures. And a much similar pattern is noted for the spit tensile strength of SCC also. Both binary and ternary mixtures developed lower chloride ion penetration properties.

Mostafa et al,2015[17] studied the influence of class F fly ash, nano silica and silica fume on the SCC with high performance. In this analysis, adequate fraction of these admixtures was replaced with the total cement content in SCC. The rheological properties, thermal properties, transport properties and the mechanical properties were detailed in this study. The fly ash content is noted to improve the rheological parameters whereas the silica nano particles and silica fume improved the transport as well as the mechanical properties. The larger portion of mineral admixtures blended with minimum fraction of the nano powders is a promising combination for a high performance self-compacting concrete.

Rahmat et al, 2012[11] predicted the fresh and hardened properties of SCC incorporated with metakaolin. A total of 15 mixes with MK content of 0%, 5%, 10%, 15% and 20% were developed by varying the w/c ratio as 0.32, 0.38 and 0.45.With no VMA, the various mixes showed up good workability and rheological properties. The compressive strength of the concrete specimens were noted to increase up to 27% upon the 14 days of curing. The compressive strength of the same were also predicted by multiple regression analysis in terms of ultrasonic pulse velocity. The tensile strength gained up to 11 % than the control specimen. Lower absorption is recorded in the SCC specimens with good electrical resistivity also. Altogether a SCC mix with 10% of metakaolin replacement is an ideal proportion for developing economically efficient concrete with better fresh and hardened concrete properties.

Beata et al, 2013[12] experimentally investigated the effects of chemical admixtures on the hydration of cement and mixture properties in SCC with very high performance. With few admixtures, micro cracking is developed whereas with the analyzed mixtures, the C-S-H gel showed improvement. The workability loss is visible according to the type of admixture that is chosen, but it does not affect the air content of very high-performance self -compacting concrete. The HRWR admixtures based cement pastes also developed bubble bridges that overtook any other fluid action of the same.

Rafat et al,2013 [18] investigated the properties of SCC developed with coal bottom ash. Fine aggregate is replaced with 10%, 20% and 30% of coal bottom ash to investigate the fresh concrete properties such as slump flow, U-funnel test, L-box test, and J-ring test, along with the hardened concrete parameters such as abrasion resistance, compressive strength, chloride permeability and sorptivity. Betterment in compressive strength is observed at i28 days whereas the chloride permeability resistance decreases for 90 days and 365 days. A co relation between the abrasion resistance and the compressive strength is developed such that both are directly proportional, since the abrasion resistance highly influenced the compressive strength of the concrete.

Ha Thanh et al,2016 [19] presented theinfluence of the super plasticizer and mineral admixtures (rice husk ash, silica fume and fly ash) on the self -compacting high- performance concrete in terms of the compressive strength of the mortar specimens. Expect the rice husk ash (RHA), both the fly ash and silica fume notedly decreased the filling and passing ability along with a hike in resistance to segregation and plastic viscosity. The bleeding in concrete is observed to mitigate with the addition of the rice husk ash. Thus, the macro-mesoporous nature of the rice husk ash can also be used as viscous modifying agent to improve the robustness of high-performance self -compacting concrete (HPSCC) at higher super plasticizer dosages. The coarser particles in RHA with large specificsurface area which thereby stipulated the attraction of the inter molecular forces and improved the compressive strength of the HPSCC. Therefore, a maximum compressive strength is visualized for SCC with 20% of fly ash and 20% of RHA.

Ali sadramomtazi et al, 2016[20] investigated the rheological, mechanical and durability characteristics of SCC developed with polyethylene terephthalatei (iPET) particles combined with the pozzolanic materials (fly ash and silica fume). These PET particles in SCC minimized the mechanical properties such as compressive strength, tensilestrength and flexural strength. To counterbalance the same, fly ash and silica fume were introduced into the mix. The lower specific gravity of these PET aggregates in concrete also reduced the density of concrete owing to the weak bonding between the PET aggregates and the cement paste. These parameters also reduced the ultra-sonic pulse velocity values due to the larger number of pores in the mix.

Wongkro et al,2014 [21] studied the effects on the compressive strength and chloride resistance of the SCC developed with high volume fly ash (HVFA) and silica fume in binary blended cement and terenary blended cement. Both HVFA and silica fume were replaced with cement at various proportions each and the tests were conducted. The binary blended cement with HVFA is noted to reduce the mechanical property (compressive strength). In contrary to the binary blended cement, the terenery blended cement improved the compressive strength. Nevertheless, both fly ash and silica fume were noted to increase the chloride resistivity at higher levels.

Navid et al, 2016 [22] analysed the influence of industrial waste (palm oil fuel ash) from power plant of palm oil industry in the SCC due to its high pozzolanic characteristics and abundance as an industrial waste. This industrial waste is replaced with OPC at 10%, 15% and i20% to investigate the mechanical and durability properties of the SCC thus developed. Both the mechanical and durability properties were noted to improve due to the reduced amount of portlandite in the system leading to the production of the C-S-H gel thereby densification of the matrix. This phenomenon also mitigated the open pores by blocking the networks.

Badogiannis et al, 2015 [23] studied the effects of metakaolin on the durability properties of self -compacting concrete. The Portland composite cement and limestone powder of high fineness were replaced with high purity metakaolin. Four mixtures were developed to find the blending nature of metakaolin with cement thereby to study the enhancement of the packing density. And another four mixes replacing metakaolin with lime -stone powder at different percentages were obtained to procure the same results. These two categories were compared with another one control mix, thus developing a total of nine mixes in this study. Crushed calcareous lime-stone aggregate and polycarboxylic based super plasticizer is involved to develop SCC at a w/c ratio of 0.6. to produce SCC. Various durability study such as open porosity, sorptivity, near surface water penetrability, Gas permeability and chloride penetrability were studied. Bases on the study the following conclusions were made. Slight improvement is found in open porosity percentage for both the replacement of cement with metakaolin and limestone powder with metakaolin. And the highest improvement is found at 14% replacement of cement with metakaolin and in limestone powder replacement as the metakaolin increases the results were getting reduced .In Sorptivity when higher order replacement metakaolin to cement and limestone powder the sorptivity value getting reduced. And the improvement in results were in between 6% to 41% for cement replaced by metakaolin and for limestone powder replaced by metakaolin the improvement was found to be in between 22% to 53%.In water permeability test as the metakaolin replacement level increase instead of cement the water permeability decreases. And for limestone powder replacement with metakaolin as the metakaolin percentage increases water permeability got reduced to some extent further increase it starts increasing the permeability value .Regarding Gas permeability slight improvement is found in both the replacement of cement with metakaolin and limestone powder with metakaolin. And the maximum reduction of 34% is obtained in higher order replacement level of cement with metakaolin. Whereas for chloride penetrability test it is found that in higher order replacement of metakaolin with cement and limestone powder resistance to chloride penetrability increases and the best results as obtained for higher order replacement of limestone powder with metakaolin.

3. Conclusions

Based on the analysis of the literatures, it is clearly evident that both chemical and mineral admixtures play a key role in the betterment of the self-compacting concrete in the following parameters.

- When silica fume is included in SCC it improved the compressive strength of the concrete whereas the workability is also noted to improve when silica fume is included in SCC along with fly ash.
- Fly ash in SCC decreased the compressive strength of the concrete, whereas fly ash along with silica fume and PET particles improved the compressive strength to a greater extend.
- The nano silica and nano fly ash in contrary to the fly ash and silica fume highly improved the compressive strength of SCC by densifying the pore structure. But the nano silica and nano fly ash reduced the workability owing to the finer nature of these mineral admixtures.
- The incorporation of admixtures such as marble powder and lime stone powder increased the water content in the concrete mix thus affecting the rheological properties of the SCC.
- The addition of mineral admixture like rice husk ash improved the mechanical properties of the SCC up to 15% of replacement with ordinary Portland cement. However the same exhibited low chloride ion resistance and developed to be a highly flowable matrix.
- The carbon nano fibers in SCC developed surface cracks along with improved mechanical properties in concrete.
- Metakaolin with SCC considerably improved the mechanical properties up to a certain percentage of replacement and also decreased the absorption of the concrete specimens. It also highly improved the durability of SCC when it is included along with lime stone powder.
- The palm oil fuel ash, a natural mineral admixture highly improved the mechanical properties as well as the durability properties of the SCC when included at adequate level.
- Coal bottom ash, an industrial by-product elevated the compressive strength of the SCC specimens. But it tends to decrease the chloride ion resistance in the SCC.
- Chemical admixtures like HRWR admixtures behaved better with the SCC by improving the C-S-H gel comparatively than the mineral admixtures in SCC.

References

[1] H A F Dehwah 2012 Corrosion resistance of self compacting concrete incorporating quarry dust powder, silica fume and fly ash *Construction and building materials* **37** 227-282.

[2] Ahmet Benli, Mehmet Karatas, Yakup Bakir 2017 An experimental study of different curing regimes on the mechanical properties and sorptivity of self-compacting mortars with fly ash and silica fume *Construction and Building Materials* 144 552–562.

[3] Cairong Lu, Hu Yang, Guoxing Mei 2015 Relationship between slump flow and rheological properties of self compacting concrete with silica fume and its permeability *Construction and Building Materials* **75** 157–162.

[4] M Mastali, A Dalvand 2016 Use of silica fume and recycled steel fibers in self-compacting concrete(SCC) *Construction and Building Materials* **125** 196–209.

[5] G Vagelis, Papadikisa 1999 Experimental investigation and theoretical modeling of silica fume activity in concrete *Construction and building materials* **29** 79-86

[6] H Y leung, J Kim, A Nadeem, Jaya prakashjaganathan, M P Anwar Sorptivity of self compacting concrete containing flyash and silica fume *Construction and building materials* **113** 369-375

[7] Divya Chopra, Rafat Siddique and Kunal 2015 Strength, permeability and microstructure of self-compacting concrete containing rice husk ash *Bio systems Engineering Volume* **130** Pages 72-80.

[8] Mehmet Gesog˘lu, Erhan Guneyisi, Mustafa E Kocabag, Veysel Bayram and Kasım Mermerdas 2012 Fresh and hardened characteristics of self compacting concretes made with combined use of marble powder, limestone filler, and fly ash *Construction and Building Materials* Volume **37** Pages 160-170.

[9] Aleksandra Kostrzanowska-Siedlarz, Jacek Gołaszewski 2015 Rheological properties and the air content in fresh concrete for self compactinghigh performance concrete *Construction and Building Materials* Volume **94** 30 Pages 555-564.

[10] Rafat Siddique 2013 Compressive strength, water absorption, sorptivity, abrasion resistance and permeability of self compacting concrete containing coal bottom ash *Construction and Building Materials, Volume* **47** Pages 1444-1450.

[11] Rahmat Madanous, S Yasinmousavi 2012 Fresh and hardened properties of self compacting concrete containing Metakolin *Construction and Building Materials* Volume **35** Pages 752-760

[12] Beata lazniewska Piekarczy 2013 The influence of chemical admixtures on cement hydration and mixture properties of very high performance self compacting concrete *Construction and Building Materials* Volume **49** Pages 643-662.

[13] Nocun, Wczelik W, Trybalska B 2007 Effect of admixtures on the rate of hydration and microstructure of cement paste *Cement WapnoBeton* p. 284–9.

[14] Güneyisi, Erhan, Mehmet Gesoglu, Asraa Al-Goody and Süleymanİpek 2015 Fresh and rheological behavior of nano-silica and fly ash blended self-compacting concrete *Construction and Building Materials* **95** 29-44.

[15] Benaicha, Mouhcine, Xavier Roguiez, Olivier Jalbaud, Yves Burtschell and Adil Hafidi Alaoui 2015 Influence of silica fume and viscosity modifying agent on the mechanical and rheological behavior of self compacting concrete *Construction and Building Materials* **84** 103-110.

[16] Rahmat Madandoust, Malek Mohammad Ranjbar, Reza Ghavidel and S Fatemeh Shahabi 2015 Assessment of factors influencing mechanical properties of steel fiber reinforced self-compacting concrete *Materials & Design* Volume **83** Pages 284-294.

[17] Mostafa Jalal, Alireza Pouladkhan, Omid Fasihi Harandi and Davoud Jafari 2015 Comparative study on effects of class f fly ash, nano silica and silica fume on properties of high performance self compacting concrete Volume **94** Pages 90-104.

[18] Siddique, Rafat 2013 Compressive strength, water absorption, sorptivity, abrasion resistance and permeability of self-compacting concrete containing coal bottom ash *Construction and Building Materials* **47** 1444-1450.

[19] Le, Ha Thanh and Horst-Michael Ludwig 2016 Effect of rice husk ash and other mineral admixtures on properties of self-compacting high performance concrete *Materials & Design* **89** 156-166.

[20] Sadrmomtazi, Ali, Sahel Dolati-Milehsara, Omid Lotfi-Omran and Aref Sadeghi-Nik 2016 The combined effects of waste Polyethylene Terephthalate (PET) particles and pozzolanic materials on the properties of self-compacting concrete *Journal of Cleaner Production* **112** 2363-2373.

[21] Wongkeo, Watcharapong, Pailyn Thongsanitgarn, Athipong Ngamjarurojana and Arnon Chaipanich 2014 Compressive strength and chloride resistance of self-compacting concrete containing high level fly ash and silica fume *Materials & Design* **64** 261-269.

[22] Ranjbar, Navid, Arash Behnia, Belal Alsubari, Payam Moradi Birgani and Mohd Zamin Jumaat 2016 Durability and mechanical properties of self-compacting concrete incorporating palm oil fuel ash *Journal of Cleaner Production* **112** 723-730.

[23] Badogiannis, Efstratios G, Ioannis P Sfikas, Dimitra V Voukia, Konstantinos G Trezos and Sotirios G Tsivilis 2015 Durability of metakaolin self-compacting concrete *Construction and Building Materials* **82** 133-141.

IOP Conf. Series: Materials Science and Engineering 1006 (2020) 012039 doi:10.1088/1757-899X/1006/1/012039

Potential of Fly Ash Polymerized Sand as an Alternative for River Sand in Concrete- A State of the Art Report

G Lizia Thankam[1], T R Neelakantan[2], S Christopher Gnanaraj[3]

[1]Research scholar, Department of Civil Engineering, Kalasalingam Academy of Research and Education, Krishnankoil,626126, India
[2,3]Faculty, Department of Civil Engineering, Kalasalingam Academy of Research and Education, Krishnankoil

E-mail: g.liziathankam@klu.ac.in

Abstract. The depletion of the natural resources and the growing demand for the natural river sand in both developing countries and developed countries paved way for the evolution of new trends in the complete and partial replacement of the natural sand as the fine aggregate. Many natural admixtures and by-products were so far analyzed for the feasibility of the replacement of the natural sand. Some of such traditional sources suggested by the researchers were foundry sand, M-sand, quarry dust, I- sand and treated sea sand. Among which M-sand tend to meet the standards and in practice too for construction in India (Author's native) and other countries without any ordeal. However, a new technique for the complete replacement of the fine aggregate was introduced which incorporates the polymerization technique in the byproducts (Fly Ash). Though the first introduced technique was quite tedious, the consequent authors proposed much simpler methodologies that develop fine aggregates with characteristics similar to that of the natural sand which is also lightweight, eco-friendly and economical as discussed in this article. These geopolymerized fine aggregates set a new era in the construction industry which in turn conserves the balance of the ecosystem.

1. Introduction

Being the second highest used material after water, concrete is vastly used enormously across the continents. Fine aggregate, a principal component and binding material in concrete, is of great demand in the current scenario. The river sand, obtained from the riverbeds, is the foremost source for the fine aggregate for decades. Over-utilization of the river sand has caused depletion of availability and increased the price [1]. Hence, to meet the burgeoning need of sand, engineers and researchers ought to discover a sustainable solution similar to the development of the M-sand which is currently in practice in the construction industry.

2. Necessity of alternative solution

Due to the increasing demand for the river sand [2] as fine aggregate, new ideas emerged in the concrete industry to replace the river sand partially or completely with no compromise in the quality of the concrete [3-14]. The restrictions by the government agencies protecting the environment and depletion of natural sources became a jump start for the evolution of new trends [15,16] for the fine aggregate in most of the developing countries. By-products and waste materials were the main sources for the researchers so that cost-effective alternative fine aggregate can be developed. Consequently, alternative fine aggregates such as manufactured sand (M-sand), iron slag sand(I-sand), crushed rock dust, glass powder, foundry sand, and others were developed in the last two decades.

3. Traditional sources for replacing river sand

3.1 Foundry sand
Foundry sand (FS) is a by-product developed during the manufacturing process of metal alloys. The rich silica content sand (silica sand) when finally becomes non-recyclable by the foundries after many cycles, it is then dumped off [17]. This foundry sand is deposited in the open area millions of tons affecting the ecosystem, though they are used in the laying of highways [18,19]. Many studies were carried out recently for the incorporation of foundry sand in concrete structures so that they can be utilized to the maximum [20-22]. The possibility of using FS in ready mix concrete [23] indicated a negative effect by lower concrete strength and higher water demand. However, in another study [24] the results indicate that replacement of fine aggregate by up to 20% of FS efficiently behaved in the concrete with minimal reduction in the compressive and flexural strengths which in turn develops better concrete without any impact on the standards of concrete diversely when the foundry sand exceeds the limit of 20%, inferior behavior is recorded in concrete mixture when compared to that of the control specimen owing to the presence of sawdust, clay, wood flour and also the fineness of foundry sand.

3.2 Crushed rock sand /M sand
The crushed rock sand was obtained from the crushed quarried stone by allowing the crushed particles to pass through a 4.75mm sieve. Better mechanical properties and a lower slump were noted upon the replacement of fine aggregate with 30% of crushed rock sand [25]. The low workability with crushed rock sand can be rectified by the usage of high range water reducers [26]. Another study [27] reported that both the mechanical and durability properties improved by 14% compared to the properties of the normal concrete. Reduction of cost is an advantage [28] while using crushed rock sand in concrete. The strength parameters of M sand with natural river were compared and revealed that similar cement mortar with natural sand can be obtained with manufactured fine aggregates also without affecting the workability, absorptivity, inelastic deformation response and strength. These parameters were also analyzed with de Larrard's Compressible Packing Model and the author corelated the packing density with the porosity of the mortars replicating as a primary property in the analysis [29]

3.3 Quarry dust
Quarry dust is a by-product obtained from the processing of stones. They are mostly not preferred in the industrial construction since the properties of the quarry dust vary based on the mineralogy of the parent rock. The compressive strength of the concrete was noted to increase up to 75% depending upon the type and quantum of the replacement [30-32]. The workability can also be improved by the usage of superplasticizers and fly ash at lower water/ cement ratio [33]. The quarry dust with mica of more than 5% is however not suitable for construction purpose since it reduces the workability [34]. Quarry dust replacement is much effective when used along with fly ash in the concrete mixture [31]; however, the properties of quarry dust depends on the mineralogy analysis before it is integrated with the concrete mixture [35].

3.4 I-Sand
A detailed study [36] was conducted by replacing fine aggregate with iron sand (I-Sand) which is a by-product obtained from the steel and iron industries. Based on the earlier studies [37-39], 30% of I-Sand in concrete as a substitute of natural sand was found to be optimum for an effective concrete without degradation in the quality [40,41]. However, in another study [42], it was observed that the

slump values decreased after 40% replacement of fine aggregate by I-sand. Though I-sand seems to be a better alternative for river sand, more rigorous research on the various properties need to be studied.

3.5 Sea Sand:
A comparison study [43] was conducted between treated and untreated sea sand analyzing their feasibility in the normal concrete as well as geopolymer concrete in view of the earlier studies[44-46]. It was observed that the treated sea sand performed better than the untreated sea sand. The chlorides and the salt content in the untreated sea sand tend to hinder the strength parameters in the incorporation of untreated sea sand in the geopolymer concrete of 8M, 10M and 12M by 13%, 13.1% and 17% respectively whereas the treated sand developed similar compressive strength to that of the natural sand.

4. Geopolymerized sand
The over usage of natural sand in India (Author's native) resulted in extensive depletion of the nature through various phenomenon. Thus, an alternative for the river sand in construction is the need of the hour almost across the globe. Only a limited amount (30%) of fly ash is used in India which were expelled out from the power plants [47,48]. The complete replacement of natural sand with fly ash were noted to reduce the compressive strength by 73% and lowered the workability by 14% [49]. However compressive strength with fly ash of marginal quantity were appreciable [50], since both the byproducts together qualified the concrete. This is achieved when the quarry dust enhances the early strength and the fly ash improves the workability of concrete. Only partial replacement of fly ash in concrete were observed to keep up the standards of concrete [51]. In order to utilize the large amount of fly ash which is dumped into the landfills from the thermal power plants [52] as well as to make the concrete more economical, fly ash is an effective supplementary cementitious material indeed[53].

4.1 Fly ash polymerized sand:
In view of the above scenario, fly ash was introduced into concrete as complete replacement of fine aggregate after geopolimerization process [54]. Thus, developed sand was termed as fly ash polymerized sand (polymerized fine aggregate). A complex mixture of fly ash and amorphous silica were dry mixed together with sodium hydroxide solution of 10M. The dry mix was then allowed to oven dry for about four hours at 50^0C, which then tends the dry mix to agglomerate as fine aggregates which lies within the zonal limits obtained upon sieving of the oven dried slurry. These fine aggregates were further heated for 100^0C for seven days which allows accomplishing the geopolymerization process. It is then washed continuously with distilled water till the electrical conductivity reaches within the limits and then oven dried for two hours and can be used as fine aggregates. These fine aggregates were tested for properties such as electrical conductivity, specific gravity, XRD patterns, pH value, grain size distribution, TDS measurements and compressive strength of the mortar samples. These fine aggregates were found to be within the zonal limits and IS Standards [55]. Also, it imparted compressive strength similar to that of the mortar specimens developed with the natural sand which enabled to stand out as an ideal substitute for natural sand as fine aggregate. These polymerized fine aggregates were modified in another study [56] by further refining the tedious process in which the fine aggregates were termed as GFS. The GFS particles were obtained by mixing together the geopolymer liquid solution(10M) and pre heated(60^0C) fly ash which is then oven dried for one hour at 100^0C after sieving with 2.36mm and 4.75mm sieve. Thus developed fine aggregate particles can be used after keeping in ambient temperature for 24 hours. Various parameters of the GFS particles were analyzed to compare its properties with the natural sand. Specific gravity, water absorption, direct shear test, alkali aggregate reaction, particle size distribution, soundness, pH value, XRF, XRD, and SEM were studied and noted to develop 93% of compressive strength similar to that of the normal river sand at 28 days. Thus, it was

concluded as an suitable alternative for the normal river sand in concrete construction works. In another study [57] by the same author, the compressive strength and flexural strength of the concrete cubes developed with GFA for M25 grade concrete was 99% of the compressive strength compared to that of the Normal river sand at 56 days denoting a gain in the compressive strength with the increase in the curing period of concrete cubes. In the third case [58] of the geopolymerized fine aggregate (GFA), the author adopted much of the procedure similar to that of the later study[56] to synthesize the GFA. Both class C fly ash geopolymerized fine aggregate(C-GFA) and class F fly ash geopolymerized fine aggregate (F-GFA) were developed and compared with the properties of manufactured sand (M Sand) as well as natural sand. After studying the physical properties such as particle size distribution, specific gravity, water absorption, pH and frictional angle, these developed fine aggregate were tested for alkali silica reaction in mortar bars as well as the compressive strength of both the mortar cubes and concrete cubes were analyzed. The mortar specimens and the concrete specimens with C-GFA were noted to attain higher compressive strength like that of the natural sand and M-Sand when compared with that of the F-GFA due to the high calcium content. The specimens thus developed showcased lower weight and higher water absorption which need to be studied further so that they can be implemented in larger scale.

5. Conclusion

Based on the analysis of the literature, few conclusions were acquired on the feasibility of polymerized fine aggregate in the construction industry

- Though many other alternatives are available for river sand in the concrete mixture, usage of fly ash is considered as the most effective one owing to the following considerations:
  - Light weight
  - Environment friendly
  - Cheaper
- Apart from the above benefits, polymerized fine aggregate exhibits much similar mechanical and durability properties of the normal river sand.
- polymerized fine aggregate particles showcased UN-reacted fly ash particles developed due to the partial break down of the fly ash particles during the production process which is also accountable for the porous nature in the surface of the polymerized fine aggregate particles.
- The high amount of water absorption thus builds out in the fly ash particles during synthesis need to be eradicated for the proper usage of polymerized fine aggregate in the construction industry.
- Altogether polymerized fine aggregate is an ideal alternative for the natural sand in the concrete structures.

6. Oncoming dimensions of polymerized fine aggregate

- Behavior of polymerized fine aggregate with other silicaceous materials can be analyzed for the enhancement of mechanical and durability properties of concrete
- Polymerized fine aggregate can be included in concrete to examine the nature of the concrete behavior under various environmental conditions (coastal structures, elevated temperatures, freezing and thawing etc,)
- Chances of implementing polymerized fine aggregate in various types of concrete like self compacting concrete and pumpable concrete in large scale construction for the reduction of cost and weight of the structures.
- Well organized complete utilization of the fly ash from the thermal power plants by the government itself for terminating the dumping of the by product (fly ash) into the landfills.

References

[1] Ilangovan R, Nagamani K 2008 Application of quarry dust in concrete construction *High performance Concrete, Federal highway Administration*.

[2] UNEP GEAS 2014. Sand rarer than one thinks. Thematic focus: Ecosystem management, Environmental governance, Resource efficiency. UNEP Global Environmental Alert Services (GEAS).

[3] Singh, Sarbjeet, Ravindra Nagar and Vinay Agrawal 2016 A review on properties of sustainable concrete using granite dust as replacement for river sand *Journal of cleaner production* **126** 74-87.

[4] Brindha D and S Nagan 2010 Utilization of copper slag as a partial replacement of fine aggregate in concrete *International Journal of Earth Sciences and Engineering* **3** no. 04 579-585.

[5] Beixing, Li, Zhou Mingkai and Wang Jiliang 2011 Effect of the methylene blue value of manufactured sand on performances of concrete *Journal of Advanced Concrete Technology* **9** no. 2 127-132.

[6] Manasseh J O E L 2010 Use of crushed granite fine as replacement to river sand in concrete production *Leonardo electronics journal of practice and technologies* **17** 85-96

[7] Elavenil S and B Vijaya. 2013 Manufactured sand, a solution and an alternative to river sand and in concrete manufacturing *Journal of engineering, computers & applied sciences* **2** no. 2 20-24.

[8] Saha, Ashish Kumer and Prabir Kumar Sarker 2017 Sustainable use of ferronickel slag fine aggregate and fly ash in structural concrete: Mechanical properties and leaching study *Journal of Cleaner Production* **162** 438-448.

[9] Vijayalakshmi M and A S S Sekar 2013 Strength and durability properties of concrete made with granite industry waste *Construction and Building Materials* **46** 1-7.

[10] Ling, Tung-Chai and Chi-Sun Poon 2014 Use of recycled CRT funnel glass as fine aggregate in dry-mixed concrete paving block *Journal of cleaner production* **68** 209-215.

[11] M Mageswari and Dr B Vidivelli 2010 The use of sheet glass powder as fine aggregate replacement in concrete *The Open Civil Engineering Journal* Vol. **4** 65-71

[12] Siddique, Rafat, Geert De Schutter and Albert Noumowe 2009 Effect of used-foundry sand on the mechanical properties of concrete *Construction and building materials* **23** no. 2 976-980.

[13] Ilangovan R and K Nagamani 2006 Studies on strength and behavior of concrete by using quarry dust as fine aggregate *CE and CR Journal, New Delhi* 40-42.

[14] Prem, Prabhat Ranjan, Mohit Verma and P S Ambily 2018 Sustainable cleaner production of concrete with high volume copper slag *Journal of Cleaner Production* **193** 43-58.

[15] Tekin, Ilker, Muhammed Yasin Durgun, Osman Gencel, Turhan Bilir, Witold Brostow and Haley E Hagg Lobland 2017 Concretes with synthetic aggregates for sustainability *Construction and Building Materials* **133** 425-432.

[16] Bilir, Turhan, Isa Yüksel, Ilker Bekir Topcu and Osman Gencel 2017 Effects of bottom ash and granulated blast furnace slag as fine aggregate on abrasion resistance of concrete *Science and Engineering of Composite Materials* **24** no. 2 261-269.

[17] Siddique Rafat, de Schutter Geert, Noumowec Albert 2013 Effect of used-foundry sand on the mechanical properties of concrete *Constr Build Mater* **46** 1–7.

[18] Javed, Sayeed and C W Lovell 1995 Uses of waste foundry sands in civil engineering *Transportation Research Record* 1486.

[19] Kleven, Jay R, Tuncer B Edil and Craig H Benson 2000 Evaluation of excess foundry system sands for use as subbase material *Transportation research record* 1714 no. 1 40-48.

[20] Kraus, Rudolph N, Tarun R Naik, Bruce W Ramme and Rakesh Kumar 2009 Use of foundry silica-dust in manufacturing economical self-consolidating concrete *Construction and Building Materials* **23** no. 11 3439-3442.

[21] Siddique, Rafat, Yogesh Aggarwal, Paratibha Aggarwal, El-Hadj Kadri and Rachid Bennacer 2011 Strength, durability, and micro-structural properties of concrete made with used-foundry sand (UFS) *Construction and Building Materials* **25** no. 4 1916-1925.

[22] Singh, Gurpree and Rafat Siddique 2012 Abrasion resistance and strength properties of concrete containing waste foundry sand (WFS) *Construction and building materials* **28** no. 1 421-426.

[23] Basar, H Merve and Nuran Deveci Aksoy 2012 The effect of waste foundry sand (WFS) as partial replacement of sand on the mechanical, leaching and micro-structural characteristics of ready-mixed concrete *Construction and Building Materials* **35** 508-515.

[24] Prabhu, G Ganesh, Jung Hwan Hyun and Yun Yong Kim 2014 Effects of foundry sand as a fine aggregate in concrete production *Construction and building materials* **70** 514-521.

[25] Celik, Tahir and Khaled Marar 1996 Effects of crushed stone dust on some properties of concrete *Cement and Concrete research* **26** no. 7 1121-1130.

[26] Mundra, Sanjay, P R Sindhi, Vinay Chandwani, Ravindra Nagar and Vinay Agrawal 2016 Crushed rock sand–An economical and ecological alternative to natural sand to optimize concrete mix *Perspectives in Science* **8** 345-347.

[27] Hameed, M Shahul and A S S Sekar 2009 Properties of green concrete containing quarry rock dust and marble sludge powder as fine aggregate *ARPN J. Eng. Appl. Sci* **4** no. 4 83-89.

[28] Ilangovan R 2014 Experimental studies on strength and behaviour of concrete by using quarry rock dust as fine aggregate

[29] Gonçalves J P, L M Tavares, R D Toledo Filho, E M R Fairbairn and E R Cunha 2007 Comparison of natural and

manufactured fine aggregates in cement mortars *Cement and Concrete Research* **37** no. 6 924-932.
[30] Pofale A D and S V Deo 2010 Comparative long term study of concrete mix design procedure for fine aggregate replacement with fly ash by minimum voids method and maximum density method *KSCE Journal of Civil Engineering* **14** no. 5 759-764.
[31] Sivakumar A and M Prakash 2011 Characteristic studies on the mechanical properties of quarry dust addition in conventional concrete *Journal of civil engineering and construction technology* **2** no. 10 218-235.
[32] Sukesh, Chandana, Katakam Bala Krishna, P Sri Lakshmi Sai Teja and S Kanakambara Rao 2013 Partial replacement of sand with quarry dust in concrete *International Journal of Innovative Technology and Exploring Engineering (IJITEE)* **2** no. 6 254-258.
[33] Anzar Hamid Mir 2015 *Int. Journal of Engineering Research and Applications* www.ijera.com ISSN : 2248-9622 Vol. **5** Issue 3 (Part -3) pp.51-58
[34] Properties of concrete A.M. Neville. Edition. 4th and final ed. Published. Harlow, Essex : Longman, 1995.
[35] Jayawardena U De S and D M S Dissanayake 2006 Use of quarry dust instead of river sand for future constructions in Sri Lanka *IAEG Paper* **38**.
[36] Manju, Unnikrishnan 2016 I-sand: An environment friendly alternative to river sand in reinforced cement concrete constructions *Construction and Building Materials* **125** 1152-1157.
[37] Cheng, An, Ran Huang, Jiann-Kuo Wub and Cheng-Hsin Chen 2005 Influence of GGBS on durability and corrosion behavior of reinforced concrete *Materials Chemistry and Physics* **93** no. 2-3 404-411.
[38] Wainwright P J and N Rey 2000 The influence of ground granulated blast furnace slag (GGBS) additions and time delay on the bleeding of concrete *Cement and concrete composites* **22** no. 4 253-257.
[39] Ganesh, Babu K and Rama Kumar V Sree 2000 Efficiency of GGBS in concrete *Cement and Concrete Research* **30** no. 7 1031-1036.
[40] Bureau of Indian Standards 2000 IS 456: Plain and Reinforced Concrete - Code of Practice
[41] IS : 2386 (Part 3) – 1963 – Method of test for aggregates for concrete (Part I) Particle size and shape.
[42] Humam, Tamara and Rafat Siddiquen 2013 Properties of mortar incorporating iron slag *Leonardo Journal of Sciences* **1** no. 23 53-60.
[43] Shinde B H and K N Kadam 2011 Strength properties of fly ash based geopolymer concrete with sea sand
[44] Huiguang, Yin, Li Yan, LvHenglin and Gao Quan 2011 Durability of sea-sand containing concrete: Effects of chloride ion penetration *Mining science and technology (China)* **21** no. 1 123-127.
[45] Dolage D A R, M G S Dias and C T Ariyawansa 2013 Offshore sand as a fine aggregate for concrete production *Current Journal of Applied Science and Technology* 813-825.
[46] Ratnayake N P, U G A Puswewala, S P Chaminda, E M T M Ekanayaka and M N Jayawardene 2014 Evaluation of the potential of sea sand as an alternative to river sand for concrete production in Sri Lanka *Journal of Geological Society of Sri Lanka*.
[47] Kumar V, R Anandkumar and M Mathur 2003 Management of fly ash in India: a perspective *Proceedings of Third International Conference on Fly Ash Utilisation and Disposal, New Delhi* vol. **1** pp. 1-18.
[48] Kumar, Vimal, Mukesh Mathur and Shashank Shekhar Sinha 2005 A case study: manifold increase in fly ash utilization in India *Fly Ash India* 1-18.
[49] Shehab, Hamdy K, Ahmed S Eisa and Ahmed M Wahba 2016 Mechanical properties of fly ash based geopolymer concrete with full and partial cement replacement *Construction and Building Materials* **126** 560-565.
[50] Ravina Dan 1998 Mechanical properties of structural concrete incorporating a high volume of class F fly ash as partial fine sand replacement *Materials and structures* **31** no. 2 84-90.
[51] Parvati V K and K B Prakash 2013 Feasibility study of fly ash as a replacement for fine aggregate in concrete and its behaviour under sustained elevated temperature *Int J of Sci &Eng Res* **4** 87-90.
[52] Ravina, Dan 1997 Properties of fresh concrete incorporating a high volume of fly ash as partial fine sand replacement *Materials and Structures* **30** no. 8 473-479.
[53] Hemalatha T and Ananth Ramaswamy 2017 A review on fly ash characteristics–Towards promoting high volume utilization in developing sustainable concrete *Journal of cleaner production* **147** 546-559.
[54] Rao S M & Acharya I P 2014 Synthesis and characterization of fly ash geopolymer sand *Journal of materials in civil engineering* ***26*(5)** 912-917.
[55] IS 2386:1963 Methods of Test for Aggregates for Concrete, Bureau of Indian Standards
[56] Agrawal U S, S P Wanjari and D N Naresh 2017 Characteristic study of geopolymer fly ash sand as a replacement to natural river sand *Construction and Building Materials* **150** 681-688.
[57] Wanjari S P, U S Agrawal and D N Naresh 2018 Geopolymer Sand as a replacement to Natural Sand in concrete *IOP Conference Series: Materials Science and Engineering* vol. **431** no. 9 p. 092011. IOP Publishing.
[58] Sharma, Anil Kumar and K B Anand 2019 Comparative study on synthesis and properties of geopolymer fine aggregate from fly ashes *Construction and Building Materials* **198** 359-367.

Feasibility Study on Paver Blocks blended with Textile Effluent Sludge and M-Sand

V Arulpandian[1], R Jagadheeswari[2], S Danielraj[3], S Muthuraman[4] and A Sakthivel[5]

[1,3]Assistant Professor, Department of Civil Engineering, National Engineering College, Kovilpatti, Thoothukudi, Tamilnadu, India.
[2]Assistant Professor, Department of Civil Engineering, K. Ramakrishnan College of Technology, Trichy. Tamilnadu, India.
[4,5]UG Student, Department of Civil Engineering, National Engineering College, Kovilpatti, Thoothukudi, Tamilnadu, India.

E-mail: arulpandi.vp@gmail.com

Abstract. Textile sludge (TS) management is a huge problem for its disposal from the textile industry. It has tremendous applications, such as walking paths, street road, and fuel stations, etc. In this manner, an innovative step has been towards the manufacture of paver blocks blended with textile effluent treatment plant (TETP) sludge to use of it in reasonable extends. A different percentage of sludge starts from 10%, 15%, 20%, 25% and 30% to be taken for this study for the effective utilization of sludge to the construction industry. This thinks about looked for to experience the potential use of TS as a binding fabric for paving blocks generation. Conventional paver block is cast with full replacement of sand by using M-sand as fine aggregate. Paver blocks consist of TS in addition to distinctive proportions was casted according to the recommendation of Indian Standards (IS) 15658 (2006), also the various results was obtained through experimentally and it was compared to the conventional paver block. The different mix combinations outcome reveals that, 30% of cement replacement by effective utilization of TS from the textile industry.

1. Introduction

The textile industry is the biggest industry unit in India. A lot of environmental issues were developed due to the rapid incremental of industrialization. The textile industry generates different types of wastes over the world. For the human community, sludge management has become a great task in recent years.

The prospective reuse of TETP sludge in construction materials was studied. The basic property of sludge which was collected from the southern part of India has been studied. The various tests have been conducted for the suitability of sludge for both structural and non-structural applications with the replacement of cement by addition of 30% sludge in the mix. Perhaps it is possible for the manufacturing of non-structural building materials by the substitution of textile sludge up-to maximum of 30% and also the setting process of the elements by the addition of sludge in the mix [1]. From the study TS proportions for different Hypo sludge combinations and TS + HS combinations taken according to the design mix. According to BIS requirements, all the combinations of Mix were satisfied [2]. The permeability of the mix was decreased by 5%

replacement of cement through the addition of TS. Beyond this substitution point, penetration properties of mortars disintegrated, owing to lesser hydration response and deficiently pozzolanic activity in cement-sludge mortars [3]. The study was compared with the standard limit through experimental investigation on fly ash unburnt bricks. The characteristics of flyash which influence a major impact in the manufacturing of unburnt bricks. Simultaneously the characterizations such as chemical, physical, and mineralogical for flysah to make sure the reactivity for the process of fabrication and it was recommended by Murugesan T et al [4]. A more than 80% of paver block quality achieved within a day after fabrication which implies the time-saving. Compared to lightweight concrete, the paver blocks less chemical assault and physical stretch due to its fast curing and less water absorption property. Paver blocks were generated from plastic waste and it's more advantageous to overcome conventional concrete due to the quick development of production [5]. The basic properties and distribution of size of foundry sand and steel slag does not meet the requirements of ASTM. The basic physical properties of all industrial wastes were obtained and it's also matched with natural sand [6]. The concrete which was made by tires rubber leads to decreases compressive and flexural strength. But at the same time, it can also supplant natural by elastic tires by kept their mechanical characteristics of concrete without elastic. The properties were improved with 10% of substitutions in the mix for making paver blocks with natural aggregates [7]. According to the IS code provision the paver blocks subjected to having more than 30 N/mm2 compressive strength as well as less than 6% of water absorption used in light and non- traffic areas such as embankment slopes, public garden, and building premises. The different mix combinations were satisfied with the IS code criteria based on their flexural strength and breaking load results [8]. Predominantly change the compressive strength of Rubber Mould Textile Sludge Paver Block (RMTSPB) at 28 days by using different mix proportions. The strength was achieved by nearly 30MPa by adding up to 20% of textile sludge in the mix. The abrasion resistance of block decreased with the percentage of sludge increased. Around 1.788mm thickness was decreased during the testing of abrasion resistance by adding 20% of textile sludge. Finally, the addition of 20% of sludge replaced as cement substitutions in the mix was recommended by Kaushal Patel et. al [9]. Form experimental investigations, the compressive strength was decreased by adding more than 40% of sludge in the mix but below this percentage, the cement was replaced by sludge and fly-ash which reveals that it doesn't make any unfavorable effect of the specimen [10].

An eco-friendly attempt is possible by adding textile sludge as composite materials in the manufacturing of paver blocks. This helps to minimize the disposal problems. There was an overwhelming future extend of the utilization of sludge was part of modern advancement of materials within the construction industry. The generation of Paver blocks blended with textile sludge results in mass transfer from industry and driving to eco-friendly attempts for the perspective situation.

2. Experimental programme

2.1 Materials used

2.1.1 Cement

The Ordinary Portland Cement (OPC) of 53 grades as per IS 1489 (1991) available from a local source was used, the specific gravity of 3.15 and fineness modulus of 3.5%.

2.1.2 Fine and coarse aggregate

M-sand was used as a Fine aggregate and collected from a nearby local source. The M-sand has a specific gravity of 2.85, fineness modulus of 5.886 and water absorption found to be 2.14%. The Coarse aggregate (CA) was collected from locally situated quarry nearby Thoothukudi district, Tamilnadu, India. The specific gravity of CA was 2.61, fineness modulus found to be 8.247, and water absorption to be 1.69%. According to IS 15658 (2006) standards, a 10 mm size was taken as the maximum size of the aggregate. For mixing and curing of paver blocks, the portable water was used..The batching of materials shown in Figure 1.The particle size distribution was clearly shown in Figure 2 for the M-sand and river sand

Fig 1(a).

Fig 1(b)

Fig 1(c)

Figure 1. a. Cement, b. M-s and and c. CA

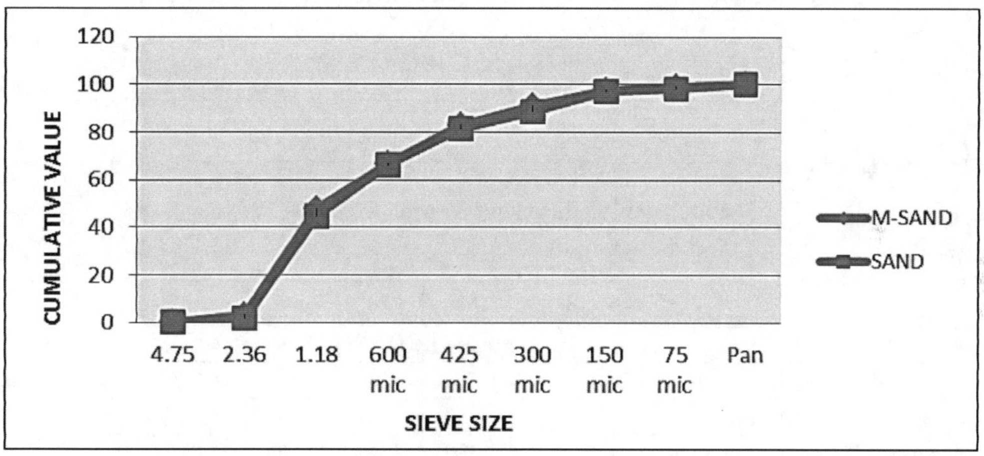

Figure 2. Sieve Analysis

2.1.3 Textile sludge

The TS was obtained from the SIPCOT, CETP, Perundurai, Erode district, Tamilnadu state, India. By random sample technique, the sludge was collected in dry conditions from the drying beds. With the help of the trowel, the dry sludge was grounded manually within room temperature. By sieving the TS obtained from industry with the help of 150- μm sieve. Due to less thickness and very little measure issues, there was taking care done for the fabrication of specimens. Its maintenance in a heap was not conceivable in a dry state comparable to cement, since it was blown absent by a slight breeze of the discuss. This sludge appears to be finer than cement in quality. It has lower specific gravity than cement. It possesses more volume and requires more water during blending. The Basic properties of specific gravity were found to be 3.16 and % of residue left in 90μ was 2.6%.The wed textile sludge and dry textile sludge as shown in figure 3 and 4 respectively.

Figure 3.Wet Textile sludge

Figure 4. Dry textile sludge

2.2 Mix combination for specimens

Paver blocks were manufactured using M40 grade concrete blended with M-sand with appropriate mix proportions. Textile sludge was added in different mix combinations for the replacement of cement. As per IS 15658 (2006) recommendations, the materials were selected. The below Table 1 and 2 shows mix proportions and mix combinations of specimens respectively,

Table.1 Mix Proportion for the paver block specimen

Grade of Concrete	Cement	FA(M-sand)	CA	W/c ratio
40	1	2.33	1.82	0.50

Table 2. Mix Combination for the specimens

S.No	Ingredients	Without Sludge	Weight of Sludge(kg)				
			10%	15%	20%	25%	30%
1	Cement	0.98	0.882	0.83	0.784	0.73	0.686
2	Textile Sludge	-	0.098	0.14	0.196	0.24	0.294
3	M-Sand	2.30	2.30	2.30	2.30	2.30	2.30
4	Coarse Aggregate	1.80	1.80	1.80	1.80	1.80	1.80

2.3 Mix design

Design mix helps to make the concrete with effectively by using desired properties of materials. This was done by choosing the appropriate combinations of constituents such as cement, aggregates, and water. The minimum strength of the paver blocks was the primary objective. As per the recommendation of IS 10262 (2009) the mix design was carried very carefully without affecting its basic properties. The mix proportions arrived for M40 grade concrete and it's shown in Table 1.

2.4 Casting of paver blocks

According to IS 15658 (2006) the size of the specimen 250 x 123x 80 mm was used in the study and it's shown in Figure 5. By adding the required amount of water in the different mix combinations the paver blocks were fabricated within the stipulated temperature without affecting the mix fresh properties. The mix was placed in the mould with an appropriate manner and vibration also was given to each layer in the mould. De-moulded the specimen after a 24 Hrs curing period.

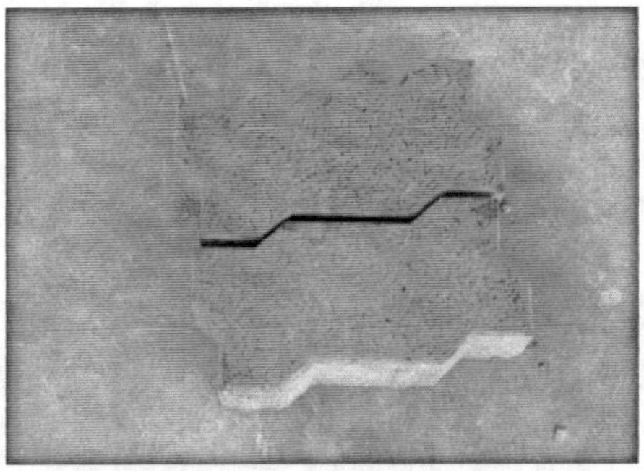

Figure 5. Paver block Specimen

3 Results and discussion

3.1 Slump test

Before usage of concrete mix, the fresh property was conducted using a slump cone and also measure the consistency of the concrete before sets. It is performed to check the workability of freshly made concrete. The slump shall be measured and recorded as True Slump and it's shown in figure 6.

Figure 6. Slump test

Figure7. Density

3.2 Density

Density is an estimation that analyzes the measure of issue an item has to its volume. An article with a lot matter in certain volume has high thickness an item with minimal issue in a similar measure of volume has a low density. Density of the specimen was carried according to the IS code and its shown in figure 7. The test results for 7 & 28 days were tabulated in Table 3 and described in figure 8.

Table. 3 Density of the specimen

S.No	%	Specimen		Weight (Kg)		Density(Kg/m^3)	Density(Kg/m^3)
		7 days	28 days	7 days	28 days	7 days	28 days
1	0	CPB1	CPB 4	5.345	5.575	2812.8	2391.26
		CPB2	CPB 5	6.120	6.190		
		CPB3	CPB 6	5.410	5.610		
2	10	TSPB1	TSPB	5.305	5.450	2171.41	2166.67
		TSPB2	TSPB	4.565	5.215		
		TSPB3	TSPB	6.155	5.040		
3	15	TSPB1	TSPB	5.030	5.630	2145.12	2140.92
		TSPB2	TSPB	5.480	5.110		
		TSPB3	TSPB	5.320	5.005		
4	20	TSPB 1	TSPB	5.555	5.335	2133.47	2168.02
		TSPB 2	TSPB	5.225	5.405		
		TSPB 3	TSPB	5.215	5.260		
5	25	TSPB 1	TSPB	5.405	5.630	2128.05	2239.84
		TSPB 2	TSPB	5.060	5.420		
		TSPB 3	TSPB	5.335	5.480		
6	30	TSPB 1	TSPB	5.350	5.350	2134.00	2199.19
		TSPB 2	TSPB	5.245	5.245		
		TSPB 3	TSPB	5.410	5.410		

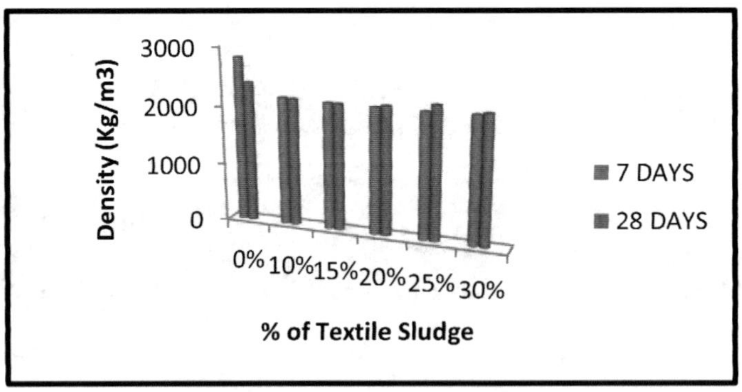

Figure 8.Densiy of the specimens

3.3 Compressive strength test

With the help of 2000 kN capacity CTM (figure 9), the compressive strength was done as per the IS code. As per IS 15658 (2006) recommendation M40 grade concrete is used for nontraffic areas as well as M55 for traffic pavements. Also it recommends that the 20 years life span of the structures by using specified concrete in the work. The various mix combinations were shown in Table 2 for fabricating of paver blocks with the TS. According to the combinations, the results are discussed and tabulated in Table 4. The results was discussed in figure 10.

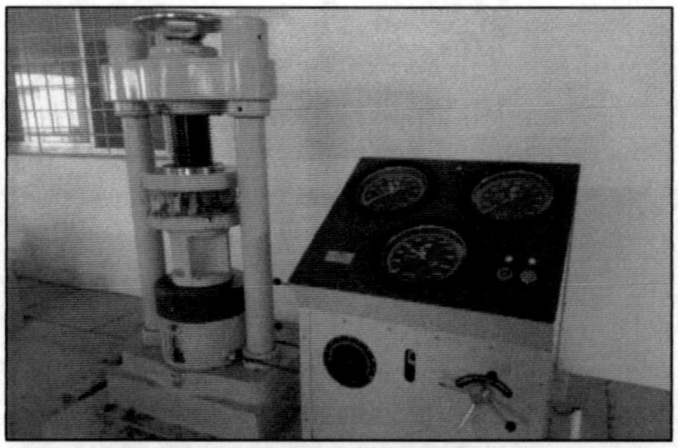

Figure 9. Compressive Strength of the Specimen

Table 4. Compressive Strength of the specimen

S.No	% of TS	Specimens		Ultimate Load(KN)		Area (mm^2)	Average Compressive Strength(N/mm^2)	
		7 days	28 days	7 days	28 days		7 days	28 days
1	0	CPB1	CPB 4	1000	1350	250X123	28.03	43.80
		CPB2	CPB 5	900	1280			
		CPB3	CPB 6	950	1410			
2	10	TSPB1	TSPB 4	770	990		25.03	34.90
		TSPB2	TSPB 5	730	1080			
		TSPB3	TSPB 6	810	1150			
3	15	TSPB1	TSPB 4	730	950		24.17	30.57
		TSPB2	TSPB 5	760	990			
		TSPB3	TSPB 6	740	880			
4	20	TSPB1	TSPB 4	720	700		26.99	24.17
		TSPB2	TSPB 5	780	750			
		TSPB3	TSPB 6	990	780			
5	25	TSPB1	TSPB 4	700	610		24.20	19.95
		TSPB2	TSPB 5	690	650			
		TSPB3	TSPB 6	750	580			
6	30	TSPB1	TSPB 4	590	650		18.86	19.62
		TSPB2	TSPB 5	540	590			
		TSPB3	TSPB 6	610	570			

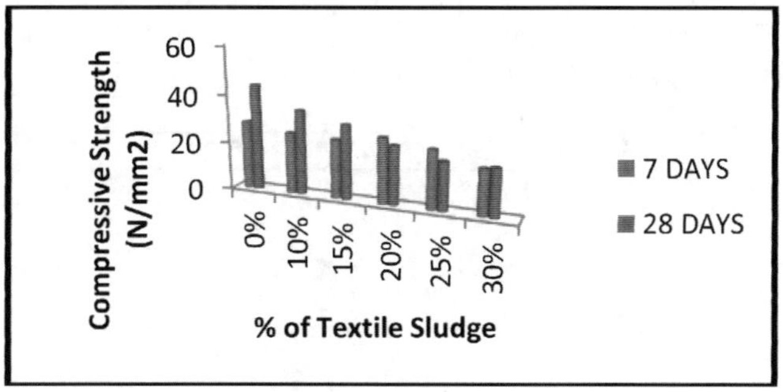

Figure 10. Compressive Strength of the specimen

3.4 Water absorption test

The water absorption, being the normal of three units, when decided in the way portrayed in Addition C, will not be in excess of 6 percent by mass and in singular examples; the water ingestion ought to be limited to 7 percent. (According to IS 15658: 2006). The average of three specimen units shall not be more than 6 percent by mass of sample, and it should be restricted to 7 percent as per IS code. Finally the dry density was used in the range of 2100 to 2165 kg/m³.The water absorption test shown in figure 11 and the results were discussed in figure 12.

Figure 11. Water absorption test

Table 5. Water Absorption of the specimen

S.No	%	Specimen of 7 Days	Specimen of 28 Days	W1 7 Days (kg)	W2 7 days (kg)	W1 28 days (kg)	W2 28 days (kg)	Water absorption for 7 Days %	Water absorption for 28days %
1	0	CPB1	CPB 4	5.575	5.635	5.545	5.635	0.93	1.8
		CPB2	CPB 5	6.190	6.240	6.120	6.240		
		CPB3	CPB 6	5.520	5.600	6.200	6.320		
2	10	TSPB1	TSPB 4	5.305	5.320	5.450	5.590	0.45	2.36
		TSPB2	TSPB 5	4.565	4.600	5.215	5.345		
		TSPB3	TSPB 6	6.155	6.175	5.040	5.150		
3	15	TSPB1	TSPB 4	5.030	5.060	5.630	5.755	0.56	2.44
		TSPB2	TSPB 5	5.480	5.510	5.110	5.240		
		TSPB3	TSPB 6	5.320	5.350	5.005	5.125		
4	20	TSPB 1	TSPB 4	5.555	5.690	5.335	5.375	2.98	0.71
		TSPB 2	TSPB 5	5.225	5.415	5.405	5.440		
		TSPB 3	TSPB 6	5.215	5.380	5.260	5.300		
5	25	TSPB 1	TSPB 4	5.405	5.505	5.630	5.670	2.02	0.99
		TSPB 2	TSPB 5	5.060	5.160	5.420	5.450		
		TSPB 3	TSPB 6	5.335	5.460	5.480	5.575		
6	30	TSPB 1	TSPB 4	5.350	5.390	5.350	5.375	0.73	0.406
		TSPB 2	TSPB 5	5.245	5.283	5.245	5.265		
		TSPB 3	TSPB 6	5.410	5.450	5.410	5.430		

*CPB-Conventional Paver Block
*TSPB – Textile Sludge Paver Block

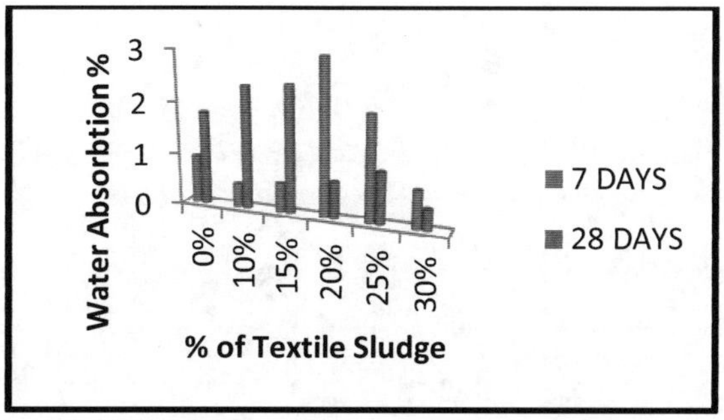

Figure 12. Water Absorption

4 Conclusion

From the experimental investigation, the preliminary properties of materials obtained and verified with IS code. The percentage of sludge varies from 5 % to 35 % for the replacement of cement and used as binding material for the manufacture of paver blocks. The following results are summarized for the manufacture of specimens in a very effective manner by utilizing textile sludge in the construction industry,

- With the addition of textile sludge and M-sand in the mix proportions, the density of the samples was decreased around 0.9 percentage of conventional paver block. This leads to making the concrete is in lightweight condition.
- For M40 grade concrete 1:2.33:1.82 mix proportions arrived and various percentage of sludge has to take for the investigation. The conventional paver blocks were cast and the compressive strength of the specimen was obtained as 44 N/mm2 using 2000kN Capacity CTM.
- Around 70 % of compressive strength of the concrete was obtained by adding up to 15 % of TS in the mix compare to CPB. Beyond the 15% of TS addition in Concrete, it leads to a decrease in the compressive strength of concrete predominantly.
- The water absorption of the TSPB was within the limit and also it satisfies the IS 15658:2006.

References

[1] Balasubramanian, P C Sabumon , John U Lazar, R Ilangovan 2006 Reuse of textile effluent treatment plant sludge in building materials *Journal of Waste Management* **26** 22–28.

[2] P Velumani & S Senthil kumar 2017 Production of sludge-incorporated paver blocks for efficient waste management *Journal of the Air & Waste Management Association* ISSN: 1096-2247 (Print) 2162-2906.

[3] Shweta Goyal ,Rafat Siddique, Sujant Jha, Devender Sharma 2019 Utilization of textile sludge in cement mortar and paste *Journal of Construction and Building Materials* **214** 169–177.

[4] Murugesan T, Bahurudeen A, Sakthivel M, Vijay R, Sakthivel S 2017 Performance evaluation of burnt clay-fly ash unburnt bricks and precast paver blocks *Materials Today: Proceedings* **4** 9673–9679.

[5] S Agyeman, N K Obeng-Ahenkora, S Assiamah, G Twumasi 2019 Exploiting recycled plastic waste as an alternative binder for paving blocks production *Journal of Case Studies in Construction Materials* **11** e00246.

[6] Manoj Kumar Dash , Sanjaya Kumar Patro b, Ashoke Kumar Rath 2016 Sustainable use of industrial-waste as partial replacement of fine aggregate for preparation of concrete – A review *International Journal of Sustainable Built Environment* **5** 484–516.

[7] Fabiana Maria da Silva et. al. 2015 Investigation on the properties of concrete tactile paving blocks made with recycled tire rubber *Journal of Construction and Building Materials* **91** 71–79.

[8] B Aneesh Narayan and J Supriya 2017 A study on textile sludge management *Anveshana's International Journal of Research in Engineering and Applied Sciences* Volume **2** ISSUE 5 , ISSN-2455-6300.

[9] Kaushal Patel et. a.l 2017 Technical feasibility study on utilization of textile sludge as a cement substitute in rubber mould paver block *International Journal of Constructive Research in Civil Engineering (IJCRCE)* Volume **3** Issue 1 PP 19-25 ISSN 2454-8693.

[10] M Indumathiet et. al 2017 Experimental investigation on partial replacement of etp sludge from textile industry on paver block *International Journal for Scientific Research & Development* Vol. **5** Issue 02 ISSN: 2321-0613.

A Comprehensive Review on the Replacement of Virgin Material using Reclaimed Asphalt Pavement (RAP) Material in Flexible Pavements

Pala Gireesh kumar[1]

[1]Associate professor and Head, Department of Civil Engineering, Shri Vishnu Engineering College for Women(Autonomous),Bhimavaram,534202,West Godavari, Andhra Pradesh, India

Email: palagireeshk@gmail.com

Abstract. With increased demand and limited aggregate and binder supply brings the use of Reclaimed Asphalt Pavement (RAP) as a valuable component in Hot Mix Asphalt(HMA) since 1970's in US. Later on, the use of RAP has evolved into a regular practice in many countries around the world and use of these materials in the past has proven to be economically and environmentally sound. Mixing of RAP in virgin materials has been greatly favored over virgin materials because of increase in cost of asphalt, scarcity of quality aggregates and asphalt. The use of RAP in pavement design is Recycling the left over Asphalt through crushing and is recycled back into asphalt which performs equal or better than the virgin mixtures. Now-a-days being more environmentally friendly is important and hence, recycling on site can offer real benefits not only through better planning and regulations, but also saving in the waste disposal charges, energy saving which is modified into the strengthen material. The use of RAP is gaining momentum in India at present and there were very limited studies carried out on the use of RAP in Warm Mix Asphalt (WMA) as well. Therefore an ample scope exists inorder to investigate the complete usage of RAP content.

1. Introduction

During the past few decades, Quarrying works have become downturned due to the increase in the emission of many poisonous gases like carbon dioxide, carbon monoxide, nitrogen dioxide etc .This results in the paucity of the origination of coarse aggregates and in the escalation of their cost[2] . In 1970's, due to Arab Oil Embargo, the petroleum and binder prices were skyrocketed. In order to overcome these circumstances, the material which was excavated during the demolition of roads can be reused.

Owing to the upsurge in the traffic conditions and perpetual mobility of vehicles, roads are getting deteriorated. So there is a need in the excavation of roads either by full depth reclamation or by surface reclamation [7]. The material which was procured during this process is called Reclaimed Asphalt Pavement. To make the project more economical, to reduce the adverse effects on the environment and also to optimize the resources the reuse of the material which was reclined on the sides of roads as a waste material through excavation can be entertained [3].

Reclaimed Asphalt Pavement, which on crushing gives aggregates and binder, in olden days was excavated into big boulders through excavators. But with the discovery of millers and bulldozers the pulverizing of RAP into nominal grade aggregates has become evident. After crushing, mixing of RAP with virgin aggregates should be done. During the past years, aged pavement was taken off and thrown away in landfills [6]. As land filling these materials has become less practical and more expensive and also the procurement of quality virgin material reduces, mixing of RAP to virgin materials has become widespread. Because the addition of RAP increases the stiffness and brittleness of pavement materials which gives high potential for cracking [16]. RAP performs better or equal to virgin aggregates [2][11]. The usage of RAP has been increasing now-a-days considering to the environmental and the economical conditions.

This article completely discuss about different works carried out by various research people and stated that the usage of RAP makes the project more economical and reduces the production of waste[5]. And also by reusing the recycled road aggregates which are generated at the same site the cost can be reduced up to 25 to 30% [4].

2. State of art for recycling of asphalt pavements

An extensive literature survey was carried out on recycling of asphalt pavements. The details regarding various methods of recycling were obtained and the laboratory conclusions which were drawn to ascertain the properties of RAP were presented below:

Brian Hill (2011) stated that the WMA technologies increase moisture and rutting susceptibility whereas RAP increases stiffness of resulting mixture. The laboratory tests conducted were Brookfield Rotational Viscometer Test, Bending Beam Rheometer Test, Hamburg Wheel Tracking Test, AASHTO T-283 Moisture Sensitivity Test , Disk-Shaped Compact Tension (DC(T)) Test. The main objective of this paper is to design virgin and high RAP HMA mixtures which satisfy super pave mix design criteria [5].

Jashanjot Singh and A.K Duggal (2015) determined the optimization of the use of RAP in surface courses such as Bituminous concrete. The usage of RAP makes the project more economical and reduces the production of waste. Several laboratory tests were conducted in order to obtain the physical and engineering properties of RAP as well as Virgin aggregates. The tests results concluded that RAP performs better or equal to Virgin aggregates and the results obtained were within the MORTH specifications [10].

Abhishek Verma et al.(2017) stated that the aggregates which are obtained from various rocks have gained more demand as construction materials for flexible pavements. Due to increase in the cost of virgin aggregates due to environmental constraints there is a need in the reuse of the material called Reclaimed Asphalt Pavement which was deteriorated. During the conduction of laboratory tests like Marshall Stability test, California Bearing Ratio test, Unconfined Compression test the stabilizers like fly ash and cement were added and concluded that the results obtained were satisfactory [1].

Krupa S Sharma, Ashok Pates(2017)determines the objective of this paper was to reuse the material which was recycled as filling material without any analysis & tests in submerged areas. To investigate the recycling mix after carrying out Marshall Stability tests. The methodology used for investigation of the materials is Marshall Stability test. It concludes that as a result of the analysis the RAP added bituminous mix given competitive as compared with fresh bituminous mix[13].

T.Anil Pradyumna and Abhishek Mittal(2013) explains that the objective of this is to study on mechanistic evaluation of hot recycle mixes with and without utilization of recycling agents and to increase performance properties of mixes. The tests like Resilience modulus test, moisture susceptibility test, Dynamic creep test, Rutting test were conducted. Based on laboratory tests conducted on fresh mixes and mixes which are having the percentage of RAP as 20 it was found that by adding RAP all the properties of bituminous mixes can be improved and RAP with 20% would behave better compare to virgin materials under same conditions[22].

Anil Kumar Yadava and Dr.Syed Aqeel Ahmad (2019) The Bituminous pavement recycling was not yet popular in India and other developing countries due to sufficient guidelines and the lack of availability of the equipment to extract the material. Different types of recycling techniques were mentioned and Hot in plant recycling technique was adopted by considering 70-100% RAP. Based on the tests conducted and mix design, analysis for future enhancement was made [2].

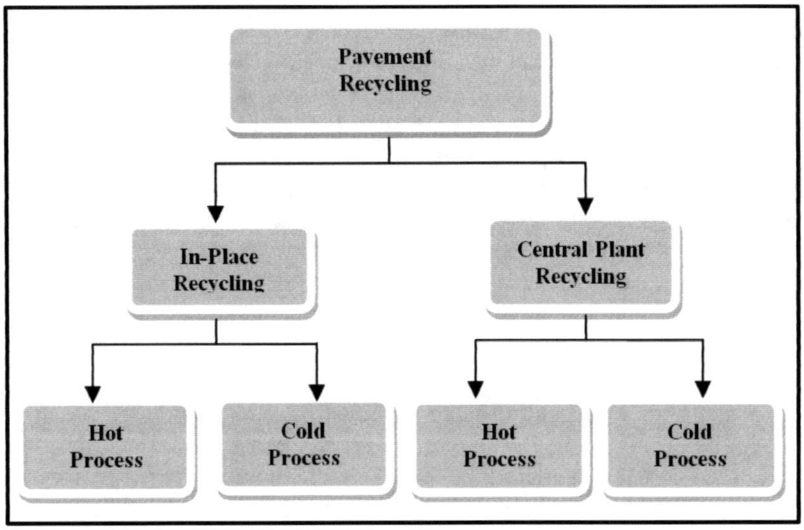

Fig.1 Classification of recycling methods based on Processes

Brajesh Mishra(2015) By reusing the recycled road aggregates which are generated at the same site the cost can be reduced up to 25 to 30 %. The materials which are recycled frequently are Reclaimed Asphalt Pavements (RAP) and Recycled concrete aggregates (RCA). Consumption of natural aggregates and binder can be made less in mixes containing asphalt by using RAP materials. The main objective is used to use RAP materials after blending in base and sub base course of materials. The laboratory tests were conducted by considering 0% RAP, 10% RAP, 20% RAP, 30% RAP, 40% RAP and 100% RAP. The usage of RAP up to
30% can be feasible and its characteristics are equal to virgin aggregates and the RAP material up to 50% is used as a replacement of granular sub base [4].

Rohan Pawar et al.(2017) The RAP material was collected and is recycled into the nominal aggregates by milling process. The objective is to determine the recycling of RAP content by the central mix plant & hot mix plant. The tests were conducted on virgin aggregates, RAP & Bitumen. All the collected materials were mixed in specified quantities and temperatures based on the requirement and conducted Marshall Stability test. It concludes that according to the Marshall Stability test conducted the values increases for 30% RAP in comparison with the conventional mixes & then reduced for 40% placement of RAP [12].

Siksha Swaroopa Kar, et al. (2018) determines that the aim of this article was to study the effects which are caued by RAP & its components on the mechanical behavior of Foamed Bituminous mixtures.The laboratory tests conducted are Indirect tensile strength, resilience modulus test. The results from the study of the Foamed Bituminous mixtures following to Indian standard conditions were analysed. The foaming qualities of fresh bitumen are used for optimum water content & the foaming temperature[21].

Imad L.AI Qadi et al.(2007) Recycling WMA material results in a mixture of the aggregates which can be reusable and asphalt binder which is called as RAP. Rutting performance typically improves by the use of RAP whereas fatigue and thermal performance has become inconsistent. The objective is to evolve the relationship between old and fresh asphalt binders in RAP[9].

Umesh Kamariya et al.(2018)determines that the objective is to determine the necessity of the mixes with and without added any admixtures and also to increase the properties of RAP materials by increasing the mixing of RAP percentages. The recycling process can be done either by hot or cold recycling process. The tests they conducted are moisture susceptibility test and rutting resistance test. It concludes that as compared to fresh mixes RAP is stiffer and brittle material and the results shows that the RAP led to increase permanent deformation and moisture damage[23].

Feipeng Xiao and Serji N. Amirkhanian (2007) explains that the objective is to measure the performance of the pavements in several design methods. The crumb rubber was mixed to the fresh binder. The purpose of this is to improve the rutting resistance properties of the rubberized asphalt mixtures.The indirect tensile strength test and rutting susceptibilitytest were also conducted to rubberize each material. It concludes that the admixture which was added along with RAP increases the resistanceagainst deformation and creep stiffness[7].

Fawaz Kaseer et al. (2019) proposed a technique to ascertain the percentage of vigorous RAP binder and to know the factors which effects the RAP material. The methodology adopted was Resilience Modulus test and the Marshall stability test. Finally the virgin binder content is mixed with recycled binder mixes as they provide a reasonable estimate of the RAP binder [6].

D.Stephen lane et al.(2015) stated that the use of RAP fits the global objective to continual development by the use of the natural resources. It was initiated to explore the potential alternatives of RAP in the higher quantities for base and sub-base. The tests conducted were Resilience modulus test which describes the material response to traffic loading. As the content of the RAP increases, the shear strength of the blend reduced below the level which is required.[19].

Martina Irene Giani et al. (2014) describes the use of RAP is the other form of the material related technology which reduces the asphalt binder, scarcity of the aggregates thus saves natural resources. The methodology used were Life cycle Assessment (LCA) and Cold In place Recycling (CIR) which analyses and describes its environmental impacts. To understand the environmental benefits which results in the use of RAP techniques, to reduce the temperature of plant production and to reconstruct the road after its usage are the advantages by adopting these techniques[14].

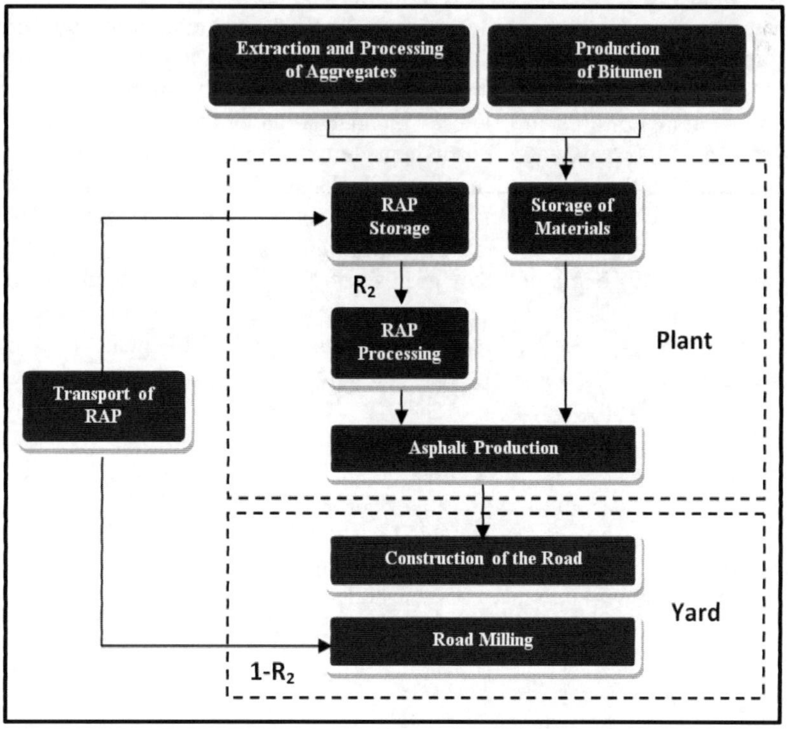

Fig.2 Processing of Asphalt pavements

Prithvi S. Kandhal (1997) the existing pavement structures can be strengthened by recycling without adding substantial overlays. The cold planning method was used which describes an automatic method for removal of asphalt pavements because recycling of asphalt surface does not increase the structural capacity of current pavement & there is no other alternative for thickness design of surface recycling[17].

S.Mangiafico et al.(2013) describes that the objective is to determine the visco elastic characteristics of binders and to examine as they amend with RAP content. Dynamic shear Rheometer and complex modulus tests were conducted. There is a necessity for analyzing performance related tests also before finalzing any recycled mix design. The tests were conducted by considering different binder contents along with varying RAP percentages [18].

Haritha Musty et al.(2012) Being economic, sustainable, and environmentally friendly, RAP can be restored for a small amount of aggregates in Hot Mix Asphalt (HMA) where standard aggregates are scare. The objective is to regulate the effects of having higher percentages of RAP and FRAP on mixture performance while meeting the present requirements of Super pave mix design. Dynamic Modulus test and Wheel cracking test were conducted by taking 20%, 40% RAP for which 20% RAP is having higher modulus[8].

Arshad Hussain and QiuYajun (2013) Within the usage of RAP, including the conservation of natural resources the cost of new asphalt mixes can also be reduced. The percentage of RAP is mostly used between 10 to 30 in hot in plant asphalt mixtures. During the construction of a pavement, environmental and financial departments are forcing to place high percentage of RAP. The laboratory tests conducted were DSR testing Rheometer testing. When the RAP binder content increases the viscosity, stiffness and critical temperature of the blend also increases[3].

K. Aravind and Animesh Das (2006) induced to plan a pavement with a hot recycled asphalt surfacing, where on conducting Marshall stability test, Creep test, Fatigue test the mix which was recycled can be finalized through a systematic mix design. The criterion gained are used in the design of pavement. Mix with 5.5% binder content is safe to produce the optimum binder content. The construction cost with recycled mix could be economical compare to virgin mix[11].

Pavel kriz et al.(2014) Performed replica of dissemination throughout realistic mix production storage and paving temperatures in order to blend RAP and fresh aggregates and to make sure pavement resistance to moist damage, Permanent deformation and cracking. 15 to 20% of RAP is commonly used in HMA/WMA in olden days in Canada and up to 40% have been recently used. The main objective is to determine an average diffusion distance or binder film thickness in a typical asphalt mix in order to calibrate the diffusion model[16].

Sara Bressiet al.(2019) explained that the use of recycled materials in bituminous pavements reduce the environmental effects and excessive use of non-renewable resources. The probable environmental effects of asphalt mixes which contains crumb rubber and RAP assuming another binder degree activation of the old binder is evaluated through life cycle analysis. The usage of small amount of crumb rubber does not justify the optimum usage of natural resources and environmental effects[20].

Munagala Sreenivasulu Reddy and Suvarna P. (2015) proposed that RAP can be used as a substitute for fine aggregates and coarse aggregates. Performance of Pavement Quality Cement Concrete (PQCC) is used as an admixture which is studied by replacing virgin aggregates with RAP and also by conducting experimental tests on RAP. Fine aggregate is replaced with 0%, 15%, 30%, 45%, 60% RAP and coarse aggregates are replaced with RAP by 0%, 15%, 30%, 45% and 60%. [15].

3. Conclusion

An extensive literature survey was carried out on recycling of asphalt pavements under different processes and by different conditions to make the economic conditions favourable and to decrease the adverse effects on the environment. Some of the conclusions drawn from the literature survey are as follows.

- It was inferred that mixes having the percentage of RAP as 20 would behave better compare to virgin mixes under same conditions.
- RAP can be used as aggregate up to concrete replacement of 30%.
- The rheological properties of the binder can be used to analyze the sample and the batch variability and their subsequent impact on the pavement performance.
- The usage of RAP makes the project more economical and reduces the optimum usage of natural resources.
- RAP can be used up to certain limit in the maintenance of roads and as an alternative to overlays.

References

[1] Abhishek Verma and Rachit Sharma et. al. 2017 Evaluation of RAP in flexible pavement layers.
[2] Anil Kumar Yadava and Dr. Syed Aqeel Ahmad 2019 A critical review of characterization and performance evaluation of RAP in road construction.
[3] Arshad Hussain and QiuYajun 2013 Effect of reclaimed asphalt pavement on the properties of asphalt binders.
[4] Brajesh Mishra 2015 Study on use of RAP materials in flexible pavements.
[5] Brian Hill 2011 Performance evaluation of warm-mix asphalt mixtures incorporating reclaimed asphalt pavement.
[6] Fawaz Kaseer and Edith Arambula et. al. 2019 A method to quantify RAP binder availability in recycled asphalt mixes.
[7] Feipeng Xiao and Hsein Juang C et. al. 2007 Rutting resistance of rubberised asphalt concrete pavements containing RAP mixtures.
[8] Haritha Musty Y and Nasim Sabahfer et. al. 2012 Use of high volume RAP for asphalt pavement rehabilitation due to increased highway truck traffic from freight transportation.
[9] Imad L.AI- Qadi and Mostafa Elseifi et. al. 2007 Reclaimed asphalt pavement - A literature review.
[10] Jashanjot Singh and Duggal A K et. al. 2015 An experimental study on RAP in bituminous concrete.
[11] Aravind K and Animesh Das 2006 Pavement design with central plant hot mix recycled asphalt mixes.
[12] Keerthi Gowda B S et. al. 2017 Recycling of bituminous pavements by RAP method.
[13] Krupa Sharma S and Ashok Patel 2018 Study on reclaimed asphalt pavements.
[14] Martina Irene Giani, Giovami Dotelli et. al. 2014 Comparative life cycle assessment of asphalt pavements using recycle asphalt, warm-mix technology and cold-place recycling.
[15] Sreenivasulu Reddy M and Suvarna P 2016 A detailed study on reclaimed asphalt pavement in pavement quality cement concrete.
[16] Pavel Kriz and Daniel Grant et. al. 2014 Reclaimed asphalt pavement-virgin binder diffusion in asphalt mixes.
[17] Prithvi S Kandhal 1997 Recycling of asphalt pavements
[18] Salvatore Mangisfico, Franciois Olard et. al. 2013 Influence of RAP content on complex modulus of asphalt binder blends and corresponding mixes.
[19] Sameer Shetty and Stephen Lane D et. al. 2015 Feasibility of RAP use as road base and sub base material.
[20] Sara Bressi and Massimo Losa et. al. 2019 A comparitive environmental impact analysis of asphalt mixtures containing crumb rubber and RAP using life cycle assessment.
[21] Siksha Swaroopa Kar and Aravind Krishna Swamy et. al. 2017 Impact of recycled asphalt pavements on properties of foamed bituminous mixtures.
[22] Anil Pradyumna T and Abhishek Mittal et. al. 2013 Characterization of RAP for use in bituminous road construction.
[23] UmeshKamariya and Amit A. Amin et. al. 2018 Utilisation of reclaimed asphalt pavement materials.

IOP Conf. Series: Materials Science and Engineering 1006 (2020) 012042 doi:10.1088/1757-899X/1006/1/012042

A Study on Effect of Reclaimed Asphalt Pavement Material over Virgin Material in Bituminous Mix

Pala Gireesh kumar[1]

[1]Associate professor and Head, Department of Civil Engineering, Shri Vishnu Engineering College for Women(Autonomous),Bhimavaram,534202,West Godavari, Andhra Pradesh, India

Email: palagireeshk@gmail.com

Abstract. With increased demand and limited aggregate and binder supply brings the use of Reclaimed Asphalt Pavement (RAP) as a valuable component in Hot Mix Asphalt(HMA) since 1970's in US. Later on, the use of RAP has evolved into a regular practice in many countries around the world and use of these materials in the past has proven to be economically and environmentally sound. Mixing of RAP in virgin materials has been greatly favored over virgin materials because of increase in cost of asphalt, scarcity of quality aggregates and asphalt. The use of RAP in pavement design is Recycling the left over Asphalt through crushing and is recycled back into asphalt which performs equal or better than the virgin mixtures. Now-a-days being more environmentally friendly is important and hence, recycling on site can offer real benefits not only through better planning and regulations, but also saving in the waste disposal charges, energy saving which is modified into the strengthen material. The use of RAP is gaining momentum in India at present and very less studies were carried out and there were no such studies on complete usage of RAP and hence this study was carried out for complete utilization of RAP content in Bituminous mixture. Experimental results show that, RAP materials has much better resistance under compressive load and impact load than virgin materials, i.e. 8.43% of impact value and 6.23% of crushing value was obtained for RAP whereas 12.83% of impact value and 9.07% of crushing value was obtained for virgin material. It was found that 30 to 50% of RAP material can be preferable for the construction of roads where as on using 100% RAP the pavement becomes weak and unstable.

1. Introduction

During the last decades, Quarrying works have become downturned due to the increase in the discharge of many poisonous gases like carbon dioxide, carbon monoxide, nitrogen dioxide etc. This results in the paucity of the origination of coarse aggregates and in the escalation of their cost[2] . In 1970's,due to Arab Oil Embargo, the petroleum and binder prices were skyrocketed. In order to overcome these circumstances, the material which was excavated during the demolition of roads can be reused.

Owing to the upsurge in the traffic conditions and perpetual mobility of vehicles, roads are getting deteriorated. So there is a need in the excavation of roads either by full depth reclamation or by surface reclamation[7]. The material which was procured during this process is called Reclaimed Asphalt

Pavement. To make the project more economical, to reduce the adverse effects on the environment and also to optimize the resources the reuse of the material which was reclined on the sides of roads as a waste material through excavation can be entertained.[3]

Reclaimed Asphalt Pavement, which on crushing gives aggregates and binder, in olden days was excavated into big boulders through excavators. But with the discovery of millers and bulldozers the pulverizing of RAP into nominal grade aggregates has become evident. After crushing, mixing of RAP with virgin aggregates should be done. During the past years, aged pavement was taken off and thrown away in landfills[6]. As landfilling these materials has become less practical and more expensive and also the procurement of quality virgin material reduces, mixing of RAP to virgin materials has become widespread. Because the addition of RAP increases the stiffness and brittleness of pavement materials which gives high potential for cracking[16]. RAP performs better or equal to virgin aggregates[2][11]. The usage of RAP has been increasing now-a-days considering to the environmental and the economical conditions.

This article mainly discusses about the RAP material which was collected near Tadepalligudem, West Godavari District, Andhra Pradesh, India and several laboratory tests were conducted on RAP material as well as on virgin material to analyze the physical and mechanical properties of both the mixes and also to determine the suitability of 100% RAP in the sustainable construction of flexible pavements. Hence an attempt has been made on the bituminous mix by considering different proportions of RAP material and varying percentages of binder content.

2. Necessity of study

Reclaimed Asphalt Pavement(RAP) is a material which is obtained during the reprocessed, reconstruction or rehabilitation of roads which can be used as a replacement for fine aggregates and coarse aggregates. The material which was recycled has more resistance to scuffing and shearing, which in turn increase the rutting resistance. Scarcity of aggregates and crude oil results in increase in the cost of asphalt binder and aggregates spurred a new interest in the reuse of RAP. Recycling solve the problem of construction waste in transportation engineering projects. Strength parameters either of same or better quality are produced by RAP when compare to virgin aggregates. Reusing of RAP reduces the consumption of fuel and reduces damages to other materials due to the transportation of minerals from quarry sites.

3. Avant-grade of reclaimed asphalt pavements

An extensive literature survey was carried out on recycling of asphalt pavements. The details regarding various methods of recycling were obtained and the laboratory conclusions which were drawn to ascertain the properties of RAP were presented below:

The usage of RAP makes the project more economical and reduces the production of waste. Several laboratory tests were performed to determine the physical and engineering properties of RAP as well as Virgin aggregates. The tests results concluded that RAP performs better or equal to Virgin aggregates and the results obtained were within the MORTH specifications [9]. The aggregates which are obtained from various rocks have gained more demand as construction materials for flexible pavements. Due to increase in the cost of virgin aggregates due to environmental constraints there is a need in the reuse of the material called Reclaimed Asphalt Pavement which was deteriorated. During the conduction of laboratory tests like Marshall stability test, California Bearing Ratio test, Unconfined Compression test the stabilizers like fly ash and cement were added and concluded that the results obtained were satisfactory[6] Abhishek Verma et al (2017).

By reusing the recycled road aggregates which are generated at the same site the cost can be reduced up to 25 to 30 %. The materials which can be recycled more are Reclaimed Asphalt

Pavements(RAP) and Recycled concrete aggregates (RCA). Consumption of natural aggregates and binder can be decreased in asphalt paving mixes by using RAP materials. The main objective is to use RAP materials after blending in base and sub base course of materials. The laboratory tests were conducted by considering 0% RAP, 10% RAP,20% RAP, 30% RAP, 40% RAP and 100% RAP. The usage of RAP up to 30% can be feasible and its characteristics are equal to virgin aggregates and the RAP material up to 50% is used as a replacement of granular sub base [4]. The methodology used for investigation of the materials is Marshall Stability test. It concludes that as a result of the analysis the RAP added bituminous mix given competitive as compared with fresh bituminous mix [1].The Bituminous pavement recycling was not yet popular in India and other developing countries due to sufficient guidelines and the lack of availability of the equipment to extract the material. Different types of recycling techniques were mentioned and Hot in plant recycling technique was adopted by considering 70-100% RAP. Based on the tests conducted and mix design, analysis for future enhancement was made [11] Anil Kumar Yadava et al(2019).

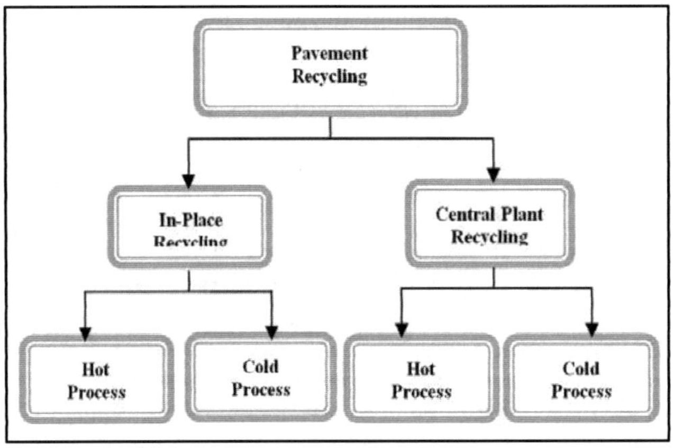

Fig.1 Classification of Recycling methods based on Processes

The RAP material was collected and is recycled into the nominal aggregates by milling process. The objective is to determine the recycling of RAP content by the central mix plant & hot mix plant. The tests were conducted on virgin aggregates, RAP & Bitumen. All the collected materials were mixed in specified quantities and temperatures based on the requirement and conducted Marshall Stability test. It concludes that According to the Marshall Stability test conducted the values increases for 30% RAP in comparison with the conventional mixes & then reduced for 40% replacement of RAP[14]. This is to increase performance properties of mixes in the tests like Resilience modulus test, moisture susceptibility test, Dynamic creep test and Rutting test. Based on laboratory tests conducted on fresh mixes and mixes which are having the percentage of RAP as 20 it was found that by adding RAP all the properties of bituminous mixes can be improved and RAP with 20% would behave better compare to virgin materials under same conditions[1].This effects the RAP & its components on the mechanical behavior of Foamed Bituminous mixtures. The laboratory tests conducted are Indirect tensile strength, resilience modulus test. The results from the study of the Foamed Bituminous mixtures following to Indian standard conditions were found. The foaming qualities of fresh bitumen are used for optimum water content & the foaming temperature[7] Siksha Swaroop Kar et al(2018).

Recycling HMA material results in a reusable mixture of aggregates and asphalt binder known as RAP. Rutting performance typically improves by the use of RAP where as fatigue and thermal performance has become inconsistent. The objective is to promote an understanding of the connection between aged and virgin asphalt binders in RAP[2]. To increase the necessity of the mixes with and without added any admixtures and also to increase the properties of RAP materials by increasing the mixing of RAP percentages. The recycling process can be done either by hot or cold recycling process. The tests they conducted are moisture susceptibility test and rutting resistance test. It concludes that as compared to fresh mixes RAP is stiffer and brittle material and the results shows that the RAP led to increase permanent deformation and moisture damage[13]. To measure the performance of the pavements in several design methods by using crumb rubber. The purpose of this is to improve the rutting resistance characteristics of the rubberized asphalt mixtures .The indirect tensile strength test and rutting susceptibility test were also conducted to rubberize each material. It concludes that the admixture which was added along with RAP increases the resistance against deformation and creep stiffness [16] Feipeng Xiao et al(2007).

A method to develop the percentage of active & available RAP binder and to know the factors which effects the RAP material. The methodology adopted was Resilience Modulus test and the Marshall stability test. Finally the virgin binder content is mixed with recycled binder mixes as they provide a reasonable estimate of the RAP binder[8]. The use of RAP fits the global objective to sustainable development by the use of the natural resources. It was initiated to explore the potential alternatives of RAP in the higher quantities for base and sub-base. The tests conducted were Resilience modulus test which describes the material response to traffic loading. As the RAP content increases, the shear strength of the blend decreased below the required level. RAP is the other form of the material related technology which reduces the asphalt binder, scarcity of the aggregates thus saves natural resources. The methodology used were Life cycle Assessment (LCA) and Cold In place Recycling (CIR) which analyses and describes its environmental impacts. To understand the environmental benefits which results in the use of RAP technologies, to reduce temperature of production in plant and to reconstruct the pavement after useful life are the advantages by adopting these techniques[22] Martina Ireneet et al(2014).

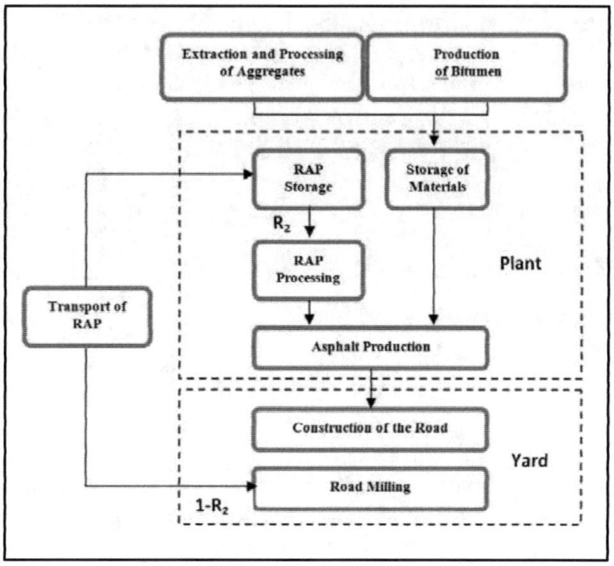

Fig.2 Processing of Asphalt Pavement

The existing pavement structures can be strengthened by recycling without adding substantial overlays. The cold planning method was used which describes an automatic method for removal of Asphalt pavements and there is no method to design of thickness for surface recycling [21]. It describes that objective is to determine the visco elastic characteristics of corresponding binders whether there is any change in reactions with RAP content or not. Dynamic Shear Rheometer and complex modulus tests were conducted in order to know about the mix reactions of different binder contents along with varying RAP percentages [20]. RAP can be restored for a small amount of aggregates in Hot Mix Asphalt(HMA) where standard aggregates are scare. The objective is to regulate the effects of having higher percentages of Reclaimed Asphalt Pavements(RAP) and Foamed Reclaimed Asphalt pavements(FRAP) on mixture performance while meetings the present requirements of Super pave method. Dynamic Modulus test and Wheel cracking test were conducted by taking 20%, 40% RAP for which 20% RAP is having higher modulus [17] Haritha musty et al (2012).

Within the usage of RAP, including the conservation of natural resources the cost of new asphalt mixes can also be reduced. The percentage of RAP is mostly used between 10 to 30 in hot in plant asphalt mixtures. During the construction of a pavement, environmental and financial departments are forcing to place high percentage of RAP. The laboratory tests conducted were DSR testing Rheometer testing. When the RAP binder content increases the viscosity, stiffness and critical temperature of the blend also increases[3].

To produce safe and optimum binder content, the analyses were done by conducting Marshall Stability test, Creep test and Fatigue test [15]. Therefore performed replica of dissemination throughout realistic mix production storage and paving temperatures in order to blend RAP and virgin aggregates and to make sure pavement resistance to moist damage, Permanent deformation and cracking . 15 to 20% of RAP is commonly used in HMA/WMA in olden days in Canada and up to 40% have been recently used. The main objective is to determine an average diffusion distance or binder film thickness in a typical asphalt mix in order to calibrate the diffusion model[19] Pavel Kriz et al (2014).

The usage of small amount of crumb rubber does not justify the optimum usage of natural resources and environmental effects but it can be used as a substitute for fine aggregates and coarse aggregates [18]. Performance of Pavement Quality Cement Concrete(PQCC) is used as an admixture which is studied by replacing virgin aggregates with RAP and also by conducting experimental tests on RAP. Fine aggregate is replaced with 0%, 15%, 30%, 45%, 60% RAP and coarse aggregates are replaced with RAP by 0%, 15%, 30%,45% and 60%. Finally the results concludes that RAP can be used as an aggregate in PQCC up to limited percentage [10] Suvarna P et al (2015).

- Summary: The above given state of art is about Reclaimed Asphalt Pavement and its properties. It is also discussed about various laboratory tests conducted on RAP. There is still some scope to investigate and to undergo research in this area.

4. Materials

It is required to determine the materials and their sources opted for this study. The materials include RAP material, Coarse aggregates, Bitumen (VG 30).

4.1 RAP material

The material which was obtained during the reconstruction and rehabilitation of roads is called Reclaimed Asphalt Pavement. RAP material which was collected for this study has undergone several laboratory tests after crushing into nominal size aggregates in order to determine its suitability in the construction of flexible pavements and to determine the physical properties like toughness, strength, hardness, quality, durability, shape and size.

4.2 Coarse aggregates

The Coarse aggregates (Virgin aggregates) of distinct sizes like 20mm,12mm, 10mm, 4.75mm, 2.36 mm were taken and different tests were conducted to determine the properties like toughness, strength, hardness, quality, durability, shape and size and to compare the results of RAP as well as Virgin aggregates.

4.3 Bitumen

Bitumen used for this study is VG- 30. It is the viscosity grade which is most suitable in the construction of flexible pavements as it has more durability and more resistance to sustain in the different temperature conditions.

5. Test results and discussions

5.1 Aggregate crushing test

This test was conducted on RAP as well as on Virgin aggregates to find out the strength.

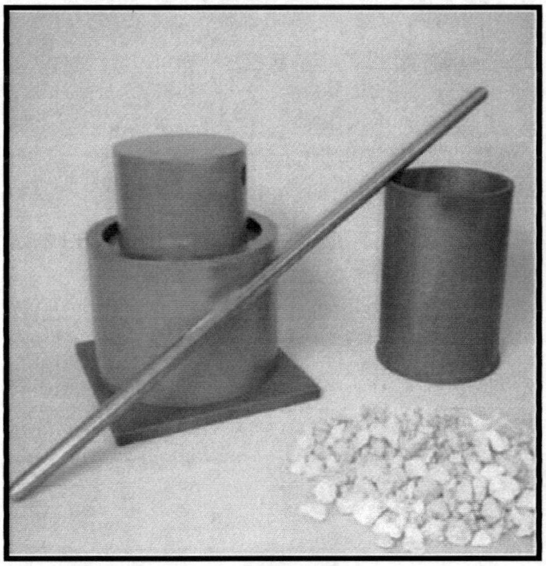

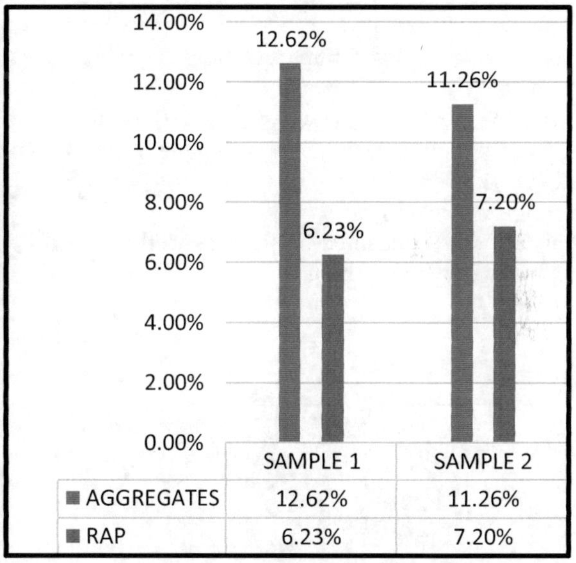

Fig 3 Crushing test apparatus

Fig 4 Analysis of crushing test

From this test it was analyzed that the value of crushing for RAP is less compare to nominal size aggregates and the value was within the MORTH limits and it was satisfactory.

5.2 Aggregate impact test

This test was conducted on RAP as well as on Virgin aggregates to find out the toughness.

IOP Conf. Series: Materials Science and Engineering 1006 (2020) 012042 doi:10.1088/1757-899X/1006/1/012042

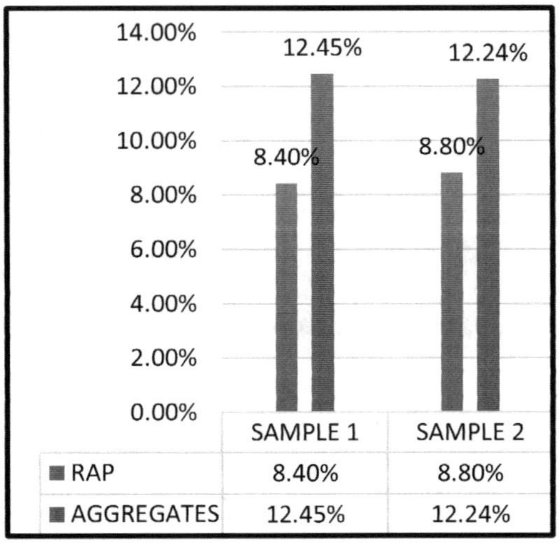

Fig 5 Impact testing machine **Fig 6** Analysis of Impact test

From the impact test it was concluded that the value of impact for RAP was less compare to nominal aggregates and the value was within the MORTH limits and it was satisfactory.

5.3 Aggregate abrasion test

This test was conducted on RAP as well as Virgin Aggregates to find out the abrasive nature between Aggregates and steel balls.

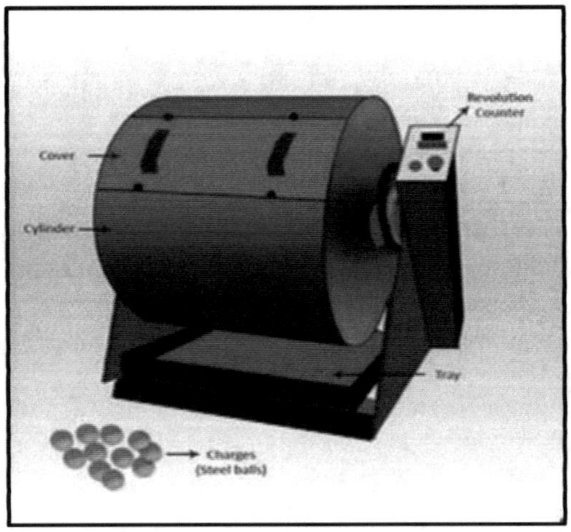

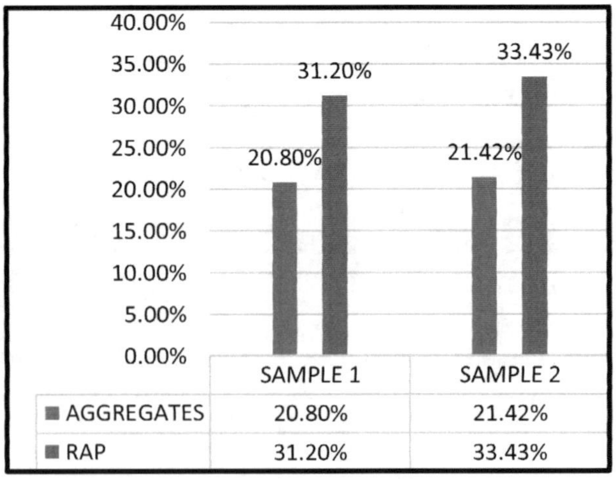

Fig 7 Abrasion testing machine **Fig 8** Analysis of Abrasion Test

From the abrasion test it was concluded that the value of the abrasion for Aggregates was less compared to RAP and the value was within the MORTH limits and it was satisfactory.

5.4 Aggregate attrition test

This test was conducted on RAP as well as for Virgin Aggregates to find out the resistance.

Fig 9 Attrition testing Machine

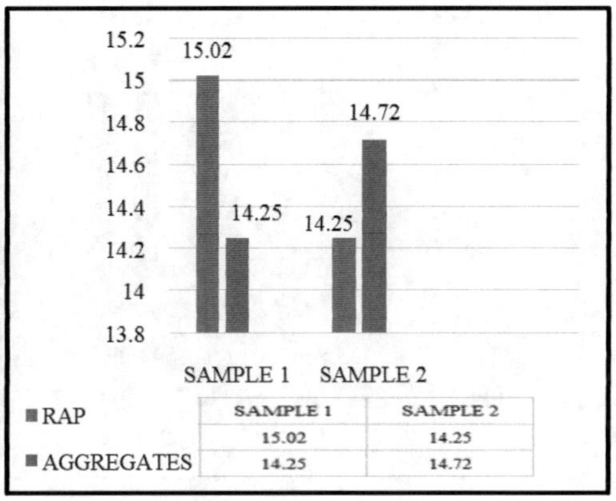

Fig 10 Analysis of Attrition test

From the attrition test it was concluded that the value of the attrition for aggregates was less compared to RAP and the values are within the MORTH limits and it was satisfactory.

5.5 Specific gravity and water absorption test

This test was conducted on Rap as well as for Virgin Aggregates to know about the quality of aggregates.

Fig 11 Specific Gravity test using wire bucket

IOP Conf. Series: Materials Science and Engineering 1006 (2020) 012042 doi:10.1088/1757-899X/1006/1/012042

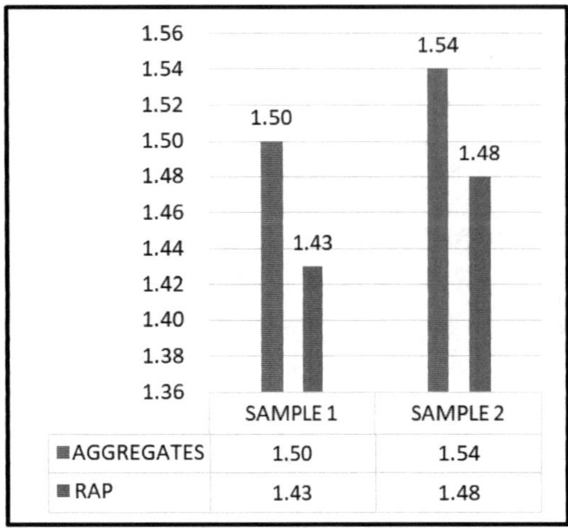

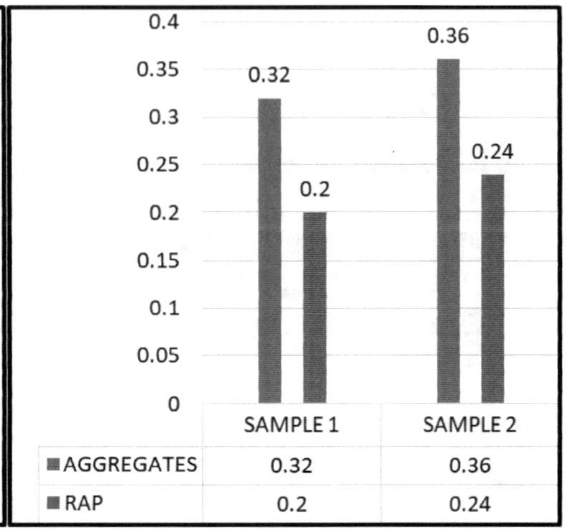

Fig 12 Analysis of Specific gravity test **Fig 13** Analysis of Water Absorption test

From the Specific gravity test it was concluded that the Specific gravity of RAP was more and water absorption of RAP was less compare to virgin aggregates and the results were satisfactory.

5.6 Results on samples of 100% rap and 100% virgin aggregate

Table 1: Results on samples of RAP and virgin aggregates

Characteristics	**100% RAP Aggregates**		**100% Virgin Aggregates**		**MORTH Specifications**	**Remarks**
	Sample 1	**Sample 2**	**Sample 1**	**Sample 2**		
Aggregate Impact value	8.40	8.80	12.45	12.24	30%	Satisfactory
Aggregate Crushing value	6.23	7.20	12.62	11.26	30%	Satisfactory
Los Angeles Abrasion value	20.80	21.42	31.20	33.43	35%	Satisfactory
Specific Gravity	1.43	1.48	1.50	1.54	2.6-3.0	Satisfactory
Water Absorption	0.2	0.24	0.32	0.36	2%	Satisfactory
Deval Attrition Test	15.02	15.67	14.25	14.72	30%	Satisfactory

6. Tests on Bitumen

6.1 Penetration test

Penetration test of bitumen was conducted to determine the hardness and consistency.

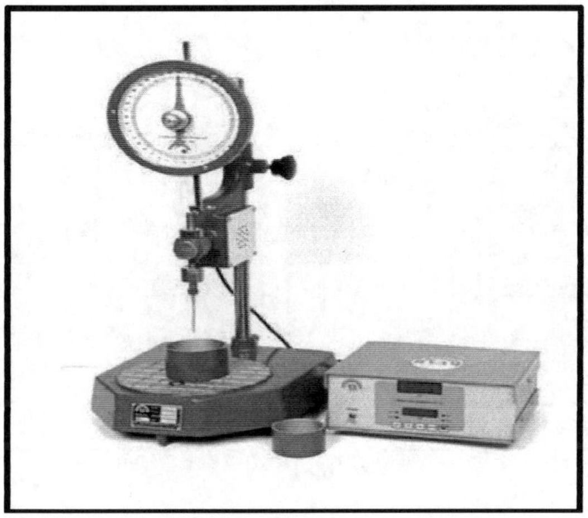

Fig 14 Penetration test

6.2 Ductility test

Ductility test of bitumen was conducted to determine the elongation nature under traffic load without getting cracks.

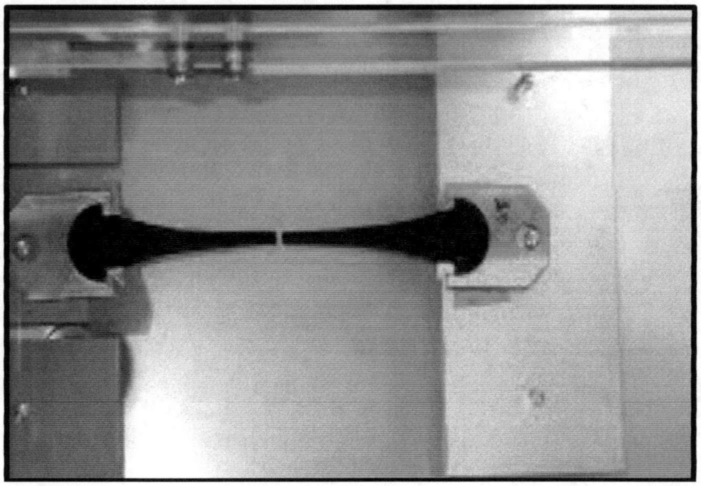

Fig 15 Ductility test by briquette mould

6.3 Softening point test

Softening test of bitumen was conducted to determine the temperature of bitumen binder heated for various road applications which is determined by ball and ring apparatus.

IOP Conf. Series: Materials Science and Engineering 1006 (2020) 012042 doi:10.1088/1757-899X/1006/1/012042

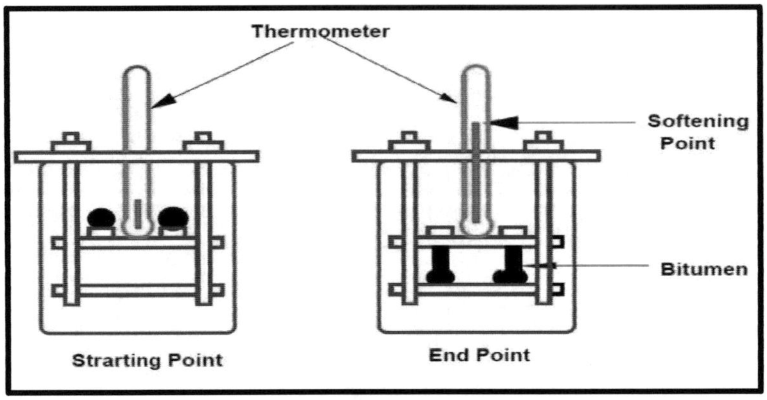

Fig 16 Softening point test apparatus

6.4 Bitumen test results

Table 2: Test results of bitumen

S NO	TYPE OF TEST	OBTAINED RESULTS	MORTH LIMITS
1	Penetration test	55	50-70
2	Softening point	40	47
3	Ductility test	38	40

7. Marshall Stability Test

7.1 Bitumen content VS marshall stability for modified and unmodified rap material

The maximum load sustained by a bituminous material at a loading rate of 50.8mm/ minute.

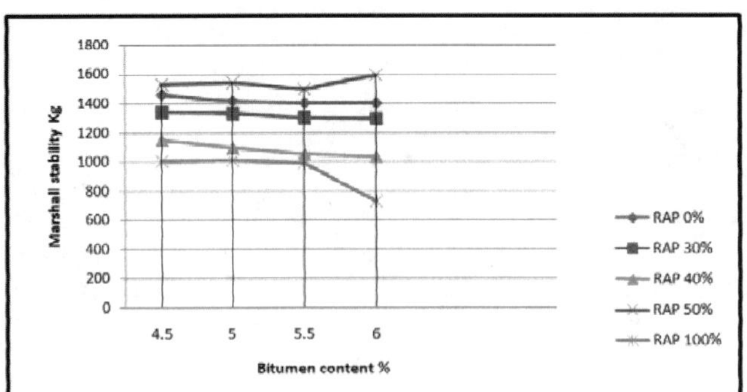

	RAP 0%	RAP 30%	RAP 40%	RAP 50%	RAP 100%
4.5	1461	1336	1148	1529	1002
5	1416	1328	1099	1543	1008
5.5	1402	1300	1052	1498	991
6	1400	1296	1035	1592	728

Fig 17 Graph of bitumen content VS marshall for modified and unmodified rap material

The minimum stability value should be 340kg.From the above graph it was observed that there is a gradual decrease in the stability value for RAP 100%, 0% with increase in the binder percentage and for the remaining percentages of RAP the results were satisfactory.

7.2 Bitumen content VS flow value for modified and unmodified rap material

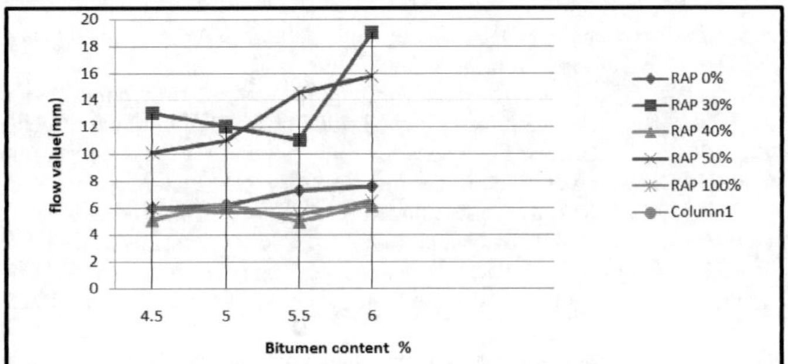

	RAP 0%	RAP 30%	RAP 40%	RAP 50%	RAP 100%
4.5	6	13	5.13	10.1	6
5	6.23	12	6.26	11	5.7
5.5	7.26	11	5	14.6	5.5
6	7.56	19	6.16	15.8	6.5

Fig 18 Graph of bitumen content VS flow value for modified and unmodified rap material

From the above graph it was observed that the results of flow value for 0% RAP, 40% RAP and100% RAP were not within the specified limits . The flow value limit is 8-17 mm.

After conducting Marshall Stability test the mechanical properties like stability value and flow value were determined. As the test was conducted only with the RAP material without adding any admixtures the test results were not satisfactory. The usage of100%RAP is not preferable according to this study.

8. Conclusion

Based on the test results obtained from this study, the following conclusions are drawn .

- Aggregates collected from RAP known as waste aggregates are suitable for bituminous mixes.
- By using the untreated RAP material, the stability of the bituminous surface may goes on decreasing due to weathering actions.
- Bituminous mix from the RAP materials with 0.5% additional bitumen content gives the maximum stability and satisfy the Marshall design criteria.
- Based on the results obtained from the laboratory tests conducted it was concluded that 100% RAP is not suitable for the construction of the flexible pavements. The preferable percentages of RAP are 30%, 50%.
- For 100% RAP (untreated), flow value which is very less than the limiting value and it is the reason of weak and unstable pavement.

References

[1] Abhishek Verma and Rachit Sharma et al 2017 Evaluation of RAP in flexible pavement layers.

[2] Anil Kumar Yadava and Dr. Syed Aqeel Ahmad 2019 *A critical review of characterization and performance evaluation of RAP in road construction.*
[3] Arshad Hussain and QiuYajun 2013 *Effect of reclaimed asphalt pavement on the properties of asphalt binders.*
[4] Brajesh Mishra 2015 Study on use of RAP materials in flexible pavements.
[5] Brian Hill 2011 *Performance* evaluation of warm-mix asphalt mixtures incorporating reclaimed asphalt pavement.
[6] Fawaz Kaseer and Edith Arambula et al 2019 A method to quantify rap binder availability in recycled asphalt mixes.
[7] Feipeng Xiao and Hsein Juang C et al 2007 Rutting resistance of rubberized asphalt concrete pavements containing RAP mixtures.
[8] Haritha Musty Y and Nasim Sabahfer et al 2012 Use of high volume RAP for asphalt pavement rehabilitation due to increased highway truck traffic from freight transportation.
[9] Imad L.AI- Qadi and Mostafa Elseifi et al 2007 Reclaimed asphalt pavement - A literature review.
[10] Jashanjot Singh and Duggal A K et al 2015 An experimental study on RAP in bituminous concrete.
[11] Aravind K, Animesh Das 2006 Pavement design with central plant hot mix recycled asphalt mixes.
[12] Keerthi Gowda B S et al 2017 Recycling of Bituminous Pavements by RAP method.
[13] Krupa Sharma S and Ashok Patel 2018 Study On Reclaimed Asphalt Pavements.
[14] Martina Irene Giani, Giovami Dotelli et al 2014 Comparitive Life Cycle Assessment of Asphalt Pavements using Recycle Asphalt, Warm-mix Technology and Cold-place Recycling.
[15]Sreenivasulu Reddy Mand Suvarna P 2016 A Detailed Study on Reclaimed Asphalt Pavement in Pavement Quality Cement Concrete.
[16] Pavel Kriz and Daniel Grant et al 2014 Reclaimed asphalt pavement-virgin binder diffusion in asphalt mixes.
[17] Prithvi S Kandhal 1997 Recycling of asphalt pavements
[18] Salvatore Mangisfico, Franciois Olard et al 2013 Influence of RAP content on complex modulus of asphalt binder blends and corresponding mixes.
[19] Sameer Shetty and Stephen Lane D et al 2015 Feasibility of RAP use as road base and sub base material.
[20] Sara Bressi and Massimo Losa et al 2019 A comparative environmental impact analysis of asphalt mixtures containing crumb rubber and rap using life cycle assessment.
[21] Siksha Swaroopa Kar and Aravind Krishna Swamy et al 2017 Impact of recycled asphalt pavements on properties of foamed bituminous mixtures.
[22] Anil Pradyumna T and Abhishek Mittal et al 2013 Characterization of RAP for use in bituminous road construction.
[23]UmeshKamariya and Amit A Amin et al 2018 Utilisation of reclaimed asphalt pavement materials.

An Exposition of Speculative and Numerical Analysis of CFST Columns

Pala Gireesh Kumar[1], Seethalam Sai Charan[2], S Reshma Sri Mani[3], Kundeti Nagarjuna[4]

[1]Associate Professor and Head, Department of Civil Engineering, Shri Vishnu Engineering College for Women (Autonomous), Bhimavaram, West Godavari, Andhra Pradesh, India.
[2,4]Assistant Professor, Gudlavalleru Engineering College
[3]Assistant Professor, Ramachandra College of Engineering

Email: palagireeshk@gmail.com

Abstract. Owing to the advantages like intense compressive strength, large ductility and enormous energy incorporation, Concrete filled steel tube column system is gaining momentum when set side by side to regular steel or reinforced concrete system. Local buckling is one such major advantage that CFST gives experimental results from various articles has proven that there is an increase in the properties of CFST over normal concrete systems. This paper aims to give brief understanding on the experiments works conducted and numerical results obtained from various articles are discussed. Experimental works exposed to view that circular concrete filled steel tubes were more appropriate than square concrete filled steel tubes. In this paper, nonlinear finite element analysis for concrete filled steel tube is effectuated through varied parameters like concrete infill grade, as short columns with diameter ratio (L/D) not exceeding 4.5 with devoid of slenderness have also been discussed. The past studies divulged that the increment of D/T ratio dwindles the load carrying capacity of CFST and moreover, the high concrete infill grade buttress the load carrying capacity. From various literature reviews it was also concluded that the confinement efficacy of concrete enlarges the resistance to axial loading of compound columns with concrete filled circular sections. Furthermorefor the CFST columns the buckling failure can be evaded, as well as the load carrying capacity can be strengthened by hauling down the slenderness ratio. Therefore ample scope exists in carrying of numerical analysis using ANSYS to find the potential of CFST columns.

1. Prelude

CFST – Concrete Filled Steel Tube structural system is basically infilled steel tubes with high-strength concrete that exploits both concrete as well as steel. They consist of hollow steel sections plugged with plain or reinforced concrete with rectangular or circular shape such a way that they are fabricated to hold up the floor load to a maximum of three to four storey height. Therefore these are implemented largely as columns as well as beams where in need of effective structural system like in high rise, multi-storey buildings and low rise industrial buildings.

To attain higher strength and superior functioning of the structure, concrete blended with supplementary cementitious material like silica fume is assimilated in disruptive CFST structures that are called as high strength concrete filled steel tubular columns. In the wake of the blended action between the key elements which promotes the high strength thereby inhibiting the inward buckling of the wall of steel. Customarily two types of compound columns were exercised in the

IOP Conf. Series: Materials Science and Engineering 1006 (2020) 012043 doi:10.1088/1757-899X/1006/1/012043

buildings; one with steel sections encased in concrete and the other with steel sections in-filled with concrete. Various researchers performed hypothetical studies as well as instigated methodical CFT models by utilizing nifty software tools to comprehend and to evaluate the compound performance of CFT. Regularly ANSYS software was employed and the models were simulated nonlinearly by the finite element method.

Column slenderness, eccentric loading embracing axial as well as with single or double bending curvature and compressive strength of concrete core were the evaluating criterion that effect the results in terms of strength and potential of concrete filled steel tubular columns. These CFST with square, rectangular and hollow circular cross sections are taking precedence over a miscellany of engineering structures.

2. Behaviour of CFST

The past studies divulged that juxtapose with cross sections, the CFST with circular sections serve better than the CFST with square sections. As a result of lesser confining pressure in the concrete core which incites the local buckling that plausibly occur in the square steel tube. In the concrete filled steel tube the structural behavior is conditioned on both the steel and concrete's Poisson ratio. Because of the higher Poisson's ratio of the steel at the early stage of loading, it does not accord a confining impact to the concrete. The steel tube happen to be bi-axially stressed and concrete core to be tri-axially stressed with the increase of longitudinal strain increases the lateral expansion of concrete than that of steel. As a consequence of this, the steel tube shifts the load from the tube to the core as it cannot hold up the normal yield stress which is similar in both the square and circular sections of CFST.

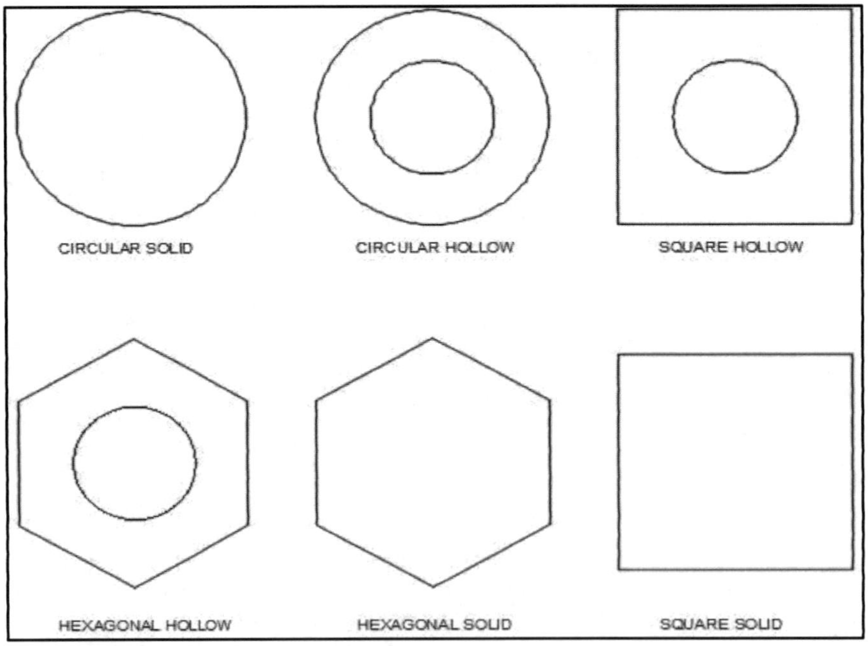

Fig 1 Varied Cross-Sections of Concrete Filled Steel Tubular Sections

3. State of the art

This section gives a brief outline of experimental and numerical workswhich are conducted by various researchers performed on Concrete Filled Steel Tubular Columns.

3.1 Concrete filled steel tubular columns: Numerical research

Neetu Devi Singh and Shubha Vaghmarey (2018) carried a non-linear analysis by employing ANSYS to the concrete filled steel tubes and deduced that the potential of circular cross-section count on the pivotal specifications like; column diameter, length, thickness, L/D and D/t ratios. Also they inferred that the slenderness ratio influences the failure mode of CFST compound column, where lesser the slenderness ratio incites the column failure at the support due to the local buckling and is crushed under direct compression. And buckling failure is seen in the column due to the larger slenderness ratio. Also the load carrying capacity of the CFST column diminishes with the increase of L/T ratio and enlarges with the increase of D/t ratio.

Alfahad B Vishnu Ramesh (2017)researched on the performance of CFST columns and contrasted the experimental and analytical results. Finite element software ANSYS was employed to analyze the CFST columns and in contrary 12 CFST column specimens with rectangular and circular sections were assessed for the failure load when subjected to axial loading. Additionally two CFST specimens with chick mesh, four cylinders filled with conventional concrete and one hollow steel tube filled with conventional concrete were also studied. Referring to the deflection values in the output, trivial dissimilarity is perceived in the experimental and numerical analysis. It is also observed that the CFST column with larger length is not as good as the shorter length CFST column. And with the increment in the height of the column, the CFST columns inclusive of wire mesh showcased load carrying capacity streets ahead of other CFST columns. Furthermore, the circular section CFST columns inclusive of wire mesh exhibited increased strength and the ductile behavior of all the CFST columns were equitably good.

Shams and SaadeghVaziricollated the results of the established three dimensional finite element model of CFST columns with the prevalent experimental values. The results showcased that the geometrical configuration of the columns and material properties of the concrete hold a great impact on the stress-strain properties of the confined concrete. Higher confinement ratio is recognized in the lower unconfined compressive strength of concrete than the higher strength concrete. Few variables like D/t ratio, unconfined compressive strength of concrete and the cross-sectional shape have a substantial effect on the increment of maximal compressive strength of concrete.

3.2 Concrete filled steel tubular columns: Experimental research

Abelmaseeh Bakos, GailanJibrael (2014)tested a total of 23 specimens and studied the axial load carrying capacity when plugged with waste glass concrete in thin walled HSS stub columns. With the replacement of fine and coarse aggregates with waste glass of 25% each displayed almost the same strength when CFST plugged with normal concrete and waste glass concrete.

Shehdeh Ghannam, N et al (2010) researched on the actual performance and the load carrying capacity of the steel columns plugged with usual and lightweight concrete. Each of four specimens with full scale rectangular cross-sections were filled with lightweight aggregate concrete and normal weight aggregate concrete respectively. The specimens were subjected to axial loads and the output delineates that the weight of the column is reduced with the use of lightweight concrete filling and on contrary a high load carrying capacity is attained simultaneously. According to the visual observations and due to the experimental failureloads, the assessed design values referring to the composite codes were surplus in almost all the columns. Moreover, the lightweight aggregate concrete filled sections sustained a squash load more than 92% and descried local failure as well as buckling failure in the column. Similarly the normal weight aggregate concrete filled sections sustained a squash load more than 87% and descried overall buckling failure at the mid height of the column. And the formulae used to calculate the squash load,

$$Nu = A_s f_{sk} / \gamma_{ms} + A_c f_{ck} / \gamma_{mc}$$

Han et al (2009) experimentally investigated a sum total of 32 specimens to study the working of concrete filled steel tubular columns under axial compression. The varied dominant specifications in the study were: square and circular sections, local area compression ratio 1.44 and

1.6, end plate thickness varying from 2 to 12 mm. The evaluation of CFST stub columns put through axial compression loading was done by finite element analysis modelling and the outcome reaffirmed a good concurrence with the test results. To examine the system of compound columns under axial compression loading, theoretical modelling was developed and studied.

Shan-Tong and Yu (1987) researched experimentally also analytically under axial compression and obtained the stress-strain relationship and strength of the concrete filled tubular specimens for discrete L/D values ranging from 2 to 5m. The authors propounded conventional test and rough methods for deducing the primary stress-strain relationship and ultimate strength of the tubes respectively. Also inferred that a plain steel tube with good ductility and energy absorption capacity is similar to the deformed behavior of a concrete filled tube.

Muhammad Naseem N et al (2006), analyzed the potential of 28 short CFST columns of circular and square sections that are axially loaded compression by varying the lengths of columns with an incremental of 500 mm from 2 m to 4 m. It was observed that the circular columns of diameter 160 mm and 111.25 mm carried a maximal axial load of 1153kN and 526kN respectively. Whereas the square columns of sides 125.66 mm and 87.38 mm carried an utmost axial load of 595kN and 450kN respectively corresponding to the ACI and LRFD codes.

J. Zeghiche and K. Chaoui (2005) have tested a total of 27 CFST columns with geometrical parameters; 5 mm wall thickness, 160 mm external diameter, varying lengths from 2 m to 4 m with an incremental of 500 mm. Three sets of columns were employed with compressive strengths of concrete 40 MPa, 70 MPa and 100 MPa. The experimental ultimate loads were varying from 126 KN for axially loaded columns filled with 40 MPa concrete to 2000kN for axially loaded columns filled 100 MPa. The experimental ultimate loads were varying from 1697kN to 963kN for single and double curvatureending.

Elremaily, Azizinamini (2002) examined the concrete-filled steel tube columns for cyclic lateral loads subjected to axial loading and under seismic loads. In regard to the interaction between the steel and concrete, analytical model was generated to anticipate the capacity of circular CFST beam-columns furthermore collated with the experimental data. The output showcased good concurrence that the CFST columns displayed better ductility and preserved their strength until its failure. They inferred that, provided the confinement of the steel tube achieves concrete strength thereby remarkably enhances the column capacity.

Lanhui, Sumiem, N et al (2007) experimentally examined the behavior of circular hollow steel tube as well as circular tubes plugged with concrete. A sum total of 24 CFST columns were tested in four series of specimens where all were loaded axially till their failure.

Shankar Jagadish (2014) studied the ultimate load carrying capacity as well as behavior of stiffened concrete filled single and double skinned thin walled steel tubular column specimens by employing finite element analysis. Ten different specimen models were analyzed for this. A conventional column with rectangular and circular cross section was first modelled and analyzed for validation. Then a circular and a rectangular column are modelled by providing outer skin only without providing conventional reinforcement and analyzed. Then in addition to the outer skin, the specimen was analyzed by providing an inner steel skin. Then the test is repeated by providing Stiffeners inside the column throughout the length and analyzed for all the cases. When the stiffened double skinned Rectangular column is compared with conventional rectangular one, the compressive strength is increasedby 309.72 %. When the stiffened double skinned circular column is compared with conventional circular one, the compressive strength is increased by 297.79 %. It is noticed that the rectangular column shows greater change than circular column when stiffener is used. Contrast to the regular specimens, the load carrying capacity is increased when the double skin is provided. As a scope of study cases of variant loadings and more complicated types of CFST specimens with various cross sections can be tried. Besides, cyclically loaded CFST beams can be studied in extension to the work.

N. Jamaluddin, N et al (2013) this journal is about the study including a sum total of 26 CFST columns of slender as well as stub members that were set up productively with the elliptical steel hollow sections at hand centralized on slenderness ratio. In order to analyze the ascendancy of slenderness on the column strength, stub column length considered two times the outer diameter and three discrete lengths were adopted for the long column specimens. Which were categorized into four series hinged on length such as series 1 delineate stub columns of length 300 mm and series 2, 3, 4 represent long columns of length 1500, 1790, 2500 respectively or in accordance with the section.Concrete Compressive strength of 30Mpa, 60Mpa and 100Mpa have been used. It was observed that the examined elliptical CFST columns had 37.5 and 40 D/t ratios and the long columns' slenderness ratio varied from 16 to 143. The axial carrying capacity ranged from 71% to 136%.

Brain Uy (2000) investigated the high strength steel box columns plugged with concrete, took up in three series of experiments to ascertain the blended response under bending and compression. All the series were deemed to be compact speaking of restrained local buckling slenderness limits nevertheless discrete plate slenderness limits. Therefore no reduction along with the local buckling was anticipated in the component plates. A cycle of tensile coupon also stub column checks were carried out under both tension and compression in order to ascertain the stress-strain behavior of the steel. The columns were casted in plates with plaster on either side in order to make certain uniform loading surface and were assessed under pure compression also under the combined bending and compression.The 28 columns were loaded at the top and the bottom of the column wielding a knife-edge concomitantly with the eccentric loading. The author judged that under pure compression, the maximal loads were derived from the curves of the specimens. Further the results as envisioned under compound actions for a column, the improved eccentricity shortened the attained largest axial load.

Morino, Tsuda (2003) worked assiduously for the past 15 years in Japan that includes "NewUrbanHousingProject", "US Japan Cooperative Earthquake Research Program" besides the work done at the annual meeting of the Architectural Institute of Japan (AIJ) by individual universities and industries. They instigated a structural system and ventilated on the advantages, latest inclinations and findings of the CFST column system. Further through profound indagation of AIJ, a rational design method for the CFST column has been initiated. They also inferred that by dint of system's construction effectiveness and the characteristics of CFST incline the system peculiarly towards high-rise and long-span structures as it saves cost of construction, time and labor.

Brain Uy (2001) performed an experimental work on the efficacy of steel plate slenderness limits and studied the potential of steel-concrete compound columns including CFST columns. Wielding the results, a numerically progressed model was augmented as well as calibrated based on rigid plastic method of analysis holding on to the international codes of practice. The study was appreciable in case of large plate slenderness values especially for larger axial forces but was not deemed for the effects of local buckling. Furthermore to conclude the author put few alterations forward in the design for the insertion of slender plated columns.

4. Peroration

In contradiction to the lengths, the CFST columns of shorter length is streets ahead of longer length and visibly less deflection is seen in increased cross sectional area of the tube.Also greater load carrying capacity is observed in these CFST columns. Juxtapose with regular CFST columns, if any enlargement in height of the column along with wire mesh have greater load carrying capacities. The steel tube's cross-sectional area has consequential upshot on ultimate axial load capacity and deformation of the column, whereas the ultimate strength is consistently improved comparatively to higher concrete areas.Also circular concrete filled steel columns have displayed increase in strength in presence of wire mesh and well-nigh all columns acquit oneself in a reasonably ductilemanner.

Finite element software ANSYS is employed to analyze the potential of steel tubular section filled with concrete (CFST) and fathom out the failure load of varied sections like circular and rectangular columns under axial loading where in consideration of both material and geometric linearity. Three-dimensional finite element models of varied cross sections have been simulated nonlinearly to delve deeper into the load transfer and under the axial compression to study the interaction betwixt steel tube and the concrete core of concrete-filled steel tubes.

References

[1] Neetu Devi Singh, Shubha Vaghmarey 2018 Study and buckling analysis of concrete filled steel tubes columns using ANSYS *International Research Journal of Engineering and Technology* vol **05** issue 12.

[2] Alani Y R A, V C Agarwal 2013 Nonlinear finite element study on circular concrete filled steel tubular columns *International Journal of Innovative Technology and Exploring Engineering* **3**: 52- 55.

[3] S Mathankumar and M Anbarasan 2016 Finite element analysis of steel tubular section filled with concrete *International Journal of Innovative Research in Science, Engineering and Technology* page1-5.

[4] M Pragna, Partheepan Ganesan 2016 Analysis of concrete filled steel tubes using ANSYS *International Journal of Latest Engineering Research and Application* Vol. **01** pp 71-78.

[5] Alfahad B, Vishnu Ramesh 2017 A study on performance evaluation of steel tubular columns infilled with concrete by ANSYS *International Journal of Advance Engineering and Research Development* Vol **4** Issue 6

[6] Abdelmaseeh Bakos Keryou and Gailan Jibrael Ibrahim 2014 Axial load carrying capacity of thin walled HSS stub columns filled with waste glass concrete *International Journal of Enhanced Research in Science Technology and Engineering* vol **3** issue 5, pp 101-111.

[7] Shedeh Ghannam, Hamid R Al-Ani and Orabi-al- Rawi 2010 Composite study of load carrying capacity of steel tube filled light weight concrete and normal concrete *Jordan Journal of Civil Engineering* vol **4** issue 2 pp 164-169.

[8] Han LH, Tao Z, Wang WD 2009 Advanced composite and mixed structures—Testing theory and design approach (Beijing,China) *Science Press (in Chinese)*

[9] Shan-Tong and Yu 1987 Experimental and analytical study on strength of concrete filled tubes 1987 *Institute for Steel Development and Growth* vol. **2** pp 25-1

[10] Muhammad Naseem Baig, FAN Jiansheng and NIE Jianguo 2006 Strength of concrete filled steel tubular columns 2006 *Tsinghua Science and Technology* vol **11** issue 6, pp 657-666.

[11] J Zeghiche and K Chaoui 2005 An experimental behavior of concrete filled steel tubular columns *Journal of Constructional Steel Research* issue 61, pp 53- 66.

[12] Elremaily and Azizinamini 2002 Behaviour of concrete-filled steel tube columns under seismic loads *Journal of Bridge Engineering* vol. **6** no. 2, pp 110-119.

[13] Lanhui Guo, Sumei Zhang, Wha-Jung Kim and Gianluca Ranzi 2007 Behaviour of hollow steel tubes and steel tubes filled with concrete *Science direct* issue 45 pp 961-973

[14] Shankar Jagadesh 2014 Analytical and numerical studies on modelling of CFST members *International Journal for Structural and Civil Engineering Research* vol **3** no. 2 pp 57-64.

[15] N Jamaluddin, D Lam, X H Dai and J Ye 2013 An experimental study on elliptical concrete filled columns under axial compression *Journal of Constructional Steel Research* issue 87, pp 6-16.

[16] Brain UY 2000 Strength of concrete filled steel box columns incorporating local buckling *Journal of Structural Engineering* vol. **126** no. 3 pp 341-352.

[17] Shams and Saadegh Vaziri 1999 Three dimensional finite element model for FST column *Journal of Structural Engineering* vol. **126** no. 3 pp 341-352.

[18] Morino and Tsuda 2003 Design and construction of concrete filled steel tube column system in Japan *Earthquake Engineering and Engineering Seismology* Vol. **4** No. 1 pp.51-73.

[19] Brain Uy 2001 Strength of short concrete filled high strength steel box columns *Journal of Constructional Steel Research* Vol. **57** pp. 113-13

AUTHOR INDEX

Institute of Physics
Dirac House, Temple Back
Bristol BS1 6BE UK

ISSN: 1757-8981
ISBN 978-1-7138-3318-5

9 781713 833185

CURRAN ASSOCIATES INC.
proceedings.com